Students and External Readers	Staff & Research

Natural Sulfur Compounds

Novel Biochemical and Structural Aspects

Natural Sulfur Compounds

Novel Biochemical and Structural Aspects

Edited by
Doriano Cavallini

University of Rome
Rome, Italy

Gerald E. Gaull

Institute for Basic Research in Mental Retardation
Staten Island, New York

and
Vincenzo Zappia

University of Naples
Naples, Italy

Plenum Press . **New York and London**

Library of Congress Cataloging in Publication Data

International Meeting on Low Molecular Weight Sulfur-Containing Natural Products,
    3d, Rome, 1979.
    Natural sulfur compounds.

    "Proceedings of the Third International Meeting on Low Molecular Weight Sulfur-
Containing Natural Products, held in Rome, Italy, June 18–21, 1979."
    Includes index.
    1. Organosulphur compounds–Physiological effect–Congresses. 2. Sulphur metabo-
lism–Congresses. I. Cavallini, Doriano. II. Gaull, Gerald E. III. Zappia, Vincenzo.
IV. Title.
QP535.S1I54 1979   574.19'214   79-28239
ISBN 0-306-40335-8

Proceedings of the Third International Meeting on Low Molecular Weight Sulfur-
Containing Natural Products, held in Rome, Italy, June 18–21, 1979.

© 1980 Plenum Press, New York
A Division of Plenum Publishing Corporation
227 West 17th Street, New York, N.Y. 10011

Printed in the United States of America

Preface

The third International Meeting on "Low Molecular Weight Sulfur Containing Natural Products,"sponsored by the International Union of Pure and Applied Chemistry, was held in the historical building of the Accademia Nazionale dei Lincei, Rome, Italy, June 18-21, 1979.

The symposium was held in order to exchange knowledge in the intriguing and complex field of sulfur biochemistry. This theme brought together scientists from many specialized areas from organic and physical chemistry to biology and medicine. The interdisciplinary nature of the meeting gave to the participants the opportunity to discuss problems of common interest approached from different scientific standpoints.

This volume contains 47 contributions presented at the meeting which mainly deal with new structural and metabolic aspects of sulfur biochemistry.

An important aspect of such a scientific meeting is the rapid publication of the proceedings. Through the cooperation of the authors in providing "camera ready" copies of their manuscripts, the good efforts of the Organizing Committee, and the Plenum Press Publishing Co., it has been possible to publish this book within a few months of the meeting.

Finally, the Editors wish to express their sincere appreciation to the Accademia dei Lincei for its generous hospitality. They also gratefully acknowledge the financial support received from the following Institutions: The University of Rome; Consiglio Nazionale delle Ricerche; Italian Society of Biochemistry; Soc. Beckman Analytical,Rome; Soc. Baldacci, Pisa; Soc. Farmitalia Carlo Erba, Milano; Soc. Gio De Vita, Rome; Soc.Tecnochimica Moderna, Rome; Soc. C. Donati, Rome

Doriano Cavallini

Gerald E. Gaull

Vincenzo Zappia

Contents

CONTENTS

Abbreviations

A = adenosine
αAA = α aminoacrilate
Acethyl-Coa = acethyl coenzyme A
Acyl-CoA = acyl coenzyme A
ACYS = adenosylcysteine
ACYM = adenosylcysteamine
Ado Hcy = S-adenosylhomocysteine
Ado Met = S-adenosylmethionine
AEC = aminoethylcysteine
Ala = alanine
Alkenyacyl-GPC = alkenyl-acyl-sn-glycero-3-phosphorylcholine
ALS = acid labile sulphide
APS = adenosylphosphosulfate
ATP = adenosine triphosphate
BAL = 2,3-mercaptoethanol
BSO = buthionine sulfoximine
Bisnor- = 4,6-dithiohexanoate or 1,2-dithiolane-3-propanoate
CBI = covalent binding index
CD = circular dichroism
CD = cysteamine dioxygenase
cis-APD = cis-2,5-bis(aminooxymethyl)-piperazine-3,6-dione
CM = carboxymethyl (cellulose)
CNS = central nervous system
CoA = coenzyme A
COMT = catechol-O-methyl transferase
ConA = concanavalin A
CSA = cysteine sulfinic acid
CSAD = cysteine sulfinic acid decarboxylase
Cycloserine = 4-aminoisoxazolidone-3
CYM = cysteamine
$CYMSO_3H$ = cysteamine sulfonate
Cys = cysteine
$CYSSO_3H$ = cysteine sulfonate
Decarboxy Ado-Met = S-adenosyl-(5')-3-methylthiopropylamine
DEAE = diethylaminoethyl (cellulose)

DephosphoCoA = dephospho coenzyme A
DFP = diisopropylphosphorofluoridrate
Diacyl-GPC = diacyl-sn-glycero-3-phosphorylcholine
DNA = deoxyribonucleic acid
DTT = dithiotreitol
3-DZA = 3-deazaadenosine
3-DZA-Hcys = 3-deazaadenosylhomocysteine
EDTA = ethylenediaminetetraacetic acid
EFA = essential fatty acid
E.M. = effective molarity
EPG = ethanolamine phosphoglycerides
ERG = electroretinogram
ESO = ethionine sulfoximine
ESP = epithio specifier protein
FAD = flavin adenine dinucleotide
FF   = fat free
GABA = γ-aminobutyric acid
GLC = gas liquid chromatography
GSH = reduced glutathione
GSSG = oxidized glutathione
GTP = guanosine triphosphate
H = hypotaurine
Hcy = homocysteine
HEPES = N-2-hydroxyethylpiperazine-N'-2-ethane sulfonic acid
Hyp = hypotaurine
IgM = immunoglobulins M
Ile = isoleucine
IR   = infrared (spectrum)
ISO-DL = isoproterenol
ISO   = DL-isoproterenol
L = lanthionine
LADH = lipoamide dehydrogenase
MAP = 6α-methyl-17α-actooxyprogesterone
2-me= 2-mercaptoethanol
MMT = S-methylmethionine
mono-TNBS salt = mono-2,4,6-trinitrobenzene sulphonate salt
3-mp = 3-mercaptopyruvate
mRNA = messenger ribonucleic acid
MS = mass spectra
MSO = methionine sulfoximine
MTA = 5'-methylthioadenosine
MTI = 5'-methylthioinosine
MTR = methylthioribose
MTR-1-P = methylthioribose 1-phosphate
MTT = 5'-methylthiotubercidin
MTX = methotrexate
NAD^+ = nicotinamide adenine dinucleotide
$NADP^+$ = NAD phosphate
NMR = nuclear magnetic resonance
NTBN = neothiobinupharidine

NPS = ninhydrin positive substances
NEM = N-ethylmaleimide
ODC = ornithine decarboxylase
O-MT = O-methylthreonine
P = pantothenate
PAP = phosphoadenosylphosphate
PAPCA = phosphoadenosylphosphocysteic acid
PAPS = phosphoadenosylphosphosulfate
PCA = perchloric acid
PCYM = pantetheine
PDE = phosphatydil-N,N-dimethyl-ethanolamine
PDH_c = pyruvate dehydrogenase complex
PHA = phytohemagglutinin
PLP = pyridoxal phosphate
PM = pyridoxamine
PME = phosphatidyl-N-monomethylethanolamine
PMP = pyridoxamine phosphate
PMR = proton magnetic resonance
PPCYM = phosphopantetheine
PPCYS = phosphopantothenylcysteine
PSH = penicillamine
PSO = prothionine sulfoximine
Ptd-choline = phosphatidylcholine
PYR = pyruvate
Pyridoxal-P = pyridoxal-phosphate
RSH = thiol
RSV = Rous Sarcoma Virus
SAH = S-adenosylhomocysteine
SAM = S-adenosylmethionine
SAM DC = SAM decarboxylase
Ser = serine
SIBA = 5'-isobutylthioadenosine
sulfane = persulfide sulfur
T = taurine
β-T = β-thiaproline
γ-T = γ-thiaproline
TAU = taurine
TCA = trichloroacetic acid
TCA cycle = tricarboxylic acid cycle
Tetranor- = 2,4-dithiobutanoate or 1,2-dithiolane-3-carboxylate
Thiocystine = bis-(2-amino-2-carboxyethyl)trisulfide
Thr = threonine
T-Ile = thiaisoleucine
TMS = trimethylsilyl
TNBS = trinitrobenzenesulphonate
TP = testosterone propionate
TPN = total parenteral nutrition
TRIS = tris(hydroxymethyl)aminoethane
tRNA = transfer ribonucleic acid
UV = ultraviolet

STEREOCHEMICAL ASPECTS OF TRANSMETHYLATIONS

OF POTENTIAL BIOLOGICAL INTEREST

A. Kjær, G. Grue-Sørensen, E. Kelstrup and

J. Øgaard Madsen

Department of Organic Chemistry, The Technical
University of Denmark, 2800 Lyngby, Denmark

INTRODUCTION

In 1954, Shive and co-workers demonstrated the occurrence
of S-methylmethionine (1) (MMT) in cabbage and several other vegetables.[1] Subsequently, 1 has been reported as a constituent of
asparagus,[2,3] jack beans,[4] pelargonium and mint leaves,[5] green
tea,[3,6] soybean meal,[7] celery,[3] tomatoes,[3,8] sweet corn,[9] po-
tatoes,[10] apples,[11] and milk.[12] The amazingly wide distribution
of 1 justifies inquiries into its chemistry, biosynthesis, metabolism, and biological significance.

Even before its discovery as a natural product, MMT (1) had
been shown to support the growth of rats on a methionine-free
diet,[13] a finding later modified to obtain only when cystine was

$$[Me]_2 \overset{+}{S}[CH_2]_2 CH(NH_3^+)CO_2^- \qquad R(Me)\overset{+}{S}[CH_2]_2 CH(NH_3^+)CO_2^-$$

$$1 \qquad\qquad\qquad 2, \text{ R = Adenosyl}$$

present in the diet.[14] Other early reports list MMT (1) as a
methyl donor in the biosynthesis of creatine in rat liver slices,[15]
though not in homogenates,[16] and as a methionine-replacing

factor, in some cases superior to methionine itself, in a number of microorganisms.[17]

Early studies on yeast[18] and bacteria,[19] in which the combined administration of homocysteine and MMT, but neither of the two alone, was shown to support the production of S-adenosylmethionine (2) (SAM)[18] or methionine,[19] led to the working hypothesis that the enzyme-catalyzed reaction:

$$[Me]_2\overset{+}{S} \cdot R + HS \cdot R \longrightarrow 2\ MeS \cdot R + H^+$$

$$(R = \text{L-}[CH_2]_2CH(NH_3^+)CO_2^-)$$

may be operating in living cells and hence possess biological significance.[20] This assumption was subsequently substantiated through more detailed studies of the ability of several cell-free extracts to catalyze the above reaction. Thus, homocysteine methyltransferases, accepting SAM and MMT as methyl donors, have been described from rat liver,[21-23] microorganisms,[20, 23-27] and plant seeds, including pea, cabbage and jack bean.[28-30] Growing plants of the last species contain MMT,[4] rendering the methyltransferase from jack bean meal[29] particularly apposite in a functional context. This enzyme, obviously different from homocysteine methyltransferase of animal or microbiological provenance, has been purified 250-fold and accepts other donors, such as dimethylpropiothetin and SAM, the latter though with only one tenth of the activity.[29]

STEREOCHEMICAL STUDIES

The monochiral S-methyl-L-methionine (1) contains prochiral, diastereotopic methyl groups which must, in principle, be transferable to a chiral or achiral acceptor molecule with different rates. An exploratory study, recently conducted in our laboratory, with (RS)-2-butyl-dimethylsulphonium ion (3) as the donor and 4-methylbenzenethiolate ion (4) as the acceptor molecule, revealed a virtually indiscriminative methyl group transfer insofar as a substantial, observed rate difference could be accounted for exclusively in terms of a remarkably large $^{12}C/^{14}C$-isotope effect.[31]

$$[Me]_2\overset{+}{S} \cdot CH(Me)Et + (p)\text{-}Me \cdot C_6H_4 \cdot S^-$$

$$(3) \qquad\qquad\qquad (4)$$

$$MeS \cdot CH(Me)Et + (p)\text{-}Me \cdot C_6H_4 \cdot SMe$$

We have extended this stereochemical exercise to the methyl group transfer from S-methyl-L-methionine to homocysteine,[29] catalyzed by the enzyme from jack bean meal; we report our results in the sequel.

Synthesis of Enzyme Substrates

  α-^{2}H-L-Methionine (5) ($\geq$ 95 % ^{2}H), produced by resolution[32] of the corresponding racemate (Stohler Isotope Chem. Inc.), was converted, upon treatment with 2-^{13}C-enriched bromoacetic acid (Stohler Isotope Chem. Inc.), into an approximately 1:1 mixture of diastereomeric salts of S-[carboxy-^{13}methyl]-α-^{2}H-L-methionine, (6) and (7) (Fig. 1). Separation of these was accomplished by recrystallization of their polyiodides, essentially following the procedure recently described by Cornforth et al.[33] The least soluble of these gives a mono-2,4,6-trinitrobenzenesulphonate salt (mono-TNBS salt) which, on analytical amino acid chromatography, possesses the longest retention time.[33,34] By x-ray diffraction, this isomer was recently demonstrated to possess the C_SS_R-configuration (7) in its cationic moiety.[33] It was found convenient in our hands to convert the two diastereomeric polyiodides into the corresponding, sparingly soluble bis-TNBS salts of

(6) and (7). By ^{1}H 270 MHz n. m. r. analysis the ^{13}C-enrichment in 6 and 7 was found to be about 79 % and 73 %, respectively; analytical amino acid chromatography revealed a content of about 7 % of 7 in 6, and about 3 % of 6 in 7, both analyzed as bis-TNBS-salts; these values, however, are to be taken as only approximate.

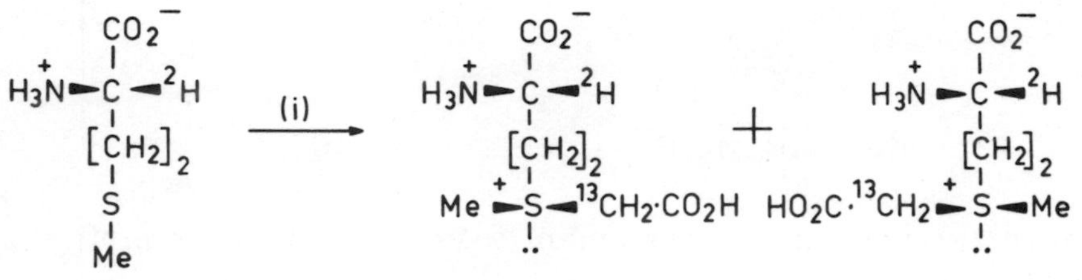

Fig. 1.   (i) Br^{13}CH$_{2}\cdot$CO$_{2}$(70.3 % ^{13}C); a-^{2}H-L-methionine ($\geq$ 95 % ^{2}H) (985 mg), H$_{2}$O (7 ml); 42 h, 20°. H$_{2}$O ad 40 ml; KI (1.20 g), KI$_{3}$ (1.0 M, 3.3 ml); 72 h at 4°; 5 x KI$_{3}$ (1.0 M, 0.66 ml) in the course of 48 h; filtration; 2.34 g (84.5 %) (least soluble isomer) (C$_{S}$S$_{R}$)^{33} (7). Filtrate: KI$_{3}$ (1 M, 1.0 ml); 12 h at 4°; 3 x KI$_{3}$ (1.0 M, 1.0 ml) in the course of 8 h; filtration; 1.15 g (41.4 %) (most soluble isomer) (C$_{S}$S$_{S}$)^{33} (6).

The anhydrous bis-TNBS-salts of 6 and 7 were subsequently subjected to decarboxylation as outlined in Fig. 2. Reaction conditions were mild enough to exclude significant epimerization due to pyramidal inversion of the sulphonium center. The resulting C$_{S}$S$_{R}$- and C$_{S}$S$_{S}$-S-[^{13}C-methyl]-a-^{2}H-methionine diastereomers, (8) and (9), were converted into their crystalline bis-TNBS-salts through a series of ion-exchange operations. ^{1}H n. m. r. analyses revealed ^{13}C contents of 69 % in 8, 76 % in 9; the specimen of 8 contained about 3.5-4 % of 9, that of 9 about 4.5-5 % of 8. Before being subjected to enzymic transfer reactions, the ions (8) and (9) were converted into their monochlorides, again by ion exchange column technique.

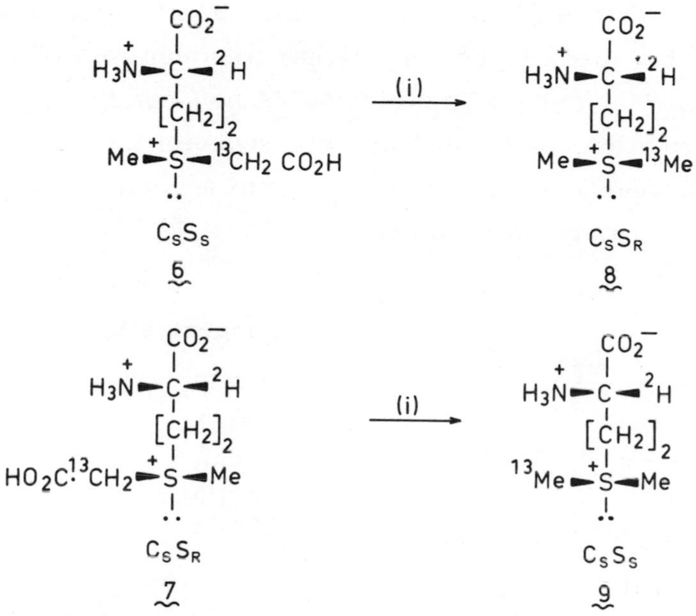

Fig. 2.   (i) Anhydrous bis-TNBS-salt (400 mg), HMPTA (3 ml), 20°, 1 h; addition of pyridine (1.2 ml), 60°, 15-17 min; ion-exchange (IR-120/Na^+), elut. with NH_3; (a) HCl (b) TNBSH, $4 H_2O$, 0°; 8 (222 mg, 53 %); 9 (202 mg, 48 %).

The monochlorides of 8 and 9, respectively, served as substrates for the transmethylation to homocysteine, catalyzed by an enzyme preparation from jack bean meal, prepared and partially purified according to literature directions.[29]

Enzymic Transmethylation

    Incubations were set up as specified in Fig. 3, resulting in
the consumption of more than 90 % of the substrates according
to n. m. r. analysis of suitable model compounds; the methionine
produced was isolated by ion exchange technique and subsequently
converted into its bis-trimethylsilyl (bis-TMS) derivatives upon
reaction with N-(trimethylsilyl)diethylamine. Control experiments,
in which the substrates, or homocysteine, were omitted from the
incubations, gave no trace of methionine. A non-catalyzed
methionine formation, detectable in the absence of the enzyme
preparation, proceeded with a rate too low to seriously compete
with the enzyme-catalyzed reaction.

Reagents	Concentration

R1:	(RS)-Homocysteine lactone, hydroiodide	30. 1 µmol/100 µl
R2:	8, or 9, as monochlorides	2. 43 µmol/50 µl
R3:	7. 5 M NaOH	1500 µmol/200 µl
R4:	1 M KH_2PO_4	2000 µmol/2000 µl
R5:	Enzyme solution	

Fig. 3.   (i) R1 + R3, N_2, 20°, 5 min; R4 added. (ii) To this
             solution (575 µl), addition of R2 (50 µl) and R5 (625 µl),
             N_2, 38-39°, 200 min.

    The bis-TMS derivatives of methionine, arising from enzymic
transmethylation of the diastereomeric substrates, (8) and (9),
Fig. 4, were separately analyzed by GLC-MS-technique.

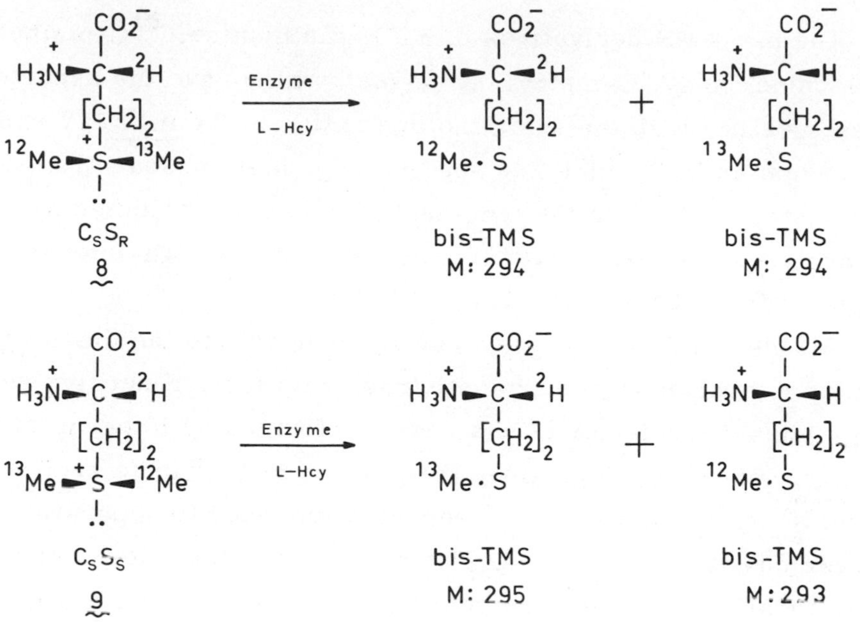

Fig. 4. Enzymic transfer of a methyl group from the diastereo-meric S-^{13}C-methyl-a-^{2}H-L-methionines to L-homo-cysteine.

Mass Spectrometric Analysis

The mass spectrum of the bis-TMS-derivative of methionine exhibits a molecular ion at m/e 293 with an intensity of about 5 % of that of the base peak at m/e 176, the latter arising by fission of the C_1-C_2-bond:

$$\text{Me} \cdot \text{S} \cdot [\text{CH}_2]_2 \text{CH}(\text{NH} \cdot \text{Si}[\text{Me}]_3) \text{CO} \cdot \text{O} \cdot \text{Si}[\text{Me}]_3$$

m/e 293 (5 %)

+

$$\text{Me} \cdot \text{S} \cdot [\text{CH}_2]_2 \text{CH}:\text{NH} \cdot \text{Si}[\text{Me}]_3$$

m/e 176 (100 %)

The bis-TMS-derivatives of α-^{2}H-methionine, ^{13}C-methyl-methionine, and ^{13}C-methyl-α-^{2}H-methionine gave the expected mass spectra, with the same isotope ratios in the m/e 177 and 178 region as in the m/e 294 and 295 region. Consequently, secondary isotope effects in the fragmentation can be neglected and MS-analyses can confidently be conducted on the high-intensity fragment ions at m/e 176, 177, and 178.

All bis-TMS derivatives were introduced into the mass spectrometer through a gas chromatograph with fast, repetitive scanning. Intensity determinations were performed by integration over 45 individual recordings along the gaschromatographic profile in order to compensate for an observable tendency to separation of the various isotopic species. The measured intensities, corrected for natural abundances, are schematically reproduced in Fig. 5.

Molecular Ions m/e	Fragment Ions m/e	% Intensity of Fragment Ions[*]	
		From $C_sS_R(8)$ m/e	From $C_sS_S(9)$ m/e
293	176	22.1	47.6
294	177	74.9	22.2
295	178	3.0	30.2

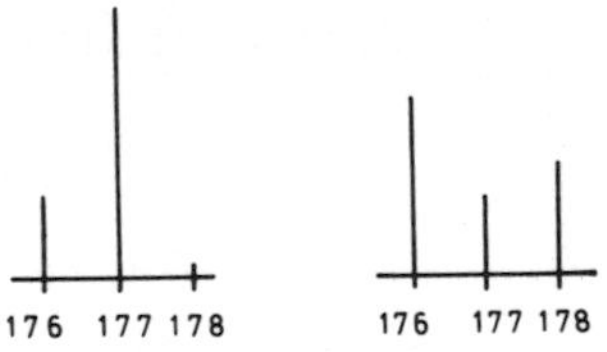

[*] Corrected for natural abundance contributions

Fig. 5.   Fragment ion intensities from enzymically produced bis-TMS derivatives of methionine.

In order to express the intensities in terms of per cent stereoselectivity, the relations in Fig. 6 were derived for the bis-TMS methionine fragment ions arising from the methionines produced in the enzyme-catalyzed transfer from (8) and its congeners ($2x$ denotes the fraction, in per cent, of (8) donating the ^{13}Me group). A similar scheme can easily be derived for the diastereomer (9).

Reactants	%		Fragment ions m/e
HCy-[H] [a] + $^{13}C^{12}C$[D] [b] $\longrightarrow$	x [c]	^{13}C[H]	177
	$50-x$	^{12}C[H]	176
	x	^{12}C[D]	177
	$50-x$	^{13}C[D]	178
HCy-[H] + $^{12}C^{12}C$[D] $\longrightarrow$	50	^{12}C[H]	176
	50	^{12}C[D]	177
HCy-[H] + $^{13}C^{12}C$[H] $\longrightarrow$	x	^{13}C[H]	177
	$50-x$	^{12}C[H]	176
	x	^{12}C[H]	176
	$50-x$	^{13}C[H]	177
HCy-[H] + $^{12}C^{12}C$[H] $\longrightarrow$	100	^{12}C[H]	176

[a] Denotes homocysteine; [b] denotes MMT with specified isotope contents in the two methyl groups and at the α-carbon; [c] $2x$ is the fraction, in per cent, of (8), donating the ^{13}Me group.

Fig. 6 Distribution of the bis-TMS fragment ions arising from L-methionine produced in the enzymically catalyzed transfer from C_SS_R-S-[^{13}methyl]-α-^{2}H-methionine (8), and its congeners, to homocysteine.

When the enrichment factors for ^{2}H and ^{13}C are designated h and c, respectively, the three equations (A)-(C) in Fig. 7 obtain. From any two of these, the sum $h+c$ can be expressed as shown and assumes the value 1.62 on insertion of the relative intensities (cf. Fig. 5) for the C_SS_R-diastereomer (8). From the known value, 0.95, for h, a ^{13}C enrichment factor of 0.67 follows. Inserting these values in any one of the equations (A)-(C), one finds a value for 2x of 90.3. It was previously established, however, that the C_SS_R-isomer (8) employed contained about 5 % of the C_SS_S-isomer (9). When corrected for this content, the value of 2x in creases to 95. Analogous calculations, performed on the isomer (9), gave the pleasingly consistent value 2x = 95.

(A) $\% \ I^{176} = \quad -hcx + 50\ hc - 50h - 50c + 100$

(B) $\% \ I^{177} = \quad 2hcx - 100\ hc + 50h + 50c$

(C) $\% \ I^{178} = \quad -hcx + 50\ hc$

It follows that:

$$h + c = \frac{100 + (I^{178} - I^{176})}{50}, \quad or \quad \frac{I^{177} + 2 \cdot I^{178}}{50}, \quad or$$

$$\frac{200 - (2 \cdot I^{176} + I^{177})}{50} = 1.62$$

h = 0.95; it follows that c = 0.67

From (A), (B), or (C): 2 x = 90.3 %

_____

Fig. 7.   Calculation of the 2x-value for the enzymic transmethylation to homocysteine from diastereomerically homogeneous C_SS_R-S-[^{13}methyl]-a-^{2}H-methionine (8).

The excellent agreement in the two experimentally indepen-dent series permits the conclusions: (i) that primary isotope effects can be neglected, and (ii) that the pro-R methyl group in MMT (10) is preferentially transferred to homocysteine in the en-zyme-catalyzed reaction. It is noteworthy that SAM, possessing the C_SS_S-configuration (11),[33] contains its methyl group in a spa-tial position different from that of the pro-R methyl group in MMT (10). The way now seems open, utilizing methyl-labelled SAM, to

clarify the stereochemical course of its S-methylation of methionine, a reaction believed to represent the biosynthesis of MMT in
jack beans.[4]

The observed stereoselectivity, defined as $2x - (100-2x)/2x + (100-2x) = 4x - 100$, thus amounts to 90 %. Its deviation from 100 % may conceivably be due to a slight competition from a non-specific, non-enzymatic transmethylation reaction, or to the presence of a few per cent of the D-enantiomer in the α-^{2}H-L-methionine utilized as the starting material. A slight epimerization of the sensitive sulphonium ion through thermal, pyramidal inversion during the various manipulations would likewise result in a diminished stereo-selectivity and cannot be entirely excluded.

CONCLUSION

The work described in this lecture was designed to answer the questions: are the two methyl groups in MMT being transfer-red to homocysteine with different rates in the enzymically cata-lyzed reaction, and, if so, which methyl group is preferentially, or exclusively, transferred? The answer to the first question is that the selectivity is 95 % or better; to the second, that the pro-R methyl group in MMT is preferentially, or perhaps even ex-clusively, transferred. A similar approach should prove useful in clarifying other stereochemical problems associated with bio-logical transmethylation reactions. Such problems are currently under investigation.

Acknowledgements

The authors are grateful to Dr. Erik Lund, The Danish Protein Institute, for his skillful performance of a series of amino acid analyses, to Dr. Klaus Bock of this Laboratory for recording a series of important 270 MHz ^{1}H n.m.r. spectra, to The Department of Technical Biochemistry of this University for placing facilities for enzyme purification at our disposal, to The Danish Council for Scientific and Industrial Research for the GC/MS-instrumentation, and, finally, to The Danish-Natural Science Research Council for generous financial support.

REFERENCES

1. R. A. McRorie, G. L. Sutherland, M. S. Lewis, A. D. Barton, M. G. Glazener, and W. Shive, J. Amer. Chem. Soc. 76 (1954) 115.

2. F. Challenger and B. J. Hayward, Chem. & Ind. (London) (1954) 729.

3. F. I. Skodak, F. F. Wong, and L. M. White, Anal. Biochem. 13 (1965) 568.

4. R. C. Greene and N. B. Davis, Biochim. Biophys. Acta 43 (1960) 360.

5. L. Peyron, Bull. Soc. Franc. Physiol. Vegetale 7 (1961) 46; C. A. 56, 12014b.

6. T. Kiribuchi and T. Yamanishi, Agr. Biol. Chem. 27 (1963) 56.

7. T. Hino, A. Kimizuka, K. Ito, and T. Ogasawara, Nippon Nogei Kagaku Kaishi, 36 (1962) 413; C. A. 61, 10889h.

8. F. F. Wong and J. F. Carson, J. Agr. Food Chem. 14 (1966) 247.

9. D. D. Bills and T. W. Keenan, J. Agr. Food Chem. 16 (1968) 643.

10. G. Werner, R. Hossli, and H. Neukom, Lebensm. -Wiss. u. Technol. 2 (1969) 145.

11. A. H. Baur and S. F. Yang, Phytochemistry 11 (1972) 2503.

12.   T. W. Keenan and R. C. Lindsay, J. Diary Sci. 51 (1968) 112;
       C. A. 68, 38277n.

13.   M. A. Bennett, J. Biol. Chem. 141 (1941) 573.

14.   J. Stekol, in 'A Symposium on Amino Acid Metabolism',
       The Johns Hopkins Press, Baltimore, Md., U. S. A.
       1955, p. 509.

15.   P. Handler and M. L. C. Bernheim, J. Biol. Chem. 150 (1943)
       335.

16.   S. Cohen, J. Biol. Chem. 201 (1953) 93.

17.   R. A. McRorie, M. R. Glazener, C. G. Skinner, and W. Shive,
       J. Biol. Chem. 211 (1954) 489.

18.   F. Schlenk and R. F. DePalma, Arch. Biochem. Biophys. 57
       (1955) 266.

19.   S. K. Shapiro, Biochim. Biophys. Acta 18 (1955) 134.

20.   S. K. Shapiro, J. Bacteriol. 72 (1956) 730.

21.   G. A. Maw, Biochem. J. 63 (1956) 116.

22.   G. A. Maw, Biochem. J. 70 (1958) 168.

23.   S. K. Shapiro and D. A. Yphantis, Biochim. Biophys. Acta 36
       (1959) 241.

24.   S. K. Shapiro, Biochim. Biophys. Acta 29 (1958) 405.

25.   S. K. Shapiro, J. Bacteriol. 83 (1962) 169.

26.   S. K. Shapiro, D. A. Yphantis, and A. Almenas,
       J. Biol. Chem. 239 (1964) 1551.

27.   E. Balish and S. K. Shapiro, Arch. Biochem. Biophys. 119
       (1967) 62.

28.   J. E. Turner and S. K. Shapiro, Biochim. Biophys. Acta 51
       (1961) 581.

29.   L. Abrahamson and S. K. Shapiro, Arch. Biochem. Biophys.
       109 (1965) 376.

30.   W. A. Dodd and E. A. Cossins, Biochim. Biophys. Acta 201
       (1970) 461.

31.   G. Grue-Sørensen, A. Kjær, R. Norrestam, and E. Wieczor-
       kowska, Acta Chem. Scand. B31 (1977) 859.

32.   G. P. Wheeler and A. W. Ingersoll, J. Amer. Chem. Soc.
        73 (1951) 4604.

33.   J. W. Cornforth, S. A. Reichard, P. Talalay, H. L. Carrell,
        and J. P. Glusker, J. Amer. Chem. Soc. 99 (1977) 7292.

34.   H. G. Gundlach, S. Moore, and W. H. Stein, J. Biol. Chem.
        234 (1959) 1761.

METHYLASE MECHANISMS: STERIC CONSTRAINTS AND MODES OF CATALYSIS

James K. Coward

Department of Pharmacology
Yale University School of Medicine
New Haven, Connecticut   06510   USA

Over the past several years, our laboratory has been studying the mechanism and regulation of biological alkyl transfer reactions.   Our approach has been to combine chemical and biological techniques in order to answer questions concerning the means by which these enzyme-mediated processes are catalyzed and controlled in living cells.   Reviews of the chemical (1) and biological aspects (2) have been published, and the present paper will deal only with more recent results of our chemical mechanism studies.   Ultimately, we would like to use this information in the design of potent, specific multisubstrate adduct ("transition state analog") inhibitors (3) of these reactions.

The enzyme catechol-O-methyl transferase (COMT, E.C. 2.1.1.6) catalyzes the transfer of an intact methyl group from S-adenosyl-methionine (SAM) to a catecholamine as shown in below.   This is representative of a group of enzymes which methylate small molecules such as the biogenic amines, norepinephrine, histamine and 5-hydroxytryptamine (4).   We have previously shown (5) by steady-

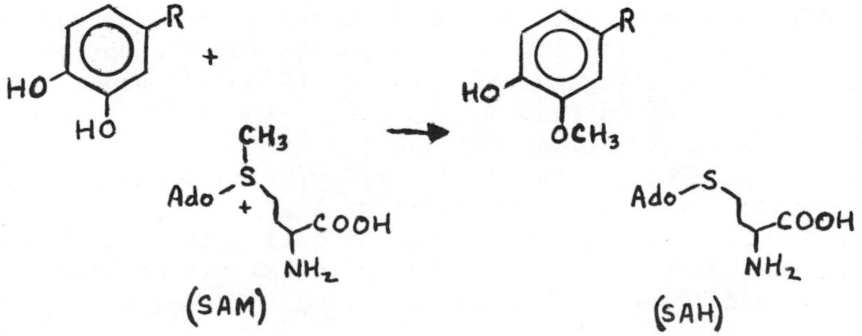

15

state kinetic analysis that the reaction proceeds by a random, sequential mechanism (6) with direct transfer of the methyl group in the ternary complex, EAB→EPQ. Recently, in collaboration with Prof. Heinz Floss and his colleagues at Purdue University, we have studied the stereochemistry of this enzyme-catalyzed methyl transfer, using SAM containing a chiral methyl group; i.e., $\overset{T}{\underset{D}{H}}\!\!>\!\!C-$ (7). The results from these studies show that the reaction proceeds with inversion of configuration at the methyl carbon, thus supporting a direct methyl transfer from SAM to the catecholamine, rather than a double-displacement ("ping-pong") (8) mechanism. These studies, together with studies using "model" reactions (1, see below) suggest very strict stereochemical constraints on S_N2 reactions, such that any deviations from a 180° array of Nuc ·······CH$_3$······· $\underset{+}{S}\!<$ in the transition state is not permitted.

A second group of SAM-dependent methylases acts on much larger substrates, such as proteins and nucleic acids (9). Only recently has progress been made on isolating purified proteins which methylate macromolecules such as tRNA (10), mRNA (11), and proteins (12). In the case of tRNA (adenine-1)-methyltransferase (E.C. 2.1.1.36) (10a) and protein carboxyl-O-methyltransferase (E.C. 2.1.1.24) (13), kinetic studies are consistent with the random sequential reaction proposed for COMT, and thus suggest a direct methyl transfer in the ternary complex. One enzyme which does not show kinetics consistent with a direct methyl transfer is histamine-N-methyltransferase (E. C. 2.1.1.8) (14). The data reported are consistent with a double--displacement mechanism of the type:

$$E + SAM \rightarrow SAH + E\text{-}CH_3 \xrightarrow{\text{histamine}} CH_3\text{-}Hist + E$$

In view of the unexpected poor nucleophilicity of imidazole towards sp^3 carbon (see below), this enzyme may require a unique mechanism to effect the required methylation of histamine. Stereochemical studies of the type done with COMT will be especially illuminating, since a double-displacement reaction (two inversions) would result in net retention of configuration at the methyl group.

In order to learn more about the chemistry of alkyl transfer reactions, we have carried out kinetic studies on several non-enzymic model reactions (1). Our most recent efforts have involved the study of reactions (1) and (2), as models for the enzyme-catalyzed methylation of 2'-hydroxyl groups of tRNA (15) and the phenolic hydroxyl groups of catecholamines (16) respectively. The rate-pH profiles for reactions (1) and (2) are shown in Fig. 1 (T=40°, μ=1.0 M). The values for k$_o$ are buffer-independent rates obtained by extrapolating linear plots of k$_{obsd}$ vs. total buffer concentration ([B$_T$]) to [B$_T$] = 0. Buffer catalysis of reactions at sp^3 carbon is rare (1), and the buffer catalysis observed in reactions (1) and (2) will be discussed in detail below.

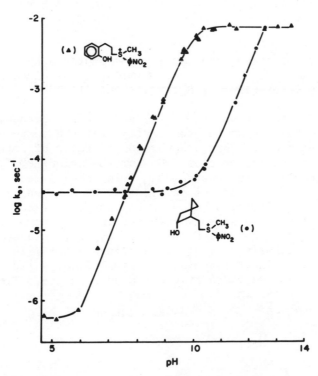

Fig. 1 pH-rate profiles for decomposition of 1 and 2a (15, 16).

   Inspection of the data in Fig. 1 reveals a low pH plateau
(k_{ROH}), followed by a rapid increase in rate (slope = 1.0) to
a new, high pH plateau (k_{RO^-}) which can be observed directly in
the case of 2a, but only inferred in the case of 1 and 2b. These
facts lead to the rate expression shown in eq. 3. A summary of

$$k_o = k_{ROH} \left(\frac{a_H}{K_a + a_H} \right) + k_{RO^-} \left(\frac{K_a}{K_a + a_H} \right) \quad (3)$$

rate constants obtained from the data shown in Fig. 1 for 1 and 2a,
and from similar data (not shown) for 2b, is given in Table I. A

Table I - Summary of kinetic data for 1 and 2[a].

Substrate	k_{ROH}, sec^{-1}	k_{RO^-}, sec^{-1}	pKapp
1	3.59×10^{-5}	31.2^{b}	16.1
2a	2.0×10^{-7}	8.58×10^{-3}	9.97
2b	2.7×10^{-6}	3.98×10^{-1b}	10[c]

[a]Best fit of eq. 3.
[b]Estimated value.
[c]Assumed value.

Bronsted plot of these data for intramolecular alkylation gives a
β value of ca. 0.3, which is very similar to β-values previously
reported for intermolecular S_N2 reactions (1).

   The reactions shown in eq. (1) and (2) are amenable to detailed
study due to their enhanced reactivity over the corresponding intermolecular reactions. Thus, comparison of rates of intermolecular
attack by the neutral species, ROH, or the anions RO^-, with rates
of intramolecular attack (k_{ROH}, k_{RO^-}) in Table I leads to an effective molarity (E.M.) (17) of ca. 10^6-10^8 M. The basis for this
type of rate acceleration has been the subject of considerable
controversy (18), and we sought to obtain the parameters, $\Delta H^{\ddagger}$,
$\Delta G^{\ddagger}$, and $\Delta S^{\ddagger}$ from the activation energies, Ea, obtained from Arrhenius plots. A summary of these data obtained for 1 and 2a, together
with the related intermolecular reaction of eq. 4 (19), is shown
in Table II. The most dramatic difference in the series is seen

Table II - Activation parameters for transalkylation[a]

Substrate, $\underline{k}$	$\underline{E}_a$ [b]	$\Delta H^{\ddagger}$ [b]	$\Delta G^{\ddagger}$ [b]	$\Delta S^{\ddagger}$ [c]
$\underline{1}$, k_{ROH}	23.1	22.5	27.0	-15
$\underline{1}$, k_{RO-}	15.7	15.1	16.1	-3.7
$\underline{2a}$, k_{RO-}	26.2	25.6	21.5	13.6
$\underline{6}$, k_{OH-}	25.6	25.0	26.3	-4.4

[a] Data obtained at six temperatures over a range of 25-40°
[b] kcal/mol
[c] cal/deg-mol.

when one compares the neutral and anionic species (k_{ROH} vs. k_{RO-}/k_{OH-}). It is clear that reactions of the anions are entropically more favored than the neutral alcohol. This is perhaps due to solvation of the anion, and associated positive $\Delta S^{\ddagger}_{solv}$ in going from the solvated initial state to the desolvated anion in the transition state.

In collaboration with Dr. Richard Schowen and colleagues at the University of Kansas, we have determined secondary kinetic isotope effects for the α-deutero analog of $\underline{1}$ (20), and have found that the ratio k_H/k_D goes from 1.27 for k_{ROH} to 0.997 for k_{RO-}, indicating a tighter transition state in the alkoxide reaction. Similar experiments, carried out with COMT, give $V_{CH_3}:V_{CD_3}=0.83$, thus indicating an even tighter transition state for the enzyme-catalyzed process (21). These data suggest that compression of the substrates in the transition-state provides a large portion of the catalytic efficiency in the COMT reaction (22).

A major finding of this research has been the establishment of general catalysis in reactions at sp^3 carbon. The reactions of the neutral species $\underline{1}$ and $\underline{2}$ (k_{ROH}) are catalyzed by added buffer bases; no catalysis is observed with the buffer conjugate acid (15, 16). However, the nature of the observed catalysis differs depending on the buffer species involved. Oxyanions, such as carbonate, phosphate, acetate, formate, etc., catalyze the reaction shown in eq. (1) and (2), while amines, such as n-butylamine, N-methylmorpholine, N, O-dimethylhydroxylamine, etc., effect the demethylation of $\underline{1}$ (and sometimes $\underline{2}$) to give the corresponding p-nitrophenylthioethers. At pH's where the anionic substrate is the predominate species present in solution, k_{RO-} is so fast that no buffer catalysis is seen over the lyate reaction; no enhanced rate is observed for the reaction of $\underline{2}$ in either n-butyl-amine or carbonate buffers (pH ca. 9-11). A Bronsted plot of the second-order rate constants obtained for general-base catalysis of reactions (1) and (2) by oxyanion buffers gives β=0.27, which is in agreement with β values obtained for both intramolecular and intermolecular nucleophilic displacement reactions, and discussed above. Interestingly, the second-order rate constants obtained for nucleophilic demethylation of 1 by amine buffers also obey the Bronsted relation, with β=0.2̄7̄. This equality of β-values for general-base and nucleophilic catalysis of reactions at sp^3 carbon may be just a coincidence, but it may also reflect a similarity in the transition state of the two processes as shown

$$
\text{B:} \cdots\text{H}\cdots \overset{\delta^+}{\text{O}} \cdots \cdots \text{CH}_2 \overset{\delta^+}{\cdots} \text{S} \begin{matrix} \nearrow \text{CH}_3 \\ \searrow \phi\text{NO}_2 \end{matrix}
$$

$$
\underset{\sim}{\text{7}}
$$

$$
\text{R--CH}_2\text{--S} \begin{matrix} \overset{\delta^+}{\cdots}\text{CH}_3 \cdots\cdots \overset{\delta^+}{\text{Nuc}} \\ \searrow \phi\text{NO}_2 \end{matrix}
$$

$$
\underset{\sim}{\text{8}}
$$

(7 vs. 8). A critical difference to note is that buffers which catalyze the ring-closure reactions (7) of eq. (1) and (2) are oxyanions, whereas buffers which effect nucleophilic displacement (8) are amines. Amine-catalyzed proton abstraction would result in charge accumulation at the amine nitrogen, whereas the observed case (7) leads to loss of charge on the oxygen atom. Thus, the results support the idea that general-base catalysis of alkyl transfer to alcohols or phenols requires loss of charge in the catalytic base in the transition state. Proton abstraction by amines to yield a charged base in the transition state apparently is untenable, and the more common route of nucleophilic demethylation is observed with amine buffers.

One exception to the general conclusions just described is found in the case of imidazole buffer. This amine, which has been implicated as a critical base in many biological reactions at sp^2 carbon and phosphorus, is non-catalytic in reactions such as eq. (1) and (2). This surprising result indicates that histidine residues cannot act as general-base catalysts in SAM-dependent methylases. In addition, it suggests that the enzyme-catalyzed methylation of histamine may require some special mechanism other than the direct transfer discussed earlier for the COMT reaction. As already mentioned, kinetic data in the literature (14) suggest a ping-pong mechanism for this reaction. Considering the extremely low reactivity of imidazole in aqueous alkyl transfer reactions, one can envision a mechanism where the enzyme-catalyzed histamine methylation is effected by a double displacement reaction. Thus, SAM might methylate a methionine residue on the enzyme, and the resulting enzyme-bound methyl sulfonium could react with histamine in a nonpolar environment to give the methylated product, N^1-methylhistamine and regenerate native enzyme. This mechanism takes advantage of the known enhanced susceptibility of methyl sulfonium salts to nucleophilic attack in non-polar media (1, 23). In addition, it allows for the methylation of a soft base, the sulfur atom of methionine, by SAM in a more polar surrounding. Soft bases are known to react readily with methylsulfonium salts in aqueous media (19). The proposed mechanism for histamine methylation is shown in Fig. 2.

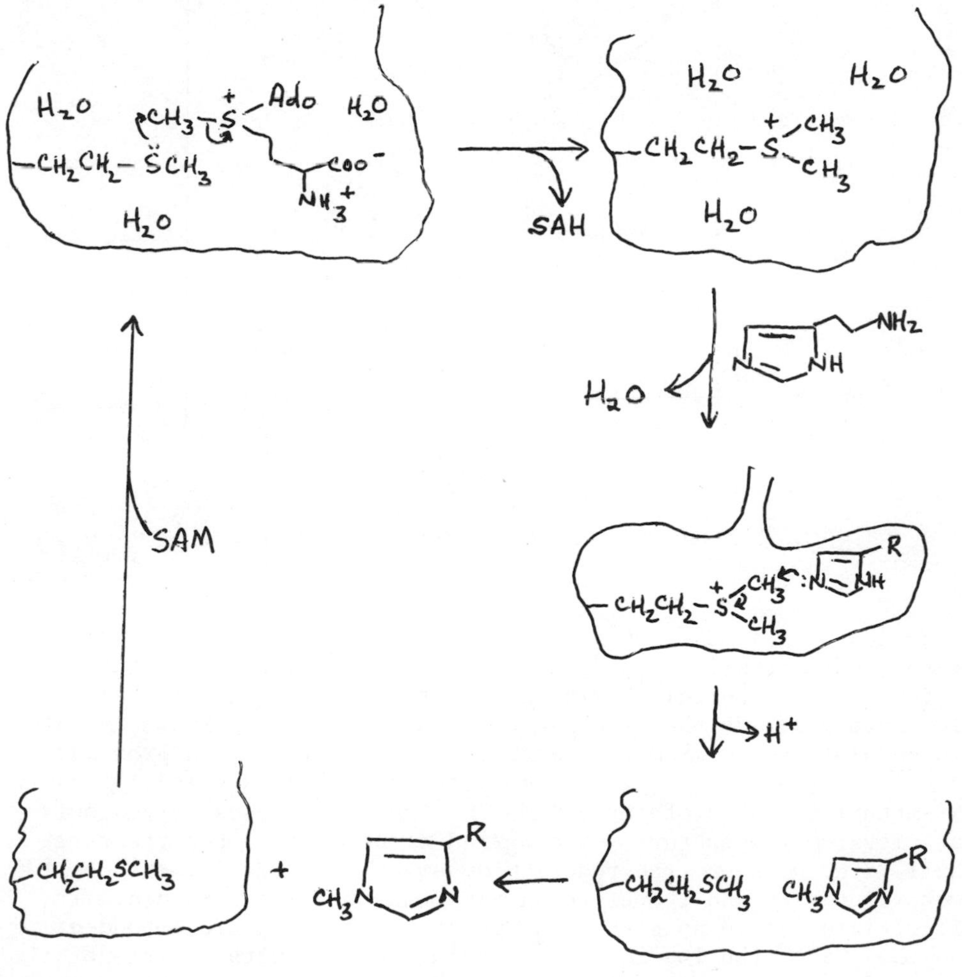

Fig. 2. Proposed role of desolvation in histamine methylation.

The role of active site polarity in enhanced enzyme-catalyzed alkyl transfers has not been investigated in detail. It is clear that indiscriminate alkylation of cellular nucleophiles is incompatible with life; therefore enzyme-catalyzed alkyl transfer must be strictly regulated in vivo. Fig. 3 shows rate data for the

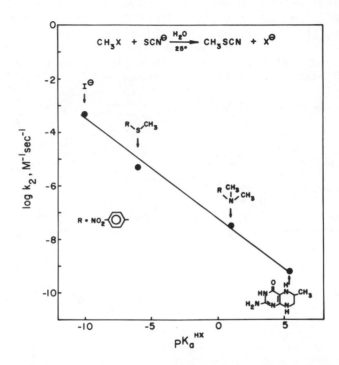

Fig. 3. Effect of leaving group on methylation rates.

reaction $CH_3X + SCN^- \rightarrow CH_3 SCN + X^-$ acquired at various temperatures, but normalized to 25° using reasonable activation parameters. The powerful alkylating agent, methyl iodide is ca. 10^2 more reactive than the p-nitrophenyl sulfonium, which in turn is ca. 10^2 and $10^{3.5}$ more reactive than ammonium and a protonated pterin, respectively (24). The latter compound is a model for N^5-methyl tetrahydrofolate, (CH_3-FH_4) and decomposes very slowly at elevated temperatures to a myriad of products. For the first half-life, however, the reaction observed is primarily the desired methyltransfer and therefore an estimate of k_2 is obtained. It is of interest to note that the rates of reaction are dependent on the pKa of the leaving group conjugate acid with a slope of ca. -0.4. One of the roles of the protein backbone of enzymes which catalyze alkyl transfer reactions might be to provide a nonpolar active site in which a more facile reaction could be effected. In the absence of these enzyme, the natural alkylating agents, SAM and CH_3-FH_4 are relatively unreactive cellular components. From data presented herein, one can calculate that proximity effects can contribute ca. 10^6 to 10^8 rate enhancement for reactions at sp^3 carbon, whereas general catalysis only enhances these reaction by ca. 10. It can be shown that the COMT reaction catalyzes the methyl transfer by ca. 10^{16} over the nonenzyme counterpart (20). Thus, combined proximity and general catalysis effects contribute only ca. 10^7-10^9 of that 10^{16} rate difference. The remaining ca. 10^8 may come either from compression of the transition-state (22), desolvation of the ground state, or both.

REFERENCES

1. J.K. Coward, in "The Biochemistry of Adenosylmethionine", F. Salvatore, et al., eds., Columbia University Press, N.Y. (1977), p. 127.
2. J.K. Coward and P.A. Crooks, in "Transmethylation", E. Usdin, et al., eds., Elsevier/North Holland, N.Y. (1979), p. 215.
3. R. Wolfenden, Ann. Rev. Biophys. Bioeng., 5:271 (1976).
4. J.M. Saavedra, Essays Neurochem. Neuropharmacol., 1:1 (1977).
5. J.K. Coward, E.P. Slisz, and F. Y-H. Wu, Biochem., 12:2291 (1973).
6. W.W. Cleland, Enzymes, 3rd Ed., 2:1 (1970).
7. R.W. Woodard, P.A. Crooks, J.K. Coward, and H.G. Floss, manuscript in preparation; c.f. H.G. Floss, et al., Ref. 1, p. 135.
8. R.T. Borchardt, J. Med. Chem., 16:377 (1973).
9. S.J. Kerr, Methods Cancer Res., 15:163 (1978).
10a. J.M. Glick and P.S. Leboy, J. Biol. Chem., 252:4790 (1977); b. J.M. Glick, V.M. Averyhart and P.S. Leboy, Biochem. Biophys. Acta, 518:158 (1978).
11. E. Barbosa and B. Moss, J. Biol. Chem., 253:7692 (1978).
12. W.K. Paik and S. Kin, Adv. Eng., 42:227 (1975).
13. M. Jamaluddin, S. Kim and W.K. Paik, Biochemistry, 14:694, (1978).
14. A. Thithapandha and V.H. Cohn, Biochem. Pharmacol., 27:263 (1978).
15. J.O. Knipe and J.K. Coward, J. Amer. Chem. Soc., 101:July, (1979).
16. J.O. Knipe, P.J. Vasquez, and J.K. Coward, manuscript in preparation.
17a. T.C. Bruice and S.J. Benkovic, "Bioorganic Mechanisms", W.A. Benjamin, N.Y. (1966), Vol. 1, p. 119; b. W.P. Jencks, "Catalysis in Chemistry and Enzymology", McGraw-Hill, N.Y. (1969), p. 11.

18a. T.C. Bruice, Ann. Rev. Biochem., 45:331 (1976); b. W.P. Jencks,
     Adv. Enz., 43:219 (1975).
19. J.K. Coward and W.D. Sweet, J. Org. Chem., 36:2337 (1971).
20. I. Mihel, J.O. Knipe, J.K. Coward, and R.L. Schowen, J. Amer.
     Chem. Soc., 101:July (1979).
21. M.F. Hegazi, R.T. Borchardt, and R.L. Schowen, J. Amer. Chem.
     Soc., 101:July (1979).
22. C.H. Gray, J.K. Coward, K.B. Schowen, and R.L. Schowen, J.
     Amer. Chem. Soc., 101:July (1979).
23. C.K. Ingold, "Structure and Mechanism in Organic Chemistry",
     2nd Ed., Cornell University Press, N.Y. (1969) p. 457.
24. J.K. Coward, unpublished results.

ADENOSYLMETHIONINE AS A PRECURSOR FOR NUCLEIC ACIDS MODIFICATION

Francesco Salvatore, Cinzia Traboni, Alfredo Colonna,
Gennaro Ciliberto, Giovanni Paolella, Filiberto Cimino

Istituto di Chimica Biologica, II Facoltà di Medicina
e Chirurgia, Università di Napoli, Via S. Pansini 5,
80131 Napoli, Italy

INTRODUCTION

Adenosylmethionine (Ado-Met) is the most important donor of
alkyl groups, which are transferred from the sulfonium pole to a
variety of nucleophilic molecular species within the cell[1]. The
reaction involves nucleophilic attack to one of the three carbon
atoms adjacent to the electron-deficient trivalent sulfur (see
Figure 1) and is catalyzed by specific enzymes[2]. Whereas the
transfer of the methyl group occurs in a large variety of reactions[1,3], the transfer of both the adenosyl moiety and of the
3-amino-3-carboxypropyl group have been described in very few
instances[4]: the acceptor molecules are enzyme proteins for the
adenosyl moiety[5-7], and transfer RNA (tRNA) for the 3-amino-
3-carboxypropyl group[8].

Due to its fundamental role as alkyl donor, adenosylmethionine
largely contributes to the post-synthetic alteration of macromolecules, both the informational (proteins and nucleic acids) and
the non-informational ones (polysaccarides, and complex phospholipids in membrane assembly). For extensive discussion concerning
the enzymology and the role of polysaccaride alteration by
adenosylmethionine we refer to a review by Ballou[9]. As far as
phospholipids and proteins are concerned we refer to other papers
of this Symposium (see Mozzi and Porcellati, and Paik and Galletti,
respectively).

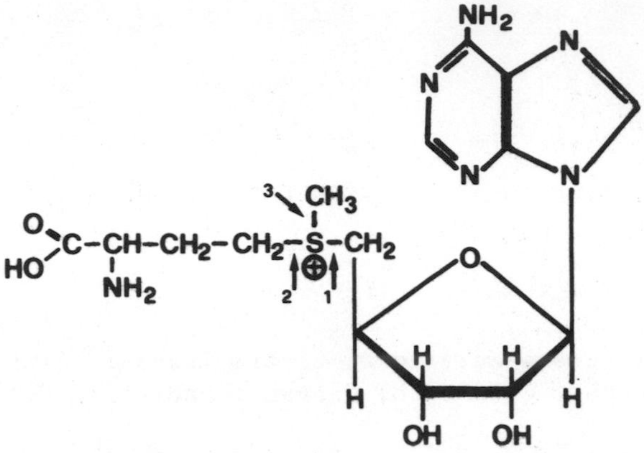

Fig. 1.   S-adenosylmethionine as a donor of alkyl groups: adenosyl
          moiety, 1; 3-amino-3-carboxypropyl group, 2; methyl
          group, 3.

POST-TRANSCRIPTIONAL METHYLATION OF NUCLEIC ACIDS

Among post-transcriptional modifications of nucleic acids,
methylation is the most frequent one; a number of methylated
nucleosides have been isolated from all types of nucleic acids[3].

DNA methylation has been studied since many years[10]. In prokaryotes, a well defined and important role has been demonstrated
for this process: methylation seems to protect specific sites from
cleavage by restriction endonucleases[11]; both 5-methylcytosine and
6-methyladenosine have been isolated as products of DNA methylation.
In eukaryotes only 5-methylcytosine is formed, and exclusively in
the sequence CpG[12]: its role has been suggested from time to time
as implicated in gene expression during cell differentiation[13-15],
also because of variations found during embryonic development[16] in
different tissues, among individuals of varying ages[17] and between
normal and transformed cells[18]. However, the evidence available
is largely correlative: in fact, no direct evidence as to the
biological functions of DNA methylation has been so far presented.

In recent years messenger RNA (mRNA) has been found to be
methylated, particularly at the 5'-end[19,20]. The type and position
of mRNA methylation is schematically depicted in Figure 2. Some
methylations are constant in all eukaryotic mRNAs, some other appear to be facultative. Whereas the methylation is mostly concen-

$m^7G(5')ppp(5')N'$. . . . . . m^6A . . $Np(polyA)-OH(3')$ "cap zero"

$m^7G(5')ppp(5')N'mp$. . . . m^6A . . $Np(polyA)-OH(3')$ "cap 1"

$m^7G(5')ppp(5')N'mpN''mp$. . m^6A . . $Np(polyA)-OH(3')$ "cap 2"

Fig. 2.   Different mRNA "cap" structures from eukaryote and viral
           systems. N',N" = 1st, 2nd nucleoside at the 5'-end;
           N'm, N"m = 2'-0-methylnucleoside of "cap 1" and "cap 2"
           structures; in some cases N'm and N"m are m^6Am, internal
           m^6A may be absent.

trated at the 5'-end, where it contributes to the formation of the
so-called "cap" structures[21], further methylation is present along
the polynucleotide chain as N^6-methyladenosine ($\sim 1/1500$ nucleotides)[22]; also present in some mRNAs are residues of 5-methylcyto-
sine[23]. The "cap" structure of mRNA appears to be required for protein synthesis; in fact, absence of capping results in decreased
ribosome binding and translation in *in vitro* systems; re-addition
of 7-methylguanosine restores the activity[24]. When methylation is
absent translation is not restored, pointing to the importance of
methyl group at the position 7 of the guanine ring; however, methyl
groups do not seem to be critical for ribosome binding to mRNA."Cap"
structure has been demonstrated to increase the stability of mRNA
toward degradation[21]. In the maturation of mRNA, the first methylation to occur is usually 7-methylguanosine; however, in some cases
ribose methylation precedes the N^7 methylation of guanosine[25].

     Another role for Ado-Met has been recently found by Shatkin[26]:
it modifies allosterically RNA polymerase, thus activating RNA initiation in a viral system, namely cytoplasmic polyhedrosis virus,
during transcription of mRNA.

     Ribosomal RNA (rRNA) is methylated mostly at the 2'-0-ribose
moiety, particularly in eukaryotes, where this type of methylation
accounts for 90% of the total methyl groups of rRNA (see review by
Maden[27]). Methylation occurs precociously during the maturation
process and before the rRNA precursor is transported to the cytoplasm[28-30]. It is interesting to note that ribose methylation
within nucleic acid has also been considered as a polysaccharide
modification because RNA resembles a polymer of phosphorylated ribose with a continuous series of purine and pyrimidine bases attached to the glycosidic groups.

TRANSFER RNA METHYLATION

A number of tRNA species (up to 50–60 species) exists within a single cell: their structural and functional individuality is due to the specific nucleotide sequence, as well as to the number of nucleoside modifications[31]. Thus, a series of modified, or minor, nucleosides are formed at post-transcriptional levels: methylation is the most frequent one, both in prokaryotes and eukaryotes, accounting for 50% of the total number of the modified nucleosides[32,33]. About 30 methylated nucleosides have been so far identified in tRNAs from different sources; however, it has been stressed that only a few (4 or 5) of these modified nucleosides constitute the bulk of all methylated nucleosides present in each type of prokaryote cell[34]. In a study on the enzymology of formation of methylated nucleosides in tRNA of *S.typhimurium* we have found that, out of eleven methylated nucleosides which have been identified in *in vivo* experiments, more than 80% of the methyl groups are localized in only 4 methylated nucleosides (5-methyluridine, 7-methylguanosine, 2-methyladenosine and 2'-O-methylguanosine)[35,33]. In other prokaryotes, as *E.coli* and *B.subtilis* (see Table 1) an analogous pattern has been observed, but the type of nucleosides are different[36,37]. This allowed to divide the methylated nucleosides in two classes on the basis of their relative abundance in tRNA of a single prokaryote[34]: the "major" methylated nucleosides, which usually are not more than 4 or 5, represent about 80% of the total methyl groups present; the "minor" methylated nucleosides, which represent all the others in the same cell type. The limited number of the "major" methylated nucleosides indicates that in each cell type a specific set of tRNA methyl transfer enzymes predominates: it is worth noting the constant presence of 5-methyluridine and 7-methylguanosine. The average number of methylated nucleosides present in a population of tRNA molecules varies from lower to higher organisms: less than 3 per cent in *Mycoplasma*, about 3 per cent in prokaryotes, and about 10 per cent in eukaryotes[38,39]. Not only does the absolute number of methylated nucleosides increase from lower to higher organisms, but their variety increases as well.

The variety of modified nucleosides which have been discovered along the primary sequence of tRNA derives from a series of post-transcriptional enzymatic modifications. In fact, the biosynthesis of tRNA takes place through a very complex process, which as a first step includes the action of RNA polymerase, which forms a precursor tRNA . Such precursor molecule, which is longer than the mature size tRNA, is in most of the cases completely lacking modified nucleosides. A not yet completely unravelled complex process of maturation[40,34] then occurs, which includes:*(i)* nucleolytic tailoring of the longer precursor; *(ii)* 3'-and 5'-end refinement (not always necessary); *(iii)* formation of modified nucleosides, which in some instances – i.e., hypermodified nucleosides – derive

Table 1. Percentage composition of methylated nucleosides in tRNA
of some prokaryotes [a]

Modified nucleosides	*S. typhimurium* (ref.35)	*E. coli* (ref.36)	*B. subtilis* [b] (ref.36)	(ref.37)
5-methyluridine	48.99[+]	43.5[+]	2.5[+]	47.80[+]
7-methylguanosine	17.58[+]	30[+]	32.2[+]	9.75[+]
2-methyladenosine	8.16[+]	13[+]	10.7[+]	1.46
2'-O-methylguanosine	7.37[+]	n.e.	n.e.	n.s.d.
1-methylguanosine	4.17	4.1[+]	10.1[+]	6.83[+]
6-methyladenosine	3.49	2.1	5.0	0.49
2'-O-methyluridine	3.35	n.e.	n.e.	n.s.d.
2'-O-methylcytidine	2.65	n.e.	n.e.	n.s.d.
mt^6A	2.12	n.d.	n.d.	n.d.
ms^2i^6A	2.11	n.d.	n.d.	0.97
3-methylcytidine	<1.0	<1.0	<1.0	n.s.d.
6-dimethyladenosine	n.d.	n.d.	3.0	n.d.
2-methylguanosine	n.d.	<1.0	5.5	6.83[+]
5-methylcytidine	n.d.	<1.0	2.0	n.s.d.
3-methyluridine	n.d.	<1.0	<1.0	n.s.d.
1-methyladenosine	n.d.	n.d.	23[+]	7.31[+]
7-methyladenosine	n.d.	n.d.	n.d.	4.87
X unidentified	---	5.7	3.2	---
Z unidentified	---	2.2	---	---
methylated pyrimidine	---	---	---	8.29[+]
ribose methylated nucleoside(s)	---	---	---	0.97
mam^5s^2U	---	---	---	4.39

a) The percentage is calculated on the basis of the total radioactivity incorporated in ^{14}C-methyl groups of tRNA nucleosides
from bacterial cultures grown in presence of [14]CH_3-methionine.
However, the data by Vold (reported in the colum, ref. 37)
include figure obtained for 5-methyluridine (O.D. measurement).
b) The discrepancies between figures from Kersten[36] and Vold[37] are
due to several factors (difference in estimation methods, difference in bacterial strains, inclusion of the methyl group derived from methyl-tetrahydrofolate in Vold's data on 5-methyl-
uridine).
+) Indicates preponderant nucleosides, called "major" methylated
nucleosides[34], which account for more than 80% of total methylated nucleosides.
mt^6A= N-[N'-methyl-N-(9-β-D-ribofuranosylpurin-6-yl)-carbamoyl]-
threonine; ms^2i^6A= $\bar{N}^6$-(Δ^2-isopentenyl)-2-methylthioadenosine;
mam^5s^2U= 2-thio-5-(N-methylaminomethyl)-uridine; n.d.= not detectable; n.e.= not estimated; n.s.d.= not specifically detected: figures are included under "methylated pyrimidine" or "ribose methylated nucleosides".

from a multistep enzymatic process. No generalization is possible
about the timing of events occurring in the maturation process of
tRNA: in fact, in some instances modifications seem to occur prior
to the processing, whereas in other cases they occur after or
during the trimming process[34,40].

Recent studies[41], particularly by X-ray crystallography, have
shown that tRNA molecules have a highly ordered tertiary structure
basically similar in different molecular species; however, it has
been also demonstrated a certain degree of flexibility for which
the modified nucleosides are considered to play a role by increasing the number and specificity of recognizable surfaces[40,41].
These are, obviously, critical for proper interaction with other
macromolecules in the multi-step process of protein biosynthesis.

A number of studies have indicated significant roles of
methylated nucleosides in some of the molecular functions of tRNA
in this process. In fact, the amino acid acceptance, the codon
response and the ribosomal binding have been on several occasions
indicated as a target for the biological role of specific methyl
groups in tRNA[32,33]. To mention some quite recent results, we
refer to the papers by Pope et $al.$[42], and by Rizzino et $al.$[43]. In
the first of these papers it has been reported that a mutant strain
of $S.typhimurium$ $(supK)$, which is able to suppress a nonsense
mutation, is defective in a tRNA methylating enzyme, which contributes to the biosynthesis of a minor methylated nucleoside, i.e. the
methylester of uridine-5-oxyacetic acid. In two tRNA species,
tRNA[Ala] and tRNA[Ser], this nucleoside is present at the first position of the anticodon, thus indicating its critical role for codon-
anticodon recognition. Evidence has also been presented that
methylated nucleosides can be essential for regulatory processes in
the cell. In the paper by Rizzino et $al.$[43] it has been reported
that in $vivo$ conditions which produce undermethylated tRNA[Ile],
tRNA[Leu] and tRNA[Val] induce a ten-fold derepression of the enzymes
involved in valine and isoleucine biosynthetic pathway, even in
presence of an excess of both these amino acids. It may be inferred
that the derepression is caused by the inability of methyl deficient tRNAs to participate properly in some cellular regulatory
functions.

For a more extensive discussion concerning the role of methylated nucleosides at the molecular level in cellular processes, we
refer to more detailed reviews[34,38,44,45] which have appeared in the
last few years. Following different lines of research, in the last decade several data have appeared in the literature which suggest a rel
tionship between the cell or tissue pattern of tRNA species and
cell differentiation or malignant transformation[46]. Differences in
tRNA patterns have been shown in several instances for malignant
tissues versus normal ones, often associated with modulation of
activity of the enzymes which methylate tRNA. Also in this case we

refer to extensive reviews[46],[47] for a detailed discussion on this
matter: here we shall only mention some recent relevant findings
which stress qualitative specific changes of malignant cells.
Kuchino and Borek[48] have found that a tRNA^{Phe} from Novikoff hepatoma and from Erlich ascites cells contains two supernumerary methylated bases: one of them, 1-methylguanine, is absent in tRNA^{Phe}
from normal mammalian liver. More rencently, Roe *et al.*[49] have
shown by complete sequence analysis that a hypermodified nucleoside,
called "Q" nucleoside, is not properly modified in the Walker mammary carcinosarcoma (this nucleoside, however, does not contain
methyl groups).

Despite all these indications, it is not yet possible to draw
any definite or generalized conclusion on the role of methylated
nucleosides in the biological functions of tRNA.

STUDIES ON tRNA METHYLTRANSFERASES

Occurrence and Localization

S-adenosylmethionine tRNA methyltransferases have been shown
to occur in all kind of cells so far studied, both prokaryote and
eukaryote. In eukaryote cells they are mostly localized in the cytosol and mitochondria: recent results show the presence of these
enzymes in nuclei of mouse L-cells[50] and of *Xenopus* oocyte[51]. These
latter results suggest that the intracellular localization of such
enzymes should be more accurately investigated by the refined techniques now available. The possible localization of these enzymes
within the nuclei is of interest if visualized in connection with
the timing and the place of tRNA maturation events.

Purification and Specificity

In spite of the difficulties inherent in the unavailability
of proper tRNA substrate, the undermethylated tRNA, partial purification of tRNA methyltransferases have been obtained from a variety of sources, both prokaryote and eukaryote cells. In some
recent cases a high degree of purification has been obtained from
eukaryote cells: we quote here only a few instances where the results
have been more successful to this aim. A combination of chromatographic analyses has been utilized in yellow lupin seeds leading to
a 300-fold purification[52]: however, the various activities were not
separated. Several purification attempts have been performed from
rat liver: the most successful preparations are those reported by
Leboy's group, who obtained high degree of purification for
tRNA(adenine-1)-methyltransferase[53] as well as for tRNA(guanine-1)-
methyltransferase and tRNA(N^2-guanine)-methyltransferase[54]. Other
less recent results are reported in the accurate review by Nau[38].

As far as prokaryote enzymes are concerned, Aschoff *et al.*[55] have
reported the purification of tRNA(guanine-7)-methyltransferase from
E.coli with the demonstration of the existence of two forms with
M_r of 100,000 and 300,000. A specific methylase involved in the
synthesis of a hypermodified nucleoside, i.e., 5-methylaminomethyl-
2-thiouridine, has also been purified almost to homogeneity by Taya
and Nishimura[56].

Purification of several enzyme activities from *S.typhimurium*
has been carried out in the authors' laboratory[35]. The procedure
is essentially based upon phosphocellulose chromatography of a cell
crude extract. The phosphocellulose, which resembles the polyanion-
ic backbone of a nucleic acid, retains most of the tRNA methylating
activities, while up to 90% of the proteins are eluted in the flow-
through. Four tRNA methylases (Peak A,B,C and D) are then separated
by eluting the column with a linear gradient of KCl ranging from 0
to 1.0 M. The specificity of each enzyme was tested by analyzing
nucleotides, nucleosides and bases, after different types of hydrolysis of undermethylated tRNA methylated *in vitro* with $^{14}CH_3$-
adenosylmethionine. On the basis of the methylated nucleosides
formed, the four enzyme preparations, respectively eluting at 0.17,
0.29, 0.39 and 0.62 M KCl, show the following products specificity:
1) N-$\boxed{\text{N'-methyl-}\underline{N}\text{-(9-β-D-ribofuranosylpurin-6-yl)-carbamoyl}}$-
threonine; 2) 1-methylguanosine; 3) 5-methyluridine, 1-methyl-
guanosine, and 3-methylcytidine (?); 4) 7-methylguanosine. Except
for enzyme preparation n. 3, the other three are free of contaminating tRNA methylases.

As it was shown also by the data reported above, tRNA methyltransferases exibit a very complex kind of specificity toward the
tRNA substrate. In fact, three requirements must be fulfilled in
order to achieve the enzymatic attachment of a methyl group to tRNA.
Each enzyme must recognize: *(i)* the proper moiety along the polynucleotide chain (either the specific base of the ribose); *(ii)* the
position of the modification at the purine, pyrimidine or furanosic
rings; *(iii)* the localized nucleotide sequence and the spatial
locus of the three-dimensional configuration in which the methylatable nucleoside is positioned. These three requirements allow us to
define three different types of specificity for tRNA methyltransferase, namely: *moiety* specificity, *ring-atom* specificity and *site*
specificity, respectively.

Many difficulties have to be faced during the purification of
tRNA methyltransferases. The major ones are: *(i)* enzyme multiplicity, which makes it cumbersome to isolate each specific reaction product, and then difficult to evaluate specific activity and extent
of purification; *(ii)* enzyme instability; *(iii)* frequent contamination by ribonuclease activity; *(iv)* possible presence of complexes with endogenous tRNA; *(v)* unavailability of a proper tRNA substrate, as we have already discussed. To point out the problem of

enzyme instability we may mention that during a preparation of cell free extract of *S.typhimurium* several tRNA methyltransferases are missed[35]. In fact, by growing *S.typhimurium* in presence of $^{14}CH_3$-methionine, eleven methylated nucleosides are identified after hydrolysis of tRNA formed during cell growth (see Table 1), while only seven methylated nucleosides derive from the hydrolysis of undermethylated tRNA methylated *in vitro* with $^{14}CH_3$-adenosyl-methionine and a crude extract of *S.typhimurium*.

Properties

tRNA methylases appear to share many similar properties[38,47]. Optimum pH is in the range between 7 and 9; reducing agents seem to be essential for their activity; monovalent anions, as carbonate, perchlorate and thiocyanate inhibit the enzyme activity; magnesium ions and polyamines, depending upon their relative concentration, stimulate the activity of eukaryotic tRNA methylases, though they seem to act at the level of tRNA conformation; monovalent cations in some cases seem to increase methylation rate.

The Michaelis constants for adenosylmethionine of tRNA methylases vary from 7×10^{-5} M for *E.coli* tRNA(guanine-1)-methyltransferase to 1.5×10^{-6} M for mammalian enzymes[38]. The four *S.typhimurium* enzymes separated in the authors' laboratory[35] show K_M values ranging from 1.5 to 3.2×10^{-5} M. While the apparent affinity constants for adenosylmethionine are very similar, those for undermethylated tRNA appear to be scattered in a wider range: for example, the K_M's found for the four enzymes isolated from *S.typhimurium*[35] range from 1.2×10^{-4} M to 6.3×10^{-5} M. This could imply that the site for adenosylmethionine is similar in all the different enzymes, whereas interaction with tRNA involves the presence of enzyme sites, which are more typical for specific methylatable sites of tRNA. It must be noted, however, that the measure of K_M for tRNA can be strongly influenced by the use of a non suitable substrate, like crude undermethylated tRNA, in which it is difficult to assess precisely what proportion of the entire preparation is the true substrate.

The molecular size of the four methylase activities isolated from *S.typhimurium* was studied[35] by comparing their sedimentation profile with that of proteins of known molecular weight. Results obtained show that four tRNA methylases have a M_r ranging from 25,000 to 65,000 daltons, except the enzyme preparation mentioned above as no. 1, which shows also an aggregate form of higher M_r (120,000). All the other tRNA methyltransferases isolated so far from other sources appear to have higher M_r[53-55,57,58].

Adenosylhomocysteine (Ado-Hcy) is the first studied and the most active inhibitor of tRNA methyltransferases, both from

prokaryote and eukaryote cells. Worthy of note is that adenosylhomocysteine is a product of transmethylation reaction, and its
intracellular concentration can be a factor of regulation of tRNA
methylases[59]. Experiments performed in our laboratory[60] (see Figure 3) show the effect of adenosylhomocysteine and some of its
analogs on the activity of tRNA(guanine-7)-methyltransferase from
S. typhimurium. These data indicate that, also for this enzyme activity, adenosylhomocysteine is a powerful inhibitor as for other
methyl transfer reactions. In Figure 4 the competitive nature of
this inhibition is shown. In Table 2 the inhibition constants of
adenosylhomocysteine and several of its analogs on this enzyme activity are reported. The analysis of the results obtained suggest
the following considerations: *(i)* the chirality at the α-carbon
atom of the amino acid chain is critical, since the D-form of the
inhibitor has a decreased inhibitory power by two orders of magnitude; *(ii)* the length of hydrocarbon chain is also critical, since
the shortening of one methylene group also lessens greatly the

S-Adenosylhomocysteine	:	$R-S-CH_2-CH_2-\underset{\underset{COOH}{\mid}}{C}HNH_2$
S-Adenosylcysteine	:	$R-S-CH_2-\underset{\underset{COOH}{\mid}}{C}HNH_2$
Methylthioadenosine	:	$R-S-CH_3$
Thioethanoladenosine	:	$R-S-CH_2-CH_2OH$
Butylthioadenosine	:	$R-S-CH_2-CH_2-CH_2-CH_3$
Isobutylthioadenosine	:	$R-S-CH_2-\underset{\underset{CH_3}{\mid}}{C}H-CH_3$
S-Inosylhomocysteine	:	$R'-S-CH_2-CH_2-\underset{\underset{COOH}{\mid}}{C}HNH_2$
Methylthioinosine	:	$R'-S-CH_3$

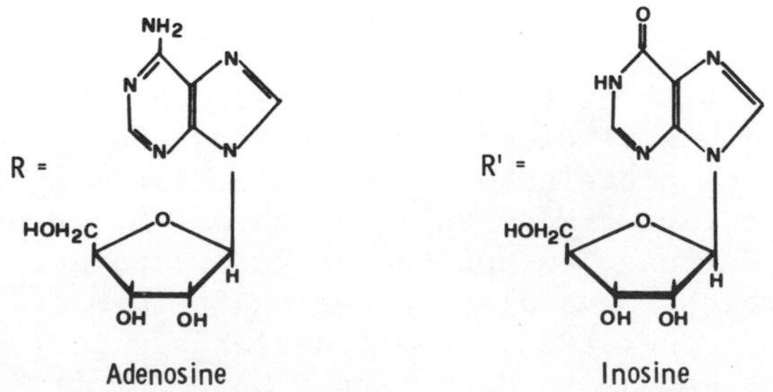

R =   Adenosine

R' =   Inosine

Fig. 3.   Structures of S-adenosylhomocysteine and several thio-
           ester analogs.

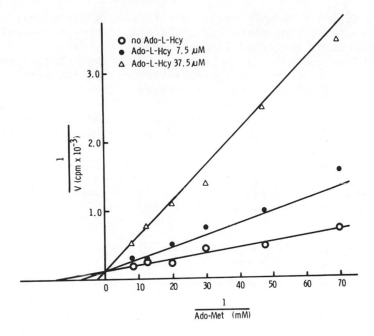

Fig. 4. Inhibition of tRNA(guanine-7)-methyltransferase by adenosyl-L-homocysteine (Ref. 60).

Table 2. Inhibition constants of S-adenosylhomocysteine and some thioester analogs on tRNA(guanine-7)-methyltransferase activity (Ref. 60).

Compound	K_i (M x 10^{-6})
S-Adenosyl-L-homocysteine	7.65
S-Adenosyl-D-homocysteine	841.17
S-Adenosyl-DL-homocysteine	64.14
S-Adenosyl-L-cysteine	545.21
Methylthioadenosine	458.57
Thioethanoladenosine	5495.61

inhibitory power (see K_i values for S-adenosylcysteine); *(iii)* the presence of only one methyl group as side chain at the sulfur atom also reduces of about 50 times the inhibitory effect (see K_i of methylthioadenosine); *(iv)* the presence of the ethyl group decreases the inhibitory power to a concentration of the inhibitor equal to 5.5 mM (see K_i of thioethanoladenosine).

As an overall conclusion it may be stated that the chain length as well as the amino and/or the carboxyl residue of the amino acid moiety appear to be critical in the binding of competitive inhibitors to the enzyme molecule. This confirms that the methyl donor substrate, Ado-Met, most likely binds through several of its chemical groups to the enzyme protein, as it has been previously indicated with other methyltransferase enzymes[61].

Table 3. Purification of tRNA(guanine-7)-methyltransferase from
S. typhimurium (Ref.62).

Step	Procedure	Total protein	Total activity	Specific activity	Purification
		mg	*units*	*units/mg protein*	*-fold*
I	Crude extract	721.9	334,740[+]	463.7	1
II	Phosphocellulose chromatography (Peak D)	6.0	156,793	26,132	56·
III	Ammonium sulfate (45-75%)	3.1	156,497	50,482	109
IV	tRNA-Sepharose chromatography	0.17	23,917	140,688	303

[+]
calculated on the basis of the amount of 7-methylguanosine formed during methylation of undermethylated tRNA incubated with $^{14}CH_3$-adenosylmethionine and *S. typhimurium* crude extract: 7-methylguanosine accounts for 41% of the total methylated nucleosides formed.

More recently, in the authors' laboratory[62] the tRNA(guanine-7)-methyltransferase from *S. typhimurium* has been further purified, essentially by affinity chromatography performed on Sepharose column with undermethylated *E. coli* tRNA attached through a six C-atoms arm (see Table 3). The enzyme resulted in 300-fold purification; analysis on SDS polyacrylamide gel electrophoresis under strong denaturing conditions showed only three protein bands.

SUMMARY

Among biochemical compounds, S-adenosylmethionine owns a unique versatility: its chemical structure, with three substituents attached to the sulfonium pole, permits its participation in several different types of biochemical reactions involving both low and high molecular weight compounds as acceptors of alkyl groups. In the case of macromolecules, both the informational and the non-informational ones, S-adenosylmethionine contributes to some of their post-synthetic modifications, thus allowing the attainment of their native configuration, which displays the specific biological function.

After briefly reviewing the modifications by S-adenosylmethionine of DNA, mRNA and rRNA, the Authors discuss the methylation of tRNA, and report the most significant data on its role in the cellular functions of tRNA. In particular, the authors summarize some results from their laboratory concerning kinetic studies on the effect of adenosylhomocysteine and several thioester analogs on tRNA(guanine-7)-methyltransferase, and concerning a fast procedure to purify this enzyme from *S. typhimurium*.

ACKNOWLEDGEMENTS

The experimental work carried out in the authors' laboratory has been supported by grants from CNR (Rome) and by an International Science Cooperative Program (CNR, Italy-National Science Foundation, USA).

REFERENCES

1. "The Biochemistry of Adenosylmethionine", F. Salvatore, E. Borek, V. Zappia, H.G. Williams-Ashman, and F. Schlenk, eds., Columbia University Press, New York (1977).
2. J.K. Coward, in "The Biochemistry of Adenosylmethionine", F. Salvatore, E. Borek, V. Zappia, H.G. Williams-Ashman, and F. Schlenk, eds., Columbia University Press, New York, p.127 (1977).
3. "Transmethylation", E. Usdin, R.T. Borchardt, and C.R. Creveling, eds., Elsevier-North Holland Press, New York-Amsterdam-Oxford (1979).

4. V. Zappia, F. Salvatore, M. Porcelli, and G. Cacciapuoti, in "Biochemical and Pharmacological Roles of Adenosylmethionine and Nervous System", V. Zappia, E. Usdin, and F. Salvatore, eds., Pergamon Press, New York and London, in press (1979).
5. S.H. Mudd, and J.D. Mann, J.Biol.Chem. 238: 2164 (1963).
6. J. Knappe, and T. Schmitt, Biochem.Biophys.Res.Commun. 71: 1110 (1976).
7. V. Zappia, and F. Ayala, Biochim.Biophys.Acta 268: 573 (1972).
8. S. Nishimura, in "The Biochemistry of Adenosylmethionine", F. Salvatore, E. Borek, V. Zappia, H.G. Williams-Ashman, and F. Schlenk, eds., Columbia University Press, New York, p. 510 (1977).
9. C.E. Ballou, in "The Biochemistry of Adenosylmethionine", F. Salvatore, E. Borek, V. Zappia, H.G. Williams-Ashman, and F. Schlenk, eds., Columbia University Press, New York, p. 435 (1977).
10. E. Borek, and P.R. Srinivasan, Ann.Rev.Biochem. 35: 275 (1966).
11. W. Arber, Prog.Nucl.Acid Res. and Mol.Biol. 14: 1 (1974).
12. J. Doskocil, and Sŏrm, Biochim.Biophys.Acta 55: 953 (1962).
13. E. Scarano, Advanc.Cytopharmacol. 1: 13 (1971).
14. R. Holliday, and J.E. Pugh, Science 187: 226 (1975).
15. A.D. Riggs, Cytogenet.Cell Genet. 14: 9 (1975).
16. R.L.P. Adams, Nature New Biology 244: 27 (1973).
17. G.D. Berdishev, G.K. Korataev, G.V. Boyarskikh, and B.F. Vanyushin, Biokhimia 32: 988 (1967).
18. E.D. Rubery, and A.A. Newton, Biochim.Biophys.Acta 324: 24 (1973).
19. A.J. Shatkin, Cell 9: 645 (1976).
20. F.M. Rottman, Int.Rev.Biochem. 17: 45 (1978).
21. A.J. Shatkin, Y. Furuichi, and N. Sonenberg, in "Transmethylation", E. Usdin, R.T. Borchardt, and C.R. Creveling, eds., Elsevier-North Holland Press, New York-Amsterdam-Oxford, p. 341 (1979).
22. R. Desrosiers, K. Frederici, and F. Rottman, Proc.Natl.Acad. Sci. USA 71: 3971 (1974).
23. D.T. Dubin, and R.H. Taylor, Nucleic Acids Res. 2: 1653 (1975).
24. S. Muthukrishnan, B. Moss, J.A. Cooper, and E.S. Maxwell, J.Biol.Chem. 253: 1710 (1978).
25. A.K. Banerjee, D. Testa, and V. Deutsch, in "Transmethylation", E. Usdin, R.T. Borchardt, and C.R. Creveling, eds., Elsevier-North Holland Press, New York-Amsterdam-Oxford, p. 331 (1979).
26. Y. Furuichi, and A.J. Shatkin, in "Transmethylation", E. Usdin, R.T. Borchardt, and C.R. Creveling, eds., Elsevier-North Holland Press, New York-Amsterdam-Oxford, p. 351 (1979).
27. B.E.H. Maden, in "Transmethylation", E. Usdin, R.T. Brochardt, and C.R. Creveling, eds., Elsevier-North Holland Press, New York-Amsterdam-Oxford, p. 381 (1979).
28. R. Perry, Ann.Rev.Biochem. 45: 605 (1976).
29. F.M. Rottman, M. Kaehler, and J. Coward, in "Transmethylation", E. Usdin, R.T. Brochardt, and C.R. Creveling, eds., Elsevier-North Holland Press, New York-Amsterdam-Oxford, p. 361 (1979).

30. F. Amalric, and J. Bachellerie, in "Transmethylation", E.
    Usdin, R.T. Brochardt, and C.R. Creveling, eds., Elsevier-
    North Holland Press, New York-Amsterdam-Oxford, p. 409 (1979).
31. P.F. Agris, and D. Söll, in "Nucleic acid-protein recognition",
    H.J. Vogel, ed., Academic Press, New York-San Francisco-London,
    p. 321 (1977).
32. F. Cimino, C. Traboni, P. Izzo, and F. Salvatore, in "Macro-
    molecules in the functioning cell", F. Salvatore, G. Marino,
    and P. Volpe, eds., Plenum Press, New York and London, p. 131
    (1979).
33. F. Salvatore, P. Izzo, A. Colonna, C. Traboni, and F. Cimino,
    in "Transmethylation", E. Usdin, R.T. Brochardt, and C.R.
    Creveling, eds., Elsevier-North Holland Press, New York-
    Amsterdam-Oxford, p. 449 (1979).
34. F. Salvatore, and F. Cimino, in "The Biochemistry of Adenosyl-
    methionine", F. Salvatore, E. Borek, V. Zappia, H.G. Williams-
    Ashman, and F. Schlenk, eds., Columbia University Press, New
    York, p. 187 (1977).
35. F. Cimino, C. Traboni, A. Colonna, P. Izzo, and F. Salvatore,
    manuscript in preparation.
36. H. Arnold, and H. Kersten, FEBS letters 36: 34 (1973).
37. B. Vold, J.Bacteriol. 127: 258 (1976).
38. F. Nau, Biochimie 58: 629 (1976).
39. E. Randerath, L.S.Y. Chia, H.P. Morris, and N. Randerath,
    Biochim.Biophys.Acta 366: 159 (1974).
40. S. Altman, ed., "Transfer RNA", MIT University Press (1979).
41. S.H. Kim, and J.L. Sussman, Horizons Biochem.Biophys. 4: 159
    (1977).
42. W.T. Pope, A. Brown, and R.H. Reeves, Nucleic Acids Res. 5:
    1041 (1978).
43. A. Rizzino, M. Mastanduno, and M. Freudlich, Biochim.Biophys.
    Acta 475: 267 (1977).
44. U.Z. Littauer, and H. Inouye, Ann.Rev.Biochem. 42: 439 (1973).
45. S.J. Kerr, Adv.Enzyme Regul. 13: 379 (1974).
46. E. Borek, C.W. Gehrke, and T.P. Waalkes, in "Transmethylation",
    E. Usdin, R.T. Brochardt, and C.R. Creveling, eds., Elsevier-
    North Holland Press, New York-Amsterdam-Oxford, p. 457 (1979).
47. S.J. Kerr, Methods in Cancer Res. 15: 163 (1978).
48. Y. Kuchino, and E. Borek, Nature 271: 126 (1978).
49. B.A. Roe, A.F. Stankiewicz, H.L. Rizi, C. Weisz, M.N. Di Lauro,
    D. Pike, C.Y. Chen, and E.Y. Chen, Nucleic Acids Res. 6: 673
    (1979).
50. A. Colonna, and S.J. Kerr, submitted for publication.
51. R. Cortese, D. Melton, T. Tranquilla, and J.D. Smith, Report
    at the Cold Spring Harbor Meeting on tRNA, Abstracts of papers,
    p. 114 (1978).
52. H. Wierzbicka, H. Jakubovski, and J. Pawelkiewicz, Nucleic
    Acids Res. 2: 101 (1975).
53. J.M. Glick, and P.S. Leboy, J.Biol.Chem. 252: 4790 (1977).

54. J.M. Glick, V.M. Averyhart, and P.S. Leboy, Biochim.Biophys. Acta 518: 159 (1978).
55. H.J. Aschoff, H. Elten, H.H. Arnold, G. Mahal, W. Kersten, and H. Kersten, Nucleic Acids Res. 3: 3109 (1976).
56. Y. Taya, and S. Nishimura, Biochem.Biophys.Res.Commun. 51: 1062 (1973).
57. G.R. Björk, and K. Kjellin-Stråby, in "The Biochemistry of Adenosylmethionine", F. Salvatore, E. Borek, V. Zappia, H.G. Williams-Ashman, and F. Schlenk, eds., Columbia University Press, New York, p. 216 (1977).
58. P. Izzo, and R.R. Gantt, Biochemistry 16: 3576 (1977).
59. G.L. Cantoni, in "The Biochemistry of Adenosylmethionine", F. Salvatore, E. Borek, V. Zappia, H.G. Williams-Ashman, and F. Schlenk, eds., Columbia University Press, New York, p. 557 (1977).
60. G. Paolella, G. Ciliberto, C. Traboni, A. Oliva, F. Cimino, and F. Salvatore, Commun. at V Congress of Italian Society of Biochemistry, Abstract of Papers, Pisa (1979).
61. V. Zappia, C.R. Zydek-Cwick, and F. Schlenk, J.Biol.Chem. 244: 4499 (1969).
62. G. Ciliberto, A. Colonna, R. Santamaria,F. Salvatore, and F. Cimino, Commun. at V Congress of Italian Society of Biochemistry, Abstract of Papers, Pisa (1979).

INVOLVEMENT OF S–ADENOSYLMETHIONINE IN BRAIN PHOSPHOLIPID METABOLISM

Rita Mozzi, Vanna Andreoli and Giuseppe Porcellati

Department of Biochemistry, The Medical School, University
of Perugia, Via del Giochetto, 06100 Perugia, Italy

INTRODUCTION

The synthesis of phosphatidylcholine (Ptd–choline) in animal
tissues is carried out chiefly by the cytidine nucleotide pathway,
although base–exchange reaction and stepwise methylation of pre-
existing phosphatidylethanolamine (Ptd–ethanolamine) also contribute
to its formation[1-7]. The N–methylation pathway, first demonstrated
in liver by Bremer and Greenberg[3] and successively described in this
tissue by several authors, has not been however unequivocally demonstrated in brain, and conflicting data have been produced in this
connection[8-10].

The problem of Ptd–ethanolamine methylation in brain was still
open in 1977, when experiments carried out in our laboratory at the
same time on prostaglandin effect upon phospholipid metabolism in
brain[11] let us think that the methylation pathway was not only occurring in brain but was somewhat dependent upon the experimental
conditions used.

In these experiments rat brain slices were incubated with some
phospholipid precursors (choline, arachidonic acid, etc.), and the
effect of prostaglandins upon their incorporation and renewal in
phospholipids was investigated. When tritiated ethanolamine was
used, as the lipid precursor, radioactivity was found not only into
ethanolamine phosphoglycerides (EPG) but also in smaller amounts

41

into Ptd–choline.   Results were not reproducible, but homogeneous
data were obtained when half hemisphere was taken as a control and
the other half was subjected to various experimental conditions.
Thus, prostaglandins did not show any appreciable effect upon methylation, whereas an increase of its rate was observed any time slices
of one hemisphere were incubated under N_2 and compared with the con-
tralateral one incubated under O_2 and CO_2.

Table 1 shows that labeled Ptd–choline was indeed synthesized
upon incubation of brain slices with tritiated ethanolamine.   The
differences found between the rate of methylation observed under
anaerobic and aerobic conditions were always noticed throughout
further experiments.   As shown in Table 2, the specific radioactivity of the synthesized Ptd–choline increased with the time of incubation, although a general decrease in respect to total lipid radioactivity was observed.

Table 1.   Radioactivity Content of Phosphatidylcholine
of Brain Slices Incubated with $[1,2-^3H]$ Ethanolamine

Addition	nCi/μmol Ptd–choline	
	A	B
none	1.6	2.1
$PGF_{2\alpha}$ (0.3 μg/ml)	1.4	1.6

Brain slices were incubated in Krebs–Ringer bicarbonate
(pH 7.4), containing in 5 ml of final volume 2.8 μCi of
$[1,2-^3H]$ethanolamine in A) 95% O_2 + 5% CO_2 or B) N_2.
Incubation for 30 min at 37°C.

Table 2.   Radioactivity Content of Phosphatidylcholine
of Brain Slices Incubated with $[1,2-^3H]$ Ethanolamine

Time	% of total radioactivity	S.R.
30'	7.2	3.4
60'	4.2	4.8

Brain slices incubated under N_2 at 37°C in Krebs–Ringer
bicarbonate (pH 7.4), containing in 5 ml of final volume
2.8 μCi of $[1,2-^3H]$ethanolamine, in the presence of 0.3
μg/ml of $PGF_{2\alpha}$.   S.R., specific radioactivity (nCi/μmol).

These findings, incidentally obtained during a study of different kind, indicated that methylation processes on lipid molecules
in brain might be somewhat dependent upon experimental conditions
and prompted a more complete study of the methylation pathway for
lipid synthesis in brain.

LIPID METHYLATION IN BRAIN HOMOGENATE

Hirata et al.[12] and Hirata and Axelrod[13] have recently reported
the presence in adrenal medulla and erythrocyte membrane of two different methyl transferases acting at different pH values and involved
in the methylation of Ptd-ethanolamine to Ptd-choline, the first
catalyzing the synthesis of phosphatidyl-N-monomethylethanolamine
(PME) from S-adenosylmethionine (SAM) at an optimum pH value of 6.5
and the second the other two methylations to produce successively
phosphatidyl-N,N-dimethylethanolamine (PDE) and Ptd-choline at an
optimum pH value of 10. On the light of these data, experiments
have been planned in order to show whether brain homogenate would
have been able to carry out such methylation reactions under similar
conditions in vitro.

For this purpose, Wistar male rats of 28-30 days of age were
sacrificed by prolonged cardiac perfusion to avoid any blood red cell
contamination of brain (as checked by assaying hemoglobin content[14]),
which would have seriously affected the final results, in relation
to previous findings[13]. Homogenized brain cortex (0.32 M sucrose)
was incubated in given amounts at different pH values (6.8 and 8.0)
with highly radioactive (15 Ci/mmol) S-adenosyl-L-[methyl-^{3}H] methionine (Radiochemical Center, Amersham, England). Radioactivity
was successively estimated in the chromatographically separated
Ptd-choline, as reported in details where appropriate. Controls,
carried out throughout the work by denaturating the product by heat
(100°C) before the addition of labeled SAM, have shown that no detectable radioactivity in lipid extracts or separated lipid was
ever observed.

Fig.1 A shows the rate of incorporation of methyl[^{3}H] groups
from radioactive SAM into brain Ptd-choline at the two mentioned
pH values. It is apparent that incorporation was much higher at
pH 8.0 than at pH 6.8 at any time of incubation, with linearity
observed up to 20 min of incubation. In another experiment (Fig.1B)
100 µg of sonicated PDE (Mann Res.Labs., New York, U.S.A.) were
added to the incubation medium ; the addition of the lipid substrate

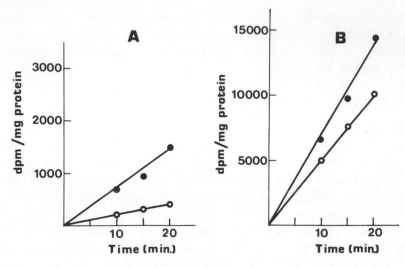

Fig.1. Time course of SAM incorporation into rat brain
Ptd-choline. Brain homogenate (0.4-0.6 mg protein) was incubated in a final volume of 80 µl
with 60 mM phosphate (pH 6.8, o-o-o-o) or 60
mM Tris-HCl (pH 8.0, ●-●-●-●), 10 mM MgCl₂
and 1.79 µM (1.72 µCi) of radioactive SAM (experiment A).   After treatment with 80 µl of cold
10 % TCA at proper intervals, the precipitated
material, washed three times after centrifugation with 4 ml of cold H₂0, was extracted with
4 ml of chloroform-methanol (2:1,v/v).   After
conventional filtration and washings, the dried
lipid phase was taken up in chloroform-methanol
(2:1,v/v) and chromatographed by TLC on silica
gel G plates in triplicate with three different
solvent systems suitable for Ptd-choline separation.   The visualized spot was counted for radioactivity content.   In experiment (B), 100 µg
of PDE were added to the incubation medium, by
sonicating for 3 min the lipid in the incubation
buffer with a 100 W MK2 ultrasonic disintegrator
(MSE, England). Activity expressed as dpm/mg protein. Each point is the mean from six experiments
with standard deviation values within 5 - 10 %.

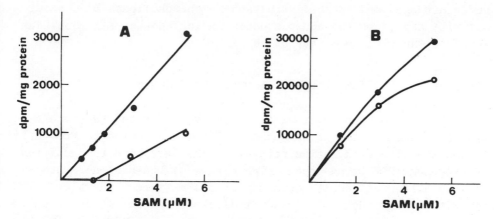

Fig.2. Incorporation of labeled SAM into rat brain
Ptd-choline at different substrate concentration.   Incubation time : 15 min.   See
Fig.1 for explanation of symbols, lettering
and expression of data.

for the second methyl transferase[12] noticeably increased the rate
of methyl incorporation at both pH values with linearity observed
up to 20 min of incubation.   These results indicate that brain
tissue also is able to carry out Ptd-ethanolamine methylation to
Ptd-choline, and that most probably the enzyme catalyzing the first
methylation step is rate-limiting, as in liver[3].

In another series of experiments the dependence of the rate of
methylation versus SAM concentration in brain homogenate was examined. Fig.2 A indicates that at the highest SAM concentration a
complete methylation of Ptd-ethanolamine to Ptd-choline is observed
also at pH 6.8, though lower than at 8.0.   The addition of PDE
(Fig.2 B) brings about an increase of methyl group incorporation
into Ptd-choline, which is no more linear with SAM concentration,
particularly at pH 6.8.

In order to identify more carefully by other means the final

product of methylation (Fig.1 and 2), the Ptd-choline spot after incubation was eluted exaustively with chloroform, the content dried
under nitrogen and treated with cabbage phospholipase D^{15}. Radioactivity assays on the aqueous phase indicated that all label was
found in the choline moiety of the lipid.

STEPWISE METHYLATION OF LIPID IN BRAIN MICROSOMES

The data of the previous section have indicated that brain
tissue is able to convert endogenous Ptd-ethanolamine to Ptd-choline
by the stepwise methylation pathway. The fact that the addition
of exogenous PDE brings about higher rate of Ptd-choline synthesis
and the different results obtained at two different pH values with
and without PDE addition already indicated that the methyl transferases acting upon membrane-bound Ptd-ethanolamine might be functioning at different pH optima.

Successive experiments have been carried out therefore with
purified rat brain microsomes incubated with radioactive SAM at different pH values, with the aim of examining the rate of labeling of
lipid intermediates, such as PME and PDE, and of Ptd-choline. As
known,[4,5,12] microsomes represent in liver and other tissues the
main subcellular fraction responsible for the metabolic pathway.

The microsomes were obtained from cerebral cortex of 30 days-
old Wistar male rats according to procedures[6], which allow a preparation of particles almost freed from contaminating myelin or other
subcellular fractions, as checked with the use of chemical or enzyme
markers[6]. After various trials and taking into consideration previous work,[12,13] the best analytical system to perform accurate and
precise separation of the monomethylated and dimethylated forms of
Ptd-ethanolamine, together with Ptd-choline and unreacted Ptd-ethanolamine, was found to be that proposed by Katyal and Lombardi,[16] as
checked with the use of sample lipids and authentic lipid standards,
followed by co-chromatography.

The effect of different pH upon lipid methylation has been examined only up to a value of 8.8, although it would be important with
successive studies to extend the range of pH up to higher values,
where pH optima have been found, at least for other tissues, for the
second methyl transferase[12,13]. On the other hand, as known, SAM
is unstable above pH 9.

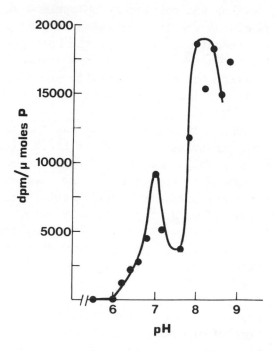

Fig.3. The effect of pH on the incorporation of methyl
groups from SAM into rat brain microsomal Ptd-
choline. Brain microsomes (0.2–0.3 mg protein)
were incubated for 20 min at 37°C in a final volume of 80 µl with 60 mM succinate–NaOH (pH 5.0–
5.5) or cacodilate (5.0–7.0) or phosphate (6.0–
7.8) or Tris–HCl (7.2–8.8) buffers and with 2.66
µM SAM (15 Ci/mmol). Treatment after incubation
was carried out as reported in Fig.1. Phospholipids were separated according to published procedures[16] with the following couple of solvents:
1), chloroform–methanol–NH_4OH (28%, w/v), 65:30:
5 (v/v/v) ; 2), n–butanol–acetic acid–H_2O, 90:
20:20 (v/v/v). Activity is expressed as dpm/
µmol total phospholipid P.

Fig.3 shows that the synthesis of Ptd-choline from endogenous Ptd-ethanolamine and added radioactive SAM is lacking at pH values lower than 6, and that it possesses an optimum pH around 8.2. Interestingly, a smaller but consistent pH optimum at about 7 is also present. On examining the incorporation rate of SAM into PDE (Fig.4 similar results were obtained although the specific activities reache were lower. The influence of pH upon the synthesis of PME is shown in Fig.5 : the smaller optimum of activity around pH 7 is no more evident and only the optimum at about pH 8.2 is present.

The data presented in Fig.3-5 show that the methyl transferase which converts Ptd-ethanolamine to PME in brain microsomes possesses an optimum pH value at 8.2, and that probably a second peak of maximal activity around 7 is presented by the successive methyl transferases. On the other hand, by carrying out similar experiments with brain microsomes from 40 days-old male rats, no peak of activity at pH 7 was detected in almost all observations, thus suggesting that the age of the animals plays an important role as regard to properties, characteristics and behaviour of lipid methyl transferases, as suggested elsewhere[9].

No clear consideration can be drawn from the results of Fig.3-5 about the pH optima of the methyl transferases in brain microsomes, since no experiments have been as yet carried out at pH values higher than 9, as mentioned previously. In a short summary related to rat synaptosomes, Crews et al.[17] have reported two pH optima, at 7.5 and 10.5, for the first and second methyl transferases, respectively.

LIPID METHYLATION IN BRAIN AT DIFFERENT AGES

The results of the previous section have shown that the age of the male Wistar rats has a noticeable effect in brain microsomes on the pH value-activity relationship, as regard to lipid methylation. Brain microsomes of 40 days-old rats do not or only slightly possess in fact the pH optimum around 7 found with similar material obtained from the 30 days-old animals (Fig.3 and 4). Parallel experiments carried out in this laboratory on 14 and 20 days-old male Wistar rats have also indicated variations in pH optima of methylation reactions occurring during maturation. These results would recommend a certain degree of caution in interpreting results of various authors about properties and characteristics of lipid methyl transferases and might explain the occurrence in the literature of conflict-

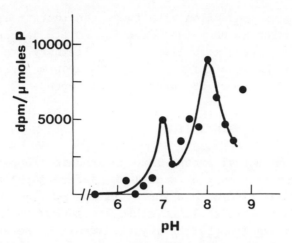

Fig.4. The effect of pH on the incorporation of
methyl groups from labeled SAM into rat
brain microsomal PDE. See Fig.3, for experimental details.

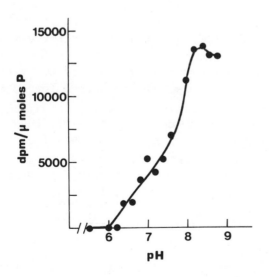

Fig.5. The effect of
pH on the incorporation
of methyl groups from
labeled SAM into rat
brain microsomal PME.
See Fig.3, for experi-
mental details.

ing   results about the characteristics of the methylation pathway
in brain.

Table 3 shows, in connection with these considerations, that
the rate of methylation at pH values from 7.8 to 8.8 decreases with
age for all the three methylated lipids.   Preliminary experiments
have also indicated that the influence of Mg^{2+} upon methylation
reactions in brain microsomes is partially age-dependent.

PLASMALOGEN SYNTHESIS BY METHYLATION

Incorporation of methyl groups into phospholipid does not take
place in brain microsomes only on diacyl-sn-glycero-3-phosphorylcho-
line (diacyl-GPC), but also on choline plasmalogen, i.e. on alkenyl-
acyl-sn-glycero-3-phosphorylcholine (alkenylacyl-GPC).   Fig.6 indicates that by isolating lysophosphatidylcholine with the above mentioned chromatographic system[16] after HCl exposure according to
standardized methodology[18], measurable activity is found in this
lipid, which represents the reaction product of alkenylacyl-GPC.
The pH value-activity relationship is similar to that observed for
the diacyl-compound, and a similar pH optimum around 8.2 was found.
The rate of incorporation, expressed as dpm/µmol total phospholipid
P, is about 1/5 of that found for diacyl-GPC (see Fig.3) ; however,
it must be mentioned that the choline plasmalogen content of brain
is very low, about 1-1.5 % of the total phospholipid content.   The
rate of methyl incorporation into choline plasmalogen therefore might
be sufficiently high, when expressed in terms of specific activity
towards its subclass, certainly higher than that observed for other
choline-containing lipids.
The results of Fig.6 are of particular interest, since they
might suggest the existence of a new and important pathway for choline plasmalogen synthesis in brain.   The mechanisms for the formation of this lipid in brain are in fact still unknown, in contrast
to what is known about ethanolamine plasmalogen synthesis.

CONCLUSIONS

The data of the present work indicate that rat brain microsomes
are able to methylate stepwise phospholipids to form phosphatidylcholine (Ptd-choline) and in addition to produce choline plasmalogen
by similar pathway.   The rate of synthesis of Ptd-choline by the

Table 3. Incorporation of Methyl Groups of S-Adenosyl-
Methionine into Rat Brain Microsomal Phospholipid
at Different Ages

Age (days)	pH used	Incorporation into		
		PC	PDE	PME
30	7.8	11,481	2756	6739
	8.2	15,534	6534	13,431
	8.8	17,554	7324	13,027
40	7.8	7600	888	4744
	8.2	9321	3547	7856
	8.8	8518	3573	5925

0.2–0.3 mg of microsomal protein were incubated in 60 mM
Tris–HCl buffer at the indicated pH values in a final volume of 80 µl. Incubation for 20 min at 37°C with labeled S–adenosylmethionine (15 Ci/mmol, 2.66 µM). Data expressed as dpm/µmol total phospholipid P.

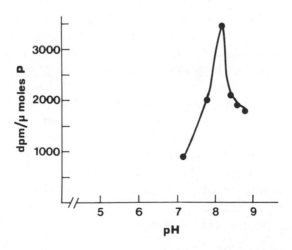

Fig.6.The effect of pH on the incorporation of methyl
groups from labeled SAM into rat brain microsomal choline plasmalogen. Activity was measured
according to the procedures reported in the text.
See Fig.3, for other experimental details.

methylation system is rather low in brain, if compared to that present in other tissues, but it is not excluded that particular brain areas or cell types may be enriched in this metabolic pathway. Moreover, only small pool of membrane Ptd-choline might be formed through this mechanism, as postulated for the base-exchange system and demonstrated[19]. This last enzymic system has been found in fact to be localized particularly in neuronal cells,[20] as compared to glia, having the first time been indicated as a neuronal marker[20]; a distribution studies of the methylation pathway in different cell types might be interesting also in this connection.

The physiological significance of this metabolic pathway in brain is unknown. Its occurrence in nervous tissue might be of some importance due to the possibility for the nerve cell to enrich its Ptd-choline fraction of polyunsaturated fatty acids,[21] particularly 20:4 and 22:6. Moreover, the Ptd-choline synthesis by the methylation pathway in red blood cells has been indicated to represent a mechanism for an enzyme-mediated flip-flop of phospholipids from the cytoplasmic to the outer surface of erythrocyte membranes,[13] producing fluidity variations in the same membrane which might affect ion movement and enzyme activities. It is possible that such implication can be drawn also for nervous membranes.

Finally, the methylation pathway could represent an enzymic system linked to the source of choline in brain. It is known that the choline moiety as such cannot pass into the brain from the blood stream, but rather a lipid-choline moiety, probably lysophosphatidylcholine,[22] might be involved in the transport of the base. The methylation pathway, however, can be visualized as another means of producing choline in brain, provided Ptd-ethanolamine is transported into the brain. Conversely, due to the noticeable presence in the brain tissue of free serine, a possible pathway for choline production in brain may be represented by an exchange reaction between free serine and endogenous brain phospholipids,[6,19,20] by the successive decarboxylation of phosphatidyl serine to Ptd-ethanolamine in the same tissue, finally followed by the described methylation reactions to produce a definite pool of active Ptd-choline.

ACKNOWLEDGEMENTS

This work was carried out by a research grant from the Consiglio Nazionale delle Ricerche, Rome (contribute n.78.01426.04/115).

REFERENCES

1. E.P.Kennedy and S.B.Weiss, The function of cytidine coenzymes in the synthesis of phospholipids, J.Biol.Chem. 222:193 (1956).
2. S.B.Weiss, S.W.Smith, and E.P.Kennedy, The enzymatic formation of lecithin from cytidine diphosphate choline and D-1,2-diglyceride, J.Biol.Chem. 231:53 (1958).
3. J.Bremer, P.H.Figard, and D.M.Greenberg, The biosynthesis of choline and its relation to phospholipid metabolism, Biochim. biophys.Acta 43:477 (1960).
4. D.Rehbinder and D.M.Greenberg, Studies on the methylation of ethanolamine phosphatides by liver preparations, Arch.Biochem. Biophys. 109:110 (1965).
5. G.A.Scarborough and J.F.Nic, Methylation of ethanolamine phosphatides by microsomes from normal and mutant strain of Neurospora crassa, J.Biol.Chem. 242:238 (1967).
6. G.Porcellati, G.Arienti, M.Pirotta, and D.Giorgini, Base-exchange reactions for the synthesis of phospholipids in nervous tissue: The incorporation of serine and ethanolamine into the phospholipids of isolated brain microsomes, J.Neurochem. 18:1935 (1971).
7. J.N.Kanfer, Base exchange reactions of the phospholipids in rat brain particles, J.Lipid Res. 13:468 (1972).
8. G.B.Ansell and S.Spanner, The metabolism of labelled ethanolamine in the brain of the rat in vivo, J.Neurochem. 14:873 (1967).
9. R.Morganstern and A.A.Abdel-Latif, Incorporation of ^{14}C-ethanolamine and ^{3}H-methionine into phospholipids of rat brain and liver in vivo and in vitro, J.Neurobiol. 5:393 (1974).
10. D.N.Skurdal and W.E.Cornatzer, Choline phosphotransferase and phosphatidylethanolamine methyltransferase activities, Int.J. Biochem. 6:579 (1975).
11. R.Mozzi, A.Orlacchio, V.Andreoli, N.Solinas, and G.Porcellati, The effect of prostaglandin $F_{2\alpha}$ on arachidonate and lipid-choline metabolism in rat brain slices in vitro, in : "Advances in Prostaglandin and Thromboxane Research", vol.3, G.Galli, C. Galli, and G.Porcellati, eds., Raven Press, New York (1978).
12. F.Hirata, O.H.Viveros, E.J.Diliberto, and J.Axelrod, Identification and properties of two methyltransferases in the conversion of phosphatidylethanolamine to phosphatidylcholine, Proc. Natl.Acad.Sci.U.S.A. 75:1718 (1978).
13. F.Hirata and J.Axelrod, Enzymatic synthesis and rapid translocation of phosphatidylcholine by two methyltransferases in erythrocyte membranes, Proc.Natl.Acad.Sci.,U.S.A. 75:2348 (1978).

14. E.Antonini, Interrelationship between structure and function in hemoglobin and myoglobin, Physiol.Rev. 45:123 (1965).

15. M.Kates and P.S.Sastry, Phospholipase D (EC 3.1.4.4) : phosphatidylcholine phosphatidohydrolase, in : "Methods in Enzymology", S.P.Kolowick and N.O.Kaplan, eds., Academic Press, New York (1969).

16. S.L.Katyal and B.Lombardi, Quantitation of phosphatidyl-N- methyl and N,N-dimethylaminoethanol in liver and lung of N-methylaminoethanol fed rats, Lipids 9:81 (1974).

17. F.T.Crews, F.Hirata, and J.Axelrod, Identification and properties of two methyltransferases that synthesize phosphatidylcholine in rat brain synaptosomes, Feder.Proc. 38:517 (1979).

18. L.A.Horrocks and G.Y.Sun, Ethanolamine plasmalogens, in : "Research Methods in Neurochemistry", N.Marks and R.Rodnight, eds., Plenum Press, New York (1972).

19. A.Gaiti, M.Brunetti, H.Woelk, and G.Porcellati, Relationships between base-exchange reaction and the microsomal phospholipid pool in the rat brain in vitro, Lipids 11:823 (1976).

20. G.Goracci, Ch.Blomstrand, G.Arienti, A.Hamberger, and G.Porcellati, Base-exchange enzymic system for the synthesis of phospholipids in neuronal and glial cells and their subfractions : a possible marker for neuronal membranes, J.Neurochem. 20 : 1167 (1973).

21. G.MacDonald and W.Thompson, Different selectivities in acylation and methylation pathways of phosphatidylcholine formation in guinea pig and rat livers, Biochim.biophys.Acta 398:424 (1975).

22. D.R.Illingworth and O.W.Portman, The uptake and metabolism of plasma lysophosphatidylcholine in vivo by the brain of squirrel monkeys, Biochem.J. 130:557 (1972).

Effect of S-Adenosyl-L-Methionine and S-Adenosyl-L-Homocysteine

Derivatives on Protein Methylation

Adriana Oliva, Patrizia Galletti and Vincenzo Zappia
Department of Biochemistry (Second Chair), University of Naples First Medical School, Naples,Italy

Woon Ki Paik and Sangduk Kim
The Fels Research Institute and Department of Bio-
chemistry, Temple University School of Medicine,
Philadelphia, Pa. 19140

A. General Description of Protein Methylation

Exactly 20 years ago, the presence of methylated amino acid ε-

N-methyllysine was first discovered in the hydrolyzate of flagella

protein of Salmonella typhimurium. During the intervening years,

it has been recognized that there exist three methylated lysines,

ε-N-mono-, ε-N-di- and ε-N-trimethyllysine. In addition, arginine,

histidine and dicarboxylic amino acid residues of certain proteins

were also found to be methylated.

These methylated amino acid residues are present ubiquitously

in nature, ranging from eukaryotic to prokaryotic organisms (1),

and their presence are extremely selective occurring in highly spe-

cialyzed proteins such as histone, flagella protein, myosin, actin,

ribosomal proteins, opsin, fungal and plant cytochrome c, calmodu-

lin and myelin basic protein. The methylation of these proteins

occurs post-translationally by a group of enzymes which are highly
specific in regard with the amino acid side chains and proteins
involved.   Thus, protein methylase I [S-Adenosylmethionine:protein-
arginine N-methyltransferase; EC 2.1.1.23] methylates the guanidino
group of arginine residues, protein methylase II [S-Adenosylmethio-
nine:protein-carboxyl O-methyltransferase; EC 2.1.1.24] methylates
free carboxyl group of aspartic and glutamic acid residues, and
protein methylase III [S-Adenosylmethionine:protein-lysine N-methyltransferase; EC 2.1.1.43] methylates the ε-amino group of lysine
residues.   Furthermore, it has been demonstrated recently that protein methylase III from calf thymus methylates the ε-amino group of
lysine residue of only histone, but not cytochrome c, while the
protein methylase III from Neurospora crassa or Saccharomyces cerevisiae methylates the ε-amino group in cytochrome c but not in histone (2,3), indicating the extreme complexicity of enzymes involved
in this post-translational side chain modification reaction.

The product of protein methylase II, carboxyl-methyl ester,
undergoes rapid and spontaneous hydrolysis into methanol at physiological pH and temperature, and the proposed physiological role(s)
of the reaction are mainly related to the neutralization of anionic
charges of the protein, which may result in conformational changes.
On the other hand, the products of protein methylase I and III
(N^G-methylated arginines and ε-N-methyllysines, respectively) are
stable in acid-hydrolysis.   These two latter reactions increase the
basicity associated with the nitrogen or increase the hydrophobicity,
thereby possibly altering the tertiary structure of the protein.

In the last few years, we have witnessed a great surge of evidence to indicate the biochemical importance of protein methylation
in cellular regulation.   Trimethylation of lysine residues of protein has been shown to provide a precursor for the biosynthesis of
carnitine, which is an important acyl carrier in mitochondrial membrane (4).   Protein carboxymethylation is strongly implicated in
leucocyte (5,6) and bacterial chemotaxis (7-10),. as well as in neu-

rosecretory processes in adrenal medulla, in the parotid and in posterior pituitary gland (6,11-15). Recent evidence from this and other laboratories demonstrated that enzymatic methylation of fungal cytochrome c facilitates the binding of this hemoprotein to mitochondria (16). Several review articles on protein methylation have been published (1,16,17), and in addition two of us (W. K. P. and S. K.) recently reviewed the subject in a book, which will be published by John Wiley & Sons, Inc., New York.

All the available evidence indicates that S-adenosyl-L-methionine is the sole biological methyl donor for the above protein methyltransferases, and the demethylated reaction product S-adenosyl-L-homocysteine is a potent competitive inhibitor for the reaction (2,18,19). Thus, there have been a few attempts in the past to investigate the effect of modification of various regions of these compounds on the reaction of protein methyltransferases. These investigations are not only important in elucidation of the reaction mechanism, but also in practical application where only a desired methylation reaction can be selectively studied, while others are inhibited and/or modulated.

In the present brief chapter, we review the effect of various derivatives and analogs of S-adenosyl-L-methionine and S-adenosyl-L-homocysteine on protein methyltransferases. In doing this, we first review protein methylase I and III reaction, and is followed by our recent data on protein methylase II reaction.

B. Protein Methylase I [S-Adenosylmethionine:protein-arginine N-methyltransferase] and Protein Methylase III [S-Adenosylmethionine:protein-lysine N-methyltransferase]

1. Ethionine

L-Ethionine, which is a potent alkylating carcinogen in rats, induces a rapid hepatic ATP deficiency and the ethylation of various cellular components (20-22). Among the latter, some proteins of the nucleus including histones are actively ethyla-

ted. After the injection of L-[^{14}C-ethyl]ethionine into rats,
Friedman et al. (23) found that a non-histone nuclear protein of
the liver, which is soluble in 0.14 M NaCl, was most heavily labeled. By amino acid analysis, they observed a single radioactivity
peak associated with N^G,N'^G-dimethylarginine (symmetric). Ethylation of arginine residues of protein proceeds most likely via
S-adenosyl-L-ethionine, which is formed by methionine adenosyltransferase [ATP:L-methionine S-adenosyltransferase; EC 2.5.1.6] (24).

Using histones as substrate, Baxter and Byvoet (25) observed
that partially purified rat liver protein methylase I could utilize
both S-adenosyl-L-methionine and S-adenosyl-L-ethionine as alkyl
donor with comparable efficiency. On the other hand, Cory et al.
(26) found that S-adenosyl-L-ethionine inhibited protein methylase
I activity by 50% at about 0.08 mM concentration. The fact that
S-adenosyl-L-ethionine can be equally utilized by protein methylase
I as S-adenosyl-L-methionine, is in contrast to the observation that
S-adenosyl-L-ethionine inhibited non-competitively the utilization
of S-adenosyl-L-methionine by rat liver protein methylase III.
However, the K_m value for S-adenosyl-L-methionine by protein methylase III was found to be 3.0 x 10^{-6} M, while the K_i value for S-
adenosyl-L-ethionine was 0.17 x 10^{-3} M, thus obliterating the biochemical significance of this analog on the in vivo enzymatic methylation of the ε-amino group of histone-lysine residues.

2. S-Adenosyl-L-homocysteine analogs.

Recently, Casellas and Jeanteur (27) investigated the effect of alteration of various regions of S-adenosyl-L-homocysteine
molecule. One interesting compound, which emerged from this investigation, was S-isobutyladenosine (SIBA). This compound has previously been shown to inhibit cell transformation induced by Rouse
sarcoma virus, mitogen-induced blastogenesis, polyoma virus replication and capping of herpes virus mRNA (28,29). Casellas and
Jeanteur (27) found that, among the many enzyme reactions tested in
vitro, the K_i value for this compound with partially purified pro-

tein methylase I was the lowest (0.15 mM). Thus, if one assumes
that hypermethylation plays a decisive role in cell transformation
(30), protein methylase I might be an attractive target for this
drug in vivo.

In extension of the above investigation, Enouf et al. (31)
studied the effect of S-adenosyl-L-homocysteine analogs on both protein methylase I and transformation of chicken embryo fibroblasts
by Rouse sarcoma virus, and found that most of the analogs derived
from alteration of the cysteine moiety were the most potent inhibitors for both functions. Furthermore, there was a direct correlation between the inhibition of protein methylase I activity and the
inhibition of virus-induced cell transformation: All the good inhibitors of protein methylase I activity strongly prevented the virus-
induced cell transformation.

The inhibitory effect of these compounds is highly specific for
protein methylase I activity; for example, although the K_i value of
SIBA with protein methylase I was 0.6×10^{-3} M with complete inhibition of the cell transformation, protein methylase III activity
was not inhibited at all at the concentration of SIBA   1.25 mM.
Similar high specificity of the inhibitory effect of SIBA on protein
methylase I has also been obtained by Casellas and Jeanteur (27).

Vedel et al. (32) recently tested the effect of sinefungin on
protein methylase I and III activity. This antifungal antibiotic,
which is a structural analog of both S-adenosyl-L-methionine and
S-adenosyl-L-homocysteine and on which the methionine or homocysteine residue is replaced with ornithine, is isolated from Streptomyces griseolus. While this antibiotic is as effective as SIBA in
inhibiting Rouse sarcoma virus-induced transformation of chicken
embryo fibroblast cells, the competitive inhibition constant K_i for
protein methylase I activity is 30-60 times lower than that of SIBA.
Thus, sinefungin appears to be one of the most potent inhibitors of
protein methylase I activity and is comparable with S-adenosyl-L-
homocysteine in regard to its inhibitory effect.

C.   Protein Methylase II [S-Adenosylmethionine:protein-carboxyl O-
     Methyltransferase]

     1.   Various sulfonium compounds as substrate for the enzyme
          Various sulfonium compounds were tested for their ability
as methyl donor for the carboxymethyl esterification of ACTH by the
action of protein methylase II. As shown in Table I, besides the
natural methyl donor, only S-adenosyl-L-(2-hydroxy-4-methylthio)-
butyric acid showed a moderate activity. The other compounds were
found to be inactive within the limits of accuracy of the analytical test.

Table I

SULFONIUM COMPOUNDS AS METHYL DONORS FOR PROTEIN METHYLASE II

Methyl donor [14C-methyl]-labeled	Concentration used (mM)	[14C]incorporated (pmol/sample)
S-Adenosyl-L-methionine	2×10^{-3}	114
	3×10^{-3}	136
	7.4×10^{-3}	175
S-Adenosyl-L-(2-hydroxy-4-methylthio)butyric acid	2×10^{-3}	1.2
	10×10^{-3}	4.8
	50×10^{-3}	16
S-Inosyl-L-methionine	10	< 1
	20	< 1
S-Inosyl-L-(2-hydroxy-4-methylthio)butyric acid	2×10^{-3}	< 1
	50×10^{-3}	< 1
S-Adenosyl-(5')-3-methylthio-propylamine	20×10^{-3}	< 1
	50×10^{-3}	< 1
Dimethylthioadenosine sulfonium salt	10×10^{-3}	< 1

The incubation mixture contained: 25 µl of citrate-phosphate buffer,
pH 6.0 (prepared by mixing 1 part of 0.5 M disodium phosphate and
0.6 part of 0.25 M citric acid), 2 mM EDTA, 50 µg of ACTH as methyl
acceptor protein, 2 µg of purified protein methylase II and [14C]-
labeled sulfonium compounds at the indicated concentrations in a
final volume of 250 µl. The incubation was carried out for 10 min.
at 37°C.

2. Various sulfonium compounds as inhibitors for the enzyme

    The sulfonium compounds were also examined for their inhibitory activity for protein methylase II in the presence of S-adenosyl-L-methionine at a final concentration of 2×10^{-6} M, corresponding to the apparent K_m value for the substrate. As listed in Table II, only 5'-dimethylthioadenosine exerts a moderate inhibition among the compounds examined, while S-adenosyl-L-(2-amino-4-carboxymethylthio)butyric acid is a very effective inhibitor.

Table II

EFFECT OF SULFONIUM COMPOUNDS ON PROTEIN METHYLASE II

Additions	Concentration used (μM)	Relative activity (%)
None	--	100
S-Adenosyl-L-(2-hydroxy-4-methylthio)butyric acid	50	100
	100	100
S-Inosyl-L-methionine	50	100
	100	100
S-Inosyl-L-(2-hydroxy-4-methylthio)butyric acid	50	100
	100	100
S-Adenosyl-L-(2-amino-4-carboxymethylthio)butyric acid	10	33
	20	18
S-Adenosyl-(5')-3-methylthiopropylamine	50	100
	100	100
S-Adenosyl-(5')-3-methylthiopropanol	50	100
	100	100
Dimethylthioadenosine sulfonium salt	50	80
	100	65

The conditions of the reaction are reported under the legend of Table I. The concentration of S-adenosyl-L-[^{14}C-methyl]methionine (95 cpm/pmol) was 2×10^{-6} M in each experiment.

3. Various thioethers as inhibitors for the enzyme

    As mentioned earlier, the transmethylation product S-adenosyl-L-homocysteine is one of the most potent inhibitors for vari-

ous transmethylation reactions, including protein methylase II.
Thus, in order to investigate the binding sites of S-adenosyl-L-
homocysteine on protein methylase II, the effect of several thioether analogs of S-adenosyl-L-homocysteine on the reaction rate has
been studied. S-Adenosyl-L-homocysteine was modified either in the
purine moiety or in the sulfur containing amino acid chain. As
shown in Table III, S-adenosyl-L-homocysteine appears to be the most
powerful inhibitor among the thioethers tested: 1.3×10^{-6} M S-adenosyl-L-homocysteine results in 50% inhibition while at a concentration of 1×10^{-5} M the inhibition is almost quantitative. The
Lineweaver-Burk plot gave K_i value of 6.5×10^{-7} M with competitive
type of inhibition.

Table III

THIOETHERS AS INHIBITORS OF PROTEIN METHYLASE II

Addition	Concentration used (μM)	[^{14}C-methyl] incorp. (pmol/sample)	Relative inhibition (%)
None	--	116	--
S-Adenosyl-L-homocysteine	10	12	90
S-Adenosyl-D-homocysteine	10	116	0
S-Adenosyl-DL-homocysteine	10	32.5	72
S-Adenosyl-L-cysteine	50	116	0
S-Inosyl-L-homocysteine	20	107.3	7.5
5'-Methylthioadenosine	20	90.5	22
n-Butylthioadenosine	100	116	0

The conditions of the reaction are shown under the legend of Table
I. The concentration of S-adenosyl-L-[^{14}C-methyl]methionine (95
cpm/pmol) was 2×10^{-6} M in each experiment.

    S-Inosyl-L-homocysteine, the 6-deaminated analog of S-adeno-
syl-L-homocysteine, gave only a negligible inhibition. The loss of
inhibitory effect by deamination enhances the importance of 6-NH_2
of the purine of the molecule in the binding to the enzyme. Among

the S-adenosyl-L-homocysteine analogs modified in the sulfur containing amino acid side chain, only 5'-methylthioadenosine (MTA) had
a significant inhibitory ability; competitive inhibition with K_i
value of 4.1×10^{-5} M. S-Isobutyladenosine (SIBA) did not exert
any inhibitory effect on the reaction. This is in great ontrast to
the observed powerful inhibitory activity of this analog on protein
methylase I, which has been described earlier.

D. Conclusion

Analogs of metabolic intermediates have been widely used to
investigate the mechanism of biochemical reactions. Along this line,
a series of sulfonium compounds and thioethers, analogs of S-adenosyl-L-methionine and S-adenosyl-L-homocysteine respectively have
been assayed as substrate and/or inhibitors of methyltransfer
reactions (33-36). However, these studies mainly concerned with
methyltransferases acting on small molecules and tRNA (33-37). In
this brief account, we have reviewed recent studies on the effect of
afore-mentioned compounds on various protein-specific methyltransferases.

The results in Table I indicate that the $2-NH_2$ and the COOH
group of methionine moiety, as well as the $6-NH_2$ group of adenine
are essential for the binding of the methyl donor to the protein
methylase II and/or to the substrate peptide. The lack of any
inhibitory effect of the sulfonium analogs (Table II) indirectly
supports the high specificity of the system herein investigated.

As shown in Table III, 5'-methylthioadenosine competitively
inhibits protein methylase II with an apparent K_i of 4.1×10^{-5} M.
This inhibition appears to be of physiological interest since the
thioether represents a product of S-adenosyl-L-methionine metabolism
and is distributed in micromolar amounts in most mammalian tissues
(38). Among all the S-adenosyl-L-homocysteine analogs, SIBA has
been the most extensively studied in vivo. Although SIBA inhibits
the protein methylase I activity with K_i value of 0.6×10^{-3} M with
complete inhibition of the cell transformation, this analog lacks

of any inhibitory activity towards protein methylase II and III
activities, thus exhibiting a high specificity in its action. Among
the various protein methylases, protein methylase I reaction appears
to be the most vulnerable target for the action of various derivatives and analogs of both S-adenosyl-L-methionine and S-adenosyl-
L-homocysteine, and sinefungin is the most potent inhibitor among
the compounds tested.

Finally, our data support the view that at least three binding
sites are required for the interaction of S-adenosyl-L-methionine
or S-adenosyl-L-homocysteine with the enzyme protein and with the
methyl-accepting polypeptide substrate. The random Bi Bi mechanism
demonstrated by Jamaluddin et al. (19) for protein methylase II
implies in fact the formation of such a ternary complex.

Acknowledgements

This work was supported by research grants AM09602 from the
National Institute of Arthritis, Metabolism and Digestive Diseases,
CA10439 and CA12226 from the National Cancer Institute, and GM20594
from the National Institute of General Medical Sciences, and from
C.N.R U.S.-Italy Cooperative Science Programm.

References

1. Paik, W. K. and Kim, S., in Advances in Enzymology (edited by
     A. Meister), Vol. 42, 227 (1975), John Wiley & Sons, New York.
2. Durban, E., Nochumson, S., Kim, S., Paik, W. K. and Chan, S.-K.,
     J. Biol. Chem., 253, 1427 (1978).
3. DiMaria, P., Polastro, E., DeLange, R. J., Kim, S. and Paik,
     W. K., J. Biol. Chem., 254 in press (1979).
4. LaBadie, J. H., Dunn, W. A. and Aronson, N. N., Biochem. J.,
     160, 85 (1976).
5. O'Dea, R. F., Viveros, O. H., Axelrod, J., Aswanikumar, S.,
     Schiffman, E. and Corcoran, B. A., Nature, 272, 462 (1978).
6. Diliberto, E. J., jr., O'Dea, R. F. and Viveros, O. H., in
     Transmethylation (edited by E. Usdin, R. T. Borchardt and C.
     R. Creveling), (1979), Elsevier North-Holland, pp. 529.
7. Kleene, S. J., Toewes, M. L., and Adler, J., J. Biol. Chem.,
     252, 3214 (1977).
8. Van Der Werf, P. and Koshland, D. E. jr., J. Biol. Chem.,
     252, 2793 (1977).
9. Adler, J., in Transmethylation (edited by E. Usdin, R. T.
     Borchardt and C. R. Creveling), (1979), Elsevier North-Holland,
     pp. 505.
10. Stock, J. B. and Koshland, D. E., jr., in Transmethylation
     (edited by E. Usdin, R. T. Borchardt and C. R. Creveling),

(1979), Elsevier North–Holland, pp. 511.

11. Diliberto, E. J., Jr., Viveros, O. H. and Axelrod, J., Proc. Natl. Acad. Sci., U.S.A., **73**, 4050 (1976).
12. Viveros, O. H., Diliberto, E. J., jr. and Axelrod, J., in Synapses (edited by G. A. Cottrell and P. N. R. Asherwood), (1977), Blackie and Son, London, pp. 368.
13. Gagnon, C., Axelrod, J. and Brownstein, M. J., Life Science, **22**, 2155 (1978).
14. Gagnon, C., Bardin, C. W., Strittmatter, W. and Axelrod, J., in Transmethylation (edited by E. Usdin, R. T. Borchardt and C. R. Creveling), (1979), Elsevier North–Holland, pp. 521.
15. Eiden, L. E., Borchardt, R. T. and Rutledge, C. O., in Trans-methylation (edited by E. Usdin, R. T. Borchardt and C. R. Creveling) (1979), Elsevier North–Holland, pp. 539.
16. Polastro, E., Schneck, A. G., Leonis, J., Kim, S. and Paik, W. K., Intn. J. Biochem., **9**, 795 (1978).
17. Paik, W. K. and Kim, S., Science, **174**, 114 (1971).
18. Lee, H. W., Kim, S. and Paik, W. K., Biochemistry, **16**, 78 (1977).
19. Jamaluddin, M., Kim, S. and Paik, W. K., Biochemistry, **14**, 694 (1975).
20. Farber, E., in Advances in Cancer Research (edited by A. Haddow and S. Weinhouse), **7**, 383 (1963).
21. Stekol, J. A., in Advances in Enzymology (edited by F. F. Nord), **25**, 369 (1963), John Wiley & Sons, New York.
22. Shull, K. H., McConomy, J., Vogt, M., Castillo, A. and Farber, E., J. Biol. Chem., **241**, 5060 (1966).
23. Friedman, M., Shull, K. H. and Farber, E., Biochem. Biophys. Res. Communs., **34**, 857 (1969).
24. Farber, E., McConomy, J., Franzen, B., Marroquin, F., Stewart, G. A. and Magee, P. N., Cancer Res., **27**, 1761 (1967).
25. Baxter, C. S. and Byvoet, P., Cancer Res., **34**, 1418 (1974).
26. Cory, M., Henry, D. W., Taylor, D. L. and Koskela, K. J., Chem. Biol. Interaction, **9**, 253 (1974).
27. Casellas, P. and Jeanteur, P., Biochim. Biophys. Acta, **519**, 255 (1978).
28. Legraverend, M., Ibanez, S., Blanchard, P., Enouf, J., Lawrence, F., Robert–Gero, M. and Lederer, E., Eur. J. Med. Chem., **12**, 105 (1977).
29. Robert–Gero, M., Lawrence, F., Farrugia, G., Berneman, A., Blanchard, P., Vigier, P. and Lederer, E., Biochem. Biophys. Res. Communs., **65**, 1242 (1975).
30. Paik, W. K., Kim, S., Ezirike, J. and Morris, H. P., Cancer Res., **35**, 1159 (1975).
31. Enouf, J., Lawrence, F., Tempete, C., Robert–Gero, M. and Lederer, E., Cancer Res., Personal Communication.
32. Vedel, M., Lawrence, F., Robert–Gero, M. and Lederer, E., Biochem. Biophys. Res. Communs., **85**, 371 (1978).
33. Zappia, V., Zydek-Cwick, C. R. and Schlenk, F., J. Biol. Chem., **244**, 4499 (1969).

34. Deguchi, T. and Barchas, J., J. Biol. Chem., <u>246</u>, 3175 (1971).
35. Coward, J. K., D'Urso-Scott, M. and Sweet, W. T., Biochem.
    Pharmacol., <u>21</u>, 1200 (1972).
36. Kerr, S. J., J. Biol. Chem., <u>247</u>, 4248 (1972).
37. Borcahrdt, R. T., in <u>The Biochemistry of Adenosylmethionine</u>
    (edited by F. Salvatore, Borek, E., V. Zappia, H. G. William-
    Ashman and F. Schlenk), (1977), Columbia University Press,
    pp. 151.
38. Rhodes, J. B. and William-Ashman, H. G., Med. Exp., <u>10</u>, 281
    (1964).

THE ROLE OF S-ADENOSYLHOMOCYSTEINE AND S-ADENOSYLHOMOCYSTEINE

HYDROLASE IN THE CONTROL OF BIOLOGICAL METHYLATIONS

Giulio L. Cantoni and Peter K. Chiang

Laboratory of General and Comparative Biochemistry
National Institute of Mental Health
Bethesda, Maryland 20205

S-Adenosylhomocysteine (AdoHcy) (Fig. 1) was first identified as one of the products of the reactions that utilize S-adenosyl-methionine (AdoMet) as a methyl donor by Scarano and Cantoni[1] in 1954. Over the last 25 years, and especially in the last three or four years, AdoHcy has attracted increasing attention because it has become well established that AdoHcy is a competitive inhibitor of many, if not most, of the reactions in which AdoMet participates [2-13].

As is well known, AdoMet is one of the most versatile compounds in biology; it can function not only as a methyl donor in a very large number of methyltransfer reactions but also as a donor of a 3-amino-3-carboxypropyl group[14], of the amino group of homocysteinyl

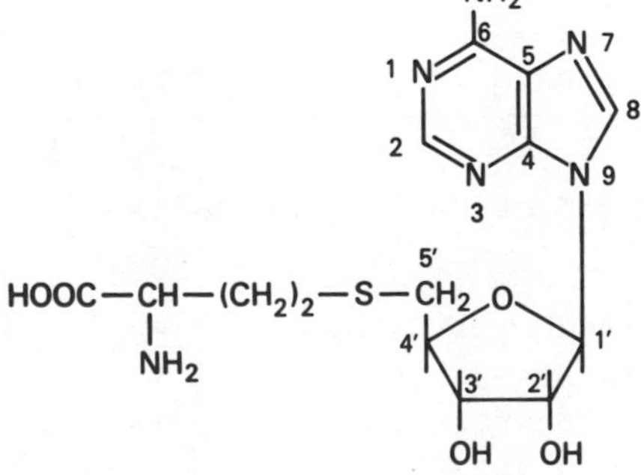

Fig. 1. S-Adenosyl-L-homocysteine.

moiety[15], and after decarboxylation, of a propylamine side chain[16].
In addition, it is cleaved enzymatically to generate thiomethyl-
adenosine and homoserine lactone[17-21], and it can function in a
variety of reactions as an allosteric effector[12, 13, 22]. AdoHcy
can inhibit competitively the great majority of these reactions, and
thus it must be considered as a key compound in the control of the
cellular utilization of AdoMet, mainly for biological methylations.

In eukaryotes, AdoHcy is metabolized through a single metabolic
pathway catalyzed by AdoHcy hydrolase, an enzyme that was first characterized by de la Haba and Cantoni (1957)[23] *. Recently AdoHcy
hydrolase from beef liver has crystallized in our laboratory by
Richards et al.[24]. The enzyme has also been purified to homogeneity
from beef liver by Palmer and Abeles[25], from lupin seed by Guranowski
and Pawelkiewicz[26], and from rat brain by Shatz et al.[27].

The reaction catalyzed by AdoHcy hydrolase is reversible and the
equilibrium is strongly in the direction of synthesis[23]. Physiologically, the reaction proceeds in the hydrolytic direction because
the products of the reaction, adenosine (Ado) and L-homocysteine
(Hcy), can be efficiently removed by a number of different enzymatic
systems (Fig. 2).

As shown in Fig. 2, an unexpected catalysis by AdoHcy hydrolase
was observed by us, namely the synthesis of S-inosylhomocysteine.
We have found recently[28] that this reaction is catalyzed by the purified beef liver and in trace amount by plant enzyme. The K_m for
inosine, however, is about 1,000 times higher than the K_m for adenosine and the V_{max} 40 times smaller. It is not yet clear, therefore,
whether under physiological conditions any S-inosylhomocysteine is
formed. We are currently trying to determine if any S-inosylhomocysteine is found in tissues, and our preliminary results would indicate that under some experimental conditions this may be the case.
It may be recalled that S-inosylhomocysteine can act in some enzymatic
systems as an inhibitor of methyltransferase reactions as shown by
Zappia et al.[5] and by Borchardt[29].

Little is known as yet about the comparative biochemistry of
AdoHcy hydrolase. We have evidence that in beef liver[24] there are
two isozymes with very similar isoelectric properties. It is of
interest that the purified enzyme from beef liver, molecular weight
of 240,000, is composed of four identical subunits, whereas the lupin

---

*
Cell-free extracts from rat kidney and liver can catalyze the
oxidative deamination of AdoHcy to yield S-adenosyl-γ-thio-α-keto-
butyrate[41]. The ketobutyrate derivative is also found in the brain
after intravenous administration of AdoHcy but it is not known what
is the physiological role, if any of oxidative deamination of AdoHcy.

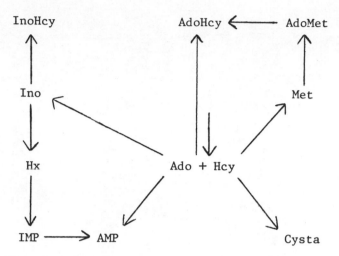

Fig. 2. Pathways for the metabolism of AdoHcy.

seed enzyme, with a molecular weight of 110,000, is composed of two identical subunits[27]. It is notable that AdoHcy hydrolase from the same tissue in different mammals, or from different tissues in the same animal, have different kinetic characteristics with respect to inhibitors.

Eloranta[30] examined the distribution of the activity of AdoHcy hydrolase in different rat tissues. He found that the activity of the hydrolase is 1,000 times greater than that of the AdoMet synthetase. In keeping with this difference, it is generally found that the concentrations of AdoHcy in the cells is 1/4 to 1/10 that of AdoMet.

From a chemical standpoint, the reaction catalyzed by AdoHcy hydrolase is quite unusual. In the synthetic direction, the reaction involves the replacement of the 5'-OH group of adenosine with a thiol nucleophile. The 5'-OH is a poor cleaving group and, since the reaction does not require ATP, a mechanism for the cleavage of the hydroxyl group is not immediately apparent. In the direction of hydrolysis, the situation is not any easier, since the reaction involves the hydrolysis of a thioether.

Palmer and Abeles[25] investigated this interesting problem and found that the enzyme contains one tightly bound NAD per subunit, a result that we have confirmed in our laboratory. The discovery that AdoHcy hydrolase contains tightly bound NAD led Palmer and Abeles to propose the complex mechanisms shown in Fig. 3. The mechanism proposed by Palmer and Abeles involves the oxidation of the 3'-hydroxyl group of AdoHcy by enzyme bound NAD. The result

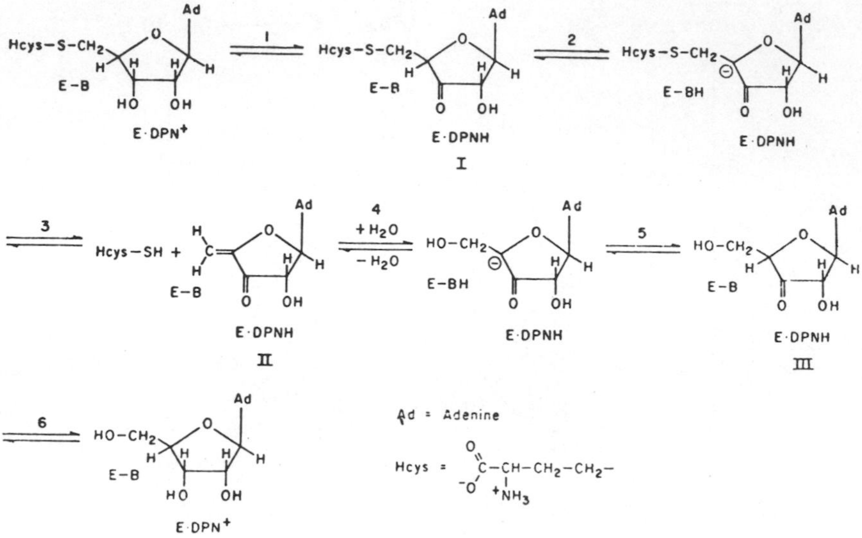

Fig. 3.   The mechanism of action of AdoHcy hydrolase.

would be the conversion of AdoHcy to 3'-keto-AdoHcy, concomitant with the reduction of NAD to NADH. Next, a base at the active site of the enzyme would remove a proton from the C-4' of the 3'-keto intermediate to form an α-ketocarbanion, which eliminates homocysteine to form 3'-keto-4',5'-dehydroadenosine. Water is then added to this intermediate in a stepwise fashion to form 3'-keto adenosine, which is finally reduced to adenosine by the enzyme bound NADH. In essence, we are dealing with a case of oxidation-reduction chemistry below the surface, as Walsh defined it[31]. The evidence in support of this mechanism is as follows: (1) Addition of adenosine to the enzyme results in a spectral increase with a maximum at 327 nm. (2) The increase is dependent on the adenosine concentration and is saturable. (3) The chromophore is bound to the enzyme and comigrates in the ultracentrifuge with the same sedimentation value as the protein component. (4) Incubation of the enzyme and adenosine in [^{3}H]-H_2O results in an exchange of the C-4' proton of adenosine with the water. In addition to the evidence advanced by Palmer and Abeles, we found the crystalline enzyme from beef liver and the purified enzyme from lupin seed exhibited a nucleosidase-like activity, causing a release of the free base, adenine, from adenosine[32]. This observation would be expected since the 3'-keto-Ado is known to decompose spontaneously with liberation of adenine[25, 32] and adds further support to the scheme proposed by Palmer and Abeles[25].

    In vertebrates, and perhaps in insects, the hydrolysis of AdoHcy provides the only source of homocysteine and of cysteine which is not an essential amino acid. Quantitatively, the formation of homocysteine and its hydrolysis are of major significance since the flow of

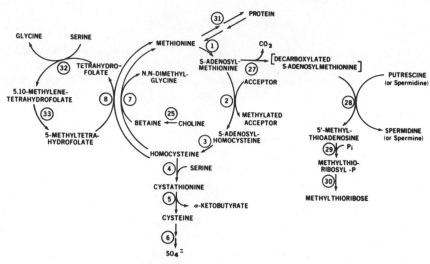

Fig. 4. Major pathways for metabolism of methionine.

methionine through this pathway is of the order of 300 μmoles/kg
body weight/day. Adult mammals and cells in culture can grow normally
with homocysteine (+ folic acid) as the sole sulfur source. However,
after malignant transformation, cells in culture become unable to
grow in the absence of methionine, although they have the biochemical
ability needed for the synthesis of methionine from homocysteine and
folic acid[33-36]. A biochemical explanation for this interesting
characteristic of certain types of transformed cells is not yet
available, and the phenomenon deserves further study also in relation
to the obvious implications for cancer chemotherapy.

With regard to the other product of AdoHcy hydrolase, adenosine,
the interest in this compound has been greatly stimulated by the discovery in 1972, that an inborne error of metabolism, adenosine
deaminase deficiency, is associated with severe immuno-deficiency
involving both B and T lymphocytes. Intracellularly, adenosine can
be formed only from AMP by 5'-nucleotidase and from AdoHcy by AdoHcy
hydrolase; probably the hydrolysis of AdoHcy is the predominant pathway. It can be removed from the cell by release into the extracellular fluid, by deamination to inosine and by reutilization to
AMP.

Kinetic considerations would lead to the conclusion that at low
concentration, the reutilization of adenosine through adenosine
kinase ($K_m \simeq 1$ μM) is much favored over deamination ($K_m \simeq 150$ μM).
Adenosine has a great variety of biological and biochemical properties. It is very likely that a number of the biological and biochemical effects of adenosine are due to its conversion to AdoHcy[37].
A comprehensive review on the role of adenosine in mammalian cells

has recently appeared in Fox and Kelley[38].

The competitive inhibition of methyltransferases by AdoHcy implies that the intracellular ratio of AdoMet/AdoHcy must be important in the regulation of biological methylation, and play a role in the determination of what we like to call the hierarchy of biological methylations.

Table 1 attempts to relate the effect of changes in the intra-cellular AdoMet/AdoHcy ratio to the relative activity of a number of transmethylases. It can be seen that the consequences of a change in the AdoMet/AdoHcy ratio is very different for the differ-ent enzymes and depends both on the K_m for AdoMet and the K_i's for AdoHcy. Generally speaking, as should be intuitively apparent, enzymes with a high affinity for AdoMet, and therefore with a K_m that is considerably lower than the intracellular concentrations of AdoMet, are more tolerant to changes in the AdoMet/AdoHcy ratio than enzymes that operate optimally at concentrations of AdoMet higher than its intracellular concentrations. It should be empha-sized, however, that the calculations shown in Table 1 are prelim-inary and tentative since they are based on the assumption that: a) the intracellular distribution of AdoMet and AdoHcy is not compart-mentalized, and b) the methyl group acceptors compound are present in concentrations that approach the saturating levels for the corres-ponding methylases. In many cases this may not be so, and we believe that it may be impossible to derive a general equation that would predict the percent enzymatic activity when the concentrations of methyl donor, methyl acceptor and inhibitor are varied and nonsatu-rating, since the kinetic behavior would vary in the different cases depending on the enzyme mechanism.

Although AdoHcy hydrolase is not found in bacteria, AdoHcy is cleaved irreversibly to adenine and S-ribosyl-L-homocysteine by AdoHcy nucleosidase[39-41]. This enzyme has been partially purified from *E. coli*, and found to catalyze also the hydrolysis of 5'-methyl-thioadenosine to adenine and methylthioribose. The enzyme is in-active towards AdoMet and a variety of other purine and pyrimidine nucleosides. The specific activity of AdoHcy nucleosidase is 1000 x greater than that of AdoHcy hydrolase emphasizing the need to remove AdoHcy, a potent inhibitor of reactions which utilize AdoMet as a substrate.

In view of the difference in the catabolism of AdoHcy in pro-karyotes and eukaryotes it would be very interesting to characterize a specific inhibitor of the bacterial AdoHcy nucleosidase. Such an inhibitor would in all likelihood be a potent and nontoxic anti-biotic.

The renewed interest in AdoHcy and its metabolism suggested to us a few years ago, that it would be useful to search for a potent

Table 1.   Sensitivity of Methyltransferases (MT) to
Change in AdoMet/AdoHcy Ratio

Enzyme	Km AdoMet	Ki AdoHcy		Km/Ki	AdoMet/AdoHcy (ratio) 4[a]	1.6[b]
		(μM)			(% maximal activity)	
	a	b		c	d	e
Epinephrine-O-MT	570	15		38	46	14
Phenylethanolamine-N-MT	10	1.4		7	39	19
Acetylserotonin-MT	14	2		7	40	20
Glycine-MT	100	35		2.9	76	46
Guanidoacetate-MT	47	17		2.8	68	40
Histone-MT	14	5.5		2.6	69	45
(Gua-7)-MT	1.4	1		1.4	74	54
Phosphatidylethanolamine-MT	1.6	1.4		1.2	82	64
Protein-MT	3	7		0.4	91	79

[a] AdoMet = 80 nmol/g, AdoHcy = 20 nmol/g liver.
[b] AdoMet - 200 nmol/g, AdoHcy = 125 nmol/g liver.

and specific inhibitor of AdoHcy hydrolase. Such an inhibitor could
be utilized to amplify the physiological antagonism between AdoMet
and AdoHcy, and we hope that its use in a variety of biological
systems would lead to a better understanding of the control of biological methylations. Our objective, therefore, was not the inhibition of a given reaction, but the desire to understand the control
mechanism operating in biological methylations.

P. Chiang et al.[42] screened a large number of compounds for
their ability to inhibit AdoHcy hydrolase and identified among them
a compound, 3-deazaadenosine (DZA), that is a very potent inhibitor
of the enzyme. The inhibition is due to the fact that 3-DZA, like
adenosine binds very tightly to the enzyme, but unlike adenosine is
not a substrate for adenosine deaminase[43] or adenosine kinase[44].
The last two enzyme reactions are responsible mainly for the metabolism of adenosine. The modification of the N-atom at the 3-position
is highly specific, because tubercidin and formycin, respectively
the 7- and 9-deazaadenosine analogs of adenosine, are neither inhibitors nor substrates for AdoHcy hydrolase.

3-DZA, however, is not an ideal inhibitor because it can also
be utilized by the enzyme place of adenosine to yield 3-DZA-Hcy.
In different tissues, for reasons that we do not understand, administration of 3-DZA can produce either the inhibition of the hydrolase,
with the resulting increase in the intracellular concentration of
AdoHcy, or its utilization as a substrate with the formation of
3-DZA-Hcy[45]. In most tissues both effects can be observed. 3-DZA,
like adenosine, is readily taken up by the cell through facilitated
diffusion, and is extremely specific in its mode of action. As
mentioned above, 3-DZA is neither a substrate nor an inhibitor for
adenosine deaminase and adenosine kinase. It is, therefore, not
incorporated in the acid soluble nucleotide pool and hence is not

incorporated in nucleic acid. As far as is known, when administered
to a variety of cellular systems or to laboratory animals, all its
effects must be explained by the inhibition of one or more methyltransferase reactions caused by the increased intracellular levels of
AdoHcy or 3-DZA-Hcy or both. This interpretation is significantly
strengthened by the fact that the biological and biochemical consequences of the administration of 3-DZA to cellular systems or laboratory animals are magnified by the simultaneous administration of homocysteine thiolactone. This compound serves intracellularly as a
source of L-homocysteine and leads to the larger accumulation of
3-DZA-Hcy. 3-DZA can, therefore, be used as a probe of the participation of biological methylations in biological process, even though
the specific biochemical target is generally not known.

     P. K. Chiang has shown that a number of methylation reactions
in rat liver are inhibited after administration of 3-DZA[45], and space
does not permit a detailed analysis of his results. There is one
reaction, however, that we wish to discuss here. We found that
administration of 3-DZA to rats will result in a drastic inhibition
in the methylation of phosphatidylethanolamine, *without* a decrease in
the total phosphatidylcholine. As is well known, in eukaryotes two
different metabolic pathways lead to the synthesis of phosphatidylcholine. In addition to the pathway that involves stepwise methylation of phosphatidylethanolamine, there is a methylation-independent
pathway that proceeds from dietary choline through phosphocholine and
cytidinediphosphocholine. The relative contribution of these two
pathways to the overall synthesis of phosphatidylcholine is not yet
fully understood, and in all probability varies in different tissues.
The finding that administration of 3-DZA results in a dramatic inhibi
tion of the incorporation of the methyl group of methionine into
phosphatidylcholine, without significantly altering the total level
of phosphatidylcholine in the liver, leads to the conclusion that in
the presence of an inhibitor of phosphatidylethanolamine methylation,
the utilization of exogenous choline is markedly stimulated. This
observation reveals a hitherto unknown control mechanism that will
require considerable further study, and promises to be very interesting.

     It should be emphasized that we do not yet know whether the
effects seen are due to the inhibition of AdoHcy hydrolase, and a
resultant increase in AdoHcy and/or to utilization of 3-DZA as a
substrate and formation of 3-DZA-Hcy, and/or, finally to the metaboli
immobilization of Hcy residues in these purinylhomocysteine compounds
We expect to have an answer to these interesting questions in the
near future.

     In addition to a variety of biochemical effects, 3-DZA exerts
many different biological effects in a number of cellular systems
through biochemical mechanisms that are as yet undefined, but that,

as mentioned earlier, must involve perturbation of biological methylations.

Zimmerman et al.[46] found that 3-DZA at very low concentrations
inhibits the lectin-dependent neutrophil-mediated cytolysis and
lymphocyte-mediated cytolysis, with concomitant increase in intracellular levels of AdoHcy. Leonard and Skeel in collaboration with
Chiang and Cantoni showed that 3-DZA at 1 μM level inhibits phago-
cytosis by resident peritoneal macrophages[47]. Chiang et al.[48]
have shown that chemotaxis by rabbit neutrophils in response to a
chemoattractant tripeptide is inhibited under conditions where the
binding of the tripeptide to the cellular receptors is not affected.
Finally, Thompson et al.[49] have shown that 3-DZA, and other inhibitors
of AdoHcy hydrolase inhibit rapidly and reversibly synaptic responses
between retinal neurons and myofibers in culture. The inhibitory
effect caused by 3-DZA was potentiated more than six times by the
addition of homocysteine thiolactone; when administered by itself
had no effect. Trager found that 3-DZA has significant activity
against *Plasmodium falciparum in vitro* (unpublished observation).

In all these cases, it might be proposed as a working hypothesis
that the biochemical event underlying these biological responses is
the alteration of normal membrane functions through inhibition of the
methylation of phosphatidylethanolamine and/or inhibition of protein
carboxymethylation.

In a variety of tissue culture systems, 3-DZA exerts a powerful
antiviral action against broad range of RNA and DNA animal viruses.
With J. Bader, we have studied in some detail the antiviral activity
of 3-DZA on Rous sarcoma virus (RSV) infection in chick embryo fibroblasts[51, 52].

The biochemical events that relate to RSV replication may be
schematically divided into two phases: there is an initial phase
during which the viral RNA is transcribed by reverse transcriptase,
probably going through a hybrid RNA-DNA formation; the infecting RNA
is then hydrolyzed by RNase H, the DNA strand is replicated, the two
DNA strands are ligated and circularized and finally integrated into
the viral genome. These steps are all catalyzed by viral enzymes,
with the exception of the joining of the sticky ends of viral DNA
that is catalyzed by host ligase. This phase lasts 8-12 hours. The
second phase involves the synthesis of viral RNA and its post-transcriptional modifications. These include polyadenylation at the 3'-
end, addition of a cap structure at the 5'-end, its stepwise methyla-
ation and methylation of internal adenine residues. There is also
ribose methylation at the 5'-cap structure. The timing of these
modification reactions is not known, and while it is generally
believed that these reactions are catalyzed by host enzymes, this
also is still uncertain[52]. 3-DZA has no effect on the initial phase
of virus replication (viral DNA synthesis). Administration of
inhibitors of DNA synthesis such as 2'-deoxyadenosine reduced focus

formation more than 10-fold[50]. In contrast, administration of 3-DZA 24 hours after infection, at a time when viral DNA is already integrated in the host genome and is transcribed into viral mRNA, produces a striking inhibition of focus formation. 3-DZA, at the effective antiviral concentration, has no effect on the growth of DNA, RNA or protein synthesis of host cells. From these experiments, we concluded, therefore, that 3-DZA interfered with the synthesis of an active species of viral RNA. Several additional experiments reinforce this tentative conclusion. Chick embryo fibroblasts can be infected with temperature-dependent RSV. In this case, the ratio of glucose/leucine uptake is low at the nonpermissive temperature, 41°, and increases about 2- to 3-fold within 6 hours after a shift of the temperature to 37°. This increase can be prevented by a variety of compounds that inhibit RNA synthesis 3-DZA is capable of preventing this increased glucose uptake induced by temperature shift, in the absence of a significant effect on RNA synthesis. Deoxyglucose/leucine uptake in normal cells is not affected by 3-DZA[50].

3-DZA also inhibits growth and replication of other RNA and DNA viruses. The effectiveness of this compound, however, varies in the different biological systems. It is now becoming increasingly clear that both the pathways and the extent of the stepwise methylation of the 5'-cap formation varies significantly in different viruses and the same is true of the enzymes involved. It is generally believed, however, that the presence of a methyl group at the 7-position of the terminal guanosine residue is important and in some cases essential for efficient translation of mRNA.

It can, therefore, be suggested as a general working hypothesis that the efficacy of 3-DZA and other AdoHcy hydrolase inhibitors and/or AdoHcy analogs as antiviral agents relates to their ability to modulate or inhibit the reactions that lead to the synthesis of a fully methylated 5'-cap structure in viral mRNAs. Considerable experimental evidence supporting this hypothesis has been accumulated, but the critical demonstration of its validity is not yet available.

REFERENCES

1. G. L. Cantoni and E. Scarano, The formation of S-adenosylhomocysteine in enzymatic transmethylation reactions, J. Amer. Chem. Soc. 76:4744 (1954).
2. A. E. Chung and J. H. Law, Biosynthesis of cyclopropane compounds. VI. Product inhibition of cyclopropane fatty acid synthetase by S-adenosylhomocysteine and reversal of inhibition by a hydrolytic enzyme, Biochemistry 3:1989 (1964).
3. J. Hurwitz, M. Gold and M. Anders, The enzymatic methylation of ribonucleic acid and deoxyribonucleic acid. IV. The properties of the soluble ribonucleic acid-methylating enzymes, J. Biol. Chem. 239:3474 (1964).
4. S. K. Shapiro, A. Almenas and J. F. Thomson, Biosynthesis of

methionine in *Saccharomyces cerevisiae*. Kinetics and mechanism of reaction of S-adenosylmethionine:homocysteine methyltransferase, J. Biol. Chem. 240:2512 (1965).

5. V. Zappia, C. R. Zydek-Cwick and F. Schlenk, The specificity of
    S-adenosylmethionine derivatives in methyl transfer reactions,
    J. Biol. Chem. 244:4499 (1969).

6. Y. Akamatsu and J. H. Law, The enzymatic synthesis of fatty acid
    methyl esters by carboxyl group alkylation, J. Biol. Chem.
    245:709 (1970).

7. T. Deguchi and J. Barchas, Inhibition of transmethylations of
    biogenic amines by S-adenosylhomocysteine, J. Biol. Chem.
    246:3175 (1971).

8. K. R. Swiatek, L. N. Simon and K.-L. Chao, Nicotinamide methyl-
    transferase and S-adenosylmethionine:5'-methylthioadenosine
    hydrolase. Control of transfer ribonucleic acid methylation,
    Biochemistry 12:4670 (1973).

9. S. J. Kerr, Regulation of tRNA methyltransferase activity, in:
    "The Biochemistry of Adenosylmethionine," F. Salvatore, E.
    Borek, V. Zappia, H. G. Williams-Ashman and F. Schlenk, eds.,
    University Press, New York (1977).

10. C. S. G. Pugh, R. T. Borchardt and H. O. Stone, Inhibition of
    Newcastle disease virion messenger RNA (guanine-7-)-methyl-
    transferase by analogues of S-adenosylhomocysteine, Biochem-
    istry, 16:3928 (1977).

11. J. K. Coward, D. L. Bussolotti and C.-D. Chang, Analogs of
    S-adenosylhomocysteine as potential inhibitors of biological
    methylation. Inhibition of several methylases by S-tuber-
    cidinylhomocysteine, J. Med. Chem. 17:1286 (1974).

12. C. Kutzbach and E. L. R. Stokstad, Mammalian methylenetetrahydro-
    folate reductase. Partial purification, properties, and
    inhibition by S-adenosylmethionine, Biochim. Biophys. Acta
    250:459 (1971).

13. G. T. Burke, J. H. Mangum and J. D. Brodie, Mechanism of
    mammalian cobalamin-dependent methionine biosynthesis, Bio-
    chemistry 10:3079 (1971).

14. S. Nishimura, Characterization and enzymatic synthesis of 3-(3-
    amino-3-carboxypropyl)-uridine in transfer RNA: transfer of
    the 3-amino-3-carboxypropyl group from adenosylmethionine,
    in:"The Biochemistry of Adenosylmethionine," F. Salvatore,
    E. Borek, V. Zappia, H. G. Williams-Ashman and F. Schlenk,
    eds., University Press, New York (1977).

15. G. L. Stoner and M. A. Eisenberg, Purification and properties of
    7,8-diaminopelargonic acid aminotransferase. An enzyme in
    the biotin biosynthetic pathway, J. Biol. Chem. 250:4029
    (1975).

16. C. W. Tabor and H. Tabor, 1,4-Diaminobutane (putrescine),
    spermidine, and spermine, Ann. Rev. Biochem. 45:285 (1976).

17. S. K. Shapiro and A. N. Mather, The enzymatic decomposition of
    S-adenosyl-L-methionine, J. Biol. Chem. 233:631 (1958).

18. S. H. Mudd, Enzymatic cleavage of S-adenosylmethionine, J. Biol.

Chem. 234:87 (1959).

19. M. Gold, R. Hausman, U. Maitra and J. Hurwitz, The enzymatic
    methylation of RNA and DNA. VIII. Effects of bacteriophage
    infection on the activity of the methylating enzymes, Proc.
    Nat. Acad. Sci. U. S. A. 52:292 (1964).
20. C. Baxter and C. J. Coscia, In vitro synthesis of spermidine in
    the higher plant, Vinca rosea, Biochem. Biophys. Res. Commun.
    54:147 (1973).
21. K. R. Swiatek, L. N. Simon and K.-L. Chao, Nicotinamide methyl-
    transferase and S-adenosylmethionine:5'-methylthioadenosine
    hydrolase. Control of transfer ribonucleic acid methylation,
    Biochemistry 12:4670 (1973).
22. G. L. Cantoni, S-Adenosylmethionine: present status and future
    perspectives, in:"The Biochemistry of Adenosylmethionine," F.
    Salvatore, E. Borek, V. Zappia, H. G. Williams-Ashman and F.
    Schlenk, eds., University Press, New York (1977).
23. G. de la Haba and G. L. Cantoni, The enzymatic synthesis of
    S-adenosyl-L-homocysteine from adenosine and homocysteine, J.
    Biol. Chem. 234:603 (1959).
24. H. H. Richards, P. K. Chiang and G. L. Cantoni, Adenosylhomo-
    cysteine hydrolase: crystallization of the purified enzyme
    and its properties, J. Biol. Chem. 253:4476 (1978).
25. J. L. Palmer and R. H. Abeles, The mechanism of action of
    S-adenosylhomocysteinase, J. Biol. Chem. 254:1217 (1979).
26. A. Guranowski and J. Pawelkiewicz, Adenosylhomocysteinase from
    yellow lupin seeds: purification and properties, Eur. J.
    Biochem. 80:517 (1977).
27. R. A. Schatz, C. R. Vunnam and O. Z. Sellinger, S-Adenosyl-L-
    homocysteine hydrolase from rat brain: purification and some
    properties, in:"Transmethylation," E. Usdin, R. T. Borchardt
    and C. R. Creveling, eds., Elsevier, New York (1979).
28. A. Guranowski, P. K. Chiang and G. L. Cantoni, to be published.
29. R. T. Borchardt, Synthesis and biological activity of analogs of
    adenosylhomocysteine as inhibitors of methyltransferases, in:
    "The Biochemistry of Adenosylmethionine," F. Salvatore, E.
    Borek, V. Zappia, H. G. Williams-Ashman and F. Schlenk, eds.,
    University Press, New York (1977).
30. T. O. Eloranta, Tissue distribution of S-adenosylmethionine and
    S-adenosylhomocysteine in the rat. Effect of age, sex and
    methionine administration on the metabolism of S-adenosyl-
    methionine, S-adenosylhomocysteine and polyamines, Biochem.
    J. 166:521 (1977).
31. C. Walsh, Chemical approaches to the study of enzymes catalyzing
    redox transformations, Ann. Rev. Biochem. 47:881 (1978).
32. A. Guranowski, P. K. Chiang and G. L. Cantoni, unpublished
    results.
33. B. C. Halpern, B. R. Clark, D. N. Hardy, R. M. Halpern and R. A.
    Smith, The effect of replacement of methionine by homocystine
    on survival of malignant and normal adult mammalian cells in
    culture, Proc. Nat. Acad. Sci. U. S. A. 71:1133 (1974).

34. R. M. Hoffman and R. W. Erbe, High *in vitro* rates of methionine biosynthesis in transformed human and malignant rat cells auxotrophic for methionine, Proc. Nat. Acad. Sci. U. S. A. 73:1523 (1976).

35. P. L. Chello and J. R. Bertino, Dependence of 5-methyltetra-hydrofolate utilization by L51789 murine leukemia cells *in vitro* on the presence of hydroxycobalamin and transcobalamin II, Cancer Res. 33:1898 (1973).

36. M. J. Wilson and L. A. Poirier, An increased requirement for methionine by transformed rat liver epithelial cells *in vitro*, Exper. Cell Res. 111:397 (1978).

37. N. M. Kredich, M. S. Hershfield and J. M. Johnston, Role of adenosine metabolism in transmethylation, in:"Transmethyl-ation," E. Usdin, R. T. Borchardt and C. R. Creveling, eds., Elsevier, New York (1979).

38. I. H. Fox and W. N. Kelley, The role of adenosine and 2'-deoxy-adenosine in mammalian cells, Ann. Rev. Biochem. 47:655 (1978).

39. J. A. Duerre, A hydrolytic nucleosidase acting on S-adenosyl-homocysteine and on 5'-methylthioadenosine, J. Biol. Chem. 237:3737 (1962).

40. C. H. Miller and J. A. Duerre, S-Ribosylhomocysteine cleavage enzyme from *Escherichia coli*, J. Biol. Chem. 243:92 (1968).

41. J. A. Duerre and R. D. Walker, Metabolism of adenosylhomo-cysteine, in:"The Biochemistry of Adenosylmethionine," F. Salvatore, E. Borek, V. Zappia, H. G. Williams-Ashman and F. Schlenk, eds., University Press, New York (1977).

42. P. K. Chiang, H. H. Richards and G. L. Cantoni, S-Adenosyl-L-homocysteine hydrolase: analogs of S-adenosyl-L-homocysteine as potential inhibitors, Mol. Pharmacol. 13:939 (1977).

43. M. Ikehara and T. Fukui, Studies of nucelosides and nucleotides. LVIII. Deamination of adenosine analogs with calf intestine adenosine deaminase, Biochim. Biophys. Acta 338:512 (1974).

44. R. L. Miller, D. L. Adamczyk, W. H. Miller, G. W. Koszalka, J. L. Rideout, L. M. Beacham, III, E. Y. Chao, J. J. Haggerty, T. A. Krenitsky and G. B. Elion, Adenosine kinase from rabbit liver. II. Substrate and inhibitor specificity, J. Biol. Chem. 254:2346 (1979).

45. P. K. Chiang and G. L. Cantoni, Perturbation of biochemical transmethylations by 3-deazaadenosine *in vivo*, Biochem. Pharmac., in press.

46. T. P. Zimmerman, G. Wolberg and G. S. Duncan, Inhibition of lymphocyte-mediated cytolysis by 3-deazaadenosine: evidence for a methylation reaction essential to cytolysis, Proc. Nat. Acad. Sci. U. S. A. 75:6220 (1978).

47. E. J. Leonard, A. Skeel, P. K. Chiang and G. L. Cantoni, The action of the adenosylhomocysteine hydrolase inhibitor, 3-deazaadenosine, on phagocytic function of mouse macrophages and human monocytes, Biochem. Biophys. Res. Commun. 84:102 (1978).

48. P. K. Chiang, K. Venkatasubramanian, H. H. Richards, G. L.

Cantoni and E. Schiffmann, Adenosylhomocysteine hydrolase and chemotaxis, in:"Transmethylation," Elsevier, New York (1979).

49. J. M. Thompson, P. K. Chiang, R. R. Ruffolo, Jr., G. L. Cantoni and M. Nirenberg, Methyltransferases are involved in neurotransmission, Society of Neuroscience Meeting, abstract (1979).

50. J. P. Bader, N. R. Brown, P. K. Chiang and G. L. Cantoni, 3-Deazaadenosine an inhibitor of adenosylhomocysteine hydrolase inhibits reproduction of Rous sarcoma virus and transformation of chick embryo cells, Virology 89:494 (1978).

51. P. K. Chiang, G. L. Cantoni, J. P. Bader, W. M. Shannon, H. J. Thomas and J. A. Montgomery, Adenosylhomocysteine hydrolase inhibitors: synthesis of 5'-deoxy-5'-(isobutylthio)-3-deazaadenosine and its effect on Rous sarcoma virus and Gross murine leukemia virus, Biochem. Biophys. Res. Commun. 82:417 (1978).

52. J. M. Bishop, Retroviruses, Ann. Rev. Biochem. 47:35 (1978).

HOMOCYSTEINE BIOSYNTHESIS IN PLANTS

John Giovanelli, S. Harvey Mudd, and Anne H. Datko

National Institute of Mental Health
Laboratory of General and Comparative Biochemistry
Bethesda, Maryland   20205, U.S.A.

INTRODUCTION

Synthesis by plants of homocysteine, the immediate precursor
of methionine, is a key reaction in biology (Allaway, 1970).
This is so because non-ruminant animals require a dietary source
of homocysteine, which is normally provided in the form of methionine. Animals metabolize these sulfur amino acids eventually
to inorganic sulfate. Plants complete the cycle of sulfur by
reductive assimilation of inorganic sulfate to methionine (and
cysteine) (Siegel, 1975), and are thus the ultimate source of
methionine in most animal diets.

In spite of its biological importance, until recently the
biosynthesis of homocysteine in plants received little experimental attention, and it was generally assumed that the process
in plants was essentially the same as in microorganisms. However
in the late 1960's it became clear that microorganisms exhibit
much diversity in the biochemistry of homocysteine biosynthesis
(Delavier-Klutchko and Flavin, 1965; Flavin and Slaughter, 1967;
Nagai and Flavin, 1967), thereby making it impossible to predict
the situation in plants. At this time the studies to be described
here were initiated. This report summarizes the major develop-
ments leading to our present understanding of homocysteine biosynthesis in plants.* It now appears that plants are unique in
certain aspects of homocysteine biosynthesis, adding even further

---

*This summary is based on a more extensive treatment of this
 subject in a review on sulfur amino acids in plants
 (Giovanelli et al., in press).

diversity to this interesting field.

EARLY STUDIES WITH PLANTS

Early studies of homocysteine biosynthesis in plants were oriented largely by what was known in microorganisms. Microorganisms were known to synthesize homocysteine by transsulfuration, the transfer of sulfur between cysteine and homocysteine via cystathionine. In bacteria, transsulfuration proceeds only from cysteine to homocysteine, whereas in fungi transsulfuration proceeds reversibly between cysteine and homocysteine (Flavin, 1975). Our early studies therefore asked whether transsulfuration occurs in plants, and if so, in what direction it proceeds.

Plants were shown to contain the two enzymes required for synthesis of homocysteine by transsulfuration. The first enzyme (Giovanelli and Mudd, 1966) catalyzes the reaction:

$$\text{Cysteine} + \text{X-O-homoserine} \longrightarrow \text{cystathionine} + \text{XOH} \qquad (1)$$

As with microorganisms, homoserine itself was not active. The nature of the homoserine ester used in Reaction (1) will be discussed below. The second enzyme (Giovanelli and Mudd, 1971) cleaves cystathionine specifically to homocysteine:

$$\text{Cystathionine} + \text{H}_2\text{O} \longrightarrow \text{homocysteine} + \text{pyruvate} + \text{NH}_3 \quad (2)$$

No cleavage of cystathionine to cysteine could be detected. Failure to detect cleavage of cystathionine to cysteine strongly suggests that transsulfuration in plants proceeds only in the direction of cysteine to homocysteine. This suggestion is supported by our recent observation (see below) that exogenous methionine, while contributing a major portion of the sulfur moiety for methionine biosynthesis, does not contribute significantly to the sulfur of cysteine.

During these early studies, an interesting finding was made in our laboratory with plants, and independently in other laboratories with microorganisms: enzymic synthesis of homocysteine occurs by a direct reaction of sulfide with a homoserine ester (Giovanelli and Mudd, 1967):

$$\text{Sulfide} + \text{X-O-homoserine} \longrightarrow \text{homocysteine} + \text{XOH} \qquad (3)$$

Homocysteine can therefore be synthesized in plants potentially by two alternative pathways which differ in the origin of the sulfur atom.

These early studies raised two major questions that will be discussed below. The first is the nature of the physiological homoserine ester (α-aminobutyryl donor) that is used for cysta-thionine synthesis [Reaction (1)] and homocysteine synthesis [Reaction (3)]. The second is the relative physiological sig-nificance of the two pathways in homocysteine biosynthesis.

THE PHYSIOLOGICAL α-AMINOBUTYRYL DONOR

"α-Aminobutyryl donor" describes any compound that can donate its α-aminobutyryl group either to cysteine to form cystathionine, or to sulfide to form homocysteine (Datko et al., 1974):

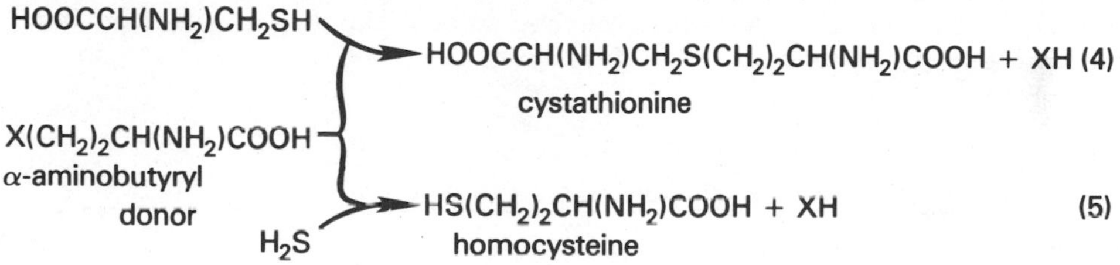

$HOOCCH(NH_2)CH_2SH$

$\longrightarrow HOOCCH(NH_2)CH_2S(CH_2)_2CH(NH_2)COOH + XH$ (4)

cystathionine

$X(CH_2)_2CH(NH_2)COOH$

α-aminobutyryl donor

H_2S

$\longrightarrow HS(CH_2)_2CH(NH_2)COOH + XH$ (5)

homocysteine

Homoserine itself is inactive in these reactions. O-Succinyl-homoserine is the physiological α-aminobutyryl donor for E. coli and S. typhimurium, while O-acetylhomoserine is used in all fungi and non-enteric bacteria examined (Flavin, 1975). Several findings, now to be discussed, show that plants use neither of these α-aminobutyryl donors, but instead use yet another O-ester of homoserine, O-phosphohomoserine:

Homoserine is Esterified by Most Plants Solely to O-Phosphohomoserine

Plants in general do not synthesize either O-succinylhomo-serine, O-acetylhomoserine, nor two other O-acylhomoserine de-rivatives, O-oxalylhomoserine or O-malonylhomoserine (Giovanelli et al., 1974). Synthesis of O-acylhomoserine derivatives was detected only in Pisum sativum and Lathyrus sativus (Giovanelli et al., 1974), plants which are unusual in accumulating O-acetyl-homoserine (Grobbelaar and Steward, 1958) or O-oxalylhomoserine (Przybylska and Pawelkiewicz, 1965), respectively. Of the above O-acylhomoserine derivatives, only O-acetylhomoserine was syn-thesized by extracts of P. sativum, and only O-oxalylhomoserine was synthesized by extracts of L. sativus.

In contrast, extracts of all plants examined (including

P. sativum and L. sativus) catalyzed the synthesis of O-phosphohomoserine (Giovanelli et al., 1974; Aarnes, 1976, 1978), according
to the reaction catalyzed by homoserine kinase (EC 2.7.1.39):

Homoserine + ATP $\longrightarrow$ O-phosphohomoserine + ADP   (6)

Plants are Unique in Using O-Phosphohomoserine as a Substrate for Cystathionine and Homocysteine Biosynthesis

O-Phosphohomoserine is used as a substrate for synthesis
of cystathionine [Reaction (4)] and homocysteine [Reaction (5)]
by extracts of all plants examined (Datko et al., 1974, 1977),
but not by any microorganism tested (See Giovanelli et al., in
press). Plants are therefore unique among the organisms studied
in using O-phosphohomoserine for cystathionine or homocysteine
synthesis.

O-Phosphohomoserine is the Only α-Aminobutyryl Donor Detected in Plants

An assay mixture containing plant cystathionine synthase
and [^{14}C]cysteine was used for detection and quantitation of
α-aminobutyryl donors in plant tissue. Under the conditions
of the assay, addition of an α-aminobutyryl donor resulted in
the conversion of [^{14}C]cysteine to [^{14}C]cystathionine at a rate
which was directly proportional to the amount of α-aminobutyryl
donor added. A crucial aspect of this assay is that it detected
not only those homoserine esters described above, but any
α-aminobutyryl donor capable of reacting with cysteine in the
presence of plant cystathionine synthase. By use of this assay,
α-aminobutyryl donor activity was demonstrated in a wide range
of plants, and was purified from several grown axenically. Only a
single endogenous α-aminobutyryl donor was detected, and was
identified as O-phosphohomoserine (Datko et al., 1974). By use
of an analogous assay containing plant homocysteine-forming sulfhydrase and radioactive sulfide, it was demonstrated that O-phosphohomoserine is also the only endogenous α-aminobutyryl donor able
to support direct synthesis of homocysteine (Datko et al., 1977).

These combined data strongly suggest that, regardless of
the pathway, O-phosphohomoserine is the physiologically important
α-aminobutyryl donor for homocysteine biosynthesis. This conclusion is consistent with in vivo experiments (Dougall and Fulton,
1967) that showed that neither O-succinyl- nor O-acetylhomoserine
was utilized in preference to homoserine for cystathionine synthesis by cells of Paul's Scarlet Rose.

PHYSIOLOGICAL SIGNIFICANCE OF THE TWO PATHWAYS

What is the relative physiological significance of transsulfuration and direct sulfhydration in plants? This question
has been answered for microorganisms mainly by combined nutrition
and enzymic studies of appropriate mutants. Transsulfuration
predominates in Salmonella, Escherichia coli and Neurospora
(Flavin, 1975). Transsulfuration also predominates in Aspergillus,
except when direct sulfhydration is derepressed by blocking cysteine biosynthesis (Paszewski and Grabski, 1974). Both pathways
have been proposed as dominant in Saccharomyces, and the relative
physiological significance of the two pathways in this organism
remains to be resolved (Flavin, 1975; Yamagata et al., 1975;
Paszewski and Grabski, 1976).

Methionine auxotrophs, which have proved so useful in assessing the relative importance of the two pathways in microorganisms, are not available in green plants. An alternative approach
of determining the kinetics of assimilation of $^{35}SO_4^{2-}$ into key

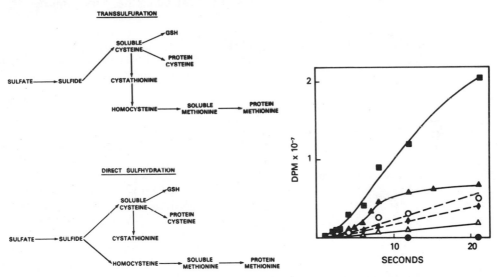

Fig. 1. (left). Models for predicting labeling patterns for
         two pathways. The chemical nature of the physiological
         inorganic sulfur precursor is not clear. The term
         sulfide is used here without intention to distinguish
         between the free and bound forms. From Giovanelli et
         al., 1978.

Fig. 2. (right). Labeling patterns of Chlorella fed $^{35}SO_4^{2-}$.
         ■, soluble cysteine; ▲, cystathionine; △, homocysteine;
         ◆, GSH; O, protein cysteine; ●, protein methionine.
         From Giovanelli et al., 1978.

sulfur amino acids, was therefore adopted. Fig. 1 outlines the
labeling patterns due to the operation of the two pathways. In
the transsulfuration pathway, radioactive sulfur must first enter
soluble cysteine, then pass through cystathionine before appearing
in homocysteine. For the direct sulfhydration pathway, cystathionine need not be labeled at early times, and label will accumulate in homocysteine independently of that in soluble cysteine.

Fig. 2 illustrates the labeling patterns observed when
Chlorella, growing photoautotrophically under steady state conditions with limiting sulfate, was incubated with $^{35}SO_4^{2-}$
(Giovanelli et al., 1978). Radioactivity accumulated first in
soluble cysteine before appearing in homocysteine. That soluble
cysteine is a necessary precursor of homocysteine is shown by
the high ratio (127:1) of soluble (^{35}S)cysteine to (^{35}S)homocys-
teine at the earliest time period of 1 sec, and the rapid decline
in this ratio to approach the equilibrium value of 4 after approximately 40 sec. These changes are those expected for transsulfuration. For the direct sulfhydration pathway the ratio would
start at 0.7 and increase with time to approach the same equilibrium value (Giovanelli et al., 1978). Furthermore, cystathionine was rapidly labeled (second only to soluble cysteine),
and rapidly reached isotopic equilibrium. Again, this is the
pattern expected for transsulfuration, which requires a rapid
turnover of cystathionine.

Quantitative analysis of these data (Giovanelli et al., 1978)
showed that transsulfuration accounts for at least 97% of homocysteine synthesis by Chlorella under the conditions specified.
These studies have now been extended to a higher plant (Lemna).
At least 90% of homocysteine biosynthesis in this plant proceeds
also through transsulfuration (P.M. Macnicol, A.H. Datko,
J. Giovanelli, and S.H. Mudd, unpublished results). The results
with these two phylogenetically distant plants suggest that transsulfuration is the predominant, perhaps exclusive, pathway for
homocysteine biosynthesis in the plant kingdom.

ENZYMES OF HOMOCYSTEINE BIOSYNTHESIS

Synthesis of cystathionine (Reaction (4)) and homocysteine
(Reaction (5)) has been demonstrated in extracts of a wide range
of plants (Datko et al., 1974, 1977). The two activities have
not been purified, and it is not known whether they are properties
of a single enzyme as in S. typhimurium, or of two separate enzymes, as in N. crassa (Flavin, 1975).

While plants are unique in using O-phosphohomoserine as
the physiological substrate, crude plant extracts can also use
other homoserine esters. For cystathionine synthesis, in general

(at near saturating concentrations) highest rates were obtained with O-malonylhomoserine, intermediate rates with O-oxalyl-, O-succinyl- and O-phosphohomoserine, and very low rates with O-acetylhomoserine (Datko et al., 1974). For the sulfhydration reaction O-phosphohomoserine was the most active of these substrates (Datko et al., 1977).

Cystathionine γ-synthase from N. crassa is inactive in the absence of N^5-methyltetrahydrofolate and is allosterically inhibited by AdoMet* (Flavin, 1975). No effect of either of these compounds was observed on the cystathionine γ-synthase activity of plants (Datko et al., 1974; Madison and Thompson, 1976).

Cystathionine β-lyase (EC 4.4.1.8), which catalyzes Reaction (2), is also widely distributed in plants, and has been purified from spinach leaves (Giovanelli and Mudd, 1971). The plant enzyme resembles the bacterial more closely than the fungal enzyme in its substrate specificity, but is more similar to the fungal enzyme, with respect to inhibition by sulfhydryl reagents.

Cystathionine is widely distributed in plant tissues, albeit at low concentrations between 0.1 to 1 μM (Datko et al., 1974a). Unless cystathionine is highly compartmented, cystathionine β-lyase operates in vivo well below saturation with its substrate.

Rhizobitoxine, an analogue of cystathionine produced by certain strains of Rhizobium japonicum (Owens et al., 1972), is a potent irreversible inhibitor of plant cystathionine β-lyase both in vivo (Giovanelli et al., 1973) and in vitro. Inhibition of the purified spinach enzyme was of the active-site-directed irreversible type (Giovanelli et al., 1971), and probably involves covalent linkage of a cleavage product of rhizobitoxine to the pyridoxal phosphate prosthetic group of the enzyme.

CONTROL OF HOMOCYSTEINE AND METHIONINE BIOSYNTHESIS

By far the most common control pattern for biosynthetic pathways is by feedback regulation by the product or products of the pathway. Dougall (1965) reported that Paul's Scarlet Rose cells growing in the presence of [^{14}C]glucose and methionine synthesized only 20% of the carbon atoms of protein methionine

---

*The following abbreviations are used: AdoMet, (S)-S-adenosyl-L-methionine (formerly designated (-)-S-adenosyl-L-methionine; AdoHcy, S-adenosyl-L-homocysteine.

from glucose. These results suggested that exogenous methionine
may decrease de novo synthesis of this amino acid, but did not
prove this point because neither incorporation of ^{14}C into soluble
methionine and its soluble derivatives (Davies, 1968), nor excretion of these latter compounds into the medium (Roberts et al.,
1955) was measured. The possibility that the methyl and 4-carbon
moieties of methionine might be regulated independently and to
different extents should also be considered.

Whether methionine regulates de novo synthesis of its sulfur
moiety has recently been examined in L. paucicostata (Giovanelli
et al., in press). Plants were grown for approximately four
generations either on $^{35}SO_4^{2-}$, or on $^{35}SO_4^{2-}$ supplemented with
2 μM methionine. Radioactivity was determined in protein cysteine
and in cystathionine and its products (homocysteine, soluble
methionine, S-methylmethionine, AdoMet, AdoHcy, and protein methionine). Only relatively small amounts of radioactive compounds
other than $^{35}SO_4^{2-}$ were found in the medium, and there was no
significant synthesis of volatile ^{35}S compounds. Growth in the
presence of exogenous methionine had little effect on the incorporation of radioactivity into protein cysteine, and did not
cause appreciable accumulation of radioactivity in soluble cysteine
and GSH. The most significant effect of exogenous methionine
was to decrease the incorporation of $^{35}SO_4^{2-}$ into cystathionine
and its products approximately 80%.

These data indicate that methionine, or a derivative thereof,
controls in vivo assimilation of sulfate into cystathionine and
its products, and therefore that the regulatory locus is at cystathionine synthesis. Furthermore, since regulation at this step
did not cause an accumulation of cysteine and its products, regulation of sulfate assimilation into cysteine is also indicated. It
has not yet been firmly established whether methionine also
controls de novo synthesis of the carbon moieties of methionine.
Such regulation of the 4-carbon moiety would be expected if exogenous methionine regulates the cystathionine synthesis step,
since it is at this step that both the sulfur and 4-carbon moieties
become committed to methionine.

Homocysteine biosynthesis requires the confluence of two
pathways, one providing the 4-carbon moiety via O-phosphohomoserine, the other providing the sulfur moiety via cysteine (Fig. 3).
Each of the converging pathways contains two branch points, and
additional branch points occur at threonine, methionine and AdoMet.
A priori, two predictions can be made regarding the control patterns of homocysteine biosynthesis in the system. One is that the
control patterns will be complex, in order that a fine balance
can be maintained between the interlocking and multiply branched
pathways. The second is that novel control patterns may occur

which may be unique to plants. The latter prediction (Datko
et al., 1974) is based on the fact that the branch point for
homocysteine and threonine biosynthesis in plants is at O-phosphohomoserine, whereas in other organisms it is at homoserine.

    Experimental support for each of these predictions is pro-
vided in the tentative scheme for control of homocysteine biosynthesis presented in Fig. 3. In this scheme, feedback regulation
of cystathionine synthesis is based on the intriguing finding
of Madison and Thompson (1976) that low concentrations of AdoMet,
while not directly affecting the activity of plant cystathionine
synthase, markedly stimulated the activity of plant threonine
synthase. O-Phosphohomoserine occurs in plants at concentrations
approximately two orders of magnitude below the Km's of cystathionine synthase and threonine synthase (See Giovanelli et al.,
in press). A strong competition by these enzymes for O-phosphohomoserine would therefore be expected in vivo. An overproduction
of methionine leading to an increased concentration of AdoMet
could stimulate threonine synthase, and thereby indirectly inhibit
cystathionine synthesis by the novel mechanism of diversion of

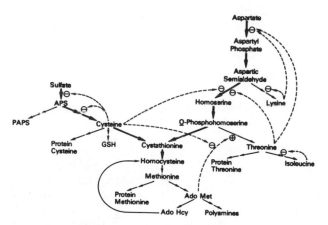

Fig. 3. Control of homocysteine and methionine biosynthesis.
    Converging pathways leading to cystathionine synthesis,
    and further conversion to homocysteine, are shown in
    heavy arrows. Relationships between effectors and their
    control sites are shown in dashed arrows. Sites of
    negative or positive control are shown by - or +, re-
    spectively. The evidence described herein suggests
    that regulation occurs at the cystathionine synthesis
    step. The physiological importance of the remaining
    control sites is less certain, and the possibility that
    other control mechanisms may operate has not been ade-
    quately explored (See Bryan, in press; Giovanelli et
    al., in press).

O-phosphohomoserine into threonine synthesis. A reduction in the flux between aspartate and O-phosphohomoserine (that is equivalent to the reduction in the flux into cystathionine) ensues by feedback regulation of homoserine dehydrogenase by threonine and cysteine, and of aspartokinase by either threonine, lysine, or both amino acids (Bryan, in press). The initial overproduction of cysteine resulting from the decreased flux of sulfur into cystathionine could similarly reduce the flux of sulfur into cysteine by a number of mechanisms that have been discussed elsewhere (Giovanelli et al., in press).

Many plant proteins are of limited nutritional value because of their low content of methionine, which is one of the first essential amino acids to become inadequate in the human diet (Allaway and Thompson, 1966). In spite of the uncertainties of the regulatory scheme shown in Fig. 3, it provides a useful working model for efforts to increase the methionine content of plants. For example, the scheme predicts that an overproduction of methionine should result from inhibition of the conversion of methionine to AdoMet, or from inhibition of the allosteric stimulation of threonine synthase by AdoMet. These proposals are currently being explored in our laboratory.

REFERENCES

Aarnes, H., 1976, Homoserine kinase from barley seedlings, Plant Sci. Lett., 7:187.
Aarnes, H., 1978, Regulation of threonine biosynthesis in barley seedlings (Hordeum vulgare, L.), Planta, 140:185.
Allaway, W. H., 1970, The scope of the symposium: Outline of current problems related to sulfur in nutrition, in: Symposium: Sulfur in Nutrition," O. H. Muth, and J. E. Oldfield, eds., Avi Publishing Co., Westport, Conn.
Allaway, W. H., and Thompson, J. F., 1966, Sulfur in the nutrition of plants and animals, Soil Sci., 101:240.
Bryan, J. K., in press, The synthesis of the aspartate family and the branched chain amino acids, in: "Biochemistry of Plants: A Comprehensive Treatise," Vol. 5, P. K. Stumpf, and E. E. Conn, eds., Academic Press, New York.
Datko, A. H., Giovanelli, J., and Mudd, S. H., 1974, Homocysteine biosynthesis in green plants. O-Phosphohomoserine as the physiological substrate for cystathionine γ-synthase, J. Biol. Chem., 249:1139.
Datko, A. H., Mudd, S. H., and Giovanelli, J., 1974a, A sensitive and specific assay for cystathionine: cystathionine content of several plant tissues, Anal. Biochem., 62:531.
Datko, A. H., Mudd, S. H., and Giovanelli, J., 1977, Homocysteine biosynthesis in green plants. Studies of the homocysteine-forming sulfhydrylase, J. Biol. Chem., 252:3436.

Davies, D. D., 1968, The metabolism of amino acids in plants,
    in: "Recent Aspects of Nitrogen Metabolism in Plants,"
    E. J. Hewitt, and C. V. Cutting, eds., Academic Press,
    New York.
Delavier-Klutchko, C., and Flavin, M., 1965, Enzymatic synthesis
    and cleavage of cystathionine in fungi and bacteria, J.
    Biol. Chem., 240:2537.
Dougall, D. K., 1965, The biosynthesis of protein amino acids
    in plant tissue culture I. Isotope competition experiments
    using glucose-U-C^{14}and the protein amino acids, Plant Physiol.,
    40:891.
Dougall, D. K., and Fulton, M. M., 1967, Biosynthesis of protein
    amino acids in plant tissue culture IV. Isotope competition
    experiments using glucose-U-C^{14} and potential intermediates,
    Plant Physiol., 42:941.
Flavin, M., 1975, Methionine biosynthesis, in: "Metabolic Pathways,"
    Vol. 7, D. M. Greenberg, ed., Academic Press, New York.
Flavin, M., and Slaughter, C., 1967, Enzymatic synthesis of
    homocysteine or methionine directly from O-succinylhomoserine,
    Biochim. Biophys. Acta, 132:400.
Giovanelli, J., and Mudd, S. H., 1966, Enzymatic synthesis of
    cystathionine by extracts of spinach, requiring O-acetylhomo-
    serine or O-succinylhomoserine, Biochem. Biophys. Res.
    Commun., 25:366.
Giovanelli, J., and Mudd, S. H., 1967, Synthesis of homocysteine
    and cysteine by enzyme extracts of spinach, Biochem. Biophys.
    Res. Commun., 27:150.
Giovanelli, J., and Mudd, S. H., 1971, Transsulfuration in higher
    plants. Partial purification and properties of β-cystathionase
    of spinach, Biochim. Biophys. Acta, 227:654.
Giovanelli, J., Owens, L. D., and Mudd, S. H., 1971, Mechanism
    of inhibition of spinach β-cystathionase by rhizobitoxine,
    Biochim. Biophys. Acta, 227:671.
Giovanelli, J., Owens, L.D., and Mudd, S. H., 1973, β-Cystathio-
    nase. In vivo inactivation by rhizobitoxine and role of
    the enzyme in methionine biosynthesis in corn seedlings,
    Plant Physiol., 51:492.
Giovanelli, J., Mudd, S. H., and Datko, A. H., 1974, Homoserine
    esterification in green plants, Plant Physiol., 54:725.
Giovanelli, J., Mudd, S. H., and Datko, A. H., 1978, Homocysteine
    biosynthesis in green plants. Physiological importance
    of the transsulfuration pathway in Chlorella sorokiniana
    growing under steady state conditions with limiting sulfate,
    J. Biol. Chem., 253:5665.
Giovanelli, J., Mudd, S. H., and Datko, A. H., in press, Sulfur
    amino acids in plants, in: "Biochemistry of Plants: A
    Comprehensive Treatise," Vol. 5, P. K. Stumpf, and E. E. Conn,
    eds., Academic Press, New York.

Grobbelaar, N., and Steward, F. C., 1958, O-Acetylhomoserine
      in Pisum, Nature, 182:1358.
Madison, J. T., and Thompson, J. F., 1976, Threonine synthetase
      from higher plants: stimulation by S-adenosylmethionine
      and inhibition by cysteine, Biochem. Biophys. Res. Commun.,
      71:684.
Nagai, S., and Flavin, M., 1967, Acetylhomoserine. An intermediate
      in the fungal biosynthesis of methionine, J. Biol. Chem.,
      242:3884.
Owens, L. D., Thompson, J. F., Pitcher, R. G., and Williams, T.,
      1972, Structure of rhizobitoxine, an antimetabolite enol-ether
      amino acid from Rhizobium japonicum, J. Chem. Soc. Chem.
      Commun., 714.
Paszewski, A., and Grabski, J., 1974, Regulation of S-amino acid
      biosynthesis in Aspergillus nidulans. Role of cysteine
      and/or homocysteine as regulatory effectors. Mol. Gen.
      Genet., 132:307.
Paszewski, A., and Grabski, J., 1976, On sulfhydrylation of
      O-acetylserine and O-acetylhomoserine in homocysteine synthesis
      in yeast, Acta Biochim. Pol., 23:321.
Przybylska, J., and Pawelkiewiez, J., 1965, O-Oxalylhomoserine,
      a new homoserine derivative in young pods of Lathyrus sativus,
      Bull. Acad. Pol. Sci. Ser. Sci. Biol., 13:327.
Roberts, R. B., Abelson, P. H., Cowie, D. B., Bolton, E. T.,
      and Britten, R. J., 1955, "Studies of Biosynthesis in
      Escherichia coli," Chapter 19, Carnegie Institution of
      Washington Publication 607, Washington, D.C.
Siegel, L. M., 1975, Biochemistry of the sulfur cycle, in:
      "Metabolic Pathways," Vol. 7, D. M. Greenberg, ed., Academic
      Press, New York.
Yamagata, S., Takeshima, K., and Naiki, N., 1975, O-Acetylserine
      and O-acetylhomoserine sulfhydrylase of yeast: Studies
      with methionine auxotrophs, J. Biochem., 77:1029.

INHIBITION OF THE SYNTHESIS OF GLUTATHIONE, GLUTAMINE, AND GLUTA-MATE BY CERTAIN METHIONINE DERIVATIVES

Alton Meister

Department of Biochemistry, Cornell University Medical College
1300 York Avenue, New York, New York 10021 U.S.A.

INTRODUCTION

In the course of studies in our laboratory on the biochemistry of glutamate and two of its principal metabolites, glutathione and glutamine, we have found that certain methionine derivatives are highly effective inhibitors of the utilization of glutamate and its derivatives. These effects are evidently explicable in terms of the structural similarities between glutamate, methionine, and their derivatives, and the consequent ability of particular methionine analogs to interact with the active sites of enzymes that normally bind and utilize glutamate or glutamine. It is of interest that a number of enzymes that act on glutamine also exhibit a relatively high affinity for methionine, and that glutamine and methionine are apparently transported across cell membranes by similar - in some instances perhaps identical - systems. Some enzymes that interact with methionine and glutamate or with derivatives of these amino acids are listed in Table I.

Antagonism between the naturally-occurring amino acids has frequently been observed in nutritional experiments. Such phenomena may probably be explained largely in terms of competition for transport or in some instances for metabolism. In contrast, protein synthesis exhibits a much higher degree of specificity. The aminoacyl tRNA synthetases seem to be highly specific for particular protein amino acids, but certain non-protein amino acid analogs are occasionally incorporated into proteins and must therefore interact with some degree of effectiveness with an aminoacyl tRNA synthetase designed for the activation of a natural protein amino acid. Although there appears to be no evidence for antagonism between glutamine and methionine for protein synthesis, these amino

TABLE I

Enzymes that Interact with Glutamate, Methionine, or with
Derivatives of these Amino Acids

Enzyme	Substrates, Inhibitors	Ref.
Glutamine transaminases	Methionine, glutamine & their α-keto acid analogs	1-4
γ-Glutamyl transpeptidase	Glutamine, methionine	5
γ-Glutamyl cyclotransferase	γ-glu-met; γ-glu-gln	6
Glutamine synthetase	Methionine sulfoximine α-Ethylmethionine sulfoximine	7-10 11
γ-Glutamylcysteine synthetase	Methionine sulfoximine Buthionine sulfoximine	12 13,14
Glutamate synthase	Methionine sulfone Methionine sulfoximine Homocysteine sulfonamide	15,16 15,16 16

acids may evidently compete for transport and in certain metabolic
reactions. For example, glutamine and methionine, as well as their
α-keto acid analogs, are good substrates of the glutamine transaminases. Glutamine and methionine are both excellent acceptor substrates of γ-glutamyl transpeptidase, and the γ-glutamyl derivatives
of glutamine and methionine are amongst the best substrates of γ-
glutamyl cyclotransferase. The methionine derivative methionine
sulfoximine interacts very effectively with glutamine synthetase
and with γ-glutamylcysteine synthetase, and thus inhibits the synthesis of both glutamine and glutathione in vivo. Methionine sulfone,
methionine sulfoximine and homocysteine sulfonamide interact with
glutamate synthase, inhibiting the utilization of glutamine for
glutamate formation by this enzyme. This paper reviews some of the
findings made in our laboratory on the inhibition of the synthesis
of glutamine, glutathione, and glutamate by methionine sulfoximine
and certain analogs of this interesting amino acid.

INHIBITION OF THE SYNTHESIS OF GLUTAMINE AND GLUTATHIONE BY METHIO-
INE SULFOXIMINE

Two major pathways of glutamate metabolism are (a) conversion
to glutamine, and (b) incorporation into the sulfur-containing tripeptide, glutathione. The enzymes that catalyze these reactions

(reactions (1)-(3)) have been purified from a number of sources and have been extensively studied[17,18/].

$$(1) \quad \text{Glutamate} + NH_3 + ATP \underset{\text{synthetase}}{\overset{\text{glutamine}}{\rightleftharpoons}} \text{glutamine} + ADP + P_i$$

$$(2) \quad \text{Glutamate} + \text{cysteine} + ATP \underset{\text{synthetase}}{\overset{\gamma\text{-glutamylcysteine}}{\rightleftharpoons}} \gamma\text{-glutamyl-}$$

cysteine + ADP + P_i

$$(3) \quad \gamma\text{-Glutamylcysteine} + \text{glycine} + ATP \underset{\text{synthetase}}{\overset{\text{glutathione}}{\rightleftharpoons}} \text{glutathione}$$

(γ-glu-cySH-gly) + ADP + P_i

Studies on the mechanisms of action of these enzymes have shown that each of the reactions involves the intermediate formation of an enzyme-bound acyl phosphate intermediate, i.e., γ-glutamyl phosphate (reactions (1) and (2)) and γ-glutamylcysteinyl phosphate (reaction (3)). There is now considerable experimental evidence, drawn from a variety of approaches, for the formation and utilization of such acyl phosphates; these findings have been reviewed[17-19/].

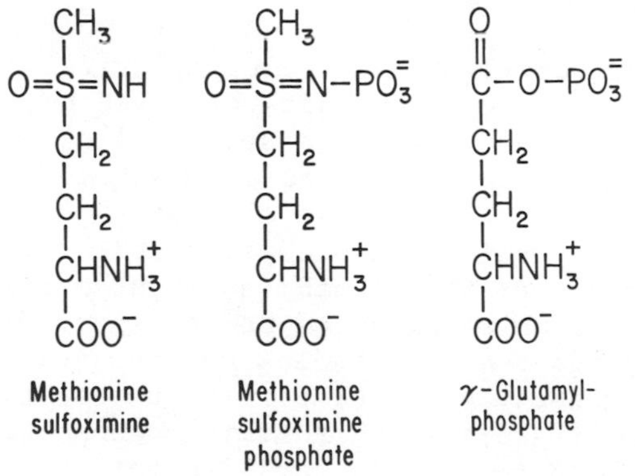

Figure 1

In studies on the inhibition of glutamine synthetase by methionine sulfoximine it was found that this compound competes with glutamate for binding to the enzyme, and that binding of methionine sulfoximine is followed by its phosphorylation by ATP to form methionine sulfoximine phosphate[7-10]. The phosphorylated compound binds tightly to the enzyme inhibiting it irreversibly. It was found that methionine sulfoximine phosphate, prepared by chemical synthesis, also inhibits the enzyme. Of the four diastereoisomers of methionine sulfoximine, only one, i.e., L-methionine-S-sulfoximine, is phosphorylated and inhibits the enzyme irreversibly, indicating that a specific configuration about the asymmetric sulfur atom is required.

Studies on the interaction of the enzyme with many substrates and non-substrates showed that L-glutamate attaches to the enzyme in an extended conformation in which the α-hydrogen atom is directed away from the active site[20,21]. Computer analysis of the active site permitted the selection of a phosphorylation site and of an ammonia binding site on the enzyme[22]. The position of the latter site was derived from the calculated position of the nitrogen atom of the tetrahedral intermediate (or transition state) presumably formed in the reaction of ammonia with enzyme-bound γ-glutamyl phosphate. Comparison of the structure of the tetrahedral intermediate with that of L-methionine-S-sulfoximine phosphate showed that these structures are very similar. L-Methionine-S-sulfoximine phosphate was found to fit as closely to the calculated enzyme sites as does the tetrahedral intermediate. These studies support the view that L-methionine-S-sulfoximine serves as an inhibitory analog of the enzyme-bound intermediate or transition state formed in the normal catalytic reaction. That methionine sulfoximine is phosphorylated also provides additional evidence for the formation of an acyl phosphate intermediate in the normal catalytic reaction.

The reaction catalyzed by γ-glutamylcysteine synthetase (reaction (2)) is analogous to that catalyzed by glutamine synthetase (reaction (1)), and there are several lines of evidence indicating that enzyme-bound γ-glutamyl phosphate is also formed in this reaction[18,23]. It is notable that only the same diastereoisomer of methionine sulfoximine that inhibits glutamine synthetase is also the only isomer that inhibits γ-glutamylcysteine synthetase; this isomer is also phosphorylated on the enzyme[12]. Thus, the mechanistic findings and the studies on inhibition by methionine sulfoximine are closely parallel for the two synthetases. The data indicate that L-methionine-S-sulfoximine phosphate is a transition-state inhibitor of both enzymes as illustrated in Figure 2.

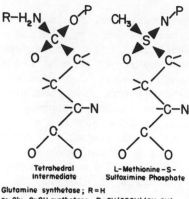

Glutamine synthetase; R=H
γ-Glu-CySH synthetase; R=CH(COOH)(CH₂SH)

Figure 2. Comparison of the structure of the tetrahedral intermediates formed in the reactions catalyzed by glutamine and γ-
glutamylcysteine synthetase and of L-methionine-S-sulfoximine
phosphate. (From 34).

SELECTIVE INHIBITION OF THE SYNTHESIS OF GLUTAMINE AND GLUTATHIONE
BY ANALOGS OF METHIONINE SULFOXIMINE

Methionine sulfoximine has long been known as a convulsant,
and it has been found that only one diastereoisomer of methionine
sulfoximine, again the L-S isomer, has convulsant activity[24]. It
thus became of significance to determine whether the convulsant
activity is associated with inhibition of glutamine synthetase or
with the synthesis of γ-glutamylcysteine (and thus of glutathione).
Furthermore, we also wished to obtain compounds that would be
useful in experimental systems for the selective inhibition of
glutamine and glutathione synthesis in vivo.

Since the mechanisms of synthesis of γ-glutamylcysteine and
glutamine are very similar and involve formation of γ-glutamyl
phosphate and analogous tetrahedral intermediates, it became evident that the design of specific one-enzyme inhibitors would require considerations based on differences in the substrate binding
sites of the two enzymes. In this work we were aided by knowledge
of the topology of the active sites of these enzymes that had
been obtained from mapping studies with various substrates and
non-substrates. We found that, although α-methylglutamate is a

substrate of glutamine synthetase, it is somewhat less active as
a substrate for γ-glutamylcysteine synthetase, indicating that
there is some steric restriction in the region of this enzyme
that lies close to the α-hydrogen atom of L-glutamate. We therefore prepared α-ethylmethionine sulfoximine[18]. This compound was
found to inhibit glutamine synthetase, but not to inhibit γ-
glutamylcysteine synthetase. α-Ethylmethionine sulfoximine was
found to induce convulsions in mice. The studies with this selective irreversible inhibitor of glutamine synthetase thus indicate
that the convulsant activity of methionine sulfoximine is most
probably due to its effect on glutamine synthetase rather than
to its effect on the synthesis of glutathione.

The reciprocal goal of obtaining a compound that inhibits
glutathione synthesis without perturbing the synthesis of glutamine was also achieved by making use of information about the
active sites of the two synthetases[13,14]. The studies on the
mapping of the active site of glutamine synthetase showed that
the S-methyl group of methionine sulfoximine attaches to the
enzyme site that normally binds ammonia. We therefore reasoned
that substitution of a bulkier group in place of the S-methyl
group of methionine sulfoximine might yield a molecule that could
not bind to glutamine synthetase. The comparable region of the
active site of γ-glutamylcysteine synthetase can bind cysteine
(and α-aminobutyrate), and would therefore be expected to bind
an analog having a group larger than a methyl group. We synthesized a series of sulfoximines including ethionine sulfoximine,
prothionine sulfoximine, and buthionine sulfoximine. Prothionine
sulfoximine and buthionine sulfoximine were found to have the
expected properties, i.e., to exhibit virtually no inhibitory
activity toward glutamine synthetase, but to inhibit γ-glutamylcysteine synthetase effectively[13,14]. Buthionine sulfoximine was
found to be about 100 times more active than methionine sulfoximine in inhibiting γ-glutamylcysteine synthetase. When injected
into mice, the kidney level of glutathione decreased to less than
20% of the control level within 1-2 hours. Buthionine sulfoximine
does not affect tissue glutamine levels, nor as expected, does
it induce convulsions in mice.

The results obtained with several sulfoximines are summarized
in Figure 3[11,13,14]. α-Methyl-methionine sulfoximine was similar
in its effects to methionine sulfoximine. The α-methyl derivatives
of prothionine sulfoximine and buthionine sulfoximine were also
synthesized and found to have properties similar to the respective
parent compounds. One would expect that the α-methyl compounds
would not undergo extensive metabolic degradation in vivo. Ethionine sulfoximine exhibits relatively less effect on glutamine synthetase, and it is a somewhat weaker convulsant as compared to
methionine sulfoximine. Buthionine sulfoximine inhibits γ-glutamyl-cysteine synthetase about 20 times more effectively than does

$$
\begin{array}{cccccc}
\boxed{\alpha\text{-et-}\atop MSO} & \boxed{\alpha\text{-me-}\atop MSO} & \boxed{MSO} & \boxed{ESO} & \boxed{PSO} & \boxed{BSO}
\end{array}
$$

α-et-MSO	α-me-MSO	MSO	ESO	PSO	BSO
					CH_3
				CH_3	CH_2
			CH_3	CH_2	CH_2
CH_3	CH_3	CH_3	CH_2	CH_2	CH_2
$O=S=NH$	$O=S=NH$	$O=S=NH$	$O=S=NH$	$O=S=NH$	$O=S=NH$
CH_2	CH_2	CH_2	CH_2	CH_2	CH_2
CH_2	CH_2	CH_2	CH_2	CH_2	CH_2
$CH_3CH_2-C-NH_3^+$	$CH_3-C-NH_3^+$	$HC-NH_3^+$	$HC-NH_3^+$	$HC-NH_3^+$	$HC-NH_3^+$
COO^-	COO^-	COO^-	COO^-	COO^-	COO^-
GLN↓	GLN↓	GLN↓	(GLN↓)		
	GSH↓	GSH↓	GSH↓	GSH↓	GSH↓↓↓
FITS	FITS	FITS	(FITS)		

Figure 3; Abbreviations: MSO = methionine sulfoximine; ESO = ethionine sulfoximine; PSO = prothionine sulfoximine; BSO = buthionine sulfoximine.

prothionine sulfoximine.

The interaction of α-ethyl-methionine sulfoximine and of buthionine sulfoximine with the respective synthetases may be considered in relation to the scheme given in Figure 4. Although both enzymes form γ-glutamyl phosphate and analogous tetrahedral intermediates or transition states, the binding sites for glutamate and for the acceptor substrates are significantly different. α-Ethyl-methionine sulfoximine can be accomodated at the active site of glutamine synthetase, but its α-ethyl group prevents it from binding to the active site of γ-glutamylcysteine synthetase. The presence of a relatively bulky group on the sulfur atom of buthionine sulfoximine effectively excludes this molecule from the active site of glutamine synthetase, but it can bind effectively to the active site of γ-glutamylcysteine synthetase which can bind an acceptor substrate which is significantly larger than a methyl group.

INHIBITION OF GLUTAMATE SYNTHASE

Glutamate synthase catalyzes the reductive amination of α-ketoglutarate according to reaction (4)[25]. The enzyme has been

(4)   Glutamine + α-ketoglutarate + NAD(P) + H$^+$ → 2 glutamate +

NAD(P)$^+$

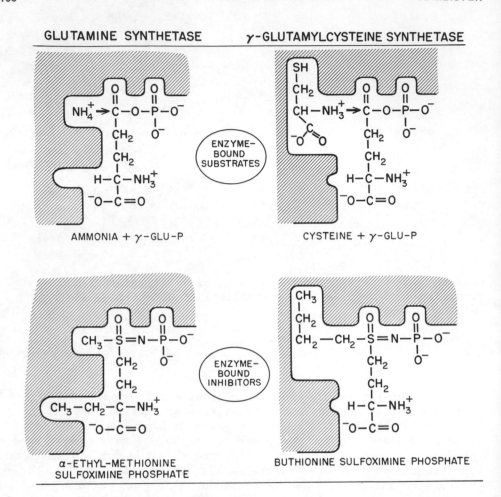

Figure 4; see the text.

obtained in highly purified form from several bacteria and from
yeast. Thus far, this enzyme activity has not been found in mammalian
tissues. In studies on the purified glutamate synthase of Aerobacter
aerogenes it was found that methionine sulfone, methionine sulfoximine, and methionine sulfoxide are effective inhibitors, competitive
with L-glutamine. The respective concentrations required for 50%
inhibition (with 0.5 mM L-glutamine) were 0.005, 0.35, and 0.7 mM[15/]
It is clear from the nature of the reaction catalyzed that the inhibition cannot be ascribed to the mechanism found for the synthetases discussed above. A further study of this phenomenon has recently been carried out in our laboratory on a highly purified and
apparently homogeneous preparation of glutamate synthase from

Saccharomyces cerevisiae[16/]. This enzyme is also inhibited by methionine sulfone and methionine sulfoximine as well as by homocysteine sulfonamide; the corresponding K_i values are 5.0, 146, and 7.3 µM[16/]. These compounds are competitive inhibitors with respect to L-glutamine. Since they each have a tetrahedral ω-sulfur atom, it is tempting to consider the possibility that they act as inhibitory analogs of the tetrahedral transition states or intermediates formed in the normal catalytic reaction in which glutamine interacts with the enzyme to form a γ-glutamyl enzyme, which is then hydrolyzed.

DISCUSSION

Figure 6 summarizes modifications of the methionine sulfoximine molecule which lead to highly selective and potent inhibitors of enzymes that catalyze the synthesis of glutamate and its utilization for the synthesis of glutamine and γ-glutamylcysteine. Homocysteine sulfonamide inhibits the synthetases competitively, but at relatively high concentrations (10 mM). Buthionine sulfoximine does not inhibit glutamate synthase appreciably except at relatively high concentrations. Since methionine sulfone inhibits all three of the enzymes, this methionine derivative would seem to be less useful as an experimental tool.

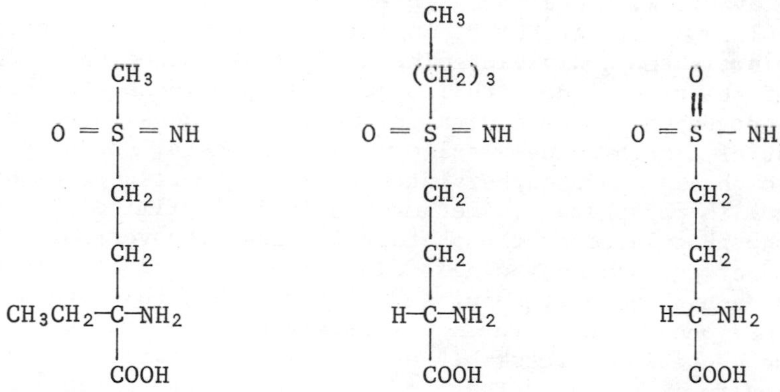

α-Ethyl-Methionine Sulfoximine
Inhibits:
Glutamine Synthetase

Buthionine Sulfoximine
Inhibits:
γ-Glutamylcysteine Synthetase

Homocysteine Sulfonamide
Inhibits:
Glutamate Synthase

Figure 6

The findings reported here give strong support to the view
that the convulsant activity of methionine sulfoximine is closely
associated with its inhibitory effect on glutamine synthetase.
Prothionine sulfoximine and buthionine sulfoximine have proven to
be effective and useful inhibitors of glutathione synthesis in
vivo, and studies in which these compounds were used have elucidated significant aspects of the metabolism of glutathione via the
γ-glutamyl cycle, the inter-organ transport of glutathione in the
mammal, and the translocation of glutathione across cell membranes
13,19,26-30/.

Finally, it should be noted that although methionine sulfoximine first came to light as a product formed in the chemical modification of proteins (i.e., in the bleaching of flour by nitrogen
trichloride[197]), there is evidence that methionine sulfoximine is
in fact a natural compound. Thus, Pruess et al[31] and Scannel et
al[32] reported the isolation of a compound with antibiotic activity
from the fermentation broth of a Streptomyces, which has the structure: L-(N^5-phosphono)methionine-S-sulfoximinyl-L-alanyl-L-alanine.
The antibiotic activity of this tripeptide containing an N-terminal
methionine sulfoximine phosphate residue against Serratia and
Bacillus subtilis was reversed by adding L-glutamine to the medium.
It is notable that the diastereoisomer present in the peptide is
identical to that which produces convulsions and which inhibits
the synthesis of glutamine and glutathione. Studies in our laboratory
carried out by Dr. W. Bruce Rowe showed that the isolated tripeptide
(kindly supplied by Dr. Arthur Stempel of Hoffman-LaRoche, Inc.),
did not inhibit sheep brain glutamine synthetase appreciably. After
treatment of the tripeptide with leucine aminopeptidase, inhibition was observed which was equivalent to that found with an equimolar amount of L-methionine-S-sulfoximine phosphate. Pruess et al[31]
were able to obtain a dephosphorylated form of the tripeptide by
treating it with phosphatase. Free methionine sulfoximine phosphate
is also dephosphorylated by phosphatases[77]. The discovery of a
natural product containing a single diastereoisomer of methionine
sulfoximine is highly interesting and suggests that this unusual
amino acid may occur elsewhere in nature. It will be interesting
to learn the mechanism, presumably an enzymatic process, by which
methionine sulfoximine is synthesized. Studies on the metabolic
breakdown of methionine sulfoximine to methane sulfinimide, methane
sulfonic acid, methane sulfinic acid, 2-imino-3-butenoic acid and
vinyl-glyoxylate have been reported[33]. Space does not permit various
speculations that might be entertained about the potential toxicity
of this amino acid to animals and its possible relationship to human
welfare.

SUMMARY

Methionine derivatives were prepared which selectively inhibit
glutamine synthetase, γ-glutamylcysteine synthetase (and there-

fore the synthesis of glutathione), and glutamate synthase. These compounds have been useful in unravelling active site topology. Studies with these compounds indicate that the convulsant effect of L-methionine-S-sulfoximine is due to its effect on glutamine synthesis rather than to an effect on glutathione synthesis. The potent and selective inhibitors of glutathione synthesis, e.g., prothionine sulfoximine and buthionine sulfoximine, have been useful in elucidating the metabolism of glutathione via the γ-glutamyl cycle, and the inter-organ transport and cellular translocation of glutathione. The natural occurrence of L-methionine-S-sulfoximine is noted and discussed.

REFERENCES

1. Meister, A., The Enzymatic Transfer of α-Amino Groups, Science 120:43 (1954).

2. Cooper, A.J.L., and Meister, A., Isolation and Properties of Highly Purified Glutamine Transaminase, Biochemistry 11:661 (1972).

3. Cooper, A.J.L., and Meister, A., Isolation and Properties of a New Glutamine Transaminase from Rat Kidney, J. Biol. Chem., 249:2554 (1974).

4. Cooper, A.J.L., and Meister, A., The Glutamine Transaminase-ω-Amidase Pathway, Critical Rev. Biochem., 4:281 (1977).

5. Tate, S.S., and Meister, A., Interaction of γ-Glutamyl Trans-peptidase with Amino Acids, Dipeptides, and Derivatives and Analogs of Glutathione, J. Biol. Chem., 249:7593 (1974).

6. Taniguchi, N., and Meister, A., γ-Glutamyl Cyclotransferase from Rat Kidney; Sulfhydryl Groups and Isolation of a Stable Form of the Enzyme, J. Biol. Chem., 253:1799 (1978).

7. Ronzio, R.A., and Meister, A., Phosphorylation of Methionine Sulfoximine by Glutamine Synthetase, Proc. Natl. Acad. Sci. U.S., 59:164 (1968).

8. Ronzio, R.A., Rowe, W.B., and Meister, A., Studies on the Mechanism of Inhibition of Glutamine Synthetase by Methionine Sulfoximine, Biochemistry 8:1066 (1969).

9. Rowe, W.B., Ronzio, R.A., and Meister, A., Inhibition of Glutamine Synthetase by Methionine Sulfoximine, Studies on Methionine Sulfoximine Phosphate, Biochemistry 8:2674 (1969).

10. Manning, J.M., Moore, S., Rowe, W.B., and Meister, A., Identification of L-Methionine-S-Sulfoximine as the Diastereo-isomer of L-Methionine-SR-Sulfoximine that Inhibits Glutamine Synthetase, Biochemistry 8:2681 (1969).

11. Griffith, O.W., and Meister, A., Differential Inhibition of Glutamine and γ-Glutamylcysteine Synthetases by α-Alkyl Analogs of Methionine Sulfoximine that Induce Convulsions, J. Biol. Chem., 253:2333 (1978).

12. Richman, P.G., Orlowski, M., and Meister, A., Inhibition of γ-Glutamyl-cysteine Synthetase by L-Methionine-S-Sulfoximine, J. Biol. Chem., 248:6684 (1973).

13. Griffith, O.W., Anderson, M.E., and Meister, A., Inhibition of Glutathione Biosynthesis by Prothionine Sulfoximine (S-n-Propyl-Homocysteine Sulfoximine), A Selective Inhibitor of γ-Glutamylcysteine Synthetase, J. Biol. Chem., 254:1205 (1979).

14. Griffith, O.W., and Meister, A., Potent and Specific Inhibition of Glutathione Synthesis by Buthionine Sulfoximine (S-n-Butyl Homocysteine Sulfoximine), J. Biol. Chem., 254:   (1979).

15. Trotta, P.P., Platzer, K.E.B., Haschemeyer, R.H., and Meister, A., Glutamine-Binding Subunit of Glutamate Synthase and Partial Reactions Catalyzed by this Glutamine Amidotransferase, Proc. Natl. Acad. Sci. U.S., 71:4607 (1974).

16. Masters, D.S., Glutamate Synthase, A Glutamine Amidotransferase from the Eucaryotic Yeast, Saccharomyces cerevisiae:Purification and Properties, Doctoral Dissertation, Cornell University Graduate School of Medical Sciences (1979).

17. Meister, A., Glutamine Synthetase, The Enzymes, (3rd ed.), 10:699 (1974).

18. Meister, A., Glutathione Synthesis, The Enzymes (3rd ed.), 10:671 (1974).

19. Meister, A., and Tate, S.S., Glutathione and Related γ-Glutamyl Compounds; Biosynthesis and Utilization, Ann. Rev. Biochem., 45:559 (1976).

20. Meister, A., On the Synthesis and Utilization of Glutamine, Harvey Lectures, Series, 63:139 (1969).

21. Meister, A., The Specificity of Glutamine Synthetase and Its Relationship to Substrate Conformation at the Active Site, Adv. Enzymol., 31:183 (1968).

22. Gass, J.D., and Meister, A., Computer Analysis of the Active
    Site of Glutamine Synthetase, Biochemistry 9:842 (1970).

23. Orlowski, M., and Meister, A., Partial Reactions Catalyzed by
    γ-Glutamylcysteine Synthetase and Evidence for an Activated
    Glutamate Intermediate, J. Biol. Chem., 246:7095 (1971).

24. Rowe, W.B., and Meister, A., Identification of L-Methionine-S-
    Sulfoximine as the Convulsant Isomer of Methionine Sulfoximine,
    Proc. Natl. Acad. Sci. U.S., 66:500 (1970).

25. Tempest, D.W., Meers, J.L., and Brown, C.M., Synthesis of
    Glutamate in Aerobacter aerogenes by a Hitherto Unknown Route,
    Biochem. J., 117:405 (1970).

26. Palekar, A.G., Tate, S.S., and Meister, A., Decrease in Gluta-
    thione Levels of Kidney and Liver after Injection of Methionine
    Sulfoximine into Rats, Biochem. Biophys. Res. Commun., 62:651
    (1975).

27. Griffith, O.W., Bridges, R.J., and Meister, A., Evidence that
    the γ-Glutamyl Cycle Functions in vivo Using Intracellular
    Glutathione; Effects of Amino Acids and Selective Inhibition
    of Enzymes, Proc. Natl. Acad. Sci. U.S., 75: 5405 (1978).

28. Meister, A., Current Status of the γ-Glutamyl Cycle, in Functions
    of Glutathione in Liver and Kidney (H. Sies and A. Wendel, eds.),
    Springer Verlag, Berlin, Heidelberg, New York; pp.43-59 (1978).

29. Griffith, O.W., and Meister, A., Translocation of Intracellular
    Glutathione to Membrane - Bound γ-Glutamyl Transpeptidase as
    a Discrete Step in the γ-Glutamyl Cycle; Glutathionuria after
    Inhibition of Transpeptidase, Proc. Natl. Acad. Sci. U.S., 76:
    268 (1979).

30. Meister, A., Griffith, O.W., Novogrodsky, A., and Tate, S.S.,
    New Aspects of Glutathione Metabolism and Translocation in
    Mammals, CIBA Foundation Symposium on The Biology of Sulfur,
    London, April, 1979.

31. Pruess, D.L., Scannell, J.P., Ax, H.A., Kellett, M., Weiss, F.,
    Demny, T.C., and Stempel, A., Antimetabolites Produced by
    Microorganisms. VII L-(N^5-Phosphono)Methionine-S-Sulfoximinyl-
    L-Alanyl-L-Alanine, J. Antibiotics, 26:261 (1973).

32. Scannell, J.P., Pruess, D.L., Demny, T.C., Ax, H.A., Weiss, F.,
    Williams, T., and Stempel, A., L-(N^5-Phosphono)Methionine-S-
    Sulfoximinyl-L-Alanyl-L-Alanine, An Antimetabolite of L-Gluta-
    mine Produced by a Streptomycete, Chem. Biol. of Peptides,
    pp. 415-421, Ann Arbor Science, Publishers (1972).

33. Cooper, A.J.L., Stephani, R.A., and Meister, A., Enzymatic
    Reactions of Methionine Sulfoximine, Conversion to the
    Corresponding α–Imino and α–Keto Acids, and to α–Ketobutyrate
    and Methane Sulfinimide, J. Biol. Chem., 251:6674 (1976).

34. Meister, A., Inhibition of Glutamine Synthetase and γ–Glutamyl-
    cysteine Synthetase by Methionine Sulfoximine and Related
    Compounds, in Enzyme-Activated Irreversible Inhibitors (N.
    Seiler, M.J. Jung and J. Koch-Weser, eds.), Elsevier-North
    Holland Biomedical Press, Amsterdam; pp. 187-211 (1978).

METHIONINE METABOLISM IN DEVELOPING NEURAL TISSUE

John A. Sturman

Developmental Neurochemistry Laboratory
Department of Pathological Neurobiology
Institute for Basic Research in
    Mental Retardation
Staten Island, N.Y. 10314, U.S.A.

The pathways of methionine metabolism in primates, including man, undergo major changes during pre- and postnatal development. The changes include activities of enzymes involved, concentrations of intermediates and end-products, and changes in the overall use of the carbon skeleton of the methionine molecule. This article will describe and discuss these changes in developing neural tissue, and compare and contrast them to those taking place in liver, when relevant.

The major developmental change which takes place in both brain and liver is the postnatal activation of the transsulfuration pathway of methionine metabolism. The net result of this pathway is the transfer of the sulfur atom from homocysteine to the carbon skeleton of serine to form cysteine. This conversion is mediated by two enzymes: cystathionine synthase (L-serine hydro-lyase {adding homocysteine}, EC 4.2.1.22) which catalyzes the β-activation of serine and the addition of homocysteine to form the thio-ether, cystathionine; cystathionase (EC 4.4.1.1) which catalyzes the γ-cleavage of cystathionine to form cysteine (Fig. 1). Both of these enzymes catalyze reactions other than those described above although their importance in vivo is uncertain (Tallan et al., 1974). In mature mammals, activities both of cystathionine synthase and of cystathionase are present in brain and liver, although cystathionase activity in

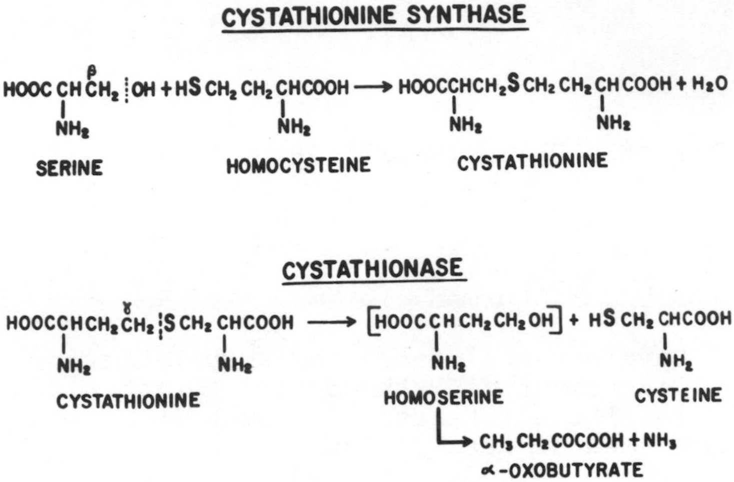

CYSTATHIONINE SYNTHASE

$$\underset{\text{SERINE}}{HOOC\,CH\,\overset{\theta}{CH_2}\,|\,OH} + \underset{\text{HOMOCYSTEINE}}{HS\,CH_2\,CH_2\,CHCOOH} \longrightarrow \underset{\text{CYSTATHIONINE}}{HOOCCHCH_2\,S\,CH_2\,CH_2\,CHCOOH} + H_2O$$

CYSTATHIONASE

$$\underset{\text{CYSTATHIONINE}}{HOOCCHCH_2\,\overset{\gamma}{CH_2}\,|\,S\,CH_2\,CHCOOH} \longrightarrow \left[\underset{\text{HOMOSERINE}}{HOOC\,CH\,CH_2\,CH_2\,OH}\right] + \underset{\text{CYSTEINE}}{HS\,CH_2\,CHCOOH}$$

$$\longrightarrow \underset{\alpha\text{-OXOBUTYRATE}}{CH_3\,CH_2\,COCOOH} + NH_3$$

Fig. 1.   Reactions of the transsulfuration pathway.

brain is quite small and contributes to the large
concentrations of cystathionine found in mature mammalian
brain (Mudd et al., 1965; Sturman et al., 1970). Cystathionine synthase and cystathionase use pyridoxal 5'-
phosphate as coenzyme, the holoenzyme of cystathionine
synthase being more resistant to a dietary deficiency of
vitamin B_6 than the holoenzyme of cystathionase (Sturman
et al., 1969). Because the transsulfuration pathway is
active the cysteine required for protein synthesis can be
synthesized from methionine, which is an essential amino
acid. In the mature mammal, therefore, cysteine is a
non-essential amino acid, although it can spare the
methionine requirement to a certain extent (Rose and Wixc
1955; Finkelstein and Mudd, 1967). This situation is
dramatically different in the human fetus, in which cyst
thionase activity is virtually absent (Sturman et al.,
1970; Gaull et al., 1972; Pascal et al., 1972). Thus fo
the human fetus, cysteine is an essential amino acid, an
must be supplied by the mother. The same is true for th
prematurely born and full term human infant since the
available data indicate that development of cystathionas
activity is a postnatal phenomenon whether the gestation
time is shortened or is of normal length(Table 1). In t

Table 1.   Cystathionase Activity in Developing Human Liver

Fetus (2nd trimester)		0^1
Neonate[2]		
680g;	24h	43
780g;	24h	0
830g;	11h	0
1000g;	8h	0
1060g;	14h	0
1260g;	3h	0
1500g;	96h	49
1650g;	216h	86
1780g;	72h	40
2250g;	24h	17
2600g;	144h	43
2730g;	96h	66
2950g;	24h	37
3450g;	7h	9
4000g;	96h	36
4250g;	72h	85
Adult		126 ± 12

From Gaull et al.,1972,1973

[1]Expressed as nmoles cysteine/mg prot/h

[2]Weight of infant at birth; time of death
   after birth

Cystathionase activity correlates significantly
with time of death after birth (r=0.776), but
not with weight at birth.   Furthermore,
correlation of cystathionase activity with time
of death after birth and weight at birth as two
independent variables does not add to the
significance of the correlation with time of
death after birth.

liver of the rhesus monkey, cystathionase activity is
present in low amounts prior to birth and then increases
rapidly and the concentration of cystathionine decreases
concomitantly (Fig. 2).   In brain, the predominant changes
during development are quite different from those in liver,
and involve an increase in the activity of cystathionine
synthase accompanied by an increase in the concentration
of cystathionine (Fig. 3).   The role of cystathionine in
brain is unknown.

LIVER

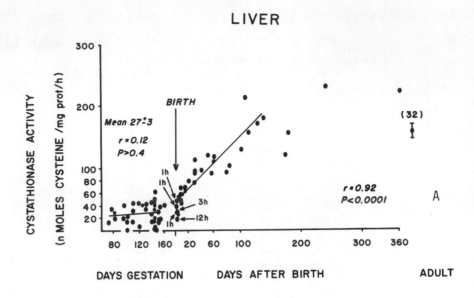

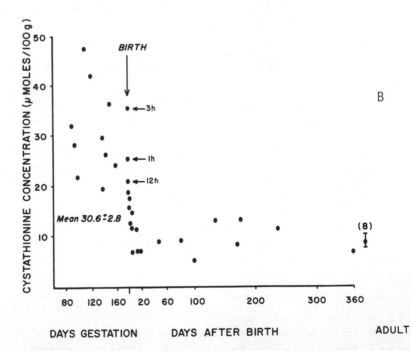

Fig. 2.   Specific activity of cystathionase (A) and
          concentration of cystathionine (B) in rhesus
          monkey liver during development.   From Sturman
          et al.,1976.

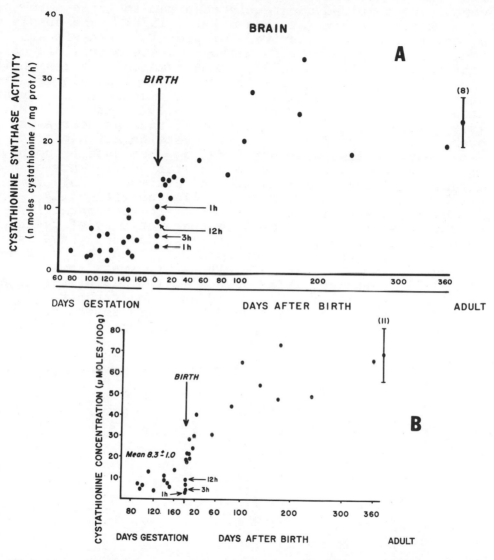

Fig. 3. Specific activity of cystathionine synthase (A) and concentration of cystathionine (B) in rhesus monkey brain during development. From Sturman et al., 1976.

The essentiality of cysteine for the fetus and newborn
may underlie their low or nonexistant ability to convert
cysteine to taurine, another low molecular weight sulfur
containing compound apparently required in large amounts
by developing brain (Sturman et al., 1978). The supply of
cysteine may all be required for protein synthesis and
none spared for taurine formation. Cysteinesulfinic acid
decarboxylase (EC 4.1.1.12), the enzyme chiefly responsibl
for taurine biosynthesis in mammals, develops slowly after
birth and reaches maximum activity in mature brain (Agrawa
et al.,1971; Pasantes-Morales et al.,1976; Rassin et al.,
1979), although the concentration of taurine decreases
over this same period (Fig. 4). Cysteinesulfinic acid
decarboxylase also uses pyridoxal 5'-phosphate as coenzyme
and is extremely sensitive to a dietary deficiency of
vitamin B_6 (Hope, 1955; Rassin and Sturman, 1975).
Cysteine and taurine must be supplied nutritionally,
therefore, for at least a portion of the postnatal life
of some mammals. It is of note that activity of cystathionase and of cysteinesulfinic acid decarboxylase is
present in tissues of the newborn rat, even though present
in smaller amounts than in adult rat tissues (Agrawal et
al., 1971; Heinonen, 1973; DeLuca et al., 1974; Loriette
and Chatagner, 1978). One would not expect the rat,
therefore, to be dependent upon a dietary source of cysteine or taurine at any stage of development. The ability

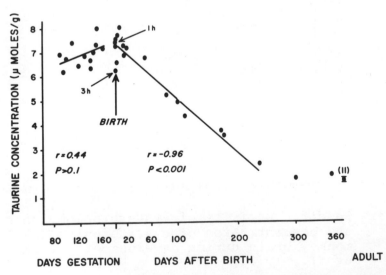

Fig. 4.   Concentration of taurine in rhesus monkey brain
          during development. From Sturman and Gaull,1975.

of some mammals, such as the primates, man and rhesus monkey, and the cat, to biosynthesize taurine may be sufficiently low, even at maturity, that taurine may be an essential nutrient throughout life (Knopf et al.,1978). The consequences of a dietary deficiency of taurine in the cat include degeneration of photoreceptor cells, disorganization of tapetal rods and loss of tapetal cells, decreased ERG responses, and eventually, blindness (Hayes et al., 1975; Wen et al., 1979). The consequences, if any, in man are not known.

The other pathway of methionine metabolism in neural tissue in which major changes occur during development is that of polyamine biosynthesis. Increased concentrations of polyamines and increased activity of the enzymes responsible for their biosynthesis, of putrescine and of ornithine decarboxylase (EC 4.1.1.17) especially, are associated with the growth process, including rapid synthesis of proteins and nucleic acids. Ornithine decarboxylase, the enzyme directly responsible for the biosynthesis of putrescine, is barely detectable in mature neural tissue, and the concentration of putrescine is low (Sturman and Gaull, 1974; 1975). Mature liver is similar to neural tissue in this respect. In rapidly growing human and rhesus monkey fetuses in the middle of gestation, however, activity of ornithine decarboxylase is easily measurable, both in brain and in liver, and the concentrations of putrescine are several fold higher than in mature tissue (Table 2, Fig. 5). The concentration of putrescine in developing monkey brain remains higher than that in adult monkey brain for somewhat longer during gestation than does ornithine decarboxylase activity. The other enzyme involved in the initiation of polyamine biosynthesis is S-adenosylmethionine decarboxylase (EC 4.1.1.50). In contrast to ornithine decarboxylase, S-adenosylmethionine decarboxylase is present in high activity in mature brain, and in lower activity in developing brain (Schmidt and Cantoni, 1973; Shaskan et al., 1973; Sturman and Gaull, 1974; 1975). The increase in activity of S-adenosylmethionine decarboxylase in brain during development is paralleled by the increase in activity of spermidine synthase and by the increase in concentration of spermidine (Table 2, Fig. 6). It should be noted that when S-adenosylmethionine is decarboxylated it is lost from the methionine-homocysteine cycle, and cannot be used for methylation reactions. S-Adenosylmethionine decarboxylase activity in liver of man and monkey decreases during development, in contrast to that in brain. Furthermore, methionine adenosyltransferase (ATP: L-methionine S-adenosyltransferase, EC 2.5.1.6), which catalyzes the formation of

Table 2. Activities of Ornithine and S-Adenosylmethionine
          Decarboxylases and Concentrations of Putrescine
          and Spermidine in Human Brain and Liver

Tissue	ODC[1]	SAMDC[1]	Putrescine[2]	Spermidine[2]
Brain				
Adult	6.2	351.0	5.5	40.7
Fetal	142.4	169.4	42.9	35.8
Liver				
Adult	2.2	12.4	6.5	21.4
Fetal	17.6	233.8	21.5	81.2

From Sturman and Gaull, 1974

[1]Expressed as pmoles CO_2/mg protein/h

[2]Expressed as umoles/100g

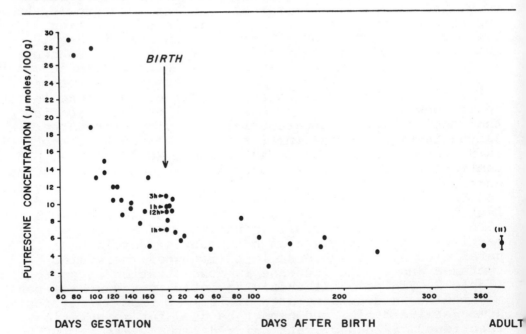

Fig. 5. Concentration of putrescine in rhesus monkey
         brain during development. From Sturman and
         Gaull, 1975.

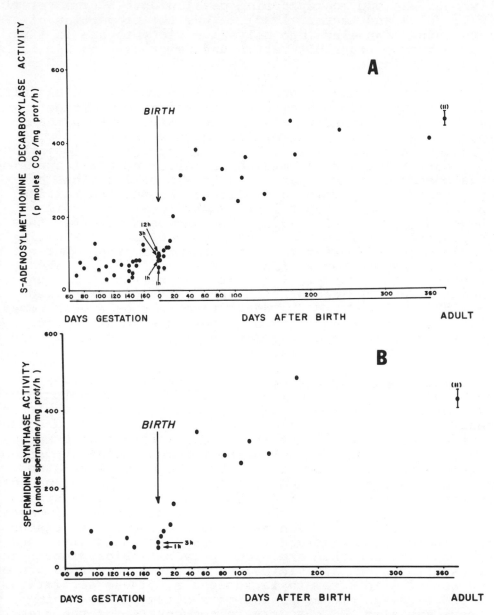

Fig. 6. Specific activities of S-adenosylmethionine
decarboxylase (A) and spermidine synthase (B)
in rhesus monkey brain during development.
From Sturman and Gaull, 1975.

S-adenosylmethionine from methionine and ATP, increases in
liver of man and monkey during development (Sturman et al.,
1970; Volpe and Laster, 1970). Thus the proportion of
methionine channeled into polyamine biosynthesis in liver,
and away from transsulfuration and remethylation is
especially large.

The most obvious conclusions from these findings with
regard to polyamines are that putrescine seems to be
especially important in developing brain, more so than
spermidine, and that spermidine may have some special
function in mature brain. Liver, in contrast, biosynthes-
izes putrescine and spermidine in larger quantities during
development than at maturity. What these functions of
putrescine and spermidine are is not known. Putrescine
is involved in the control of RNA synthesis in brain
(Reynolds and Russell, 1973). It has been reported that
putrescine is transported axonally in optic axons during
development and during regeneration after crush, but not
in intact mature nerves (Ingoglia et al., 1977). These
studies suggest that actively growing nerves during development or during regeneration have putrescine supplied
from the cell body to the nerve terminals or growth cones.
This result is potentially of importance in attempting to
answer the question of why mature mammalian central nerves
have little or no plasticity.

The conclusion reached from these findings from the
point of view of methionine metabolism in developing brain
is that involvement of S-adenosylmethionine in methylation
reactions is more important than its involvement in synthesis of polyamines. Furthermore, in its role as a methyl
donor it is converted to S-adenosylhomocysteine by various
methyltransferases, and the carbon skeleton is retained
within the methionine-homocysteine cycle. The absence of
an active transsulfuration pathway at these early stages
of development supports the concept that the most importan
metabolic role of methionine in developing neural tissue i
methylation. Thus metabolic pathways which involve loss o
the methionine carbon skeleton, polyamine biosynthesis and
transsulfuration, are minimal or absent in order to promot
and conserve this remethylation cycle. These alterations
in the pathways of methionine metabolism in developing
tissue may also be necessary to compensate for the greater
utilization of methionine for protein synthesis at this
time.

REFERENCES

Agrawal, H.C., Davison, A.N., and Kaczmarek, L.K., 1971,

Subcellular distribution of taurine and cysteinesulph-inate decarboxylase in developing rat brain, Biochem. J., 122:759-763.

DeLuca, G., Ruggeri, P., and Macaione, S., 1974, Cysta-thionase activity in rat tissues during development, Ital. J. Biochem., 23:371-379.

Finkelstein, J.D., and Mudd, S.H., 1967, Trans-sulfuration in mammals: The methionine-sparing effect of cystine, J. Biol. Chem., 242:873-880.

Gaull, G.E., Sturman, J.A., and Raiha, N.C.R., 1972, Development of mammalian sulfur metabolism: Absence of cystathionase in human fetal tissues, Pediat. Res., 6:538-547.

Gaull, G.E., von Berg, W., Raiha, N.C.R., and Sturman, J.A., 1973, Development of methyltransferase activities of human fetal tissues, Pediat. Res., 7:527-533.

Hayes, K.C., Carey, R.E., and Schmidt, S.Y., 1975, Retinal degeneration associated with taurine deficiency in the cat, Science, 188:949-951.

Heinonen, K., 1973, Studies on cystathionase activity in rat liver and brain during development, Biochem. J., 136:1011-1015.

Hope, D.B., 1955, Pyridoxal phosphate as the coenzyme of the mammalian decarboxylase for L-cysteine sulphinic and L-cysteic acids, Biochem. J., 59:497-500.

Ingoglia, N.A., Sturman, J.A., and Eisner, R.A., 1977, Axonal transport of putrescine, spermidine and sperm-ine in normal and regenerating goldfish optic nerves, Brain Res., 130:433-445.

Knopf, K., Sturman, J.A., Armstrong, M., and Hayes, K.C., 1978, Taurine: An essential nutrient for the cat, J. Nutr., 108:773-778.

Loriette, C., and Chatagner, F., 1978, Cysteine oxidase and cysteine sulfinic acid decarboxylase in develop-ing rat liver, Experientia, 34:981-982.

Mudd, S.H., Finkelstein, J.D., Irreverre, F., and Laster, L., 1965, Transsulfuration in mammals: Microassays and tissue distributions of three enzymes of the pathway, J. Biol. Chem., 240:4382-4392.

Pasantes-Morales, H., Mapes, C., Tapia, R., and Mandel, P., 1976, Properties of soluble and particulate cysteine sulfinate decarboxylase of the adult and the develop-ing rat brain, Brain Res., 107:575-589.

Pascal, T.A., Gillam, B.M., and Gaull, G.E., 1972, Cystathionase: Immunochemical evidence for absence from human fetal liver, Pediat. Res., 6:773-778.

Rassin, D.K., and Sturman, J.A., 1975, Cysteine sulfinic acid decarboxylase in rat brain: Effect of vitamin B_6 deficiency on soluble and particulate components, Life Sci., 16:875-882.

Rassin, D.K., Sturman, J.A., and Gaull, G.E., 1979, Source
    of taurine and GABA in cerebrum of developing rhesus
    monkey, Trans. Am. Soc. Neurochem., 10:161.
Reynolds, A.G., and Russell, D.H., 1973, Stimulation of
    ^{14}C-uridine incorporated into RNA by intracisternal
    injection of putrescine, Fed. Proc., 32:429.
Rose, W.C., and Wixom, R.L., 1955, The amino acid require-
    ments of man. XIII. The sparing effect of cystine
    of the methionine requirement, J. Biol. Chem., 216:
    763-773.
Schmidt, G.L., and Cantoni, G.L., 1973, Adenosylmethionine
    decarboxylase in developing rat brain, J. Neurochem.,
    20:1373-1385.
Shaskan, E.G., Haraszti, J.H., and Snyder, S.H., 1973,
    Polyamines: Developmental alterations in regional
    disposition and metabolism, J. Neurochem.,20:1433-1452.
Sturman, J.A., and Gaull, G.E., 1974, Polyamine biosynthe-
    sis in human fetal liver and brain, Pediat. Res., 8:
    231-237.
Sturman, J.A., and Gaull, G.E., 1975, Polyamine metabolism
    in the brain and liver of the developing monkey,
    J. Neurochem., 25:267-272.
Sturman, J.A., and Gaull, G.E., 1975, Taurine in the brain
    and liver of the developing human and monkey, J.
    Neurochem., 25:831-835.
Sturman, J.A., Cohen,P.A., and Gaull, G.E., 1969, Effects
    of deficiency of vitamin B$_6$ on transsulfuration,
    Biochem. Med., 3:244-251.
Sturman, J.A., Gaull, G.E., and Niemann, W.H., 1976,
    Cystathionine synthesis and degradation in brain,
    liver and kidney of the developing monkey, J.
    Neurochem., 26:457-463.
Sturman, J.A., Gaull, G.E., and Raiha, N.C.R., 1970,
    Absence of cystathionase in human fetal liver: Is
    cystine essential? Science, 169:74-76.
Sturman, J.A., Rassin, D.K., and Gaull, G.E., 1970,
    Distribution of transsulphuration enzymes in various
    organs and species, Int. J. Biochem., 1:251-253.
Sturman, J.A., Rassin, D.K., and Gaull, G.E., 1978, Taurin
    in the development of the central nervous cystem, in
    "Taurine and Neurological Disorders," A. Barbeau, an
    R.J. Huxtable, eds., Raven Press, New York.
Tallan, H.H., Sturman, J.A., Pascal, T.A., and Gaull, G.E
    1974, Cystathionine γ-synthesis from homocysteine an
    cysteine by mammalian tissue, Biochem. Med., 9:90-10
Volpe, J.J., and Laster, L., 1970, Trans-sulfuration in
    primate brain: Regional distribution of methionine-
    activating enzyme in the brain of the rhesus monkey
    at various stages of development, J. Neurochem., 17:
    413-424.

Wen, G.Y., Sturman, J.A., Wisniewski, H.M., Lidsky, A.A., Cornwell, A.C., and Hayes, K.C., 1979, Tapetum disorganization in taurine-depleted cats, <u>Invest. Ophthalmol. Vis. Sci.</u>, in press.

ANTAGONISTS OF FOLIC ACID AND VITAMIN B$_6$ IN REGULATION OF S-ADENOSYLMETHIONINE AND S-ADENOSYLHOMOCYSTEINE CONTENT IN NORMAL AND MALIGNANT TISSUES

Yuriy V.Bukin and Eugeniy N.Orlov

Cancer Research Center of U.S.S.R. Academy of Medical

Sciences, Moscow, U.S.S.R.

It is known that S-adenosylmethionine (SAM)[a] is the propyl-amino-group donor in the biosynthesis of polyamines (spermidine and spermine)[1,2] and universal methyl-group donor in all reactions of biological methylation except those involved in the biosynthesis of L-methionine[3,4]. Some biochemical functions of SAM may be critical for the viability of malignant cells. The polyamines which arise from decarboxylated products of SAM and L-ornitine perform specific function in reduplication of DNA and in cell division[5]. As it was shown by Corti et al.[6] ,the antitumor activity of methylglyoxal bis(guanilhydrazone) and related compounds is due to the inhibition of SAM decarboxylase (EC 4.1.1.50) in malignant tissue. The antitumor activity of cycloleucine (the conformational analogue of L-methionine) was suggested to be due to inhibition of L-methionine conversion to SAM catalyzed by SAM synthetase (EC 2.5.1.6)[7].

Enzymatic methylation reactions of proteins and nucleic acids in presence of SAM are of fundamental for cell division, differenciation and the normal genome expression. SAH which is the product of SAM demethylation possesses the properties of natural inhibitor of the majority of transmethylases[8,9] and apparently may exert in vivo a direct action on SAM metabolism and physioligical functions of cells. The high rate of tRNA turnover which is characteristic for malignant tumors is apparently due to abnormally high activity of

----

a Abbreviation: SAM, S-adenosylmethionine; SAH, S-adenosylhomocysteine; pyridoxal-P, pyridoxal 5'-phosphate; MTX, methotrexate; cis-APD, cis-2,5-bis(aminooxymethyl)-piperazine-3,6-dione.

tRNA methylases in tumor cells[10,11],moreover some tRNA methylases
of tumors are relatively more sensitive to SAH[12] . Recently Robert-
Gero and Lederer reported[13] that structural analogue of SAH,
5'-deoxy-5'-S-isobutyl-adenosine, possessing antimitigenic and
antiviral activity in cell culture inhibits also in in vivo experiments Friend virus leukemia in mice. Chiang et al. found that the
analogue of SAH, 5'-deoxy-5'-(isobutylthio)-3-deazaadenosine (a non-
competitive inhibitor of SAH hydrolase /EC 3.3.1.1/) has a selective
antiviral activity against Rouse sarcoma virus in chick embrio cells,
and against Gross murine leukemia virus in mouse embrio cells, and
reverses under certain condition the malignant transformation induced
by oncogenic virus[14].

Thus development of various approaches to regulation of SAM
and SAH metabolism in tumors may provide methods for control of
malignant growth.

According to our data presented below one of the approaches to
regulation of biosynthesis of SAM and SAH   in tissues may be
based on the use of antimetabolites of folic acid and vitamin B_6
blocking in vivo the conversion of folate to 5-CH_3-H_4folate by
inhibition of dihydrofolate reductase (EC1.5.1.3) and serine transhydroxymethylase (EC 2.1.2.1).

Dihydrofolate reductase catalyzes in presense of NADPH (or
NADH) the reduction of folate and H_2folate to H_4folate that is the
acceptor and carrier of active one-carbon units, the main source of
which are glycine and L-serine[15]. Pyridoxal-P -dependent serine
transhydroxymethylase catalyzes conversion L-serine to glycine
coupled with the conversion of H_4folate to 5,10-CH_2-H_4folate[15,16].
This enzyme plays a key role in formation of active one-carbon
units in cells. Alternative metabolic pathways of 5,10-CH_2-H_4folate
include its participation in thymidin monophosphate biosynthesis and
its conversion to 5,10-CH-H_4folate or 5-CH_3-H_4folate[15,16]. Since
the content of SAM in tissues is determined to considerable degree
by L-methionine concentration in cells[1,7,17]and since the resynthesis
of L-methionine is dependent on activity of 5-CH_3-H_4folate - L-homocysteine methyltransferase (EC 2.1.1.13)[15,18-20] we assumed that
blocking of the neogenesis of the methyl groups would limit the SAM
formation and decrease content of SAM in cells.

In order to achieve blocking of the neogenesis of methyl groups
we used MTX (which is a classical inhibitor of dihydrofolate reductase [15,21] and inhibits also the transport of 5-CH_3-H_4folate into
cells[22,23]) and some antagonists of vitamin B_6 and antimetabolites
of pyridoxal-P which, in accordance with our data, inhibit serine
transhydroxymethylase in vivo: D-cycloserine[24,25], cis-APD[25-27],
hydrazine (sulphate or hydrochloride)[24], L-penicillamine and decaborane ($B_{10}H_{14}$)[28]. SAM and SAH were assayed in trichloroacetic acid
supernatans of tissues by the phosphocellulose-column-chromatographic

method of Eloranta et al.[29] The method for estimation of serine transhydroxymethylase activity in tissues and characteristics of some of the antimetabolites were indicated previously[25].

The effect of MTX on SAM and SAH content in liver tissue of mice and rats is shown in Fig.1. The level of SAM and SAH in liver tissue of the animals injected i.p. 6 hr before sacrifice by moderate doses of MTX (2.5 mg/kg of body weight) is decreased by 40 -50% as compared with control. The decrease in SAM content in the tissue produced by MTX was accompanied by a corresponding decrease in SAH content thus the ratio SAH/SAM practically did not change.

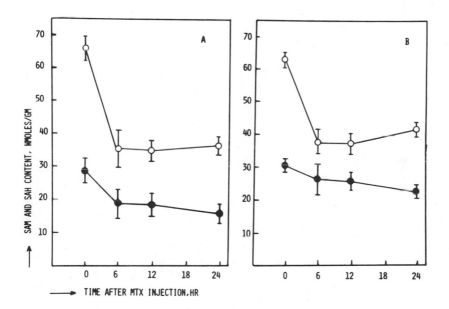

Fig. 1.   SAM and SAH content in liver tissue of mice and rats treated with MTX.
MTX was injected into mice (A) and rats (B) i.p. in a dose of 2.5 mg/kg of body weight, and the animals were killed at specified time intervals for SAM and SAH estimation. Symbols: (—O—),SAM; (—●—), SAH. Circles and vertical lines represent the average values of SAM and SAH content (nmoles/gm of wet weight of tissue) and their 95% confidense limits; there were 4-8 animals in each group.

In separate experiments we found that MTX administration into mice or rats in relatively high doses (12.5 mg/kg of body weight) leads to a decrease in initial SAM and SAH content in liver tissue no more than 50%. Within 3 days after the administration of MTX the content of SAM and SAH in liver of mice was still lower than in control, but it became normal in liver of rats.

The effect of D-cycloserine and cis-APD on SAM and SAH content in liver tissue of rodents is presented in Fig.2. A single injection of D-cycloserine into mice (4 gm/kg of body weight) producing within 4 hr moderate 40-50% decrease in activity of serine transhydroxymethylase in liver tissue, leads to gradual 40% reduction of SAM level in the tissue. As the initial SAH content in the tissue within 4 hr after D-cycloserine administration does not change, the SAH/SAM ratio increases from about 0.45 to 0.75. Within 12 hr after D-cycloserine administration the activity of serine transhydroxymethylase and SAM and SAH content in liver tissue of mice return to normal.

In rats cis-APD after single i.p. injection of 1.0 gm/kg of body weight produced 50-60% inhibition of serine transhydroxymethylase in liver tissue during only about 2 hr which was not accompanied by remarcable decrease in SAM content in the tissue (the data are not shown in Fig.2). However, more dramatic and prolonged decrease in the enzyme activity in liver tissue of rats in the course of repeated injections of cis-APD led to relatively sharp and prolonged decrease in SAM content in the tissue which was accompanied also by a decrease in SAH content (see Fig.2,B). Although the enzyme activity in this case was restored within 12 hr after the first injection of cis-APD the level of SAM within 24 hr was still decreased as compared with control. The delayed decrease in SAM content in the tissue may be the result of some metabolic shok produced by previous inhibition of serine transhydroxymethylase. Indeed, the decrease in the rate of $5,10-CH_2-H_4$folate formation from H_4folate catalyzed by the enzyme, may lead to preferable conversion of H_4folate to relatively unreactive polyglutamate derivatives of H_4folate or to $10-CHO-H_4$folate which possesses properties[30] of a dihydrofolate reductase inhibitor.

It shoud be noted that decaborane, the most active inhibitor of serine transhydroxymethylase in in vivo experiments, decreases the SAM and SAH content in liver tissue of rats no more than MTX does, namely by 50-55% as compared with the control (see Fig.3). In separate experiments simultaneous injections of MTX and cis-APD into mice or rats lead to maximal decrease in SAM and SAH level in liver tissue by 55-60%. The data obtained suggest that H_4folate- as well as pyridoxal-P - dependent neogenesis of methyl groups plays an important role in biosynthesis of SAM and in maintaining its high level in liver tissue. Under conditions of blocking of the neogenesis of methyl groups the new steady state concentrations of SAM in the tissue became about one-half as compared with the normal content. This relatively low ("basal") level of SAM

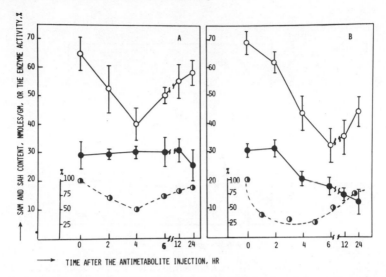

Fig. 2.   SAM and SAH content in liver tissue of mice
          and rats treated with D-cycloserine and cis-APD.
D-Cycloserine was injected i.p. into mice (A) in dose a 4.0
gm/kg of body weight. cis-APD was injected i.p. into rats (B)
three times with intervals of 2 hr, each time in a dose of
1.0 gm/kg of body weight. At the specifed time intervals after
the injections of D-cycloserine or first injection of cis-APD
the animals were killed for assay of SAM and SAH content and
activity of serine transhydroxymethylase. Symbols: (—O—),SAM;
(—●—),SAH; (—◑—), activity of the enzyme in percent to
control, mean values. See also legend to Fig.1.

content in liver is controlled obviously by concentration of preformed L-methionine in cells and also by the rate of L-methionine
resynthesis catalyzed by betaine- L-homocysteine methyltransferase
(EC 2.1.1.5) and dimethylthetin-L-homocysteine methyltransferase
(EC 2.1.1.3).

     In comparative experiments on rats we have found that all the
tested antagonists of vitamin B$_6$ and pyridoxal-P antimetabilites,
which inhibit serine transhydroxymethylase in liver within 6 hr
after i.p. injection, produce also a remarcable decrease in content
of SAM in the tissue. Among them, besides those mentioned above,
are hydrazine sulphate and DL-penicillamine which beeng injected
into rats in a dose of 0.1 gm/kg of body weight, decrease within
6 hr the content of SAM in the liver tissue by 55% and 15%, respectively. Unlike these antimetabolites 4-deoxypyridoxine, D-isomer
of penicillamine and isonicotinic acid hydrazide did not inhibit
serine transhydroxymethylase in rat liver and practically did not
decrease SAM content in the tissue within 6 hr after the administration. However, within 24 hr after the isonicotinic acid hydrazide
injection (0.3 gm/kg of body weight) the SAM content in rat liver

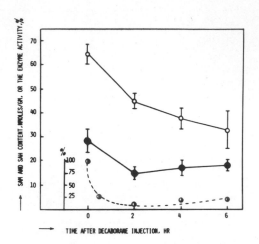

Fig. 3.   SAM and SAH content in liver tissue of rats
            treated with decaborane ($B_{10}H_{14}$).
Decaborane was injected into rats i.p. in a dose of 20 mg/kg of
body weight in 0.25 ml of olive oil. Symbols: (—O—),SAM;
(—●—), SAH; (—◑—), activity of serine transhydroxymethylase in percent to control,mean values. See also legend
to Fig.1.

decreased by more than 30%. It should be mentioned that a single
i.p. injection of cis-APD (3.0 gm/kg of body weight) or DL-penicillamine (0.12 gm/kg of body weight) into rats lead to pronounced,
40-45% decrease in SAM content in liver within 24 hr after the
administration, but within initial 6 hr period associated with
serine transhydroxymethylase inhibition, the decrease in SAM content
in the tissue was only about 15%.   The mechanisms of these "late
effects" may be different in each case and require further studies.

     Some of pyridoxal-P antimetabolites mentioned above reveal in
experiments on rodents antitumor activity comparable with that of
MTX, if injected daily in relatively high doses. Thus according to
our data[31,32] cis-APD injected into mice or rats (i.p. twice a day
in a 1.0 gm/kg of body weight each time during 10-15 days) produces
the inhibition of the growth of mouse solid form sarcoma 180 and rat
Pliss limphosarcoma by 60-80%. As it was shown by Littman et al.[33],
DL-penicillamine (preferably L-penicillamine), injected daily into
mice in a dose of 0.06 gm/kg, retarded the growth of sarcoma 180
in mice.

     Comparative data on the effect of MTX, cis-APD and DL-penicill-
amine on SAM and SAH content in liver and tumor tissue of mice or
rats are presented in Table 1. Single injection of MTX into mice
bearing sarcoma 180 (2.5 mg/kg of body weight) decreases within 24 hr
after the administration the SAM and SAH level in liver and sarcoma
180 by about 20% and 30%, respectively. In rats bearing Pliss

Table 1.
Comparative effect of MTX, cis-APD and DL-penicillamine on SAM and SAH content in liver and tumor tissue of rodents

Mice bearing sarcoma 180 (tumor weight 0.5-0.65 gm) and rats bearing Pliss lymphosarcoma (tumor weighw 5-6 gm) were injected i.p. with saline (control) or with the antimetabolites. MTX was injected into mice (2.5 mg/kg of body weight) and rats (0.8 mg/kg of body weight) 24 hr prior to sacrifice. cis-APD was injected into mice and rats twice (1.0 gm/kg of body weight each time) 24 hr and 12 hr prior to sacrifice. DL-Penicillamine (0.120 mg/kg of body weight) was injected into mice 24 hr prior to sacrifice. There were 4-6 rats or 6-8 mice in each group.

Tissues and experimental conditions	Content of SAM and SAH in nmoles/gm of wet weight of tissue and in percent to control			
	SAM		SAH	
	nmoles/gm[a]	%	nmoles/gm[a]	%
Mouse liver				
Control	64,4 ± 6,1		21,2 ± 2,5	
MTX	45,2 ± 5,2	70	15,2 ± 1,6	72
cis-APD	46,9 ± 4,0	73	18,2 ± 3,1	86[b]
Penicillamine	40,8 ± 5,6	63	10,0 ± 2,2	47
Sarcoma 180				
Control	45,6 ± 3,1		16,0 ± 2,8	
MTX	27,0 ± 4,2	59	13,6 ± 3,7	85[b]
cis-APD	36,0 ± 2,9	79	19,0 ± 1,9	85[b]
Penicillamine	30,6 ± 4,1	65	7,6 ± 1,4	48
Rat liver				
Control	59,9 ± 8,9		26,2 ± 10,5	
MTX	35,6 ± 13,4	66	18,3 ± 3,9	70[b]
Control	64,8 ± 7,8		21,6 ± 3,9	
cis-APD	32,4 ± 8,3	57	16,2 ± 4,7	80
Pliss lymphosarcoma				
Control	54,1 ± 11,3		24,3 ± 9,6	
MTX	28,6 ± 9,7	53	16,5 ± 4,1	68[b]
Control	56,6 ± 6,9		20,3 ± 3,7	
cis-APD	29,3 ± 11,9	51	17,8 ± 9,7	88[b]

[a] Average values ± 95% limits are indicated.
[b] Difference from control (based on Studen's test) is not significant; in all other cases the difference in SAM and SAH content is significant as compared with control ($p < 0.05$).

limphosarcome, which is very sensitive to MTX, comparatively low
dose of the drug (0.8 mg/kg of body weight) cause the decrease of
SAM content in liver and tumor tissue correspondingly by 34% and 47%,
in average. The data obtained suggest that the decrease in SAM
content in tumor cells may be a part of mechanism of antitumor
action of MTX as well as the disturbances in DNA and RNA synthesis
generally assumed[21, 34] or in ATP synthesis discussed recently by
Kaminskas[35].

        Administration of pyridoxal-P antimetabolite, cis-APD, into
rodents leads during 24 hr to decrease in SAM content in mouse sarcoma 180 and rat Pliss limphosarcoma by about 20% and 50%,
correspondingly. Single injection of DL-penicillamine into mice
decreases the SAM content in sarcoma 180 by 35%. The decrease in
SAM content was accompanied in some cases by the decrease in SAH
content in tumors (see Table 1). According to our data cis-APD and
DL-penicillamine (preferably L-isomer) inhibit serine transhydroxymethylase in tumor tissues. The decrease in rate of $5,10\text{-}CH_2\text{-}H_4$fo-
late formation , which is essential for thymidine monophosphate
synthesis, interfere with the DNA synthesis in tumor cells. Together
with this effect the decrease in SAM content in the tumor cells due
to restriction in $5\text{-}CH_3\text{-}H_4$folate synthesis from $5,10\text{-}CH_2\text{-}H_4$folate
may be significant for the antitumor effect of cis-APD or DL-peni-
cillamine.

        Finally, we shall briefly discuss the possible role of pyri-
doxal-P-dependent cystathionine synthase (EC 4.2.1.22) in regulation
of SAH level in tissues. El Oranta et al.[29] found recently that the
concentration of SAH increased 3-5 fold in liver of rats fed
vitamin B_6-deficient diet, and suggested that the effect is due to
an increase in L-homocysteine content in the tissue as a result of
metabolic block in the trans-sulphuration pathway. Taking into
consideration the above data and the fact that growth of sarcoma 180
in mice maintained on vitamin B_6-deficient diet is inhibited[36] we
suggested that the decrease in cystathionine synthase activity and
possible increase in L-homocysteine and SAH content in sarcoma 180
may play certain role in the regression of the tumor. However, the
corresponding experiments did not confirm this assumption. Despite
the fact that cystathionine synthase activity in liver and sarcoma
180 of mice fed vitamin B_6-deficient diet was significantly decreased we observed the 2-fold increase in SAH content only in liver
tissue whereas in the tissue of sarcoma 180, the growth of which
was inhibited, the SAH content was the same (about 20 nmoles/gm of
wet tissue weight) as in fast growing sarcoma 180 in mice fed diet
supplemented with vitamin B_6. Moreover, in experiments on rats
treated with decaborane (20 mg/kg of body weight) when cystathionine
synthase was inhibited in liver by 70-80%, the i.p. injection of
L-methionine and glycine (each in a dose of 100 mg/kg of body
weight) did not cause more pronounced increase in SAH content as
compared with rats injected only with L-methionine and glycine.
It was not be excluded that under the above conditions the biosynthe-

sis of SAH from L-homocysteine in the reversible reaction catalyzed
by SAH hydrolase is limited by adenosine concentration in tissue.

The data obtained suggest that the increase in SAH content in
liver of vitamin B$_6$-deficient mice and rats is not due only to
decrease in cystathionine synthase activity. Molecular basis for
SAH metabolic disturbances under condition of vitamin B$_6$-deficency
remain to be clarified.

REFERENCES

1. J. B. Lombardini and P. Talalay,Formation, functions and
    regulatiory importance of S-adenosyl-L-methionine, in:
    Advances in Enzyme Regulation, G. Weber, ed., 9:349,
    (1971).
2. D. M. Grinberg, Utilization and dissimilation of methionine,
    Chapter 13 in: Metabolic Pathways, D. M. Grinberg, ed.,
    vol. 7 (Metabolism of Sulfur Compound),Academic Press,
    New York (1975).
3. G. L. Cantoni, Biological methylation: selected aspects,
    Ann. Rev. Biochem., 44:435 (1975).
4. G. L. Cantoni, S-Adenosylmethionine:present status and futur
    perspectives, p. 557 in: The Biochemistry of S-Adenosyl-
    methionine, F. Salvatore, E. Borek V. Zappia, H. G.
    Williams-Ashman, and F. Schlenk, eds., Columbia Univer-
    sity Press, New York (1977).
5. C.W. Tabor and H. Tabor, 1,4-Diaminobutane (putrescine),
    spermidine, and spermine, Ann. Rev. Biochem., 45:285
    (1976).
6. A. Corti, C. Dave, H.C. Williams-Ashman, E. Mihich and
    A. Schenone, Specific inhibition of the enzyme decar-
    boxylation of S-adenosylmethionine by methylglyoxal bis
    (guanilhydrazone) and related substances, Biochem. J.,
    139:351 (1974).
7. Ting-Chao Chou, A. W. Coulter, J. B. Lombardini, J. R.
    Sufrin, and P. Talalay, The enzymatic synthesis of adeno-
    sylmethionine: mechanism and inhibition, p. 18 in:
    The Biochemistry of S-Adenosylmethionine, F. Salvatore,
    E. Borek, V. Zappia, H.G. Williams-Ashman, and F.
    Schlenk, eds., Columbia University Press, New York
    (1977).
8. T. Deguchi and J. Barchas, Inhibition of transmethylation
    of biogenic amines by S-adenosylhomocysteine, J.Biol.
    Chem., 246:3175 (1971).
9. V. Zappia, C. R. Zydek-Cwick, and F. Schlenk, The speci-
    fiticy of S-adenosylmethionine derivatives in methyl
    transfer reactions, J.Biol. Chem., 244:4499 (1969).
10. E. Borek, and S. J. Kerr, Atypical transfer RNA's and their
    origin in neoplastic cells, Advan. Cancer Res., 25:163
    (1972).

11. S. J. Kerr, Interaction of normal and tumor transfer RNA
     methyltransferases with ethionine-induced methyl-defi-
     cient rat liver transfer RNA, Cancer Res., 35:2969
     (1975).
12. V. A. Deev, T. V. Venkstern, and A. A. Bayev, Comparative
     investigation of kinetic constants of some methylases
     in normal and malignant tissues, Doklady Akad. Nauk
     S.S.S.R., 236:755 (1977).
13. M.Robert-Gero and E. Lederer, Antiviral, antimitogenic and
     antimalarial effect of synthetic analogues of S-adeno-
     sylhomocysteine, p. 111 in: Bioorganic Chemistry and
     Molecular Biology: Results and Perspectives, Yu. A.
     Ovchinnikov and M. N. Kolosov, eds.,"Nauka", Moscow
     (1978).
14. P. K. Chiang, G. L. Cantoni and J. P. Bader, Adenosylhomo-
     cysteine hydrolase inhibitors: Synthesis of 5'-deoxy-
     5'-(isobutylthio)-3-deazaadenosine and its effect on
     Rous sarcoma virus and Gross murine leukemia virus,
     Biochem. Biophys. Res. Commun.,82:417 (1978).
15. R. L. Blakley, The Biochemistry of Folic Acid and Related
     Pteridines, North-Holland Publishing Company, Amsterdam
     (1969).
16. Yu. V. Bukin, Folic Acid and Vitamin B_6 Metabolism in Leu-
     kemia, Chapter 6 in: Role of Endogenous Factors in
     Leukemia Development, M. O. Raushenbakh, ed.,"Medizina",
     Moscow (1974).
17. J.B. Lombardini, M.K. Burch and P. Talalay, An enzymatic
     derivative double isotop assay for L-methionine,
     J. Biol. Chem.,246:4465 (1971).
18. J. D. Finkelstein, W. E. Kyle, and B. J. Harris, Methionine
     metabolism im mammals. Regulation of homocysteine
     methyltransferases in rat tissue, Arch. Biochem. Biophys.
     146:84 (1971).
19. J.D. Finkelstein, Enzyme defects in sulfur amino acid
     metabolism in man, Chapter 15 in: Metabolic Pathways,
     D. M. Greenberg, ed.,vol. 7 (Metabolism of Sulfur Com-
     pound), Academic Press, New York (1975).
20. V. Herbert and K. C. Das, The role of vitamin B_{12} and folic
     acid in hemato- and other cell-poiesis, Vitamins and
     Hormones, 34:1 (1976).
21. J. R. Bertino, Folate Antagonists, p. 469 in: Handbook of
     Experimental Pharmacology, Part 2, vol. 38, A. C. Sarto-
     relli and D. G. Johns, eds., Springer-Verlag, Berlin
     (1975).
22. G. B. Henderson, A. W. Schrecker, C. Smith, M. Gordon,
     E. M. Zevely, K. S. Vitols and F. M. Huennekens,
     Transport of methotrexate and other folate compounds:
      components, mechanism and regulation by cyclic nucleo-
     tides, Advances in Enzyme Regulation, G. Weber, ed.,
     15:141 (1977).

23. D. W. Horne, W. T. Briggs and C. Wagner, Transport of 5-methyltetrahydrofolic acid and folic acid in freshly isolated hepatocytes, J. Biol. Chem.,253:3529 (1978).

24. Yu. V. Bukin, A. V. Sergeev and M. O. Raushenbakh, Selective inhibition of serine hydroxymethyltransferase in mice liver and spleen by pyridoxine antagonists, p. 445 in: Chemistry and Biology of Pteridines, K. Iwai, M. Akino, M. Goto, Y. Iwanami, eds.,International Academic Printing Co., Tokyo (1970).

25. Yu. V. Bukin, V.A. Draudin-Krylenko and W. Korytnyk, Potentiating action of 4-vinylpyridoxal on inhibition of serine transhydroxymethylase by D-cycloserine and its dimer, Biochem. Pharmacol., 28: (in press, 1979).

26. Yu. V. Bukin and V. A. Draudin-Krylenko, Some properties of the enzyme-inhibitor complexes of D-cycloserine or its dimer and serine transhydroxymethylase, p. 44 in: Third All-Union Symposium: Structure and Functions of Enzyme Active Centers (Abstracts, Pustchino-na-Oke, 10-13 August, 1976), Publ. "Nauka", Moscow (1976).

27. V. A. Draudin-Krylenko, Molecular bases of antileukemic action of D-cycloserine and its dimer, Khim.-farm. Zh., (Chem. Pharm.J.,U.S.S.R.,Moscow), 10:3 (1976).

28. V. A. Draudin-Krylenko, A. A. Levchuk, E. N. Orlov and Yu. V. Bukin, Irreversible inhibition and disturbances in induced formation of some vitamin B$_6$-dependent enzymes in rat liver and lymphosarcoma after parenteral administration of boron hydrides into rats, p.55 in: New Problems of Vitaminology (Abstracts of All-Union Meeting, 20-21 April, 1978, Moscow), vol. 1, M. F. Nesterin, ed. (1978).

29. T. O. Eloranta, E. O. Kajander and A. M. Raina, A new method for the assay of tissue S-adenosylhomocysteine and S-adenosylmethionine. Effect of pyridoxine deficiency on the metabolism of S-adenosylhomocysteine, S-adenosylmethionine and polyamines in rat liver, Biochem. J., 160:287 (1976).

30. P. B. Rowe and G.P. Lewis, Mammalian folate metabolism. Regulation of folate interconversion enzymes, Biochemistry, 12:1962 (1973).

31. Yu. V. Bukin, Regulation of the metabolism of vitamin B$_6$ and folic acid coenzymes: theoretical and practical aspects, p.21 in: New Problems of Vitaminology (Abstracts of All-Union Meeting,20-21 April, 1978, Moscow), vol. 1, M. F. Nesterin, ed. (1978).

32. Yu. V. Bukin and V. A. Draudin-Krylenko, The regulation of serine transhydroxymethylase avtivity in normal and tumor cells and in cells of host organism, Vestnik Akad. Med. Nauk SSSR, in press (1979).

33. M. L. Littman, T. Taguchi, Y. Shimizu, Acceleration of
     growth of sarcoma 180 with pyridoxamine and retardation
     with penicillamine, Proc. Soc. Exp. Biol. Med.,113:667
     (1963).
34. W. M. Hryniuk, Purineless death as a link between growth
     rate and cytotoxicity by methotrexate, Cancer Res.,
     32:1506 (1972).
35. E. Kaminskas, Inhibition of sugar uptake by methotrexate
     in cultured Ehrlich ascites cells, Cancer Res.,
     39:90 (1979).
36. F. Rosen, E. Mihich and C. A. Nicol, Selective metabolic
     and chemotherapeutic effects of vitamin B_6 antimetabo-
     lites, Vitamins and Hormones, 22:609 (1964).

RECENT STUDIES ON THE METABOLISM OF 5'-METHYLTHIOADENOSINE

Vincenzo Zappia[x], Maria Carteni-Farina[x], Giovanna Cacciapuoti[x], Adriana Oliva[x] and Agata Gambacorta[xxx]

[x]Department of Biochemistry, Second Chair, University of Naples First Medical School, Naples, Italy and [xxx]Laboratory for the Chemistry of Molecules of Biological Interest, C.N.R., Naples, Italy

INTRODUCTION

5'-Methylthioadenosine (MTA) represents one of the main products of S-adenosylmethionine (Ado-Met) metabolism and is distributed ubiquitously in micromolar amounts in several prokaryotes and eukaryotes[1,2]. Although the chemical structure of this thioether was elucidated in 1924[3], its biological role as product of methionine metabolism was demonstrated by Schlenk in 1952[4], even before the discovery of its precursor Ado-Met[5].

As reported by Klee and Mudd[6], the thioether exhibits in solution a predominant extended <u>anti</u> conformation (see fig.1): i.e.

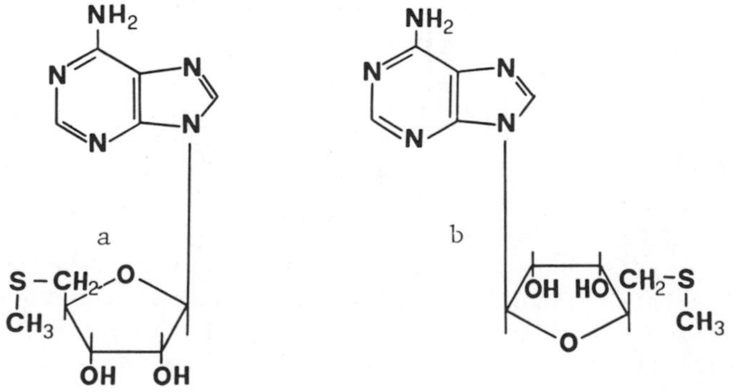

Fig. 1. Structure of 5'-methylthioadenosine in (a) <u>syn</u> and (b) <u>anti</u> conformation

133

carbon 2 of the base is directed away from carbon 5' of the ribose.

The biological relevance of the molecule has been elucidated only in recent years and a large variety of functions has been attributed to MTA. Moreover a number of evidences suggests that this thioether is also involved in the control of cell growth and division[7,8].

SYNTHESIS AND ANALYTICAL PROCEDURES

The synthesis of MTA has been accomplished by Baddiley[9] starting from adenosine: the nucleoside is converted to the 2',3'-isopropylidene derivative and then a toluene-p-sulfonyl group is introduced in 5' position. This latter is replaced by a methylthio group according to Raymond[10]. An alternative and easier procedure, involving the acid hydrolysis of Ado-Met, is generally employed for preparative purposes[11].

The purification of MTA is easily performed by column ion exchange chromatography[11]. For analytical purposes the thioether can be separated from the related sulfur compounds by paper and thin-layer chromatography on silica gel[12,13] or by H.V. electrophoresis.

A new analytical procedure, recently developed in our Laboratory, involves high pressure liquid chromatography[14]. Fig.2 reports a representative chromatogram of MTA and related adenosyl-sulfur compounds. The relative retention times permit an accurate separation of the compounds with a sensitivity limit of 50 pmoles, thus

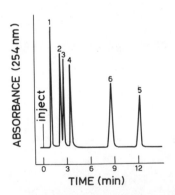

Fig. 2.   High performance liquid chromatography separation of MTA and related adenosine-sulfur compounds. 1,adenosylhomocysteine; 2, adenine; 3, 5'-methylthioadenosine; 4, S-adenosylmethionine; 5, S-adenosyl(5')-3-methylthiopropylamine; 6, S-adenosyl(5')-3-methylthiopropanol (i.s.). Column: Partisil 10 SCX, room temperature, eluent: ammonium formate 0.5M pH 4.0; flow rate: 3.0 ml/min (From Zappia et al.[14]).

allowing the estimation of MTA in biological materials.

The enzymatic reactions leading to MTA from Ado-Met (see Fig. 3) can also be assayed spectrophotometrically by using adenosine deaminase (EC 3.5.4.4) , which is very effective in deaminating MTA, whereas Ado-Met and S-adenosyl(5')-3-methylthiopropylamine (decarboxy-Ado-Met) are unmodified. The amount of the MTA formed is calculated from the decrease in optical density at 265 nm after addition of an excess of adenosine deaminase[15-16].

GENERAL METABOLIC PATHWAYS

In mammalian tissues the conversion of Ado-Met into MTA is operated through three independent pathways as summarized in fig.3: pathway 1, which is considered the most relevant quantitatively, involves transalkylation reactions between decarboxy-Ado-Met and putrescine or spermidine[17,18]. Two mol of MTA are released/mol of spermine and 1 mol/mol of spermidine.

The direct cleavage of Ado-Met into the thioether and α-amino-γ-butyrolactone, which is then converted non-enzymatically into homoserine, is catalyzed by a lyase distributed in prokaryotes and eukaryotes[19,21] (pathway 2).

A third enzymatic route (pathway 3) leading to MTA involves the transfer of 3-amino-3-carboxypropyl group from Ado-Met to uridine in tRNA[22]: the biological significance of this modification of tRNA is still under discussion. The reported enzymatic routes are virtually irreversible. Another pathway of MTA formation has been reported in <u>Escherichia coli</u> by Stoner and Eisenberg[23];

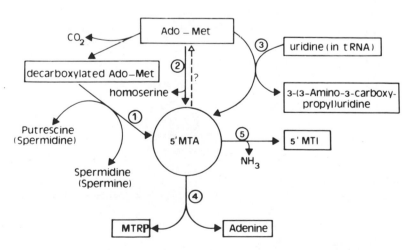

Fig. 3. Pathways of biosynthesis and degradation of MTA

Ado-Met is first converted, through transamination, into S-adenosyl-
2-oxo-4-methylthiobutyric acid which in turn is cleaved into MTA
and 2-oxo-3-butenoic acid. These reactions represent a critical step
in biotin biosynthesis[23].

MTA phosphorylase is the main enzyme responsible for the degradation of the thioether in mammalian tissues. It prevents the cellular
accumulation of MTA and plays a primary role in purine salvage: the
reaction (pathway 4) involves the phosphorolytic cleavage of the
glycosidic bond of MTA and the release of methylthioribose-1-phos-
phate (MTR-1-P) and adenine.

Two classes of MTA nucleosidases have been reported in the Literature: a hydrolytic nucleosidase that cleaves MTA into adenine
and methylthioribose (MTR) has been described in Aerobacter aerogenes[19] and in Escherichia coli[21-25], and a phosphorolytic nucleosidase
has been purified from several mammalian tissues: i.e. rat ventral
prostate[1], rat lung[26], human prostate[27] and human placenta[28].
The occurrence of MTA phosphorylase in Caldariella acidophila[29,30],
a thermophilic bacterium growing optimally at 87°C, represents the
only example of a phosphorolytic cleavage of MTA in prokaryotes.

The deamination of MTA into 5'-methylthioinosine (MTI) (pathway 5) has been reported in bacterial systems[31]. The reaction is
probably operative also in eukaryotes since 5'-isobutylthioadenosine
(SIBA), an analogue of MTA (see fig. 11), is actively deaminated in
chick embryo fibroblasts[32].

Fig. 4 summarizes the pathways of MTA catabolism involving MTR-
1-P as a common intermediate both in eukaryotes and prokaryotes.
Ferro et al.[33] have recently described in Aerobacter aerogenes

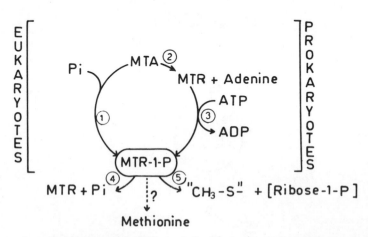

Fig. 4.   Pathways of MTA catabolism in eukaryotes and in
          prokaryotes.

an enzyme catalyzing the ATP-dependent phosphorylation of MTR (pathway 3); in turn MTR-1-P can be cleaved into ribose-1-P and "methylthio" groups by the action of methylthiolase[7,8]; alternatively it is dephosphorylated by alkaline phosphastase[26,30]. The recycling of the thiomethyl moiety of MTR-1-P into methionine has been demonstrated in prokaryotes[34] and eukaryotes[11,35], though the single reactions and the relative enzymes involved are still unknown.

NEW POLYAMINE BIOSYNTHESIS

Besides spermine and spermidine, the new symmetrical polyamines sym-nor-spermidine and sym-nor-spermine have been detected in thermophilic bacteria[36,37] and in several arthropods[38]. The biosynthetic pathway appears to be unusual in that the carbon skeleton of the new polyamines is entirely derived from Ado-Met, while the tetramethylene moiety of spermine and spermidine originates from putrescine. Furthermore, large amounts of MTA are released in the course of the biosynthesis: two mol of MTA are formed per mol of sym-nor-spermidine and three mol/mol of sym-nor-spermine. Whether the large amounts of MTA formed in thermophilic bacteria are related to the thermal stability of these organisms is still matter of investigation.

MTA PHOSPHORYLASE: MOLECULAR PROPERTIES, KINETICS AND SPECIFICITY

MTA phosphorylase purified from human placenta[28] and human prostate[27] requires reducing agents for maximal activity and is inhibited by thiol-blocking reagents, thus suggesting the presence of essential thiol groups. The enzyme has been purified 400-fold from human placenta and the use of the affinity chromatography by Hg-Sepharose has been found particularly advantageous[28]. As reported in Fig. 5, the molecular weight of the enzyme, estimated by gel filtration with a Sephadex G-150, is 95,000 daltons. Ferro et al. reported a mol. weight of 31,000 for the hydrolytic MTA nucleosidase purified from E. coli[25]. The apparent Km values for MTA calculated for the enzyme from human prostate (25 µM) and rat prostate (300 µM) indicated that the cellular concentrations of the thioether are saturating with respect to the enzyme. Oppositely, the cellular levels of phosphate are of the same order of magnitude as the calculated K_m values for phosphate ions, thus implying a regulatory role of the cellular phosphate concentration on the reaction rate.

In Fig. 6 is compared the effect of temperature on the reaction rate of MTA phosphorylase purified from Caldariella acidophila and human placenta. As observed in many other enzymes from thermophilic microorganisms, the phosphorylase from Caldariella acidophila shows a remarkable thermophilicity with an optimum temperature at 95°C while the same enzyme from human tissues displays an optimum of activity at 65°C.

The kinetics as a function of the two substrates and the inhibition patterns observed with adenine and ribose-1-P are consistent

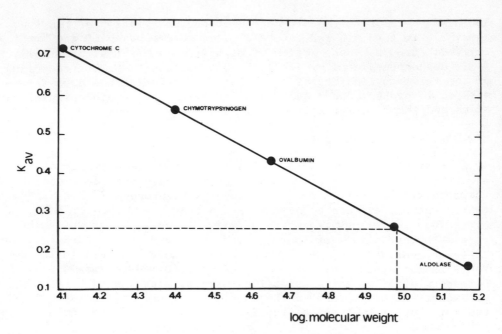

Fig. 5. Determination of the molecular weight of MTA nucleosidase by gel filtration. The dotted line indicates the K_{av} of the nucleosidase (From Cacciapuoti et al.[28]).

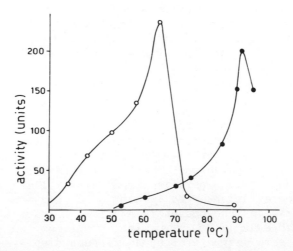

Fig. 6. Effect of temperature on reaction rate of MTA phosphory- lase from C. acidophila (•) and from human placenta (o) (From Cartenì-Farina et al.[30]).

with an equilibrium-ordered mechanism, involving a ternary complex
between MTA, phosphate and the enzyme according to Fig. 7[26].

    The specificity of human 5'-methylthioadenosine phosphorylase
is rather strict if compared with that of the enzyme purified from
E. coli [25]. The replacement of the sulfur atom of 5'-methylthioade-
nosine by selenium and the replacement of the methyl group by an
ethyl one are the only substrate modifications compatible with enzymic activity. The rate of breakdown of 5'-methylselenoadenosine
equals that of 5'-methylthioadenosine (see Fig. 8). This finding
agrees with the generally accepted view that the enzyme systems
that normally utilize sulfur metabolites also convert their selenium analogues, i.e. the interchangeability of methionine and sele-
nomethionine has been demonstrated in protein synthesis[39] as well
as that of S-adenosylmethionine and Se-adenosylselenomethionine in
polyamine biosynthesis.

    The ethyl derivative shows 60% of the activity of an equimolar
concentration of 5'-methylthioadenosine. Conversely, the replacement of N-7 of adenine by a methinic radical resulted in almost
complete loss of activity. The resistance of 5'-methylthiotuberci-
din to enzymic hydrolysis has also been observed by Coward[40] with
the enzyme purified from rat prostate.

    Replacement of the bivalent sulfur in the thioether confor-
mation by a charged sulphonium group results in a loss of activity:
the positively charged group probably prevents the catalytic interaction with the enzyme. On the other hand, the binding of the sulphonium group of 5'-dimethylthioadenosine sulphonium salt to non-
catalytic sites of the enzyme protein could explain the non-compe-
titive inhibition exerted by this molecule[27].

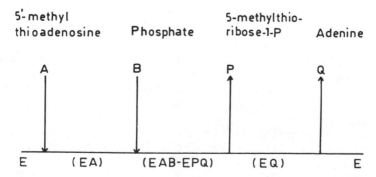

Fig. 7.   Proposed kinetic mechanism of 5'-MTA phosphorylase purifi-
       ed from rat lung (From Garbers[26]).

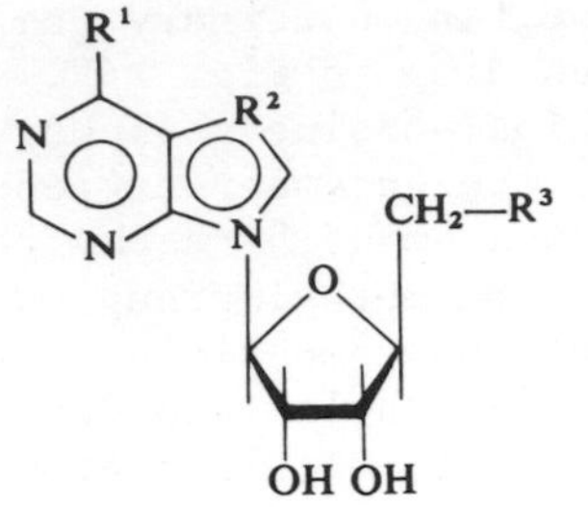

	R^1	R^2	R^3	
	$-NH_2$	$-N=$	$-S-CH_3$	(I)
	$-NH_2$	$-N=$	$-Se-CH_3$	(II)
	$-NH_2$	$-CH=$	$-S-CH_3$	(III)
	$-OH$	$-N=$	$-S-CH_3$	(IV)
	$-NH_2$	$-N=$	$-\overset{+}{S}\overset{CH_3}{\underset{CH_3}{<}}$	(V)
	$-NH_2$	$-N=$	$-S-C_2H_5$	(VI)
	$-OH$	$-N=$	$-S-C_2H_5$	(VII)

Substrate analogues	Concn. (mM)	Enzyme activity (units/mg)	Relative activity (%)
5'-Methylthioadenosine (I)	0.4	5.7	100
5'-Methylselenoadenosine (II)	0.4	5.4	95
5'-Methylthiotubercidin (III)	2.0	N.D.	—
5'-Methylthioinosine (IV)	0.4	0.5	8.9
5'-Dimethylthioadenosine sulphonium salt (V)	1.0	N.D.	—
5'-Ethylthioadenosine (VI)	0.4	3.42	60
5'-Ethylthioinosine (VII)	1.0	N.D.	—

Fig. 8.   Substrate specificity of 5'-MTA phosphorylase (From Zappia et al.[27]).

BIOLOGICAL ROLES, REGULATORY ASPECTS AND TRANSPORT

As reported in Table 1, MTA exerts a significant inhibitory effect on a variety of biological systems: in most instances the inhibitory concentrations are of the same order of magnitude as the assumed cellular levels of the thioether[2,41]. However, only an accurate measurement of the MTA concentration in the various tissues could permit a correct evaluation of the physiological significance of the effects reported.

Recent data reported by Raina[41] on the inhibition of MTA on spermine synthase from bovine brain are particularly interesting in that indicate that the nucleoside may play a role in the regulation of spermine synthesis in animal tissues. They also suggest that the thioether or its nucleosidase resistant derivatives, e.g. methylthiotubercidin and dimethylthioadenosine (see Fig. 8), could exert their effect also in vivo, therefore acting as possible antiproliferative agents.

Of special interest is also the regulatory effect of MTA on protein carboxymethylation since this reaction is involved in several cellular functions, e.g. bacterial[48] and mammalian chemotaxis[49] and neurosecretion mechanisms[50]. Finally, it has to be mentioned the effect of MTA on the enzyme Ado-Met lyase, which directly converts Ado-Met into MTA and homoserine lactone. Most probably the phy-

Table 1. Inhibitory effect of methylthioadenosine on
biological systems

Concentration used	Inhibited system	Authors and year of finding
4.0 mM	Ado-Met lyase (S.cerevi-siae)	Mudd[20] (1959)
0.5 mM	Histamine and acetylse-rotonin methylation	Zappia et al.[12] (1969)
0.2 mM	RNA synthesis in salivary glands of D.melanogaster	Law et al.[44] (1969)
19.0 μM	DNA methylation	Linn et al.[42] (1977)
10.0 μM	Restriction endonuclease reaction	Linn et al.[42] (1977)
35.0 μM (K_i)	Protein(arginine) N-methylation	Casellas & Jean-teur[43] (1978)
0.3 μM (K_i)	Spermine synthase (bovine brain)	Pajula & Raina[41] (1979)
0.4 mM	PHA and ConA stimulated lymphocyte transformation	Ferro[2] (1979)
0.2 mM	Ado-Met lyase (rat liver)	Zappia et al.[45] (1979)
1.0 mM	Spermine synthase (rat ventral prostate)	Pegg & Hibasami[47] (1979)
41.1 μM	Protein O-carboxymethyl-ation	Oliva et al.[46] (1979)

siological significance of this reaction is related to the synthesis
of MTA rather than to the control of Ado-Met concentrations, as it
has been previously proposed.

The proposed regulatory roles of MTA are summarized in Fig. 9.

Despite its biological importance, the mechanism of transport
of the thioether in the cell has been investigated only in few instances. A saturable transport system has been described in the pro-
tozoan Ochromonas malhamensis[35]; the apparent K_m value calculated
for MTA is 0.30 mM, significantly lower than the K_m of Ado-Met and
methionine,thus indicating a specificity of the system towards MTA.

The transport of MTA in mammalian cells has been investigated
only in isolated and perfused rat liver[51]. Fig. 10 shows that the
thioethers methionine and MTA are incorporated at a much higher rate
than Ado-Met, probably because of its charged sulphonium pole. The
existence of a high affinity permease specific for thioethers and
a low affinity system specific for sulphonium compounds has been
postulated.

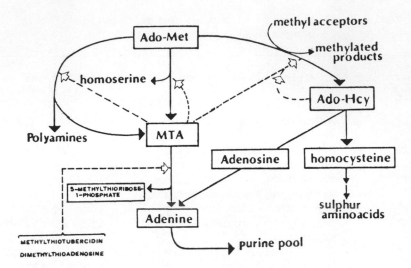

Fig. 9. Regulatory role of MTA on the metabolism of adenosine-sulfur   compounds. Solid lines indicate the metabolic pathways,
broken lines indicate the inhibitory effect.

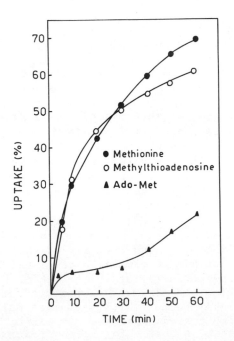

Fig. 10. Uptake of methionine, MTA and Ado-Met by isolated and
perfused rat liver (From Zappia et al.[51]).

ANALOGUES OF MTA WITH ANTIPROLIFERATIVE ACTIVITY

Two structural analogues of MTA are endowed with antimitotic activity: the 7-deaza analogue, methylthiotubercidin (MTT) and 5'-isobutylthioadenosine (SIBA). MTT exerts a potent irreversible inhibition on lympocytes stimulated by various mitogens[2]. This analogue is also a competitive inhibitor of MTA phosphorylase purified from human[27] and rat prostate[40].

The antiproliferative effects of SIBA have been studied more extensively: this compound inhibits cell transformation induced by oncogenic viruses[52], the growth of transformed mouse mammary cells[53], the mitogen-stimulated blastogenesis of lymphocytes[54] and the capping of herpes virus mRNA[55].

The mechanism of action of this drug is still under investigation . It has been postulated that the antiproliferative effect could be related to the inhibition exerted by SIBA on tRNA methylases[56], on protein methylase I[43] and on S-adenosylhomocysteine hydrolase[57]. These studies have been performed in view of a hypothesized structural similarity between SIBA and Ado-Hcy (see Fig. 11).

In our opinion, however, the structure of SIBA is more closely related to MTA than to Ado-Hcy and therefore the enzymes involved in MTA metabolism could be the most probable targets for the drug action. It is interesting to mention, in this respect, that SIBA is actively metabolized in mammalian tissues by MTA phosphorylase which converts the molecule into isobutylthioribose-1- phosphate and adenine[58]. The affinity of the drug towards this enzyme is even ın the natural substrate.

On the basis of these results, the cytostatic effect as well as the other pharmacological actions of the drug could be ascribed to a competition in vivo between SIBA and MTA. Moreover the formation of 5 -isobutylthioribose-1-phosphate from SIBA could interfere with the synthesis of methylthio groups essential for cell division.

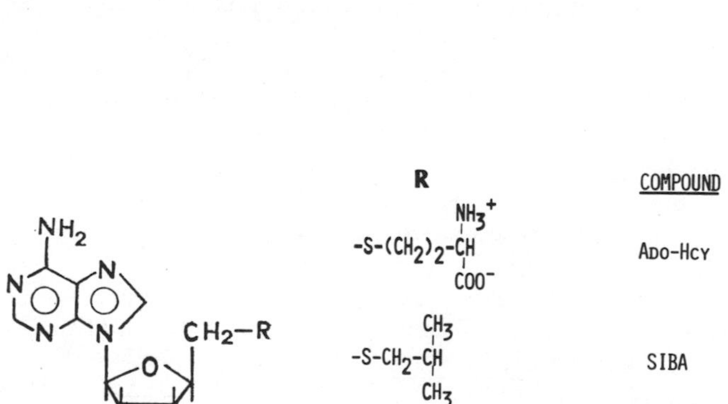

Fig. 11. Structure of SIBA and related thioethers.

AKNOWLEDGMENTS

   The   authors thank their colleagues who contributed to some of
the reported experiments and whose names appear in the list of references. The C.N.R., Rome, Italy, project "Control of Neoplastic
Growth" has provided support.

REFERENCES

1. A. E. Pegg and H. G. Williams-Ashman, Phosphate stimulated
       breakdown of 5'-methylthioadenosine by rat ventral prostate,
       Biochem. J. 115:241 (1969).
2. A. J. Ferro, Function and metabolism of 5'-methylthioadenosine,
       in:"Transmethylation", E. Usdin, R. T. Borchardt, and C. R.
       Creveling, eds., Elsevier/North-Holland, New York (1979).
3. U. Suzuki, S. Odake, and T. Mori, Uber einen neuen schwefelhal-
       tigen Bestandteil der Hefe, Biochem. Z. 154:278 (1924).
4. L. Raymond, L. Smith, and   F. Schlenk, Determination of adeni-
       ne thiomethylriboside and 5-thiomethylribose, and their dif-
       ferentiation from methionine, Arch. Biochem. Biophys. 38:159
       (1952).
5. G. L. Cantoni, S-Adenosylmethionine: a new intermediate formed
       enzymatically from L-methionine and adenosine triphosphate,
       J. Biol. Chem. 204:403 (1953).
6. W. A. Klee, and S. H. Mudd, The conformation of ribonucleosides
       in solution. The effect of structure on the orientation of
       the base, Biochemistry 6:988 (1967).
7. J. I. Toohey, Methylthio group cleavage from methylthioadenosine
       Description of an enzyme and its relationship to the methyl-
       thio requirement of certain cells in culture, Biochem.
       Biophys. Res. Commun. 78:1273 (1977).
8. J. I. Toohey, Methylthioadenosine phosphorylase deficiency in
       methylthio-dependent cells, Biochem. Biophys. Res. Commun.
       83:27 (1978).
9. J. Baddiley, Adenine 5'-deoxy-5'-methylthiopentoside (Adenine
       thiomethyl pentoside): a proof of structure and synthesis,
       J. Chem. Soc. (London). 1348 (1951).
10. A. L. Raymond, Thiosugars, J. Biol. Chem. 107:85 (1934)
11. F. Schlenk, and D. J. Ehninger, Observations on the metabolism
       of 5'-methylthioadenosine, Arch . Biochem. Biophys. 106:95
       (1964).
12. V. Zappia, C. R. Zydek-Cwick, and F. Schlenk, The specificity
       of S-adenosylmethionine derivatives in methyl transfer
       reactions, J. Biol. Chem. 244:4499 (1969).
13. V. Zappia, P. Galletti, A. Oliva, A. De Santis, New methods for
       preparation and analysis of S-adenosyl(5')-3-methylthio-
       propylamine, Anal. Biochem. 79:535 (1977).
14. V. Zappia, P. Galletti, M. Porcelli, C. Manna, and F. Della
       Ragione, High performance liquid chromatographic separation

of natural adenosyl-sulfur compounds (Manuscript in preparation).

15. V. Zappia, P. Galletti, M. Cartenì-Farina, and L. Servillo, A coupled spectrophotometric enzyme assay for methyltransferases, Anal. Biochem. 58:130 (1974).

16. V. Zappia, M. Cartenì-Farina, and P. Galletti, Adenosylmethionine and polyamine biosynthesis in human prostate, in"The Biochemistry of Adenosylmethionine", F. Salvatore, E. Borek, V. Zappia, H. G. Williams-Ashman, and F. Schlenk, eds., Columbia University Press, New York (1977).

17. H. Tabor, and C. W. Tabor, Biosynthesis and metabolism of 1,4-diaminobutane, spermidine, spermine and related amines, Adv. Enzymol. 36:203 (1972).

18. H. Tabor and C. W. Tabor, 1,4-Diaminobutane (putrescine), spermidine and spermine, in "Ann. Rev. Biochem.", E. E. Snell, P. D. Boyer, A. Meister, C. C. Richardson eds., Annual Review Inc., Palo Alto California, vol. 45 (1976).

19. S. K. Shapiro, and A. N. Mather, The enzymatic decomposition of S-adenosylmethionine, J. Biol. Chem. 233:631 (1958).

20. S. H. Mudd, The mechanism of the enzymatic cleavage of S-adenosylmethionine to α-amino-γ-butyrolactone, J. Biol. Chem. 234:1784 (1959).

21. V. R. Swiatek, L. N. Simon, and K. L. Chao, Nicotinamide methyltransferase and S-adenosylmethionine: 5'-methylthioadenosine hydrolase. Control of transfer ribonucleic acid methylation, Biochemistry 12:4670 (1973).

22. S. Nishimura, Y. Taya, Y. Kuchino, and Z. Ohashi, Enzymatic synthesis of 3-(3-amino-3-carboxypropyl)uridine in Escherichia coli phenylalanine transfer RNA: transfer of the 3-amino-3-carboxypropyl group from S-adenosylmethionine, Biochem. Biophys. Res. Commun. 57:702 (1974).

23. G. L. Stoner, and M. A. Eisenberg, Purification and properties of 7,8-diaminopelargonic acid aminotransferase. An enzyme in the biotin biosynthetic pathway, J. Biol. Chem. 250:4029 (1975).

24. J. A. Duerre, A hydrolytic nucleosidase acting on S-adenosylhomocysteine and on 5'-methylthioadenosine, J. Biol. Chem. 237:3737 (1962).

25. A. J. Ferro, A. Barrett, and S. K. Shapiro, Kinetic properties and the effect of substrate analogues on 5'-methylthioadenosine nucleosidase from Escherichia coli, Biochim. Biophys. Acta 438:487 (1976).

26. D. L. Garbers, Demonstration of 5'-methylthioadenosine phosphorylase activity in various rat tissues, some properties of the enzyme from rat lung, Biochim. Biophys. Acta 533:82 (1978).

27. V. Zappia, A. Oliva, G. Cacciapuoti, P. Galletti, G. Mignucci and M. Cartenì-Farina, Substrate specificity of 5'-methylthioadenosine phosphorylase from human prostate, Biochem. J.

175:1043 (1978).

28. G. Cacciapuoti, A. Oliva, and V. Zappia, Studies on phosphate-
    activated 5'-methylthioadenosine nucleosidase from human
    placenta, Int. J. Biochem. 9:35 (1978).

29. A. Gambacorta, M. De Rosa, M. Cartenì-Farina, G. Napolitano,
    and G. Romeo, Studies on 5'-methylthioadenosine phosphory-
    lase from Caldariella acidophila, an extreme thermoacido-
    philic microorganism, The Special FEBS Meeting on Enzymes
    Dubrovnik-Cavtat, Abstr. no. S7-11 (1979).

30. M. Cartenì-Farina, A. Oliva, G. Romeo, G. Napolitano, V. Zappia,
    M. De Rosa, and A. Gambacorta, Methylthioadenosine phospho-
    rylase from Caldariella acidophila. Purification and pro-
    perties (Manuscript in preparation).

31. F. Schlenk, C. R. Zydek-Cwick, and N. K. Hutson, Enzymatic
    deamination of adenosine sulfur compounds, Arch. Biochem.
    Biophys. 142:144 (1971).

32. F. Lawrence, M. Richou, M. Vedel, G. Farrugia, P. Blanchard,
    and M. Robert-Géro, Identification of some metabolic pro-
    ducts of 5'-deoxy-5'-S-isobutylthioadenosine, an inhibitor
    of virus induced cell transformation, Eur. J. Biochem.
    87:257 (1978).

33. A. J. Ferro, A. Barrett, and S. K. Shapiro, 5-Methylthioribose
    kinase a new enzyme involved in the formation of methionine
    from 5-methylthioribose, J. Biol. Chem. 253:6021 (1978).

34. S. K. Shapiro, The function of S-adenosylmethionine in methio-
    nine biosynthesis, in"Transmethylation and Methionine Bio-
    synthesis", S. K. Shapiro, and F. Schlenk, eds., The Univer-
    sity of Chicago Press, Chicago (1965).

35. Y. Sugimoto, T. Toraya, and S. Fukui, Studies on metabolic
    role of 5'-methylthioadenosine in Ochromonas malhamensis
    and other microorganisms, Arch. Microbiol. 108:175 (1976).

36. M. De Rosa, S. De Rosa, A. Gambacorta, M. Cartenì-Farina, and
    V. Zappia, Occurrence and characterization of new polyamines
    in the extreme thermophile Caldariella acidophila, Biochem.
    Biophys. Res. Commun. 69:253 (1976).

37. M. De Rosa, S. De Rosa, A. Gambacorta, M. Cartenì-Farina, and
    V. Zappia, The biosynthetic pathway of new polyamines in
    Caldariella acidophila, Biochem. J. 176:1 (1978).

38. V. Zappia, R. Porta, M. Cartenì-Farina, M. De Rosa, and A.
    Gambacorta, Polyamine distribution in eukaryotes: occurrence
    of sym-nor-spermidine and sym-nor-spermine in arthropods,
    FEBS Letters 94:161 (1978).

39. J. L. Hoffman, K. P. McConnell, and D. R. Carpenter, Aminoacy-
    lation of Escherichia coli methionine t-RNA by selenomethio-
    nine, Biochim. Biophys. Acta 199:531 (1970).

40. J. K. Coward, N. C. Motola, and J. D. Moyer, Polyamine biosyn-
    thesis in rat prostate. Substrate and inhibitor properties
    of 7-deaza analogues of decarboxylated S-adenosylmethionine
    and 5'-methylthioadenosine, J. Med. Chem.20:500 (1977).

41. R. L. Pajula, and A. Raina, Methylthioadenosine, a potent inhi-

bitor of spermine synthase from bovine brain, FEBS Letters
99:343 (1979).

42. S. Linn, B. Eskin, J. A. Lautenberger, D. Lackey, and M. Kimball,
Host-control modification and restriction enzymes of Esche-
richia coli B and the role of adenosylmethionine, in"The
Biochemistry of Adenosylmethionine", F. Salvatore, E. Borek,
V. Zappia, H. G. Williams-Ashman, and F. Schlenk, eds.,
Columbia University Press, New York (1977).

43. P. Casellas, and P. Jeanteur, Protein methylation in animal cells.
II Inhibition of S-adenosyl-L-methionine protein(arginine)
N-methyltransferase by analogs of S-adenosyl-L-homocysteine,
Biochim. Biophys. Acta 519:255 (1978).

44. R. E. Law, R. M. Sinibaldi, M. R. Cummings, and A. J. Ferro,
Inhibition of RNA synthesis in salivary glands of Drosophila
melanogaster by 5'-methylthioadenosine, Biochem. Biophys.
Res. Commun. 73:600 (1976).

45. V. Zappia, M. Porcelli, and F. Della Ragione, S-Adenosylmethionine lyase: specificity and mechanism of reaction (Manuscript
in preparation).

46. A. Oliva, P. Galletti, V. Zappia, W. K. Paik, and S. Kim, Substrate specificity of S-adenosylmethionine: protein-O-methyl-
transferase from calf brain, The Special FEBS Meeting on
Enzymes, Dubrovnik-Cavtat, Abstr. no. S7-82 (1979).

47. A. E. Pegg, and H. Hibasami, The role of S-adenosylmethionine
in mammalian polyamine synthesis, in"Transmethylation", E.
Usdin, R. T. Borchardt, and C. R. Creveling, eds., Elsevier/
North-Holland, New York (1979).

48. J. B. Stock, and D. E. Koshland, Jr., Identification of a methyltransferase and a methylesterase as essential genes in bacterial chemotaxis, in"Transmethylation", E. Usdin, R. T. Borchardt, and C. R. Creveling, eds., Elsevier/North-Holland,
New York (1979).

49. E. J. Diliberto, Jr., R. F. O'Dea, and O. H. Viveros, The role
of protein carboxymethylase in secretory and chemotactic
eukaryotic cells, in"Transmethylation", E. Usdin, R. T. Borchardt, and C. R. Creveling, eds., Elsevier/North-Holland,
New York (1979).

50. L. E. Eiden, R. T. Borchardt, and C. O. Rutledge, Protein carboxymethylation in neurosecretory processes, in"Transmethylation", E. Usdin, R. T. Borchardt, and C. R. Creveling, eds.,
Elsevier/North-Holland, New York (1979).

51. V. Zappia, P. Galletti, M. Porcelli, G. Ruggiero, and A.
Andreana, Uptake of adenosylmethionine and related sulfur
compounds by isolated rat liver, FEBS Letters 90:331 (1978).

52. A. Raies, F. Lawrence, M. Robert-Géro, M. Loche, and R. Cramer,
Effect of 5'-deoxy-5'-S-isobutyladenosine on polyoma virus
replication, FEBS Letters 72:48 (1976).

53. C. Terrioux, M. Crépin, F. Gros, M. Robert-Géro, and E. Lederer,
Effect of 5'-deoxy-5'-S-isobutyl-adenosine (SIBA) on mouse

mammary tumor virus, <u>Biochem</u>. <u>Biophys</u>. <u>Res</u>. <u>Commun</u>. 83:673
(1978).

54. C. Bona, M. Robert-Géro, and E. Lederer, Inhibition of mitogen
induced blastogenesis by 5'-deoxy-5'-S-isobutyladenosine,
<u>Biochem</u>. <u>Biophys</u>. <u>Res</u>. <u>Commun</u>. 70:622 (1976).

55. B. Jacquemont, and J. Huppert, Inhibition of viral RNA methylation Herpes simplex virus type 1 infected by 5'-S-isobutyl
adenosine, <u>J</u>. <u>Virol</u>. 22:160 (1977).

56. J. Hildesheim, R. Hildesheim, E. Lederer, and J. Yoh, Étude de
l'inhibition d'une t-ARN N_2-guanine méthyl transférase du
foie de lapin par des analogues de la S-adenosyl-homocysteine,
<u>Biochimie</u> 59:989 (1972).

57. G. L. Cantoni, H. Richards, and P. K. Chiang, Inhibitors of S-
adenosylhomocysteine hydrolase and their role in the regulation of biological methylation, <u>in</u>"Transmethylation", E.
Usdin, R. T. Borchardt, and C. R. Creveling, eds., Elsevier/
North Holland, New York (1979).

58. M. Cartenì-Farina, F. Della Ragione, G. Ragosta, A. Oliva, and
V. Zappia, Studies on the metabolism of 5'-isobutylthioade-
nosine (SIBA). Phosphorolytic cleavage by methylthioadenosine
phosphorylase, <u>FEBS</u> <u>Letters</u> in press (1979).

TWO DIFFERENT MODES OF SYNTHESIS AND TRANSFORMATION OF CYSTEINE (AND RELATED AMINO ACIDS) BY PYRIDOXAL-P-DEPENDENT LYASES

A.E. Braunstein and E.V. Goryachencova

Institute of Molecular Biology, USSR Academy

of Sciences. Moscow B-334, U.S.S.R.

A variety of amino acids with electrophilic substituents (RO, RS, indolyl, etc.) in positions ß or γ are of general biological significance as protein constituents and metabolic intermediates, or in more restricted roles as detoxication products or secondary nitrogen reserves (in plants). In the biosynthesis and metabolism of such amino acids essential steps are catalysed by a group of widespread pyridoxal-P (PLP) dependent enzymes, classified as lyases (EC class 4, see Ref.[6]), which effect, more or less selectively, elimination and/or replacement reactions of the X-substituent, according to equations I - IV :

α,ß-Elimination

$$X\overset{\beta}{C}HR.\overset{\alpha}{C}H\overset{+}{N}H_3.COO^- + H_2O \rightleftharpoons XH + NH_4^+ \quad \overset{\beta}{R}CH_2.\overset{\alpha}{C}O.COO^- \quad (I)$$

ß-Replacement

$$\overset{\beta}{X}CHR.\overset{\alpha}{C}H\overset{+}{N}H_3.COO^- + YH \rightleftharpoons XH + Y\overset{\beta}{C}HR.\overset{\alpha}{C}H\overset{+}{N}H_3.COO^- \quad (II)$$

ß,γ-Elimination

$$\overset{\gamma}{X}CHR.\overset{\beta}{C}H_2.\overset{\alpha}{C}H\overset{+}{N}H_3.COO^- + H_2O \longrightarrow XH+NH_4^+ + \overset{\gamma}{C}H_3.\overset{\beta}{C}H_2.\overset{\alpha}{C}O.COO^- \quad (III)$$

γ-Replacement

$$\overset{\gamma}{X}CHR.\overset{\beta}{C}H_2.\overset{\alpha}{C}H\overset{+}{N}H_3.COO^- + YH \rightleftharpoons XH + Y\overset{\gamma}{C}HR.\overset{\beta}{C}H_2.\overset{\alpha}{C}H\overset{+}{N}H_3.COO^- \quad (IV)$$

Extensive comparative studies of several PLP-dependent lyases involved in the synthesis and transformations of cysteine, serine and their natural or synthetic analogues have been conducted during the last decennium by us and a group of associates (see Refs.[2-4,11-14] and Table 1). Such lyases may have

149

narrow substrate specificity and be strictly selective for one
reaction type, but mostly there occur relative substrate speci-
ficities and plurifunctional activities in more than one reaction type; lyases with reversible and with apparently unidi-
rectional catalytic action are known in each subgroup (see
examples in Table 1).

Within the frame of accepted theory of PLP-enzyme action,
the mechanisms of lyase-catalysed reactions of types I-IV were
interpreted in similar, but not quite identical ways by Soviet-
Russian and American investigators. Both our school [2-5, 11-13]
and American enzymologists[7a-10, 14] assume, on a broad experimental basis, that the initial step in all such reactions
is dissociation of the α-H atom in enzyme-bound PLP-substrate
aldimines. According to E.E. Snell and his followers, [7-10,14]
the next step in all PLP-lyases is prototropic conversion of
these intermediates to pyridoxamine-P ketimines, with further
changes leading either via $\Delta^{\beta,\gamma}$unsaturated substrate-coenzyme
imines to elimination or replacement of γ-substituent (reactions III and IV), or by way of $\Delta^{\alpha,\beta}$-imines (aminoacrylate-coen-

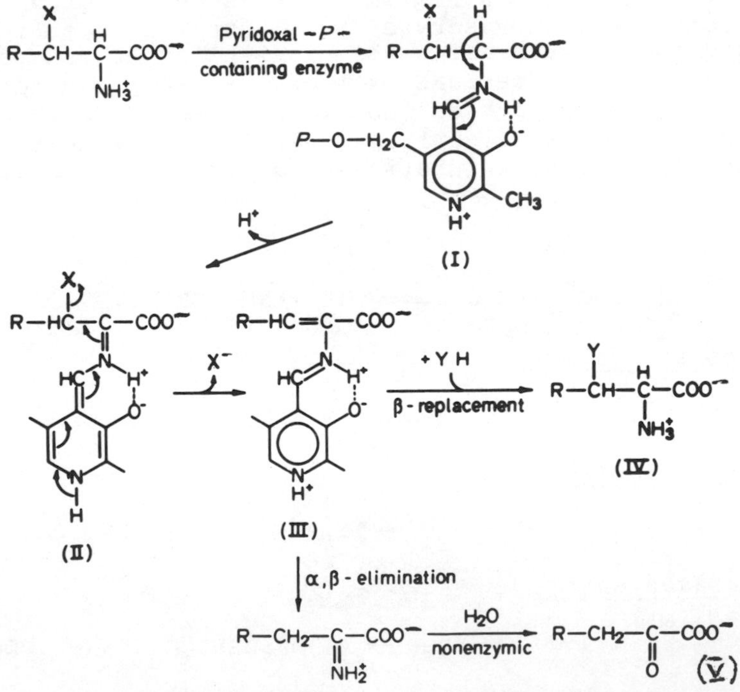

Fig. 1. Scheme of the mechanisms of α,β-elimination
and β-replacement reactions catalysed by PLP-
dependent lyases (after L.Davis & D.Metzler [9]).

Table 1. PLP-Dependent Lyases Studied [4, 13]

MECHANISMS OF PLP-DEPENDENT ß-REPLACEMENT REACTIONS

Enzymes [a] (Classification,[2] name, source)	Reaction types	Primary substrates	Replacing agents (cosubstrates)
A 2 b. _ß-Replacing_			
1. Cysteine lyase (chicken yolk-sac)	II →	Cys	HSO_3^-, Cys,AlkSH, H_2S
2. Serine sulfhydr- ase (chicken liver, yeast)	II →	Ser,Ser(OAcyl), Cys,Cys(SAlk), Ala(Cl),Ala(CN)	Hcy,H_2S, AlkSH, $NH_2(CH_2)_2SH$, $HO(CH_2)_2SH$, H_2O
3. Cystathionine ß-synthase (rat liver;allelozyme of lyase 2)	⇌ II	Ser,Ser(OAcyl), Cys,Cys(SAlk), Ala(Cl),Ala(CN)	same as for lyase 2
4. ß-Cyanoalanine synthase (lupine seedlings)	II →	Cys, Ala(Cl), Ala(SCN)	HCN, H_2S, MeSH, EtSH, $HO(CH_2)_2SH$, $NH_2(CH_2)_2SH$
A 2.a. _α,ß-Eliminating_			
5. Alliinase (garlic)	I →	Alliine (and its analogues)	
6. Serine dehydr- atase (rat liver)	I →	Ser, Thr, erytro-_ßHO-Asp_	
A 2 c. _α,ß-Eliminating_ **and _ß-Replacing_** **(ambifunctional)**			
7. Tryptophanase (E.coli, other microorganisms)	⇌ ⇌ I, II	Trp, Ser,Cys, Cys(SAlk) _et al._	3- and 5-Ind, H_2O, H_2S, Alk(SH) _et al._
A 1 d.ß,γ- and α,ß- **-Specific (pluri-** **functional)**			
8. γ-Cystathionase[b] (rat liver, micro- organisms)	→ → I, II(?) III, ⇌ IV	Cystathionine, Hse, Cys, Hcy(?) Cys Cys	[b]

[a] Lyases 1 — 5 and 8 were purified to 95—100 % homogeneity by techniques developed in our laboratory.

[b] Evidence concerning reactions of types II and IV is scanty, see ref.[14].

zyme imines) to elimination or replacement in position ß, i.e.,
- reactions I and II. We entirely agree in regard to reaction types III and IV - the elimination or replacement of substituents at C-γ; their mechanism will not be further considered here. Our views are also essentially in accord as regards
the mechanism of reactions of types I and I+II (ß-replacement
via α,ß-elimination) catalysed by eliminating and ambifunctional ß-specific lyases. As illustrated in Fig.1, this mechanism postulates conversion of PLP-substrate aldimines (1) by
α-H elimination to the deprotonated intermediate (quinonoid
tautomer of type 2), followed by release of X^- and conversion to $\Delta^{\alpha,\beta}$-substrate-PLP aldimine (3). Intermediate 3 may release ammonia and α-oxo acid (5) by twofold hydrolysis (α,ß-
elimination, Eq. I) or add the cosubstrate, YH, at the double
bond and undergo hydrolysis to the new amino acid (4),
YCHR·CHNH$_3$·COO$^-$, and PLP-enzyme (ß-replacement, Eq.II). The
reaction sequences of Fig.1. were substantiated for elimina-
tion-specific (5, 6) and plurifunctional (7, 8 et al.) lyases by a large body of experimental evidence, considered
below. [a]

    However, the situation is, apparently, different for one
subgroup among the PLP-dependent lyases extensively studied
in recent years in our laboratory on comparative lines,cf.
surveys.[3,4,12, 13] When highly purified, lyases belonging to
this subtype, namely, cysteine lyase, serine sulfhydrase,
cystathionine ß-synthase, and ß-cyanoalanine synthase (enzymes 1 - 4 in Table 1; 2 and 3 obviously are species-specific
allelozymic forms of one lyase [3]) exclusively catalyse ß-replacement (Eq. II). Under whatever conditions tested, they
fail to promote αß-elimination (Eq. I) or γ-specific reactions (Eqs.III or IV), and do not comply to any of the criteria characterizing lyases of the other subtypes[4,13], and in
particular - to those criteria which support the scheme for
ß-replacement presented in Fig.1.

    What follows is, essentially, a synopsis of similar and
distinctive features revealed by comparison of exclusively
ß-replacing lyases with the other subtypes. This comparison
seems to indicate that lyases of the former type act in a way
that differs significantly from the reaction mechanism outlined in Fig.1, see Refs.[4,13]. In our early papers on the
theory of pyridoxal catalysis,[1,3]we pointed out that, owing
to supplementary inductive effects of the electrophilic ß-X
group and of adjacently bound cosubstrate YH (or its anion Y^-)
α-H exchange and replacement of ß-X might proceed in the active site of lyases directly via ß-X-α-aminoacid⇌PLP-aldimines

[a]   Throughout this article, the individual lyases (1 - 8)
and reaction types (I - IV) are designated by the same nu-
merals as in Table I.

with polarized α-C—H bond, without formation of intermediate
PMP-ketimines and of $\alpha\beta$-unsaturated imines (coenzyme-amino-
acrylate Schiff bases).

Steric course of ß-X substitution. The scheme illustrat-
ing this suggestion (Scheme L in Ref.[1]) and reproduced in our
recent publications[4,11,13,23] implies steric inversion at ß-C
and hence is untenable. It conflicts with the fact that replacement of ß-X by lyases 1 - 4 (Table 1) proceeds with retention of configuration, as observed earlier by us[11] and by
others[15,16]; more recently, extensive studies by H. Floss and
associates have conclusively demonstrated inversionless steric course both for ß-replacement [17,18] and for α,ß-elimina-
tion reactions [19](i.e, finally, substitution of H for ß-X), catalysed by PLP-dependent lyases. Hence, a descriptive scheme
carrying no implication as to the mechanism of stereospecific
inversionless substitution should provisionally be used for
the action of replacement-specific β-lyases. e.g.:

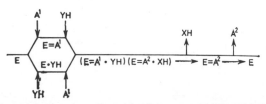

Steady-state kinetic mechanisms. The kinetic mechanism,
according to Cleland's classification,[20]should be "Uni Sequential Tri" for α,ß-elimination, as indicated by the reaction
steps in the scheme on Fig.1, and [ordered] "Ping Pong Bi Bi"
for ß-replacement by elimination-addition (I+II). Ping-pong
kinetics was reported for the case of O-acetyl-serine sulfhydrase (EC 4.2.99.8).[21] Yet, for Ala(CN) synthase (lyase 4;
substrate, cysteine; cosubstrate, KCN)[22] and for serine sulfhydrase from yeast (lyase 2; substrate, cysteine; cosubstrate, 2-mercaptoethanol),[23] our group, in collaboration with
L.V. Kozlov, has recently demonstrated "Random Bi Sequential
Bi" kinetics, implying formation of a ternary complex (amino-
substrate⇌PLPenzyme•cosubstrate) by mutually dependent binding of the two substrates (factor α>1):

Michael reactions (nucleophilic addition at double bonds).
In elimination-mediated ß-replacement (I+II) by plurifunctional lyases, such as tryptophanase, ß-tyrosinase, threonine
synthase, the step in which the cosubstrate's anion (Y⁻) adds
at the ß-C of the $\Delta^{\alpha\beta}$-aminosubstrate-coenzyme Schiff base is
a Michael reaction (nucleophilic addition).[13] Several nuc-

leophilic inhibitory reagents can serve as active-site-speci-
fic Michael addends.[9,13,14] Thus, M. Flavin et al.(see
Refs.[9,14]) devised a test demonstrating formation of enzyme-
bound aminoacrylate or aminocrotonate imines. They used N-
ethylmaleimide (^{14}C-labeled NEM), which traps such unsaturated intermediates by nucleophilic Michael addition at the
double bond: it condenses stereospecifically at the ß-C atom,
yielding labeled adducts of the structure shown in Table 2, and
thereby reduce the output of α-oxo acid, the major normal
elimination product. E. Goryachenkova (1978, unpublished) applied this test to some ß-active lyases; see Table 2. With
enzymes active in elimination reactions (lyases 6, 8), findings similar to those of Flavin were obtained. NEM cannot be
used, because of its rapid spontaneous reaction with SH groups,
in the usual assays for ß-replacement-specific lyases (1→4),
where either aminosubstrate or cosubstrate or both are thiols.
Fortunately, Ala(CN) synthase (lyase 4) allowed, by its capacity to catalyse β-replacement between chloroalanine and
cyanide, to perform the test in the absence of thiol-containing substrate (Table 2). The reagent did not interfere with
this reaction — a result in accord with our view that no unsaturated Schiff-base intermediate is involved. Yet it should
be kept in mind that negative findings never constitute defi-
nitive proof: failure to detect a $\Delta^{\alpha,\beta}$ Schiff base by the NEM
test may be due to rapidity of its further conversion to product (cf. Ref.[9], p.70). Similar caution must be observed in
evaluating concordant evidence recently reported by Abeles
and associates.[24,25] They found that polyhalo-alanines, e.g.,
Ala(Cl$_2$) and Ala(F$_3$), act as "suicide inactivators" of γ-cys-
tathionase (lyase 8) and some other eliminating PLP-enzymes:

Table 2. Action of PLP-dependent lyases in presence
of [[14]-C] -N-ethylmaleimide (NEM)

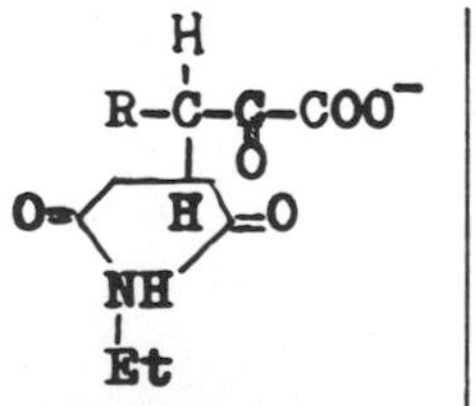 R-C(H)-C(O)-COO⁻ structure with O=ring, NH, Et	Lyases and substrates	Assays		Adduct trapped by NEM
		Control	With [14-C]-NEM	
		End-pro-duct, μmol	Inhibition, per cent	
Products formed from NEM and — a Hse, Thr, Cystathionine: "KEDB", R = CH$_3$; b Serine: "KEDP", R = H.	8 + Hse	26,2 [a]	63	KEDB: +
	6 + Thr	13,5 [a]	35	KEDB: +
	6 + Ser	18,0 [b]	70	KEDP: +
	4 + Ala(Cl)+	++	0	none
	+ NaCN	++		

[a]α-oxobutyrate; [b]pyruvate; [c] Ala(CN)

they bind to the coenzyme as pseudosubstrate aldimines and eliminate Hal-H; the resulting ß-halogenated $\Delta^{\alpha,\beta}$amino-acid--coenzyme Schiff base (an activated Michael acceptor) is attacked at ß-C by a nucleophilic group of the protein; its covalent binding inactivates the lyase.[24] By a related, more intricate multi-step mechanism, involving alternating deprotonations at α-C, ß-C, and tautomerization steps, another "suicidal" pseudosubstrate, propargylglycine, is converted in the active site of γ-cystathionase into another highly activated Michael acceptor, a 2-aminopenta-$\Delta^{3,4}$,$\Delta^{4,5}$-dienoate-coenzyme Schiff base, which is attacked and covalently bound at its (allenic) γ-C atom by a second nucleophilic group of the enzyme protein, thus inactivating the lyase (see Ref.[25]). In accordance with our view that the action of replacement-specific ß-lyases does not involve formation of α,β-unsaturated substrate-coenzyme imines, Silverman and Abeles[24] found that cystathionine ß-synthase (our lyase $\underline{4}$) is not subject to inactivation by typical "suicide" substrate analogues, such as the Michael-acceptor precursors $Ala(Cl_2)$ and $Ala(F_3)$ - they merely cause moderate competitive inhibition of this lyase.

Spectral features in the visible range. Like other PLP-enzymes, the γ- or ß-active lyases have typical spectral absorption bands in the 410–430 nm range (mostly associated with positive CD peaks), which are assigned to hydrogen-bonded "internal" PLP-lysine aldimines.[9] In the replacement-specific ß-lyases ($\underline{1}$, $\underline{2}$, $\underline{4}$) this absorption band is non-dichroic;[13] its position is not substantially shifted on addition of aminosubstrate (without cosubstrate), although its size may decrease - e.g., in lyase $\underline{1}$ with cysteine - without appearance of new peaks in the 325–335 nm range (PMP-ketimine or substituted aldamine forms) or above 470 nm (Tolosa,1968). In serine sulfhydrase ($\underline{2}$) no spectral change is caused by serine, but on interaction with a thiol-containing amino- or cosubstrate (Cys, Hcy) the peak is sharpened and slightly shifts to the right (430 $\longrightarrow$ 437 nm).[26] S•E and pseudoS•E complexes of elimination-active lyases ($\underline{7}$, $\underline{8}$ et al.) usually display spectra with a conspicuous absorption peak in the 490--550 nm region, associated with a deep and narrow negative CD extremum; similar peaks were observed with some pseudoS•E complexes of other PLP-enzymes. Such maxima, indicating an extended conjugated π-system, are attributed to deprotonated S•E and I•E imines in the tautomeric quinonoid form (like $\underline{2}$ in Fig.1) or - in the 455–480 nm range - to α,β-unsaturated PLP-aldimines;[9,20] they are transient in spectra of complexes with adequate substrates and may be persistent in pseudoS•E complexes. We failed to detect such long-wave maxima in S•E or pseudoS•E spectra of the replacement-specific ß-lyases $\underline{1}$ - $\underline{4}$, see Refs.[2,4,13] These results are likewise in accord with non-occurrence of quinonoid Schiff bases or ketimines as intermediates in the catalytic cycle. Yet mind the caution about inconclusiveness of negative evidence.

Isotopic exchange reactions in positions α and β. In the end-products of ß-replacement reactions catalysed in presence of cosubstrate, YH, by PLP-lyases (either replacement-specific or plurifunctional), the α-H atom is replaced by labeled hydrogen from the aqueous medium. In absence of cosubstrate YH, elimination-active lyases (7, 8, and others) catalyse fairly rapid α-H exchange and exchange of ß-H atoms at varying rates - extremely rapid with lyase 8 (Refs.[8,9,14]) and chiral in substrates or allosubstrates (with no ß-substituent) having a C_4 or longer carbon chain. With selectively ß-replacing lyases [4](1, 2, 4),[11,27] α-H exchange is very slow or nil in absence of YH (in non-reacted aminosubstrate); in the presence of adequate YH, excess aminosubstrate exchanges its α-H at a rate commensurate with that of the overall replacement reaction. In the case of cysteine lyase (1), L-serine — a non-substrate and inhibitory analogue with K_i equal to K_m of cysteine — will, in the presence of appropriate cosubstrate, exchange its α-H at a rate similar to that of cysteine.[11] These findings imply that specific binding of cosubstrate, i.e., formation of a ternary (allo)substrate⇌PLPenzyme·cosubstrate complex, is an essential step in the complete reaction cycle.[13]

Interactions with enantiomeric cycloserines. The antibiotic D-cycloserine (4-aminoisoxazolidone-3) irreversibly inhibits, in bacteria, certain enzymes metabolizing D-amino acids. In eucariotes, the L-enantiomer (and DL-cycloserine) are the potent inactivators of transaminases and some other PLP-enzymes acting on structurally analogous L-amino acids. As demonstrated earlier by our colleagues M.Karpeisky, R.Khomutov, E. Severin et al.,[27-29] the following steps appear to be in-

α-Amino   acids   Cycloserine and its
                                         5-substituted derivatives

volved in the inactivation of transaminases by cycloserine and its 5-R derivatives. The agent - a sterically rigid pseudosubstrate - forms an aldimine intermediate with enzyme-bound PLP; rearrangement of this imine to the labile tautomeric PMP-ketimine is followed by decyclization of the isoxazolidine ring and blocking, by the resulting reactive acyl residue, of an essential nucleophilic group of the catalytic site; see Refs.[30,4,13](PMP-ketimines and pyridoxamine were isolated from suitably disintegrated transaminase-cycloserine complexes[27,28]). In this laboratory and elsewhere, DL-cycloserine was found to strongly inhibit other PLP-enzymes whose

catalytic mechanism comprizes aldimine—ketimine rearrangement
of intermediate Schiff bases, e.g., aspartate ß-decarboxylase,
kynureninase, γ-cystathionase, cf.[4,13]. Conversely, PLP-enzymes whose reaction cycle presumably does not involve ketimine intermediates, e.g., α-decarboxylases of glutamate[2,5]
and dopa (Turano et al.,1974) proved insensitive to cycloserine[29,4]. As a tentative generalization, we inferred that typical inactivation by aminoisoxazolidone pseudosubstrates might
be a feature proper to only those PLP-enzymes whose normal reaction mechanism includes an essential PMP-ketimine step.[2]

   We applied this test to ß-specific PLP-dependent lyases
which, in our opinion, might not require this step. As already reported,[4,13] lyases 1 - 4 are refractory to inhibition by
L-cycloserine (cSer). Slight affinity to the enzymes could be
demonstrated at high cSer concentrations ($\geqq$ 0.1 M): filtrates of denatured cSer-enzyme mixtures contained coenzyme-inhibitor imines (reducible with NaB^3H_4 to secondary amines);
in acid solution the imines decomposed, releasing PLP and pyridoxal, but no PMP or pyridoxamine (lyases 2 and 4; Goryachenkova, Rabinkov et al.,1978, see[13] and Figs.4,5). These findings seem to support the view that PMP-ketimines are probably
not formed in the catalytic cycle of selectively ß-replacing
lyases. In contrast, the elimination-active lyases have for

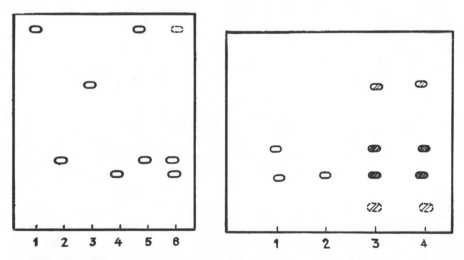

Fig. 4. Chromatography of L-cycloserine complexes from
Ser sulfhydrase (2) and γ-cystathionase (8). 1,PL, 2,PLP;
3,PM; 4, PMP; 5, I·coE complex from lyase 2; 6, I·coE complex from lyase 8 .
   Fig.5. Chromatograph of L-cycloSer complexes from β -re-
placing lyases upon reduction with Na-borotritide.Radioactive
spots hatched. 1,Pxl-cycloSer; 2,pyridoxaminooxyalanine;
3, complex from Ala(CN)synthase;4,complex from γ-cystathio-
nase (8).

L-cSer K_i values of the order of $10^{-4} - 10^{-5}$ M (for the D-isomer affinities are 20-100 fold lower). L-cycloserine is not
bound covalently by the lyase proteins (as opposed to the
transaminases): from the adducts gel-filtration will release
native (labile) apoenzymes and coenzyme-inhibitor imines, yielding mainly PMP and PM on acidification (Fig.4). If our cri-
terium based on the occurrence of abortive half-transaminati-
on from cSer to PLP is valid, the fairly high sensitivity of
lyases 5, 6 and 8 to inhibition by L-cSer (though atypical in
comparison with transaminase inactivation) indicates the probable involvement of PMP-ketimines (or tautomeric quinonoid
 Schiff bases, see Fig.1)in enzymic α,β-elimination reactions.

    Effects of 3- and 4-mercapto-2-amino acids. It has long
been known that 1,2- and 1,3-aminothiols, and especially the
substrate-like mercapto-amino acids - penicillamine, homocysteine et al. - strongly inhibit many PLP-enzymes(transaminases, α-decarboxylases, γ-cystathionase, and others) by conden-
sing with the carbonyl group of PLP in the active site to stable heterocyclic compounds (thiazolidines, resp., thiazanes),
see Refs.[1,30,31,32]
In our recent studies, the exclusively β-replacing lyases,
1 - 4, proved completely refractory to inhibition by D- or
DL-penicillamine, as well as by L-cysteine or L-homocysteine
at high concentrations — 10^{-3} - 10^{-1} M, i.e., far in excess
of K_s for Cys or K_{cos} for Hcy. In contrast, the activities
of eliminating lyases 5 — 8 were sensitive to competitive inhibition by these compounds in similar concentrations; the
I·E complexes could readily be resolved (e.g., by gel-filtration) with release of native apoenzymes; see[2,3,4, 13].

    In the overviews just quoted we attempted to explain the
dissimilar behaviour towards mercapto-amino acids and, at the
same time, to account for the contrasting reaction specifici-
ties of the β-replacement-specific lyases, as opposed to the
elimination-active ones, by a seemingly plausible assumption
about the geometries of substrate binding. We assumed that
the reactions catalysed by α,β-eliminating lyases proceed by
a trans mechanism of 1,2-elimination, well established
for many similar chemical and enzymic reactions. It was postulated that in selectively α,β-eliminating enzymes the S·E
complexes contain substrate-PLP aldimines in fixed geometry,
with the β-substituent in orientation anti (trans)to the α-H
atom,i.e.,syn to the α-N atom. This conformation, regarded by
us as favouring elimination, is also appropriate for condensation of a β-mercapto-amino acid with PLP in the active site
to a thiazolidine. We accordingly assumed that in replacement-
specific β-lyases the substrate's β-substituent is fixed in
the inverse orientation, namely, syn to the α-H (anti to α-N).
Such a conformation, precluding thiazolidine formation with
$C^{4'}$ of the coenzyme, was considered unfavourable for an elimination reaction (cf. Fig. 4 in Ref.[2] or Fig. 9 in Ref.[4]). For

ambifunctional ß-lyases we postulated freedom of rotation
around the ∝-C—ß-C bond. However, in the light of newer evidence (see below) this interpretation appears oversimplified
and is in need of revision.

Topochemical aspects of the molecular mechanism in reactions catalysed by PLP-dependent lyases. As mentioned before,
H. Floss et al.[17-19]recently demonstrated conclusively that
in ß-replacement reactions catalysed by PLP-dependent lyases[*]
and likewise in enzymic ∝,ß-elimination, producing an oxo
acid in which the substrate's ß-X group is stereospecifically replaced by an H-atom, the substitutions at the ß-C atom
occur with retention of configuration. They also presented
evidence indicating that in elimination-active and in selectively ß-replacing lyases alike the amino-acid substrates are
fixed in the active sites in similar orientation - with the
ß-substituent group syn to the ∝-H. The authors' inference is
that all interactions at C-atoms ∝, ß (and evidently ɣ, as
well) in PLP-lyase-catalysed reactions take place on one side only of planar coenzyme-substrate complexes[19],in accordance with concepts set forth by H.Dunathan[10].

Their findings and conclusions are manifestly conflicting with the tentative interpretation, presented above,of the
sum of our experimental data. Some of the contradictions may
find solution, as indicated by Floss et al.[19],if the inversionless ß-replacement results from a double-displacement process
mediated by a suitably positioned anionic group of the protein,
or - a suggestion more plausible, in our opinion - if conformational change of the enzyme during the catalytic process
(e.g., on binding of the cosubstrate) reorients the leaving
ß-X and the incoming substituent relative to the π-plane. In
fact, there is evidence, spectrophotometric and other - indicating occurrence of conformational changes (reorientation of
the coenzyme plane) in transaminases in the course of the catalytic cycle (cf. Ref.[30]). Inspection of space-filling atomic models of PLP-substrate aldimines in ß-lyase-catalysed
reactions shows that, for example, in ∝,ß-elimination the orientation of ß-X, e.g., in cysteine-PLP aldimine,must be different for the catalytic act (HS—ß-C bond normal to the π-
plane), and for condensation to thiazolidine with PLP (skew-
ed by approx. 25[°]); thus, binding of the amino-acid substrate
should be sufficiently loose to allow ß-X to assume either
the productive (perpendicular) or the inhibitory (inclined)
orientation. In the case of a replacement-specific lyase the
binding must then be so tight as to exclude such wobbling of
the ß-X—C bond. Obviously, this is not the only possible speculative reaction model.

---

[*] Ala(CN) synthase of approx. 10 per cent purity was the
only ß-replacement-specific lyase used in these studies.

For the time being, a number of perplexing questions are still unsolved. For example - why are exclusively ß-eliminating lyases, such as serine (threonine) dehydratase, alliin lyaseş etc., incapable to effect a replacement step (Michael addition); — what is the difference in dynamic topography (stereochemistry) between this enzyme subtype and the ambifunctional ß-lyases, displaying the same behaviour towards L-cycloserine or aminothiol inhibitors,etc. Further investions are needed to shed light on these and other obscure aspects of the problem.

REFERENCES

1. A.E. Braunstein (1960) in The Enzymes, 2nd ed.Vol.2,Acad. Press, N.Y., pp. 113-184.
2. A.E. Braunstein (1972) in Enzymes: Structure and Function FEBS Symposium No 29, North-Holland, Amsterdam, pp.135-150.
3. A.E. Braunstein, E.V. Goryachenkova, E.A. Tolosa, I.H. Willhardt, & L.L.Yefremova(1971)Biochim.Biophys.Acta, 242, 247-260.
4. A.E. Braunstein & E.V. Goryachenkova (1976) Biochimie (Paris) 58, 5-17.
5. A.E. Braunstein & M.M. Shemyakin(1953) Biokhimiya(USSR) 18,393-411.
. Enzyme Nomenclature 1978. Recommendations of IUB. Academ. Press Inc. 1979.
7. E.E. Snell(1978) Vitamins,Hormones 16, 77-105.
7a.E.E. Snell & S.J. Di Mari(1970) in The Enzymes, 3rd ed., Vol.2, Academic Press, N.Y.,pp. 335-370.
8. E.E. Snell(1973) Advanc.Enzymol. 42, 287-395.
9. L. Davis & D.E. Metzler(1972) in The Enzymes, 3rd ed., Vol.7, Academ.Press,N.Y.,pp.33-74.
10. H.C. Dunathan(1971) Advanc.Enzymol. 35, 70-131.
11. E.A. Tolosa, R.N. Maslova, E.V. Goryachenkova,I.H. Willhardt & L.L. Yefremova (1975) Eur.J. Biochem. 53, 429-436.
12. T.N. Akopyan, A.E. Braunstein & E.V. Goryachenkova(1975) Proc.Natl.Acad.Sci.U.S.A. 72, 1617-1621.
13. A.E. Braunstein, E.V. Goryachenkova, R.A. Kazaryan, L.A. Polyakova & E.A. Tolosa (1979) "The ß-Replacing and ,ß-Eliminating Pyridoxal-P-dependent Lyases: Enantiomeric Cycloserine Pseudosubstrates and the Catalytic Mechanism",in Frontiers in Bioorganic Chemistry and Molecular Biology (Yu.A. Ovchinnikov & M.N. Kolosov,Eds.), Chapt.9, Elsevier-North-Holland, Amsterdam.
14. M. Flavin,"Methionine Biosynthesis", in Metabolism of Sulfur Compounds, vol.7 of Metabolic Pathways, Acad.Press, N.Y,,pp. 457-503.
15. F. Chapeville & P. Fromageot(1961)Biochim.Biophys.Acta 49, 328-234.
16. A. Sentenac & P. Fromageot(1964) Biochim.Biophys.Acta 81, 289-300.

17. H.G. Floss, E. Schleicher & R. Potts(1976)J.Biol.Chem. 251, 5478-5481.
18. M.-D. Tsai, J. Weaver, H.G. Floss, E. Conn, R.K. Creveling & M. Mazelis(1978)Arch.Biochem.Biophys. 190,553-559.
19. M.-D. Tsai, E. Schleicher, R. Potts, G.E. Skye & H.G. Floss(1978)J.Biol.Chem.253, 5350-5354.
20. W.W. Cleland(1963)Biochim.Biophys.Acta 67,104-196;(1970) in The Enzymes, 3rd ed.,Vol.2,Academ.Press,N.Y.,pp.1-65.
21. P.F. Cook & R.T. Wedding(1976) J.Biol.Chem.251,2023-2029.
22. E.A. Tolosa, L.V. Kozlov, A.G. Rabinkov & E.V. Goryachenkova (1978) Bioorganich. Khimiya(USSR) 4, 1334-1340.
23. E.A. Tolosa, I.H. Willhardt, L.V. Kozlov & E.V. Goryachenkova (1979) Biokhimiya(USSR) 44, 453-459.
24. R.B. Silverman & R.H. Abeles (1976) Biochemistry 15, 4718-4723; (1977) Ibidem 16, 5515-5520.
25. W.Washtien & R.H. Abeles(1977) Biochemistry 16,2485-2491.
26. L.L. Yefremova & E.V. Goryachenkova (1975) Biokhimiya (USSR) 40, 505-512.
27. M.Ya. Karpeisky & Yu.N. Breusov (1965)Biokhimiya(USSR) 30, 153-160.
28. R.M.Khomutov, E.S. Severin, G.K. Kovaleva, N.N.Gulyaev, N.V. Gnuchev & L.P. Sashchenko(1968)in Pyridoxal Catalysis: Enzymes and Model Systems (2nd IUB Symposium, Moscow 1966) Interscience Publs.,N.Y. pp.631-650.
29. L.P. Sashchenko, E.S. Severin & R.M. Khomutov(1968) Biokhimiya(USSR), 33, 142-147.
30. A.E. Braunstein(1973) in The Enzymes, 3rd ed.,Vol.9,Acad. Press,N.Y., pp 379-481.
31. S.R. Mardashev & Chao Tsien-Rey(1960) Doklady AN SSSR 133, 2809-2812.
32. A. Pestaña, I.V. Sandoval & A. Sols (1971) Arch.Biochem. Biophys. 146, 373.

POSSIBLE PHYSIOLOGICAL FUNCTIONS OF TAURINE IN MAMMALIAN SYSTEMS

Jørgen G. Jacobsen, M.D.

Medical Department M,
Odense University Hospital
DK-5ooo Odense C, Denmark

INTRODUCTION

Taurine is an aminosulfonic acid of great antiquity in phylogeny, occurring ubiquitously throughout the Animal Kingdom from Mytilus to Man[1]. Since 1968, when the biochemistry and physiology of taurine and taurine derivatives was first extensively reviewed[1], interest in this compound has increasedrapidly, as testified to by several recent reviews[2-6]. Two international meetings, devoted to various aspects of taurine metabolism and function, have been held, the proceedings of which have been published[7,8]. Studies, exploring the role of taurine in development and nutrition[9], in mammalian heart[10,11], and in the retina[12,13], are reported elsewhere at this meeting.

It is beyond the scope of the present communication to consider in detail current evidence for physiological functions, proposed for taurine. Instead, selected topics, reflecting the interests of the author as a clinician, are touched upon and some speculative suggestions put forth regarding physiological and pathophysiological mechanisms, in which future research may implicate taurine.

The physiological functions, which have been proposed for taurine, are listed in Table I.

Table I. Physiological functions proposed
for taurine

1. Participation in bile acid conjugation
2. Osmoregulator function
3. Neurotransmitter function
4. Neuromodulator function
5. Membrane stabilising function.

1. PARTICIPATION IN BILE ACID CONJUGATION

The conjugation of taurine with bile acids is known to
occur in a large number of species above the selachians[1].
The importance of conjugated bile acids for digestion
should not obscure the fact, that excretion of bile acids
conjugated with taurine and/or glycine constitutes an important means of removing cholesterol from the body. Considerable species differences exist with regard to which
of these two amino acids is preferentially used for conjugation, some, in particular herbivores, using only glycine,
others, in particular carnivores, using exclusively taurine
Recent studies of the effect of dietary taurine deficiency
in the cat[14] and the kitten[15] have shown, that in this
species, in which bile acids are conjugated exclusively
with taurine[1] and hepatic synthesis of taurine from cysteine via cysteine sulfinic acid is severely limited[16],
taurine may be an essential nutrient, the deficiency of
which may i.a. lead to retinal degeneration[17]. It would
be of interest to know whether serum cholesterol levels
in the taurine-deficient cat shows any tendency to increase.

Human liver resembles that of the cat in having only a
negligible capacity for taurine synthesis from cysteine via
cysteine sulfinic acid and contains less taurine than most
species, using taurine for conjugation with bile acids[18],
but differs from that of the cat in being able to use also
glycine for conjugation. Dietary supplementation markedly
increases the fraction of total bile acids, conjugated with
taurine[19], while supplementary glycine does not influence
this fraction. The daily intake of taurine in normal man
has been estimated to be 40-400 mg[20], but variable amounts
of taurine may be lost from taurine-containing foods during
preparation[20,21]. Hence, taurine deficiency might occur
not only in the cat, but perhaps also in Man. This possibility has already received some attention in the neonate[22,2.
but should also be looked for in Adult Man. Several studies
have in fact been carried out, in which an effect of taurine
administration on serum cholesterol levels has been looked

for, but the results were inconclusive[1]. To obtain further
evidence on this question, a study was carried out in 1967-
1970 at the Department of Cardiology, Rigshospitalet, Co-
penhagen in collaboration with Dr. Ole Færgemann, in which
14 patients with familial type II hyperlipidemia and greatly elevated serum cholesterol levels were given 1.5 to 3.0
grammes of taurine daily as a dietary supplement for periods from 5 to 36 months (average 9.3 months). No significant reduction in serum cholesterol levels could, however,
be demonstrated, either during the initial period or over
the ensuing months. In view of more recent studies, which
have shown that patients with familial type II hyperlipidemia are deficient in their ability ro remove cholesterol
from circulating lipoproteins due to the absence of the
normal lipoprotein receptor in their cells[27], the failure
of taurine to influence the serum cholesterol levels
through an action at a point more distal to the block
in cholesterol metabolism in these patients, is inconclusive. Further studies in patients with non-familial hypercholesterolemia are needed to decide whether a limited
availability of taurine in Man may influence the rate of
cholesterol removal from the body via bile acid conjugation.

2.OSMOREGULATOR FUNCTION

In invertebrates, taurine appears to participate in
osmoregulation, in particular in marine animals[1,3]. In
mammals, the high concentration of taurine in lens has
been thought to have an osmoregulatory function[25-27].
The high concentration of taurine in the lens of the
kitten remains almost unchanged during taurine deficiency,
suggesting an important role for taurine in this organ[12].
With age, taurine concentrations in lenticular tissue decline[26], but the relationship of this to the development
of cataract is unknown, as is the the effect of taurine
supplementation on the changes occurring in the lens with
advancing age.

3. NEUROTRANSMITTER FUNCTION

While taurine may not fulfill all the criteria for
being a neurotransmitter[28,29], strong evidence, reviewed
elsewhere[2-6,29], has accumulated over the past years implicating taurine as an inhibitory neurotransmitter in
certain parts of the central nervous system. The most
convincing evidence is probably the work by McBride & Frederickson[30], who in a series of elegant experiments have
shown with a high degree of probability that taurine is the

inhibitory transmitter released by the stellate cells of
the outer molecular layer of the rat cerebellum onto the
dendrites of the Purkinje cells. Some reservations still
remain, however, with regard to a more generalized transmitter function for taurine in rat brain, in view of the
demonstrated slow rate of synaptosomal taurine uptake[31].
    The cerebellum being responsible for the control and
regulation of motor function, it has been suggested, that
certain cerebellar disorders, in particular Friedreich's
ataxia, might be the result of taurine deficiency in
specific areas of the brain[32-34]. In one study[34], the renal excretion of taurine in patients with Friedreich's
ataxia was found to be elevated compared with a normal
control group. Increased renal taurine excretion has also
been reported in cerebellar dyssynergia[35], and in a patient
with hyperβ-alaninemia, β-aminoaciduria, γ-aminobutyric
aciduria somnolence and seizures[36,37]. β-alanin excretion
was also increased in the group of patients with Friedreich's ataxia, and plasma β-alanin levels were stated to
be slightly elevated[34], although the actual values are
in the normal range as reported elsewhere[37]. An accumulation of β-alanin would be an attractive explanation for
the findings in patients with Friedreich's ataxia, since
β-alanin could inhibit taurine uptake not only in the renal tubule, but also elsewhere, e.g. the cerebellum and
the retina, thereby causing disturbance in normal function
of these organs, neurological symptoms and retinitis.
Furthermore, an accumulation of β-alanin could also inhibit
glial uptake of GABA[38], thereby disposing to seizures,
sometimes occurring (as the Ramsay-Hunt syndrome) in patients with Friedreich's ataxia[32]. It is of interest,
that the patient described by Scriver et al.[36,37], with
hyperβ-alaninemia also excreted excessive amounts of
GABA in the urine and had seizures. Further studies on
patients with cerebellar disorders and other neurological
diseases, notably epilepsy, in whom measurements of the
concentrations of taurine, GABA and β-alaninein body fluids
are reported, would be of considerable interest.

    The role of taurine in the retina has been thoroughly
reviewed elsewhere[27] and is the subject of several papers
at this meeting[12,13]. It is of interest that taurine uptake
into platelets from patients with retinitis pigmentosa has
recently been shown to be decreased[39]. Platelets are considered to be useful models for synaptosomes in the central
nervous system[40] and have been much used in studies of the
uptake mechanisms of various biogenic amines[41]. Since retinitis pigmentosa may also be found in patients with Friedreich's ataxia, it would be of interest to examine the
transport and content of taurine in platelets from patients

3. NEUROMODULATOR FUNCTION

The fact that only 17% of the total taurine content of rat brain is associated with the synaptosomal fraction,[42] and that the rate of taurine uptake into synaptosomes is rather slow even with the high affinity uptake system[36], suggests that the major pool of taurine in brain may well have a function different from that of a neurotransmitter. Perhaps taurine should be considered a neuromodulator, although it does not completely meet all the criteria for such a factor[28], or perhaps taurine in brain as elsewhere has a membrane stabilising function, to be discussed below. The major pool of taurine may be associated with glial cells, this appears to be the case at least in the retina[43]. In analogy with GABA[44], it may be suggested that taurine, after release from synaptic vesicles, is removed from the synaptic cleft by uptake into adjacent glial cells, where the bulk of taurine is stored. In glial cells, GABA is converted to glutamine and later returned to the synaptic vesicles. No such mechanism for shuttling taurine has at present been described.

5. MEMBRANE STABILISING FUNCTION

Outside the central nervous system, taurine in high concentrations is present in the heart, in striated muscle, in the adrenal gland, and in platelets and lymfocytes, but all tissues contain some taurine.

The function of taurine in the mammalian heart has been the subject of considerable research in recent years[45-48], and is discussed elsewhere at this meeting[10-12]. The hypothesis has been developed that taurine in heart tissue, and presumably elsewhere, acts as a membrane stabilising factor by virtue of an interaction with intracellular calcium ions.[46,48] Taurine is thought to increase the binding of calcium ions to some intracellular structure, possibly mitochondrial membranes, and thus to counteract calcium-induced potassium efflux from the cells[46]. The uptake of taurine by heart muscle is stimulated by β-adrenergic activation, which also increases calcium influx into the heart[47]. In cardiac insufficiency, intracellular taurine levels are increased, presumably secondary to increased β-adrenergic stimulation[47,48]. The uptake and content of taurine in platelets from spontaneously hypertensive rats has also been shown to be increased[49]. While catecholamines thus appear to promote taurine uptake into cells, taurine conversely inhibits the release of catecholamines from nerve ganglia[50], as well as from adrenal

chromaffin granules, resulting in a diminution of the
stress-induced rise in blood sugar[51].

From these studies it would appear, that taurine is
a naturally occurring, ubiquitously present, catecholamine
antagonist, inhibiting both the release and in some measure the calcium-induced effects of catecholamines, thereby stabilising the membranes of excitable tissues. By
extension, taurine may also antagonise other exogenous
stimuli, affecting cells by increasing calcium influx.

FUTURE PERSPECTIVES IN TAURINE RESEARCH

While a clear understanding of the physiological
functions of taurine has not yet been achieved, the work
reviewed above suggests that important information may
emerge from further studies of the interrelationship between taurine, intracellular calcium, and factors affecting
both. It would be of interest to know more about the effect
of catecholamines on taurine levels, not only in the heart,
but also in platelets, the aggregation of which is stimulated by catecholamines[52]. Does taurine counteract this
effect? Many years ago, taurine added in vitro was reported to delay the coagulation of blood[53]. More information
is also needed about taurine levels in in patients with
cardiac arrhythmias, cardiac insufficiency during treatment with digitalis drugs, in digitalis intoxication and
such toxic conditions as alcoholic cardiomyopathy, in
which mitochondrial accumulation of calcium may be inhibited[54]. A third area for further study is of the
reported effect of taurine in patients with alcohol withdrawal[55].

With time, such studies, and many others, may perhaps
permit the hypothesis, that taurine, throughout evolution,
has served as a cellular buffer, protecting the cell
against excessive and potentially harmful effects of exogenous stimuli, whether it be light, osmotic changes,
electric impulses, circulating hormones, drugs or toxic
substances. It this hypothesis could be substantiated, the
question arises, whether modern Man, resembling the cat in
having only a limited capacity for synthesising taurine
from cysteine via cysteine sulfinic acid[16,18] and ingesting
foods, from which taurine to an unknown degree has been
lost[20,21], may be living precariously on the brink of
taurine deficiency, not only during the neonatal period,
as previously discussed, but also with advancing age. If s
what would be the consequences? Table II attempts to show
the effects, that sustained taurine deficiency might have.

Table II. Hypothetical effects of taurine deficiency
in adult Man

1. Decreased excretion of cholesterol,disposing to
       hypercholesterolemia
       atherosclerosis
2. Decreased protection against catecholamines,
   disposing to
       hypertension
       hyperglycemia
       cardiac arrhythmias
       increased platelet aggregation
       thrombotic episodes
3. Decreased availability of taurine in cerebellum,
   retina, lens, disposing to
       cerebellar disorders
       retinal degeneration
       lenticular changes, cataract
4. Decreased protection against drugs and toxic
   substances, disposing to e.g.
       digitalis intoxication
       alcoholic cardiomyopathy
       alcoholic changes of the CNS

It is clear that only further studies on the uptake,
synthesis, content and disposal of taurine in human
tissues in a variety of situations and at various ages
can confirm or reject the suggestion proposed here, that
taurine may be an essential nutrient in Man, and taurine
deficiency, therefore, a potential health hazard.

1. J.G.Jacobsen and L.H.Smith, Jr., Biochemistry and
      physiology of taurine and taurine derivatives,
      Physiol. Rev. 48:424 (1968).
2. A. Barbeau, N. Inoue, Y. Tsukada, and R. Butterworth,
      The neuropharmacology of taurine, Life Sci. 17:669
      (1975) .
3. S.I. Basken, A. Leibman, and E.M. Cohen, Advances
      Biochem. Psychopharmacol. 15:153 (1976).
4. P. Mandel and H. Pasantes-Morales, Advances Biochem.
      Psychopharmacol. 15:141 (1976).
5. K.C. Hayes, A review on the biological function of
      taurine, Nutr. Rev. 34:161 (1976).
6. P. Mandel and H. Pasantes-Morales, Taurine in the ner-
      vous system, Rev. Neuroscience 3:157 (1978).
7. "Taurine", R.J. Huxtable and A. Barbeau, eds., Raven
      Press, New York (1976).
8. "Taurine and Neurological Disorders", A. Barbeau and
      R. J. Huxtable, eds., Raven Press, New York (1978).

9. G.E. Gaull, Taurine in development and nutrition, in: "Low Molecular weight sulfur-containing natural products", Plenum Press, New York (1979).

1o. R.J. Huxtable, The regulation and function of taurine in the heart: relationship to the adrenergic system, in:"Low molecular weight sulfur-containing natural products", Plenum Press, New York (1979).

11. J. B. Lombardini, Effects of ischemia on taurine levels in cardiac tissue, in:"Low molecular weight sulfur-containing natural products", Plenum Press, New York (1979).

12. H. Pasantes-Morales, Effect of taurine on calcium transport in excitable tissues, in:"Low molecular weight sulfur-containing natural products, Plenum Press, New York (1979).

13. S. Schmidt, Taurine in retinal degeneration, in:"Low molecular weight sulfur-containing natural products, Plenum Press, New York (1979).

14. K. Knopf,J. A. Sturman, M. Armstrong, and K.C. Hayes, Taurine: an essential nutrient in the cat, J. Nutr. 1o8:773 (1978).

15. J. A. Sturman, D. K. Rassin, K.C. Hayes, and G.E. Gaull, Taurine deficiency in the kitten: Exchange and turnover of (^{35}S) taurine in brain, retina and other tissues, J. Nutr. 1o8:1462 (1978).

16. J.G.Jacobsen, L.L. Thomas and L.H. Smith, Jr., Properties and distribution om mammalian L-cysteine sulfinate carboxy-lyases, Biochim. Biophys. Acta 85:1o3 (1964).

17. K.C. Hayes, R.E. Carey, and S.Y. Schmidt, Retinal degeneration associated with taurine deficiency in the cat, Science 188:949 (1975).

18. J.G. Jacobsen and L.H. Smith, Jr., Comparison of decarboxylation of cysteine sulphinic acid-1-C^{14} and cysteic acid-1-C^{14} by human, dog and rat liver and brain, Nature 22o:575 (1963).

19. A.S. Truswell, S. McVeigh, W.D. Mitchell, and B. Bronte Stewart, Effect in man of feeding taurine on bile acid conjugation and serum cholesterol levels, J. Atherosclerosis Res. 5:526 (1965).

2o. D. A. Roe and O. Weston, Potential significance of free taurine in the diet, Nature 2o5:287 (1965).

21. N.R. Jones, Free amino acids in fish. I. Taurine in the skeletal muscle of codling (Gadus callarias), J. Sci. Food Agr. 6:3 (1955).

22. G.E. Gaull, D.K. Rassin, K. Hemonen, and N.C.R. Räihä, Milk prote n quantity and quality in low birth-weight infants II. Effects on selected aliphatic amino acids in plasma and urine, J. Pediat. 9o:348 (1977).

23. J. Rigo and J. Senterre, Is taurine essential for the neonates? Biol. Neonate 32:73 (1977).

24. M. S. Brown and J.L. Goldstein, Familial hypercholesterolemia: Defective binding of lipoproteins to cultured fibroblasts, associated with impaired regulation of 3-hydroxy-3-methylglutaryl coenzyme A reductase activity. Proc. Natl. Acad. Sci. 71:788 (1974).

25. G.W. Barber, Free amino acids in senile cataractous lenses. Possible osmotic etiology. Invest. Ophthal. 7:564 (1968).

26. S.I. Baskin, E.M. Cohn, and J.J. Kocsis, Taurine changes in visual tissues with age, in:"Taurine", R. J. Huxtable and A. Barbeau, eds., Raven Press, New York (1976).

27. A. I. Cohen, Retinal organization and function:Possible roles for taurine, in:"Taurine and Neurological Disorders", A. Barbeau and R.J. Huxtable, eds., Raven Press, New York (1978).

28. J.D. Barchas, H. Akil, G.R. Elliott, R.B. Holman, and S.J. Watson, Behavioral neurochemistry: neuroregulators and behavioral states, Science 200:964 (1978).

29. S.S. Oja and P. Kontro, Neurotransmitter actions of taurine in the central nervous system, in:"Taurine and Neurological Disorders", A. Barbeau and R.J. Huxtable, eds., Raven Press, New York (1978).

30. N.J. McBride and R.C.A. Frederickson, Neurochemical and neurophysiological evidence for a role of taurine as an inhibitory neurotransmitter in the cerebellum of the rat, in:"Taurine and Neurological Disorders", A. Barbeau and R.J. Huxtable, eds., Raven Press, New York (1978).

31. P. Kontro and S.S. Oja, Taurine uptake by rat brain synaptosomes, J. Neurochem. 30:1297 (1978).

32. A. Barbeau, Taurine in Friedreich's ataxia, in: "Taurine and Neurological Disorders", A. Barbeau and R.J. Huxtable, eds., Raven Press, New York (1978).

33. Quebec Cooperative Study on Friedreich's Ataxia. Phase I, A. Barbeau, ed., Canad. J. Neurol.Sci. 3:269 (1976).

34. B. Lemieux, A. Barbeau, V. Beroniade, D. Shapcott, G. Breton, G. Geoffroy, and B. Melancon, Amino acid metabolism in Friedreich's ataxia, Canad. J. Neurol. Sci. 3:373 (1976).

35. C.D. Hall, F.R. Stowe, and G.K. Summer, Familial cerebellar dyssynergia and myoclonus epilepsy associated with defect of amino acid metabolism. Neurology 24: 375 (1974).

36. C.B. Scriver, S. Puerschel, and E. Davies, Hyper-β-alaninemia associated with β-aminoaciduria and γ-aminobutyric aciduria, somnolence and seizures, New Engl. J. Med. 274:635 (1966).

37. C.R. Scriver, W. Nutzenadel, and T.L. Perry, Disorders of β-alanine and carnosine metabolism, in:"The Metabolic Basis of Inherited Disease", J.B. Stanbury, J. B. Wyndgaarden and D.S. Frederickson, eds., 4th ed., McGraw-Hill, New York (1978).

38. L.L. Iversen and J.S. Kelly, Uptake and metabolism of GABA by neurons and glial cells, Biochem. Pharmacol. 24:375 (1974).

39. E. M. Airaksinen, M.M. Airaksinen, P. Sivhola, M. Sivhola, and E. Tuovinen, Uptake of taurine by platelets in retinitis pigmentosa, Lancet I:474 (1979).

40. J.M. Sneddon, Blood platelets as a model for mono-amine-containing neurons, Prog. Neurobiol. 1:151(1973)

41. "Platelets, a Multidisciplinary Approach", G. de Gaetano and S. Garattini, eds., Raven Press, New York (1978).

42. D.K. Rassin, J.A. Sturman, and G.E. Gaull, Taurine in the developing rat brain: subcellular distribution and association of (^{35}S) taurine in maternal, fetal, and neonatal rat brain, J. Neurochem. 28:41 (1977).

43. B. Ehinger, Glial uptake of taurine in the rabbit retina, Brain Res. 60:512 (1973).

44. P.L. McGeer, J.C. Eccles, and E.G.McGeer, Molecular Neurobiology of the Mammalian Brain, Plenum Press, New York (1978).

45. R.J. Huxtable, Metabolism and function of taurine in the heart, in:"Taurine", R.J. Huxtable and A. Barbeau eds., Raven Press, New York (1976).

46. R.J. Huxtable, Regulation of taurine in the heart, in: "Taurine and Neurological Disorders", A. Barbeau and R.J. Huxtable, eds., Raven Press, New York (1978).

47. J. Chubb and R.J. Huxtable, Transport and synthesis of taurine in the stressed heart, in:"Taurine and Neurological Disorders", A. Barbeau and R.J. Huxtable eds, Raven Press, New York (1978).

48. P. Dolara, F. Ladda, A. Mugelli, L. Mantelli, L. Zilletti, F. Franconi, and A. Grotti, Effect of taurine on calcium,inotropism, and electrical activity of the heart, in:"Taurine and Neurological Disorders A. Barbeau and R.J. Huxtable, eds. Raven Press, New York (1978).

49. M. K. Paasonen, J.-J. Himberg, and E. Solantunturi, Taurine in pletelets and heart tissue, in:"Platelets a Multidisciplinary Approach", G. de Gaetano and S. Garattini, eds. Raven Press, New York (1978).

50. K. Kuriyama, M. Muramatsu, K. Nagakawa, and K. Kahita, Modulatingrole of taurine on release of neurotransmitters and calcium transport in excitable tissues, in:"Taurine and Neurological Disorders", A. Barbeau and F.R.J.Huxtable, eds.,Raven Press, New York (1978)

51. K. Kuriyama and K. Nagakawa, Role of taurine in adrenal gland: A preventive effect on stress-induced release of catecholamines from chromaffin granules, in: "Taurine", R.J. Huxtable and A. Barbeau, eds., Raven Press, New York (1976).

52. E.C.Rossi, Interactions between epinephrine and platelets, in:"Platelets, a Multidisciplinary Approach", G. de Gaetano and S. Garattini, eds., Raven Press, New York (1978).

53. J. H. Sterner and G. Medes, The effect of certain sulfur compounds on the coagulation of blood. Am. J. Physiol. 117:92 (1936).

54. R.J. Bing, Cardiac metabolism: Its contributions to alcoholic heart disease and myocardial failure, Circulation 58:965 (1978).

55. H. Ikeda, Effects of taurine on alcohol withdrawal, LancetII:5o9 (1977).

CYSTEINE DIOXYGENASE OF RAT LIVER: SOME PROPERTIES AND ITS

SIGNIFICANCE ON CYSTEINE METABOLISM OF RAT

Kenji Yamaguchi

Department of Medical Chemistry
Osaka Medical College
2-7, Daigaku-machi, Takatsuki
Osaka, Japan

INTRODUCTION

Cysteine dioxygenase[EC.1.13.11.20] is a soluble enzyme catalyzing the oxygenation of cysteine to cysteine sulfinate(CSA), which
has been considered as a key intermediate of cysteine metabolism to
hypotaurine, taurine, isethionic acid, pyruvate and sulfate in mammalian tissues[1-3]. Recently, rat liver supernatant has been shown
to contain an enzyme catalyzing the oxidation of cysteine to CSA[4-8].
The enzyme responsible for the catalysis was shown to dioxygenase by
Lombardini et al[9] and was named cysteine dioxygenase[10]. However,
the significance of this enzyme on the cysteine metabolism and
taurine biosynthesis from cysteine has been doubted because of
following reasons: (1) The CSA forming enzyme activity of rat liver
homogenate was extremely low[7,8], and (2) most of the attempt to
purify the enzyme were unsuccessful because of it great instability
with standard purification procedures.

Recently two important findings in this connection have been presented from our laboratory[10-12]: (1) The CSA forming enzyme activity
of rat liver homogenate was markedly enhanced by the preincubation
with 10 mM L-cysteine under nitrogen gas at 37° to reach 10-20 fold.
In addition, the enzyme was converted to an inactive form during
purification procedure, and the inactive enzyme was activated by
preincubation with L-cysteine under anaerobic condition. (2) The
activated enzyme was rapidly and irreversibly inactivated during
aerobic assay condition and this inactivation was prevented by a distinct cytoplasmic protein in rat liver, called protein-A. On the
basis of these findings, the enzyme from cytosol of rat liver was
purified to homogeneity by Yamaguchi et al.[13]. In the present report is discussed the significance and the regulatory role of cysteine

dioxygenase in the overall cysteine metabolism." Some properties of purifed cysteine dioxygenase and of partially purified protein-A will also be discussed.

SOME PROPERTIES OF PURIFIED CYSTEINE DIOXYGENASE OF RAT LIVER

Cysteine dioxygenase was purified from rat liver cytoplasm to homogeneity. A summary of the purification procedure is given in Table 1.

Analysis of activation of the enzyme

The CSA forming enzyme activity was not detected in the preparations after acetone precipitation without preactivation and the inactive enzyme was activated by the anaerobic incubation in the presence of L-cysteine. As shown in Fig. 1A and B, the activation was dependent on the substrate concentration and the incubation temperature. The maximum activation was observed when the enzyme was preincubated with 10 mM L-cysteine at 37° for 30 min. Simillar anaerobic activation of the enzyme by L-cysteine was observed either in the presence of or in the absence of protein-A, suggesting that protein-A did not associate with the enzyme activation.

Table 1. Summary of purification of rat liver
cysteine dioxygenase

Step	Volume (ml)	Total Units	Total protein (mg)	Specific Activity	Yield (%)
1. Homogenate	1414	43187	39261	1.1	100
2. Acetone precipitation	934	44923	14369	3.1	104
3. 1st DEAE-cellulose	533	22493	773	29	52
4. 2nd DEAE-cellulose	175	9286	175	53	22
5. Sephadex G-100	95	5014	4.41	1138	12
6. Hydroxyapatite	26	2551	1.67	1528	6
7. DEAE-Sephadex A-25	25	1598	0.76	2103	4
8. Sephadex G-75	7	710	0.25	2840	2

Assay of enzyme activity was performed under standard assay conditions with anaerobic activation procedure as reported by Hosokawa et al.[15]. A 0.5-mg of protein-A was added into assay mixture using the preparation after Step 3. Units are in μmol of CSA produced per hour. Specific activity is in units per mg protein. (Yamaguchi et at., ref. 13, with permission)

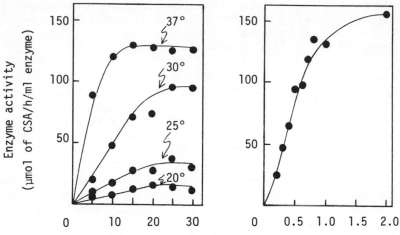

Fig. 1.   Effects of temperature and of the concentration of L-
          cysteine on the anaerobic activation of cysteine dioxy-
          genase. A, Cysteine dioxygenase was preincubated with 10 mM
          L-cysteine under anaerobic condition at 10, 20, 30 and 37°,
          and the volume of preincubation mixture was 0.05 ml. B,
          Cysteine dioxygenase was preincubated under normal acti-
          vation conditions except the concentration of L-cysteine
          given in the figure, and the volume of preincubation mixture
          was 0.05 ml. After mixture volume was adjusted to 1.0 ml,
          the enzyme activity was assayed under standard assay con-
          citions. (From Yamaguchi et al., ref. 13, with permission)

    The activation effect of analogues of L-cysteine such as S-
methyl-L-cysteine, DL-homocysteine, cysteamine and N-acetyl-L-
cysteine were examined. Since some compounds among of them exhibited
a inhibitory effect on the enzyme activity, the preactivation of
enzyme was carried out in 0.05 ml of reaction mixture, and then the
mixture was diluted to 1 ml with prewarmed buffer solution containing
1 mM L-[^{35}S]cysteine and protein-A, followed by the aerobic incubation under standard conditions. Under these conditions no significant inhibition by the additions was observed. The enzyme was activated by all analogues of L-cysteine. Carboxymethyl-L-cysteine and
carboxyethyl-L-cysteine were most effective (13).

Possible conformational change of cysteine dioxygenase by activation

    In order to examine the possibility of conformational change of
the active site of enzyme by activation procedure, the following experiments were carried out. As shown in Fig. 2, the activity of enzyme without the preactivation procedure was markedly decreased by

heat treatment at higher temperature than 50° and no enzyme activity
was detected after heat treatment at 60°. When the same experiment
was carried out after the preactivation procedure, no significant
loss of enzyme activity was observed after heat treatment even at 65°.

As another criterion of possible conformational change of enzyme
during preactivation, the susceptibilities of enzyme to the inhibitio
by chelating agent and to the trypsin inactivation were examined
after and before preactivation procedure. It is noteworthy that the
dramatical decrease of susceptibility of enzyme to the inhibition by
chelating agent was observed after preactivation. As shown in Table
2, no significant inhibition by any chelating agent so far tested was
observed even at 1 mM of concentration. In addition, tryptic digestion of the enzyme could be almost totally prevented by the activation of enzyme as shown in Table 2. Since protein-A was added into
every reaction mixture in these experiments, the changes of the enzyme properties caused by the activation procedure was not due to
Protein-A.

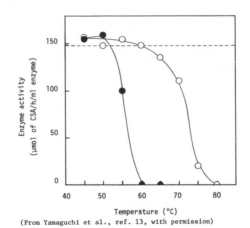

(From Yamaguchi et al., ref. 13, with permission)

Fig. 2. The effect of preactivation of cysteine dioxygenase on the
thermolability. The heat treatment of activated and inactivated enzyme was carried out under anaerobic condition to
avoid the substrate consumption and under aerobic condition
to avoid the anaerobic activation of inactive enzyme, respectively. Heat treatment was carried out for 3 min at
given temperature in this figure in 0.1 ml of reaction mixture containing 10 mM L-cysteine. After heat treatment, both
the activated and inactivated enzyme were subjected to the
usual activation procedure and the volume of mixture was
diluted to 1 ml with buffer containing L-[^{35}S]cysteine as an
isotopic tracer for assay of enzyme activity. Assay was
performed under standard assay conditions except that the
substrate concentration was 1 mM. Plotted are enzyme activity of activated(O) and of nonactivated enzyme(●).
Dotted line is the enzyme activity without heat treatment.

Table 2.   Inhibition by chelating agent and trypsin inactivation
           of activated cysteine dioxygenase

Treatment	Relative enzyme activity (%)	
	before activation	after activation
None		100
0.1 mM o-Phenanthroline	0	98
0.1 mM Bathophenanthroline sulfonate	12	98
0.1 mM EDTA	3	90
Trypsin Treatment	20	90

Anaerobic activation of the enzyme was performed by normal activation
procedure. Trypsin (10 μg) was added into standard assay mixture and
then incubated at 37° for 10 min. After trypsin treatment, trypsin
inhibitor (100 μg) was added into the mixture and the enzyme activity
was assayed under standard assay conditions. Trypsin treatment of
the enzyme before and after activation was performed under aerobic
and anaerobic condition respectively. Protein-A was added into every
mixture. (From Yamaguchi et al., ref. 13, with permission)

Nonlineality of enzyme assay and stabilizing mechanisms of protein-A

As shown in Fig. 3, if protein-A was omitted from assay mixture,
the plots of the cumulative amount of CSA produced (P) versus time
is indeed nonlinear and is visibly curving from the earliest observable time point. Thus the instantaneous reaction velocity (P/min=
Vt) at given time (t min) is decreasing during the course of the
assay. It should be mentioned that the substrate concentration is
still saturating at the end of experimental time. If the inactivation of the enzyme during the reaction were the first order decay
process, then the instantaneous reaction velocity(V_t) should obey an
equation of the form $V_t = Pe^{-kt}$ and a plots of log Vt versus time should
be linear with a negative slope as reported by Knoonts and Shiman[14].
The successive 1-min changes of product were subtracted yielding an
average velocity over 1-min period through out the reaction. These
velocities were plotted against time on logarithmic scale as shown in
Fig. 3. The points clearly fit a straight line, and therefore imply
a first order process. From this slope, the points at which the
velocity was decreased to 50% of the initial velocity(V_o), i.e. the
half-life of the enzyme, can be determined. Under the normal conditions employed here, the reaction has a half-life about 4 min.
The true initial velocity(V_o) of the reaction can also be calculated
by extrapolation of the plots of log(Vt) versus time back to zero
time. Using Vo values, the apparent Km for L-cysteine in the absence
of protein-A was obtained to be $6.7 \times 10^{-4}M$. Since this Km value did
not significantly differ from that in the presence of protein-A,
protein-A did not affect the affinity of enzyme for cysteine. Since
the irreversible inactivation during assay occurred in the

presence of both oxygen and substrate, a stable enzyme-substrate complex may be formed in the presence of oxygen, and protein-A may alter the mode of susceptibility of enzyme to substrate or oxygen.

Other properties of cysteine dioxygenase

On the basis of the experimental results obtained from the gel filtration using Sephadex G-75, SDS-polyacrylamide gel electrophoresis and isoelectrofocussing using 1% ampholyte(pH3-10), purified cysteine dioxygenase is a single subunit protein having a molecular weight of around 23,000 with a pI of 5.5. The purified enzyme contains 1 atom of iron per mol of enzyme protein but not copper in significant quantity. The optimum pH of purified enzyme was 8.5-9.0. Cysteine dioxygenase activity was found in livers of mouse, rat, pig, dog and rabbit. It is of interest that no enzyme activity was observed in the malignant tumor cells such as rat hepatoma cell (AH 2440, AH 109A) and mouse Ehrlich ascites tumor cell, and fetal or neonatal rat liver. No exogenous cofactor such as hydrogen carriers was not required for the enzyme activity but the addition of Fe^{2+}, Fe^{3+} or Cu^{2+} inhibited the enzyme activity[12].

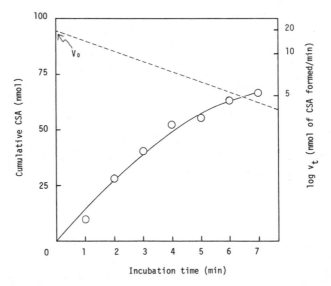

Fig. 3. The change in activity during a normal assay of cysteine dioxygenase. The substrate concentration during activation and assay was 10 mM and 1 mM respectively. Plotted are cumulative amount of CSA versus time (O), and log Vt(nmol of CSA formed/min) versus time (------). The plots of log Vt versus time were obtained by least square analysis of the observed value of Vt. (From Yamaguchi et al., ref. 13, with permission)

Some properties of partially purified protein-A

Protein-A was partially purified from rat liver cytosol by acetone fractionation, DEAE-cellulose, CM-cellulose and Sephadex G-200 column chromatography[16]. The molecular weight of protein-A was calculated as around 80,000 from the results of gel filtration using Sephadex G-200 superfine column. The pI value of protein-A was 7.8. A 0.2 µg of partially purified protein-A was enough to stabilize 0.5 µg(10 units) of purified cysteine dixoygenase. The stabilization of active cysteine dioxygenase during assay caused by protein-A was not affected by chelating agents such as EDTA in high concentration. This fact suggests that the stabilizing effect of protein-A is not associated with any metals[16]. In contrast to cysteine dioxygenase, protein-A is widely distributed in rat tissues such as liver, brain, heart, kidney and spleen[16].

REGULATION OF CYSTEINE DIXOYGENASE ACTIVITY IN RAT LIVER

Induction of cysteine dioxygenase in rat liver by methionine and cysteine

We have found that cysteine oxidase activity in rat liver, which was assayed by NAD-depending assay system[8], was induced by either cysteine or methionine in the presence of permissive amounts of glucocorticoids. Recently, we have reexamined the effect of cysteine or methionine injection on the cysteine dioxygenase activity assayed by the new assay method as mentioned above. As shown in Table 3, the hepatic cysteine dioxygenase activity was also remarkably increased by the intraperitoneally injected cysteine or methionine. On the other hand, the increase of enzyme activity mediated by cysteine or methionine was not observed in adrenalectomized rat.

Table 3. Effects of hydrocortisone, L-cysteine, L-methionine on the rat liver cysteine dioxygenase activity

Injection	Specific enzyme activity Units/mg protein S.E.	
0.9% NaCl	1.319 ± 0.030	(6)
L-Cysteine (1g/kg body weight)	2.769 ± 0.109	(4)
L-Methionine (0.5g/kg body weight)	2.800 ± 0.318	(4)*
Hydrocortisone acetate (0.1g/kg body weight)	2.349 ± 0.129	(4)

Injections were carried out 4 h prior to sacrifice. The number of animals are given in parentheses. (From Hosokawa et al., ref. 15 and * from Yamaguchi, ref. 18, with permission)

It is of interest that the hepatic cysteine dioxygenase activity
of intact rat is highly responded to the dietary protein contents.
Fig. 4 summarizes the effect of dietary protein contents on the activities of cysteine dioxygenase and cysteine desulfhydrase, and the
urinary taurine excretion. Cysteine dioxygenase activity increased
slightly with the dietary protein content up to 20% and dramatically
at protein contents higher than 20%. The activity in the liver of
rats fed on 30% and 40% protein diets was respectively about 10 and
20 times that of rats fed on the 2% protein diet. The enzyme activities reached a plateau at dietary protein contents of over 40%. The
high hepatic activity of cysteine dioxygenase was observed as long as
the animals were maintained on the high protein diets for 2 weeks at
least. On the other hand, the activity of cysteine desulfhydrase,
another enzyme involved in the degradation of L-cysteine in mammals,
was much lower than that of cysteine dioxygenase and did not significantly respond to the dietary protein. It is of interest that the
urinary taurine excretion was increased with the increase in dietary
protein contents as shown in same figure. These findings suggest
that the excess cysteine derived from diet was mostly metabolized in
the liver to taurine via CSA through the adaptively increased activity of cysteine dioxygenase.

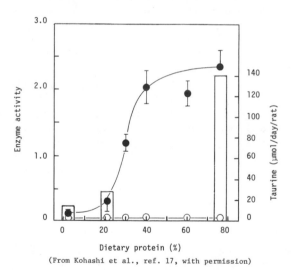

(From Kohashi et al., ref. 17, with permission)

Fig. 4. The effects of dietary protein on the activities of cysteine
dioxygenase, cysteine desulfhydrase in liver and the urinary
taurine excretion of intact rat. Plotted are cysteine dioxygenase activity(●), Cysteine desulfhydrase activity
(○), and urinary taurine contents(▭). Results are
expressed as the mean ± S.D.(represented by vertical line) of
six and three animals fed on basal diet (20% protein diet)
and other diets, respectively. Animals were fed on experimental diets for 2 days prior to sacrifice.

CYSTEINE DIOXYGENASE ACITIVITY OF RAT LIVER DURING DEVELOPMENT

Fig. 5 summarizes the relationship between cysteine dioxygenase activity in rat liver with or without hydrocortisone injection and the hepatic cysteine contents during development. No significant activity was detected in fetal and postnatal rat liver younger than 4-day-old, and first appeared in detectable levels at the 8th post-natal day. The enzyme activity gradually increased to 13% of the adult level by 12-day-old and then markedly increased to reach the adult level at the 28th postnatal day.

It is noteworthy that the response to hydrocortisone mediated induction of cysteine dioxygenase activity was also lack in the liver of 2-day-old neonatal rat and in the liver from fetal, of which mother rat in 18-pregnant-day was injected with hydrocortisone(0.1 mg/ g body weight). The induction mediated by hydrocortisone was first observed in a 4-day-old rat liver as shown in Fig. 5. The response to the hydrocortisone mediated induction of the enzyme was increased with age to reach adult level by 16-postnatal-day. It is unlikely that the lack and appearance of hydrocortisone-mediated induction of rat liver cysteine dioxygenase during development is not due to the lesion of hydrocortisone into nucleus, since the hydrocortisone receptor in an adult levels is observed immediately after birth[21] and some enzyme such as tyrosine aminotransferase in rat liver younger than 4-day-old including fetal rat liver are induced by hydrocortisone[22]. Therefore, it is likely that the transcription rate of the gene coding mRNA for cysteine dioxygenase may be regulated by a specific negative regulators during development. Killewich and Feigelson have reported that the similar appearance of tryptophan pyrrolase activity in rat liver during development and that the de novo synthesis of tryptophan pyrrolase resulted from the appearance of its mRNA during development and the glucocorticoid induction of the enzyme in the developing rat liver occured through the increased levels of its mRNA[23]. Fig. 5 also shows the hepatic cysteine contents in developing rat. The cysteine contents was much higher in 4-day,and 2-day-old young rat(0.37 μmol/g liver weight) than in adult rat(0.16 μmol/g liver weight). The contents was decreased with the appearance of cysteine dioxygenase activity and of the responce to hydrocortisone mediated induction. These findings suggest that the tissue taurine of rat in early developing period is derived from exogenous sourse such as maternal-fetal transport and mother milk. In this connection, the findings reported by Sturman et al. that the concentration of taurine in rat milk is large for the first few days after birth, is of interest.

The effect of cysteine on the cysteine dioxygenase activity of 8-day-old rat liver, in which the enzyme activity was nascently induced by hydrocortisone as shown in Fig. 5, was also examined. The enzyme activity was not increased by the intraperitoneally injected cysteine(50 μg/g body weight), by which the enzyme activity in adult

rat liver was increased by 210%. This fact suggest that the regulation of the cysteine dioxygenase activity mediated by cellular
cysteine level was lack in the early stage of development and it may
explain the high cysteine contents in fetal and neonatal rat liver.

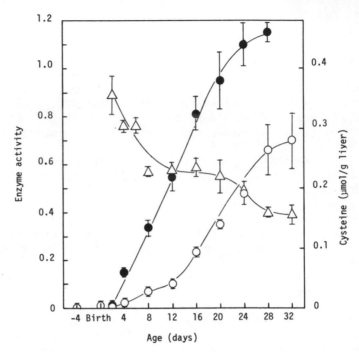

Fig. 5. The appearance of enzyme activity and response to hydrocortisone mediated induction of cysteine dioxygenase, and
the change of cysteine contents in rat liver during development. Rats were injected with hydrocortisone acetate(0.1
mg/g body weight) 25h, 20h, 15h, 10h prior to sacrifice at
2 or 4-, 8-, 12-, and over 16-postnatal-day respectively,
since the time reaching the maximum induction is differ
from each other depending on the age. Enzyme activity with
(●) or without(○) hydrocortisone injection. Cysteine
contents in rat liver(△). Results are expressed as the
mean ±S.D.(represent by vertical line) of five animals.
(From Hosokawa et al., ref. 19, 20, with permission)

ACKNOWLEDGMENTS

    I am grateful to my collaborators, in particular Drs. Y. Hoso-
kawa, N. Kohashi, Y. Kori and S. Sakakibara, with whom most of this
work was carried out. My thanks also go to Prof. I.Ueda, Osaka
Medical College, for his continuous encouragement and useful discussion. The author thanks Prof. Y. Sakamoto, Institute of

Cancer Research, Osaka University Medical School, for his useful advice and encouragement. The early part of this work was carried out in Prof. Sakamoto's laboratory. This study was supported in part by a grant from the Scientific Research fund of Ministry of Education and Culture of Japan, from TANABE Amino Acid Research Fund and from KANAE Shinyaku Kenkyu-Kai Fund.

REFERENCES

1. J. G. Jacobsen and L. H. Smith Jr., Biochemistry and physiology of taurine and taurine derivatives, Physiol. Rev. 48: 424-511 (1968)

2. F. Chapeville and P. Fromageot, La formation de l'acide cystéine-sulfinique à partir la cysteine chez le rat, Biochim. Biophys. Acta, 17: 275 (1955)

3. M. Tavachnik and H. Tarver, The conversion of methionine-^{35}S to cystathionine-^{35}S and taurine-^{35}S in the rat, Arch. Biochem . Biophys., 56: 115-121 (1955)

4. B. Sörbo and L. Ewetz, Enzymatic oxidation of cysteine to cysteine sulfinate in rat liver, Bichem. Biophys. Res. Commun., 18: 359-363 (1965)

5. C. De Marco, R. Mosti and D. Cavallini, Sulla ossidazione della cisteina e della cisteamina a derivati solfinice, catalizata dal fegato di ratto, Boll. Soc. Ital. Biol. Sper., 42, 94-96 (1966)

6. L. Ewetz and B. Sörbo, Characteristics of the cysteinesulfinate-forming enzyme system in rat liver, Biochim. Biophys. Acta, 128: 296-305 (1966)

7. J. B. Lombardini, P. Turini, D. R. Biggs and T. P. Singer, Cysteine oxygenase: 1. General properties, Physiol. Chem. & Physics, 1: 1-23 (1969)

8. K. Yamaguchi, S. Sakakibara, K. Koga and I. Ueda, Induction and activation of cysteine oxidase of rat liver. I. The effects of cysteine, hydrocortisone and nicotinamide on hepatic cysteine oxidase and tyrosine transaminase activities of intact and adrenalectomized rats, Biochim. Biophys. Acta, 237: 502-512 (1971)

9. J. B. Lombardini, T. P. Singer and P. d. Boyer, Cysteine oxgenase: II. Studies on the mechanism of the reaction with ^{18}oxygen, J. Biol. Chem., 244: 1172-1175 (1969)

10. K. Yamaguchi, S. Sakakibara, J. Asamizu and I. Ueda, Induction and activation of cysteine oxidase of rat liver. II. The measurement of cysteine metabolism in vivo and the activation of in vivo activity of cysteine oxidase, Biochim Biophys. Acta, 297: 48-59 (1973)

11. S. Sakakibara, K. Yamaguchi, Y. Hosokawa, N. Kohashi, I. Ueda and Y. Sakamoto, Two components of cysteine oxidase in rat liver, Biochem. Biophys. Res. Commun., 52: 1093-1099 (1973)

12. S. Sakakibara, K. Yamaguchi, Y. Hosokawa, N. Kohashi, I. Ueda and Y. Sakamoto, Purification and some properties of rat liver

cysteine oxidase(cysteine dioxygenase), Biochim Biophys. Acta, 422: 273-279 (1976)

13. K. Yamaguchi, Y. Hosokawa, N. Kohashi, Y. Kori, S. Sakakibara and I. Ueda, Rat liver cysteine dioxygenase(cysteine oxidase), J. Biochem., 83: 479-491 (1978)

14. W. A. Knoontz and R. Shiman, Beef kidney 3-hydroxyanthranilic acid oxygenase-Purification, characterization, and analysis of the assay, J. Biol. Chem. 251: 368-377 (1976)

15. Y. Hosokawa, K. Yamaguchi, N. Kohashi, Y. Kori and I. Ueda, Decrease of rat liver cysteine dioxygenase(cysteine oxidase) activity mediated by glucagon, J. Biochem. 84: 419-424 (1978)

16. Y. Hosokawa, K. Yamaguchi and N. Kohashi, Study on rat liver cysteine dioxygenase, Sulfur-containing Amino Acids, 1: 251-263 (1978)

17. N. Kohashi, K. Yamaguchi, Y. Hosokawa, Y. Kori, O. Fujii and I. Ueda, Dietary control of cysteine dioxygenase in rat liver, J. Biochem. 84: 159-168 (1978)

18. K. Yamaguchi, Cysteine oxidase, Seikagaku, 47: 241-263 (1975)

19. Y. Hosokawa, K. Yamaguchi, N. Kohashi, Y. Kori and I. Ueda, The regulation mechanisms of cysteine oxidase, Seikagaku, 48: 698 (1976)·

20. Y. Hosokawa, K. Yamaguchi, N. Kohashi, Y. Kori and I. Ueda, The cysteine dioxygenase activity in postnatal rat liver, Seikagaku, 50: 918 (1978)

21. G. Giannopoulos, Ontogeny of glucocorticoid receptors in rat liver, J. Biol. Chem., 250: 5847-5851 (1975)

22. O. Greengard, Enzymic differentiation in mammalian liver-injection of fetal rats with hormones causes the premature formation of liver enzymes, Science, 163: 891-895 (1969)

23. L. A. Killewich and P. Feigelson, Developmental control of messenger RNA for hepatic tryptophan 2,3-dioxygenase, Proc. Natl. Acad. Sci. USA, 74: 5392-5396 (1977)

CYSTEAMINE PATHWAY OF TAURINE BIOSYNTHESIS

Giorgio Federici, Giorgio Ricci, Luigi Santoro,
Antonio Antonucci and Doriano Cavallini

Institute of Biological Chemistry, University of Rome
and Centre of Molecular Biology, C.N.R. Rome, Italy

Three main routes have been discovered in animals able
to produce taurine (TAU), called the cysteine sulfinic acid (CSA)
pathway, the cysteamine pathway and the inorganic pathway.
The CSA pathway is generally claimed to be the main route through which cysteine is changed into TAU in animal body
(1-4). This claim is supported by the discovery of hypotaurine ,
thiotaurine and TAU as cysteine metabolites and by the occurrence of cysteine oxygenase and CSA decarboxylase activities in tissues (5-11). However some questions are in opposition to this
claim. In fact CSA decarboxylase is very active in rat, mouse
and dog liver and in dog kidney, but most tissues of these and
other animals, including man, show much lower activity of this
enzyme or are even lacking this activity (3, 10). Heart muscle
where TAU has a very high content and where it is actively produced in vitro from cysteine and cysteamine (12, 13) appears to
lack this decarboxylase (10, 12, 14); neverthless rats fed a pyri-
doxin deficient diet decrease, without estinguish, TAU biosynthesis. This has been recently confirmed by Yamaguchi et al. (15)
who conclude that other routes, not involving pyridoxal phosphate,
must exist for the production of TAU even in rat tissues, where
the CSA pathway seems the most important one. Cysteine oxygenase activity, essential for CSA pathway, has been detected in
high amount only in rat liver (10), and for this reason it is very
difficult to reconcile the absence of this enzyme with the active
conversion of cysteine into TAU by rat heart and brain in vitro.

187

Moreover CSA is a good substrate for the widely distributed Aspartate aminotransferase (17). This will certainly decreases the probability of CSA being used or to be transferred to other tissues for the TAU production.

In conclusion the irregular distribution of enzymes and other data makes questionable wether the CSA pathway is the main route for the production of TAU.

The occurrence of cysteamine pathway has been established by the detection of hypotaurine and TAU as cysteamine metabolites both in vivo and in vitro (18-23), by the occurrence of cysteamine as common metabolite in animals (12, 24) and by the presence of cysteamine oxygenase activity in tissues (25). This enzyme adds an oxygen molecule to the sulfhydryl group of cysteamine to produce hypotaurine. It is therefore a dioxygenase according to the terminology introduced by Hayaishi (26), and behaves as an oxygenase at low concentrations of substrate and as an oxidase at higher concentrations (27). The presence of catalytic amounts of sulfide, selenium or certain redox dyes depresses the oxidasic activity in favour of the oxigenasic one (28, 29). Cysteamine oxygenase is specific for cysteamine and other few sulfhydryl compounds, but not for cysteine (30).

The part of cysteamine pathway going from cysteine to cysteamine is still under active investigation. At present only one way is clearly known to convert cysteine into cysteamine, operating in the course of the biosynthesis of phosphopantheteine and Coenzyme A. Since free cysteamine is split from phosphopantheteine by pantheteinase (2, 31), which is also a widely distributed enzyme, cysteamine could be produced as a branch way of the Coenzyme A synthesizing system (2, 32). The fast exchange rate of phosphopantheteine found by Tweto et al. (33) suggests an appreciable contribution of cysteamine pathway to the production of TAU. Other alternative hypothetical mechanism for cysteamine production, namely use of lanthionine and aminoethylcysteine as intermediate or cysteine with reversibly blocked sulfhydryl group , i.e. cysteine-S-sulfonate, are under active investigation.

A third pathway for TAU production from serine and inorganic sulfate was proposed by Martin, involving the activation of sulfate in the form of phosphoadenosylphosphosulfate (34, 35). The distribution of enzymes involved in this route is unknown and by the data calculated from a Martin's paper a rough estimation indicates that after an injection of 20 uCi of labelled sulfate no more than   one molecule among 40.000 of TAU extracted from liver was labelled. Furthermore this pathway is strongly depressed by cysteine in favour of the other routes (36). These data suggest a very minor contribution of this path to the production of TAU in animals.

In this context it is important to examine the tissue levels of cysteine oxygenase and cysteamine oxygenase activities in order to distinguish, if possible, the relative contribution of CSA pathway and cysteamine pathway in TAU biosynthesis.

Cysteine oxygenase activity was determined in the standard assay conditions reported by Yamaguchi (10) with 20 μmoles of L-^{35}S-cysteine, 4 μmoles NAD^{+}, 0.5 μmoles Fe(NH$_4$)$_2$(SO$_4$)$_2$ 10 μmoles hydroxylamine chloridrate, 100 μmoles phosphate buffer pH 6.8 and tissue homogenate in 0.14 M KCl. The reaction was started with substrate and terminated by the addition of trichloroacetic acid (TCA) at 10% final concentration. After centrifugation a 0.3 ml aliquot of supernatant was applied to a Dowex 50 x 8 H^{+} form column (0.5 x 5 cm) and reaction products from cysteine were eluted with 10 ml of water.

Cysteamine oxygenase activity was determined as previously reported (30) with 2 μmoles of ^{35}S-cysteamine, in the presence or absence of 0.2 μmoles of phenazine methosulfate as cofactor, with 100 umoles of phosphate buffer pH 7.6, in a final volume of 2 ml. After treatment with TCA aliquots were subjected to chromatography in Whatman n°4 paper strip by use of water-saturated phenol as solvent. Radioactivity was determined with a Packard 7201 radiochromatogram scanner and quantitative values were calculated by integration of radioactive peaks.

Table 1 shows the comparative activities of cysteine oxygenase and cysteamine oxygenase in some animal organs. Cysteamine oxygenase is assayed in the presence of 0.2 μmoles of phenazine methosulfate. It is evident that in all tissues tested, except for rat liver, cysteamine oxygenase activity is present in the same order of magnitude, or higher, with respect to the cysteine oxygenase activity. In these results are remarkable the differences observed between rat, cat and man liver. In rat liver cysteine oxygenase activity is higher than cysteamine oxygenase, while in cat and man liver the reverse is observed. This fact is in good agreement with the known distribution of CSA decarboxylase. Actually this enzyme is present in high amount in rat liver, but has very low activity in cat and man liver (37).

Fig. 1 reports the chromatographic separation of cysteamine products by incubating 30 mg of human liver homogenate with

TABLE I

CYSTEINE DIOXYGENASE AND CYSTEAMINE DIOXYGENASE ACTIVITY IN SOME ANIMAL ORGANS

	HEART		LIVER		KIDNEY	
	CYS	CYM	CYS	CYM	CYS	CYM
RAT	0.75	6.50	13.4	8.1	0.95	4.6
MOUSE	0.82	0.95	5.4	5.9	1.95	2.1
RABBIT	1.85	1.95	1.7	6.2	0.5	1.8
CAT	--	--	0.6	8.7	1.1	6.4
MAN	--	--	1.4	13.2	--	--

CYS = Cysteine dioxygenase , CYM = Cysteamine dioxygenase. Values in micromoles of products per gram of fresh tissue per hour. Each value is the average of three different experimental results which varied by about 10%.

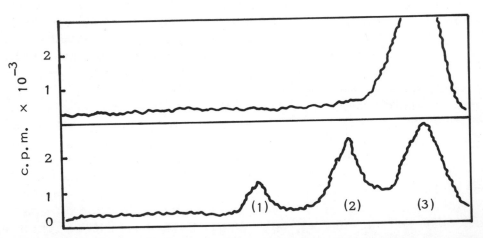

Fig. 1   CYSTEAMINE OXYGENASE ACTIVITY IN HUMAN LIVER 30 mg of human liver homogenate, obtained by biopsy, were incubated, as reported in the text, with ^{35}S-cysteamine in the presence of cofactor. A = zero time, B = 4 hours incubation. The radioactive peaks correspond to taurine (1), hypotaurine (2) and unreacted cysteamine (3).

[35]S-cysteamine, in the presence of phenazine methosulfate. After 4 hours incubation about 50% of cysteamine is oxidized to hypo-taurine and taurine. The same result is obtained if the activity is performed in the absense of phenazine. This result is unexpected since usually the activity in the presence of cofactor is between three or five times higher than in the absence. This particular aspect merit to be investigate further in order to examine whether a natural compound is present in human liver acting as cofactor for cysteamine oxygenase. Moreover, as shown in fig. 1, a con-sistent portion of hypotaurine is recovered oxidized to TAU.

      An interesting question is pointed out by considering the cat. When this animal is fed with a TAU-free diet containing ca-sein as protein source, a retinal degeneration is observed after 23 weeks of diet. The retinal and plasma levels of TAU decrease up to 4% of the basal in plasma and at about 60% in retina. This decrease is partially reversed in the retinal tissue by addition of cysteine to the diet, While a smaller effect is observed on plasma TAU level (38, 39). The retinopathy can be also prevented or re-versed by substituting casein in the diet with lactalbumin or egg albumin (40). The fact that cysteine, supplemented as free amino-acid together with cysteine-poor-protein, is unable to regenerate the normal TAU levels is of interest. This behaviour could be ex-plained by the presence in the cat urine of very high amounts, up to one hundred milligrams for day, of the rare substance "felinine" which is formed by addition of cysteine to an aliphatic chain (41). It follows that the biosynthesis of this unusual aminoacid traps an high amount of cysteine. The possibility of the use of cysteine for TAU synthesis in the cat, either through the CSA or the cystea-mine pathway, is therefore impaired.

      In conclusion the preliminary results reported above in-dicate that, at least in cat and man, the cysteamine pathway could represent an alternative route for TAU production, and that the role of TAU as an essential aminoacid is still questionable.

Acknowledgements

      We are indebted to Prof. M. Messini and to Prof. S. Spada for the samples of human liver.

REFERENCES

1) Awapara, J., in Taurine, Raven Press - New York, page 1-9, (1976)

2) Cavallini, D., Scandurra, R., Duprè, S., Federici, G., Santoro, L., Ricci, G. and Barra, D., in Taurine, Raven Press - New York, page 59-66, (1976)

3) Chatagner, F., Lefauconnier, J.M. and Portemer, C., in Taurine, Raven Press - New York, page 67-71, (1976)

4) Singer, P.T., in Metabolic pathways - vol. VII, Academic Press - New York, page 535-546, (1975)

5) Awapara, J. and Berg, M., in Taurine, Raven Press - New York, page 135-143, (1976)

6) Cavallini, D., De Marco, C. and Mondovì, B., J. Biol. Chem. 234 : 854-857 (1959).

7) Cavallini, D. and Mondovì, B., Giornale di Biochimica, 3 : 170-183 (1951).

8) Lombardini, J.B., Turini, P., Biggs, D.R. and Singer, T. P., Physiol. Chem. & Physics, 1 : 1-23 (1969).

9) Sakakibara, S., Yamaguchi, K., Hosokawa, Y., Kosashi, N. Ueda, I. and Sakamoto, Y., Biochim. Biophys. Acta, 422 : 273-279 (1976).

10) Jacobsen, J.C., Thomas, L.T. and Smith, L.H., Biochim. Biophys. Acta, 85 : 103-116 (1964).

11) Lin, J.C., De Maeio, R. and Metrione, R.M., Biochim. Biophys. Acta, 250 : 558-567 (1971).

12) Huxtable, R. and Bressler, R., in Taurine, Raven Press - New York, page 45-57 (1976)

13) Read, W.O. and Welty, J.D., J. Biol. Chem., 237 : 1521 - 1522 (1962).

14) Huxtable, R., in Taurine, Raven Press - New York, page 99-119 (1976).

15) Yamaguchi, K., Shigeisha, S., Sakakibara, S., Hosokawa, Y. and Ueda, I., Biochim. Biophys. Acta, 381 : 1-8 (1975).

16) Kohashi, N., Yamaguchi, K., Hosokawa, Y., Kori, Y., Fujii, O. and Ueda, I., J. Biochem., 84 : 159-168 (1978).

17) Leinweber, F.J. and Monty, K.J., Anal. Biochem., 4 : 252-256 (1962).

18) Cavallini, D., De Marco, C. and Mondovì, B., Enzymologia 23 : 101-110 (1961).

19) Cavallini, D. Mondovì, B. and De Marco, C., Ricerca Scientifica, 24 : 2649-2651 (1954).

20) Eldjarn, L. Scand. J. Clin. Lab. Invest., 6 Suppl., 13 : 7-96 (1954)

21) Eldjarn, L., Pihl, A. and Sverdrup, A., J. Biol. Chem., 223 : 353-358 (1954).

22) Mondovì, B. and Tentori, L., Italian J. Biochem., 10 : 436-445 (1961).

23) Cavallini, D., Federici, G., Ricci, G., Duprè, S., Antonucci A. and De Marco, C., FEBS lett., 56 : 348-351 (1975)

24) Cavallini, D., De Marco, C. and Mondovì, B., Ricerca Scientifica, 25 : 2901-2903 (1955).

25) Cavallini, D., Scandurra, R. and Duprè, S., in Biological and chemical aspects of oxygenases, Maruzen Co., Tokyo, page 73-92 (1966).

26) Hayaishi, 0. in Proceedings of the Plenary Session, Sixth International Congress of Biochemistry, New York (1964).

27) Cavallini, D., Fiori, A., Costa, M., Federici, G. and Marcucci, M., Physiol. Chem. & Physics, 3 : 175-180 (1971)

28) Wood, J.L. and Cavallini, D., Arch. Biochem. Biophys., 119 : 368-372 (1967).

29) Cavallini, D., Scandurra, R. and De Marco, C., Biochem. J. 96 : 781-786 (1965).

30) Cavallini, D., Federici, G., Ricci, G., Duprè, S., Antonucci, A. and De Marco, C. FEBS lett., 56 : 348-351 (1975).

31) Cavallini, D., Duprè, S., Graziani, M.T. and Tinti, M.G., FEBS lett. 1 : 119-121 (1968).

32) Cavallini, D. Scandurra, R., Duprè, S., Santoro, L. and Barra, D., Physiol. Chem. & Physics, 8 : 157-160 (1976).

33) Tweto, J., Liberati, M. and Larrabee, A.R., J. Biol. Chem. 246 : 2468-2471 (1971).

34) Gorby, W.G. and Martin, W.G., Proc. Soc. Expt. Biol. Med. 148 : 544-549 (1975).

35) Martin, W.G., Sass, N.L., Hill, L., Tarka, S. and Truen R., Proc. Soc. Expt. Biol. Med., 141 : 632-633 (1972).

36) Yamaguchi, K., Seikagaku, 47 : 241-263 (1975).

37) Sturman, J.A., in Human Vitamin B_6 Requirements, Proc. of a Workshop of Natl. Acad. Sci., page 37-60 (1976).

38) Schmidt, S.Y., Berson, E.L. and Hayes, K.C., Investigative ophthalmology, 15 : 47-52 (1976).

39) Berson, E.L., Hayes, K.C., Rabin, A.R., Schmidt, S.Y. and Watson, G., Investigative Ophthalmology, 15 : 52-58 (1976).

40) Hayes, K.C., Carey, R.E., Schmidt, S.Y., Science, 188 : 949-951 (1975).

41) Westall, R.G., Biochem. J., 55 : 244-252 (1953).

KEY   WORDS

Cysteamine oxygenase

Cysteine oxygenase

Taurine

Taurine Biosynthesis

Hypotaurine

Cysteine sulfinic acid

Cysteine

METABOLISM OF HYPOTAURINE IN SOME ORGANS OF THE RAT

Yves Pierre, Claude Loriette and
Fernande Chatagner

Laboratoire de Biochimie, 96 boulevard Raspail,
75006 Paris

INTRODUCTION

Although various physiological roles were suggested for taurine in different tissues like brain, heart,
retina, the origin of this substance in these tissues
is still unclear. Taurine could be provided by the diet
and could also be synthesized. Until now, three main
pathways were described or postulated in mammals for
this biosynthesis [1] :
- the "inorganic pathway" untilizing inorganic
  sulfate,
- the "cysteamine pathway",
- the "cysteine sulfinic pathway".

It is often difficult to evaluate the contribution
of taurine from the diet or synthesized by one of these
pathways to the establishment of taurine level in a tissue ; indeed, taurine synthesized in a tissue could be
transported by the plasma, and for some tissues the uptake of taurine from the plasma is not neglectable [2].

However, as far as the biosynthesis is considered,
the cysteinesulfinic (CSA) pathway and the cysteamine
pathway are considered as prevalent [3], each tissue showing a preference for the one of the other route [4]. In
this respect, it is of interest that in these both pathways the intermediate metabolite formed is hypotaurine [5].
The metabolism of hypotaurine is not clearly known,
although the oxidation of hypotaurine into taurine is

required as the final step of the prevalent pathways of
taurine biosynthesis. Surprisingly this step is still
poorly understood.

According to a recent report [6] hypotaurine is oxidized into taurine by ultraviolet irradiation, which
suggests that a non enzymatic oxidation could occur ;
on the other hand, an enzymic oxidation of hypotaurine
into taurine, catalysed by hypotaurine oxidase, was described in various tissues of the rat [7] and in the retina
of different species [8]. In view of the poor knowledge
which still exists about the metabolism of hypotaurine
and especially about its enzymic oxidation into taurine,
we decided to initiate experiments in order to contribute to a better understanding of this last step.

MATERIAL AND METHODS

Albino rats of our local breeding were used. They
were fed a commercial diet (UAR). In some preliminary
experiments, the rats received a synthetic diet containing either 18% or 60% of casein. Various tissues of
rats of definite age were homogeneized in different buffers (Tris buffer, phosphate buffer, or in a buffered
sucrose solution). The age of the rat appears as important as it has been reported [7] that the enzymic activity changes depending upon the age of the rat. The determination of the hypotaurine oxidase activity was performed as previously described [7] or according to the modifications we brought to this method. The determination
of taurine concentration was carried out according to
the method of Anzano et al.[9] after deproteinisation of
the incubated mixtures with sulfosalicylic acid.

The products employed were purchased from
Calbiochem (Hypotaurine), from Koch Light (BAL : 2,3-
mercaptopropanol), from NBC, USA (DTT : dithiotreitol),
and from Boehringer Mannheim (NAD[+]). All other reagents
were of analytical grade.

RESULTS

In a first set of experiments we determined the
production of taurine from hypotaurine, under the conditions previously described [7], in various tissues
(brain, liver, kidney, spleen, heart, lung, muscle) of
rats of different ages (5, 10, 18, 50, 120 days-old).
The conclusions we could draw out from these observations were that in every case the enzymic activities

were very low. In some tissues and in some conditions
(spleen of 120 days-old rats, brain of 50 days-old rats)
the figures we obtained (respectively 3.3 μmoles of taurine and O μmoles of taurine/g of wet tissue) are in
keeping with those already published [7], whereas the activity of kidney was lower and in liver, whatever the
age of the rat, the production of taurine was unsigni-
ficant or nil.

    In other experiments we used mainly spleen as the
activity was the highest in this tissue. We observed
that in spleen, like in retina [10] the enzymic activity
was located in the cytosol obtained by centrifugating
at 105.000 g for an hour an homogenate prepared in phos-
phe buffered sucrose and that the pH optimum for this
enzyme was around 7, which means similar to that observed for the enzyme from retina [8].

    In addition, we observed that similar results were
obtained when the homogenate of spleen was prepared in
0.1 M Tris buffer pH 7.4 and in 0.1 M or 0.2 M phosphate buffer pH 7.5. As we observed that no production of
taurine was obtained with an homogenate of spleen or of
kidney from an adult rat when homogeneisation was performed in phosphate buffer and the assays were carried
out without addition of BAL, we questionned the role of
BAL. If the effect of BAL is mainly to maintain the enzyme in a reduced state - which is the most likely -
one can envisage that similar results will be obtained
when the homogeneisation of the tissue is performed in
a buffer containing DTT. Therefore we measured the production of taurine by homogenates of spleen, heart,
liver made in 0.2 M phosphate buffer pH 7.5 containing
10 mM DTT. Under those conditions, the molar concentration of DTT in the assay is similar to the concentration
of BAL under the conditions described [7]. We observed
that, on three rats (male, 120 days-old), the concentration of taurine in spleen was 16.13 ± 0.34 μmole/g wet
tissue and the production of taurine was 4.2 ± 0.37, in
the heart the figures were respectively 22.96 ± 0.24
and 3.4 ± 0.75, and in the liver 11.90 ± 0.51 and O.
On the other hand, when the assays were performed on
these homogenates containing DTT, with addition of BAL,
the production of taurine was decreased. One can thence
question whether a too high concentration of reducing
substances in the assay is unfavorable to the enzymic
activity.

    Finally, as we observed[11] that the concentration of

taurine in the liver and in the kidney of rats fed on
a 60% casein diet were higher than those observed in similar tissues of rats fed on a 18% casein diet, these
results led us to suggest that the activity of hypotaurine oxidase could be higher in tissues of rats fed on
a high protein-containing diet. Although the concentration of taurine in the spleen of these rats were not
determined, we performed experiments on spleen and kidney of rats fed on the two diets for at least one week.
As no activity of hypotaurine oxidase was noted, in
our conditions the liver was excluded from these measu-
res. The homogenates were prepared in 0.2 M phosphate
buffer pH 7.5 containing DTT, and the assays were carried out without addition of BAL.

The results are reported in table 1.

Table 1

Concentration of taurine (a) and activity of
hypotaurine oxydase (b) in kidney and spleen
of rats fed on a 18% casein diet (A) and on
a 60% casein diet (B).

		Kidney	Spleen
A	a	6.67 ± 0.33	10.67 ± 0.88
	b	0	1.77 ± 0.87
B	a	10.00 ± 0.88	11.22 ± 1.17
	b	2.6 ± 0.96	1.93 ± 0.40

The values are the mean ± SEM of 3 experiments in
each case. The concentration of taurine is expressed in
μmoles of taurine/g wet tissue and the activity of hypotaurine oxidase in μmoles/h/g wet tissue.

From these figures it appears that, whatever the
diet, the concentration of taurine was the same in the
spleen whereas it was increased in the kidney of rats
fed on the 60% casein diet, which is in keeping with our
previous observations [11]. Furthermore the formation of
taurine from hypotaurine remained approximatively the
same whatever the diet in the spleen while a clear increase was observed in the kidney of the 60% casein fed
rats.

DISCUSSION

The results so far obtained and which are only preliminary led us however to suggest that :
- hypotaurine oxidase is an SH-dependent enzyme,
- the activity is mainly present in the spleen of
rats. It remains to explain why, under our conditions,
no activity was noted in the liver of the rats,
- the activity of the enzyme was unchanged in the
spleen, whatever the diet, whereas in the kidney, a high
protein-containing diet increases the activity.

REFERENCES

1. J.G. Jacobsen and L.H. Smith Jr, Biochemistry and
    Physiology of Taurine Derivatives, Physiol. Rev.
    48:424 (1968).
2. R.J. Huxtable, Regulation of Taurine in the Heart
    in Taurine and Neurological Disorders, A. Barbeau   .
    and R.J. Huxtable, ed., Raven Press, New York (1978).
3. J. Awapara, The Metabolism of Taurine in the
    Animal, in Taurine,R.J.Huxtable and A. Barbeau, ed.,
    Raven Press, New York (1976).
4. D. Cavallini, R. Scandurra, S. Dupré, S. Santoro
    and D. Barra, A New Pathway of Taurine Biosynthesis,
    Physiol. Chem. Phys. 8:157 (1976).
5. F. Chatagner and B. Bergeret, Décarboxylation enzy-
    matique in vivo et in vitro de l'acide cystéine
    sulfinique dans le foie des animaux supérieurs,
    C.R. Acad. Sci. 232:448 (1951).
6. G. Ricci, S. Dupré, G. Federici, G. Spoto, R.M.
    Matarese and D. Cavallini, Oxidation of Hypotaurine
    to Taurine by Ultraviolet Irradiation, Physiol.
    Chem. Phys. 10:435 (1978).
7. S.S. Oja, M.L. Karvonen and P. Lähdesmäki, Biosyn-
    thesis of Taurine and Enhancement of Decarboxylation
    of Cysteine Sulphinate and Glutamate by the Electri-
    cal Stimulation of Rat Brain Slices, Brain. Res.
    55:173 (1973).
8. R.M. Di Giorgio, G. De Luca and S. Macaione, Hypo-
    taurine Oxidase Activity in Retina, Bull. Mol. Biol.
    and Med. 3:115 (1978).
9. M.A. Anzano, J.O. Naewbanij and A.J. Lamb, Simpli-
    fied two-step Column-chromatographic Determination
    of Taurine in Urine, Clin. Chem. 24:321 (1978).
10. R.M. Di Giorgio, S. Macaione and G. De Luca, Sub-
    cellular Distribution of Hypotaurine Oxidase Acti-
    vity in Ox Retina, Life Sci. 20:1657 (1977).

11. C. Loriette, H. Pasantes-Morales, C. Portemer and
    F. Chatagner, Dietary Casein Levels and Taurine
    Supplementation ; Effects on Cysteine Dioxygenase
    and Cysteine Sulfinate Decarboxylase Activities
    and Taurine Concentration in Brain, Liver and
    Kidney of the Rat, Nutr. Metab. (in press).

HYPOTAURINE OXIDATION BY MOUSE TISSUES

P. Kontro and S.S. Oja

Department of Biomedical Sciences
University of Tampere
Box 607, SF-33101 Tampere 10, Finland

INTRODUCTION

There are possibly two main biosynthesis pathways of taurine in mammalian tissues. Taurine is formed either from cysteine via cysteine sulphinic acid and hypotaurine intermediates (Jacobsen and Smith, 1968) or from phosphopantothenylcysteine via cysteamine and hypotaurine (Cavallini et al., 1976; Scandurra et al., 1977). The key intermediate in both metabolic routes is hypotaurine, 2-amino-ethanesulphinic acid. The oxidation of hypotaurine to taurine has not yet been adequately demonstrated. Cavallini et al. (1954) were unable to detect any oxidation, whereas Sumizu (1962) reported that liver homogenates could oxidize hypotaurine in the presence of NAD^+. Later, Fiori and Costa (1969) were unable to confirm his observation, but Oja et al. (1973) found some activity in tissues of developing rats. Fiori and Costa even suggested that hypotaurine is oxidized by the trace amounts of H_2O_2 produced by cellular metabolism. Recently, Di Giorgio et al. (1977; 1978) have demonstrated hypotaurine oxidation in the retinal subcellular fractions of different animals using the method of Oja et al. (1973). The properties of the oxidation reaction have not, however, been investigated in detail, even if the hypotaurine: NAD^+ oxidoreductase (EC 1.8.1.3) enzyme has been prematurely listed and assigned a number on the basis of the poorly controlled short communication of Sumizu (1962). In particular, the true enzymatic nature of hypotaurine oxidation has never been satisfactorily confirmed.

We have now studied the oxidation of hypotaurine to taurine in adult and immature mouse tissues. Our aim was to establish unequivocally the enzymatic nature of the reaction, to characterize

its properties and to find out the optimal assay conditions for
hypotaurine oxidation.

METHODS

Preparation of [^{35}S]hypotaurine

[^{35}S]Hypotaurine was prepared from [^{35}S]cystamine (sp. act.
0.13-4.4 TBq/mol in different batches, The Radiochemical Centre,
Amersham) as suggested by Scandurra et al. (1969). In our modified
method the reaction mixture (15 ml), containing 0.05-1.5 mmol [^{35}S]-
cystamine (0.2 GBq), 60 μmol Cu^{2+} ions and 150 μmol NaOH, was
incubated for 3 h in a shaking water bath at 311 K. The mixture
was then acidified with acetic acid and layered on Dowex 50 H$^+$-
columns (200-400 mesh). Taurine was first eluted out with H_2O and
hypotaurine thereafter with 1 mol/l NH_4OH. Pooled hypotaurine fractions were evaporated to dryness in a waterbath at 323 K under reduced pressure and then dissolved in water. This chromatographic
process was repeated 3-5 times. The radiochemical purity of [^{35}S]-
hypotaurine was assessed by two-dimensional thin layer chromatography
on DC-Alufolien Kieselgel 60 plates. Our [^{35}S]hypotaurine preparations contained 1-4% of [^{35}S]taurine and traces of [^{35}S]cystamine
as radiochemical impurities.

Conversion of hypotaurine to taurine in vivo

A small dose of [^{35}S]hypotaurine (0.4 μmol, 0.25 MBq) in saline
was injected into the femoral muscle of adult mice (wt. 20-25 g).
The animals were killed by decapitation from 10 min to 10 h after
the injection. Samples from various tissues were homogenized in
4 volumes of cold 0.6 mol/l perchloric acid (PCA) and centrifuged.
The supernatants were neutralized with 4.8 mol/l KOH, centrifuged
again and samples from the final supernatants subjected to thin
layer chromatography and liquid scintillation counting. Only
the spots corresponding to taurine, hypotaurine and cystamine contained radioactivity. These spots were well separated from each
other on silica gels by either phenol-water (4:1, w/v) or ethanol-
water (7:3, v/v). They were scratched out of the plates, mixed
with 5 ml of scintillation cocktail (4 g PPO, 0.1 g POPOP, 50 ml
ethanol, 350 ml Triton X-100 and xylene to make 1 l) and counted
(LKB-Wallac, model 81000). Any quenching was corrected for with
the external standard-channels ratio method. The average counting
efficiency was 90%.

Oxidation of hypotaurine in vitro

Tissue samples were homogenized in 4 volumes of 0.2 mol/l HEPES
buffer (pH 7.4) at 278 K. Samples of the homogenate (0.1 ml) were
immediately incubated with [^{35}S]hypotaurine in a shaking water bath

for 4 h at 310 K and pH 7.4 in open tubes under air in a final volume of 0.11 ml of the above HEPES buffer (if not otherwise specified). The incubations were terminated with 20 μl of cold PCA (0.6 mol/l) and the samples were deproteinized, chromatographed and counted as above. The samples in which PCA was already added together with [^{35}S]hypotaurine in the beginning of incubation served as zero controls.

Subcellular distribution of hypotaurine oxidation

Tissue samples were homogenized in 5 volumes of cold 10 mmol/l HEPES buffer (pH 7.5) containing 0.25 mol/l sucrose and 3 mmol/l MgCl$_2$. The liver was fractionated according to Fleischer and Kervina (1974) and the brain according to Whittaker and Barker (1972). Samples of tissue homogenate, and nuclear, mitochondrial, microsomal and cytoplasmic fractions were incubated as above with [^{35}S]hypotaurine. Protein contents of the subcellular fractions were determined by the method of Lowry et al. (1951).

RESULTS AND DISCUSSION

The injected [^{35}S]hypotaurine was readily converted to [^{35}S]-taurine in vivo in adult mice. For instance, more than 90% of the radioactivity was already recovered in [^{35}S]taurine in the liver and kidney 60 min after the [^{35}S]hypotaurine injection (Fig. 1). Only

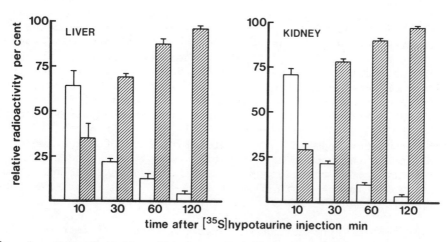

Fig. 1. Relative distribution of radioactivity in hypotaurine (clear columns) and taurine (hatched columns) in the liver and kidney of adult mice in vivo at varying intervals after an intramuscular injection of [^{35}S]hypotaurine.

traces of [35S]hypotaurine were present in the samples obtained
still later. It seems that hypotaurine is a precursor of taurine
in vivo, as Cavallini et al. (1954) suggested. [35S]Methionine
is also converted to hypotaurine and taurine in the rat (Peck and
Awapara, 1967). Measurable amounts of hypotaurine have indeed been
found in the rat brain in vivo (Perry and Hansen, 1973). However,
such experiments do not reveal anything of the nature and sequence
of the reaction(s) involved.

In order to characterize more closely the oxidation reaction of
hypotaurine, we incubated homogenates of different tissues of adult
mice with [35S]hypotaurine. These first incubations were carried
out at 310 K and pH 7.4 for up to 4 h in open tubes without cofactors. We were unable for some time to convince ourselves whether or
not there occurs any conversion of hypotaurine to taurine under
such conditions. We continued tenaciously with our experiments,
however, because hypotaurine was so rapidly converted to taurine in
vivo. Only after numerous trials could we infer that there occurred
some conversion also in vitro, but its detection was just at the
sensitivity limits of our assay. It is thus no wonder that Cavallini
et al. (1954) had earlier failed to detect any. Only 2-3% of the
added hypotaurine were oxidized by homogenates of the liver and kidney, about 1% by the heart, spleen, muscle and lung, and still less
by the brain. These traces of activity disappeared when the tissue
samples were boiled or incubated at 273 K. The activity was slightly
enhanced by NAD^+, $NADP^+$ or oxidized glutathione, whereas reduced
glutathione, NADH and NADPH had no effect.

There are some indications that hypotaurine oxidation may be
faster in homogenates prepared from tissues of developing animals
(Oja et al., 1973). Here, too, this was found to be the case; the
activity was generally higher in tissues of developing mice, being
highest in tissues excised from about one-week-old mice. The properties of the reaction were therefore further studied with the liver
and brain of developing mice. The activities of other enzymes along
the biosynthesis pathways of taurine, viz. cysteine dioxygenase (EC
1.13.11.20) (Misra and Olney, 1975) and cysteine sulphinate decarboxylase (EC 4.1.1.29) (Agrawal et al., 1971; Oja et al., 1973) increase during development. The diminishing rate of oxidation of
hypotaurine in the course of maturation of the liver and brain shows
a strikingly better correlation with the decreasing concentration of
taurine in developing brain (Oja et al., 1968) and liver (Macaione
et al., 1975) than the activities of the above two other enzymes.

Oxidation of hypotaurine in liver homogenates from developing
mice was significantly increased by both nicotinamide dinucleotides,
whereas in the brain $NADP^+$ was apparently ineffective (Table I).
FAD^+ did not cause a statistically significant enhancement in the
present experiments. When the time course of oxidation was studied

Table I.   Effects of certain cofactors on hypotaurine oxidation by homogenates from developing mice

Cofactor	Hypotaurine oxidation (% of control)	
	Liver	Brain
None (control)	100 ± 12 (20)	100 ± 10 (15)
NAD^+	$220 \pm 26^\dagger$ (20)	$139 \pm 9^\dagger$ (15)
$NAD^+ + Cu^{2+}$	$292 \pm 36^\dagger$ (12)	$184 \pm 10^\dagger$ (7)
$NADP^+$	$206 \pm 20^\dagger$ (8)	102 ± 5 (3)
FAD^+	103 ± 3 (3)	123 ± 5 (8)

Samples from 6 to 8-day-old mice were incubated with 0.05 mmol/l [^{35}S]hypotaurine (0.2 GBq/l). The concentrations of the dinucleotides were 50 mmol/l and that of Cu^{2+} ions 50 μmol/l. Means ± S.E.M. are given. Number of experiments in parentheses. Significance of the differences from the corresponding controls: $^\dagger p < 0.01$.

Table II   Effects of certain cations on hypotaurine oxidation by homogenates from developing mice

Cation	Hypotaurine oxidation (% of control)	
	Liver	Brain
None (control)	100.0 ± 12.0 (20)	100.0 ± 12.0 (9)
Fe^{2+}	86.5 ± 6.8 (6)	–
Fe^{3+}	124.8 ± 7.5 (5)	74.3 ± 5.0 (3)
Cu^+	123.6 ± 11.0 (5)	–
Cu^{2+}	$133.1 \pm 9.5^*$ (12)	130.2 ± 18.2 (7)
Mg^{2+}	94.1 ± 5.5 (3)	–
Mn^{2+}	74.9 ± 6.0 (3)	–
Zn^{2+}	71.5 ± 6.3 (3)	–
Mo^{5+}	85.5 ± 13.1 (3)	–

Samples from 6 to 8-day-old mice were incubated with 0.05 mmol/l [^{35}S]hypotaurine (0.2 GBq/l). The concentrations of the metal chlorides were 50 μmol/l. Means ± S.E.M. are given. Number of experiments in parentheses. Significance of the differences from the corresponding controls: $^* p < 0.05$.

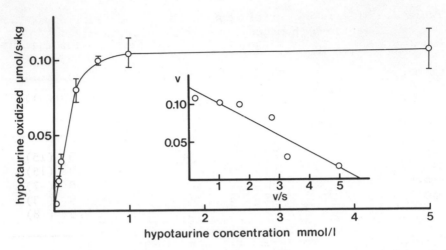

Fig. 2.   Oxidation of hypotaurine by liver homogenates from 6-day-
           old mice at varying concentrations of hypotaurine. The
           samples were incubated with 50 μmol/l Cu^{2+} ions, 50 mmol/1
           NAD^+ and 0.2 GBq/1 [^{35}S]hypotaurine. Means ± S.E.M.
           (n=3) are shown. In the smaller graph the data are
           plotted in a v vs. v/s plot.

with liver homogenates, NAD^+ appeared definitely more effective than
$NAPD^+$ during the first two hours of incubation, i.e. before the reaction rate gradually started to slow down. The rate of oxidation by
the liver was significantly further increased when Cu^{2+} ions were
added (Table II). A number of other cations known as oxidase acti-
vators (e.g., Fe^{2+}, Mg^{2+}, Mn^{2+}, Zn^{2+} and Mo^{5+}) had no apparent effect. Cu^+ and Fe^{3+} caused no significant enhancement either. The
reaction rate increased nearly 3-fold in the liver and 2-fold in
the brain when both Cu^{2+} ions and NAD^+ were present (Table I). The
activation by Cu^{2+} ions was also dependent on the copper ion concentration, the optimum being of the order of 50 μmol/1.

     The oxidation rate was also dependent on the concentration of
hypotaurine in the incubation medium. The reaction appeared to obey
simple Michaelis-Menten kinetics (Fig. 2). The kinetic constants
were estimated from a linear transformation of the Michaelis-Menten
equation in a v against v/s plot. The apparent Michaelis constant
(K_m) was about 0.2 mmol/1 and the maximal velocity (V) about
0.1 μmol/s x kg. In our crude liver homogenate the apparent K_m
for hypotaurine oxidation was of the same order of magnitude as
K_m for the partially purified L-cysteine sulphinate decarboxylase
(Jacobsen et al., 1964), the preceding enzyme in the biosynthesis
pathway.

Table III. Effects of inhibitors on hypotaurine oxidation by homogenates from developing mice

Inhibitor tested	Inhibitor concentration (mmol/l)	Hypotaurine oxidation (% of control)		
		Liver		Brain
None (control)	–	100.0 ± 4.4	(6)	100.0 ± 17.8 (3)
$HgCl_2$	0.1	103.4 ± 3.1	(3)	–
	10.0	93.5 ± 8.7	(3)	$23.8 \pm 4.8^*$ (3)
$AgNO_3$	0.1	108.4 ± 0.2	(3)	–
	10.0	$79.4 \pm 7.5^*$	(3)	$14.0 \pm 1.8^\dagger$ (3)
$CdCl_2$	0.1	106.8 ± 1.0	(3)	–
	10.0	$50.5 \pm 9.1^\dagger$	(3)	–
ICH_3COOH	0.1	95.2 ± 5.4	(3)	–
	10.0	$37.1 \pm 1.6^\dagger$	(3)	$14.3 \pm 2.0^\dagger$ (3)
NaF	0.1	106.4 ± 2.2	(3)	–
	10.0	$76.5 \pm 12.2^*$	(3)	57.2 ± 8.6 (3)
NaCN	0.2	84.9 ± 12.2	(4)	–
BAL	4.0	$51.3 \pm 8.4^\dagger$	(11)	–

Samples from 8 to 11-day-old mice were incubated with 50 mmol/l NAD^+, 50 µmol/l Cu^{2+} ions and 0.05 mmol/l [^{35}S]hypotaurine (0.2 GBq/l). Means + S.E.M. are given. Number of experiments in parentheses. Significance of the differences from the corresponding controls: $^*P < 0.05$, $^\dagger P < 0.01$.

Enzyme inhibitors, $HgCl_2$, $AgNO_3$, $CdCl_2$ and iodoacetate at a concentration of 0.1 mmol/l had no significant effects on hypotaurine oxidation in liver homogenates (Table III). Only at a very high concentration (10 mmol/l) did $AgNO_3$, $CdCl_2$ and iodoacetate significantly inhibit the reaction. Moreover, an addition of sulphydryl groups in the form of 2,3-dimercapto-1-propanol (BAL) significantly reduced the oxidation activity in the liver – an observation at variance with the claims of Sumizu (1962). A possible involvement of free sulphydryl groups in hypotaurine oxidation thus remains unsettled. On the other hand, the oxidation of hypotaurine by brain tissue was much more sensitive to certain of the above agents. The inhibition caused by NaF suggests that the possible enzyme is activated by metal ions or is a metalloenzyme. Metal-chelating agents also to some extent reduced the formation of taurine (Table IV), which may reflect the requirement of cupric ions in the reaction. The oxidation was again relatively more sensitive in brain than in liver homogenates.

Table IV. Effects of chelating agents on hypotaurine
           oxidation by homogenates from developing mice

Chelating agent (0.1 mmol/1)	Hypotaurine oxidation (% of control)	
	Liver	Brain
None (control)	100.0 ± 1.5	100.0 ± 10.0
o-Phenanthroline	105.2 ± 3.6	63.3 ± 9.3
Hydroxylamine	95.8 ± 1.9	$48.3 \pm 8.8^*$
α,α'-Dipyridyl	$68.3 \pm 5.3^\dagger$	$55.0 \pm 10.0^*$
Diethyldithiocarbamate	$75.8 \pm 1.6^\dagger$	$50.0 \pm 7.6^*$
Bathocuproine sulphonate	$88.2 \pm 0.6^\dagger$	130.0 ± 10.4
Bathophenanthroline sulphonate	$87.4 \pm 0.5^\dagger$	86.7 ± 12.0
8-Hydroxyquinoline	95.3 ± 1.9	$51.7 \pm 14.2^*$

Samples from 7 to 9-day-old mice were incubated with 50 mmol/1
NAD^+ and 0.05 mmol/1 [^{35}S]hypotaurine (0.2 GBq/1). Means $\pm$
S.E.M. (n=3) are given. Significance of the differences from
the corresponding controls: $^*P<0.05$, $^\dagger P<0.01$.

Table V. Subcellular distribution of hypotaurine oxidation
          in 7-day-old mice

Subcellular fraction	Hypotaurine oxidation (% of total)	
	Liver	Brain
Nuclear	12.5 ± 1.5	10.0 ± 2.3
Mitochondrial/P_2	13.0 ± 1.8	13.2 ± 3.7
Microsomal	30.5 ± 3.4	2.0 ± 0.4
Soluble cytoplasmic	42.5 ± 4.9	75.4 ± 5.2

Samples of different fractions were incubated for 2 h with
50 µmol/1 Cu^{2+} ions, 50 mmol/1 NAD^+ and 0.05 mmol/1 [^{35}S]-
hypotaurine (0.2 GBq/1). Means $\pm$ S.E.M. of triplicate
samples of 4 separate fractionations are given.

A great deal of the enzyme activity was lost during the sub-cellular fractionation procedures. The activity recovered in the liver was only 33.6 ± 3.2 per cent and in the brain 68.7 ± 11.2 per cent (means ± S.E.M., n=12) of the total original activity of tissue homogenates. This loss resulted solely from the denaturation and/or inactivation of the enzyme during the time-consuming isolation and incubation procedures, since all the protein in tissue homogenates was recovered in the subcellular fractions. Most of the oxidation of hypotaurine occurred in the cytoplasmic and microsomal fractions in the liver (Table V). About one tenth of the recovered activity was found in the mitochondrial and nuclear fractions. In the brain most of the activity was in the soluble cytoplasma, but microsomes were almost devoid of activity. The enzyme activity in the ox, pig, horse and wether retina is also mainly associated with the soluble fractions (Di Giorgio et al. 1977; 1978). Cysteine sulphinate decarboxylase has been recovered predominantly in the soluble cyto-plasmic and crude mitochondrial fractions in the adult rat brain Agrawal et al., 1971; Rassin and Gaull, 1975; Pasantes-Morales et al., 1977). Most of taurine is in the soluble fractions in both the adult and immature rat brain (Agrawal et al., 1971; Rassin et al., 1978) and the liver (Macaione et al., 1975).

The oxidation of hypotaurine in both the liver and brain was gradually more and more reduced when the incubation temperature was lowered to 278 K. The Q_{10} value was about 1.6 within the tempera-ture interval from 293 to 310 K for the reaction in the liver. The activation energy, calculated from the Arrhenius plots, was about 60 kJ/mol. When the incubation mixtures were thoroughly oxygenated with pure O_2, the reaction rate was approximately doubled. It was 191.3 ± 9.7 per cent (mean ± S.E.M., n=5) of a control which also contained both NAD^+ and Cu^{2+} ions but was incubated only under air in open tubes. This result indicates that the partial pressure of oxygen in the medium under air is not high enough to saturate the enzyme system(s) responsible for hypotaurine oxidation. The finding is at variance with the comment of Sumizu (1962) that hypotaurine oxidation is not coupled with the consumption of atmospheric oxygen. Inorganic sulphite obviously does not compete with hypotaurine for the enzyme since the reaction rate was at the 0.05 mmol/l hypo-taurine concentration uninfluenced by the presence of 0.5 mmol/l Na_2SO_3.

The pH dependence curve of the oxidation reaction was rather flat-topped (Fig. 3). The pH optimum was about 9.0. In all pre-vious studies on hypotaurine oxidation the pH of the incubation mixture has been adjusted to 7.4–7.8 (Cavallini et al., 1954; Su-mizu, 1962; Oja et al., 1973; Di Giorgio et al., 1977). In con-trast to the present results Di Giorgio et al. (1978) reported a pH optimum of 7.0 for hypotaurine oxidation in the retina. The too acidic pHs tested may account for the frequent difficulties in

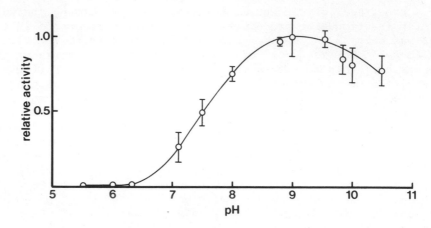

Fig. 3.   Oxidation of hypotaurine by liver homogenates from 8-day-
          old mice at varying pH. The samples were incubated either
          in 0.2 mol/1 HEPES buffer (pH 5 to 9) or in 0.1 mol/1
          glycine-NaOH buffer (pH 9 to 11). The true pH of each
          incubation mixture was checked at 298 K. The samples were
          then incubated for 4 h at 310 K with 50 μmol/1 Cu^{2+} ions,
          50 mmol/1 NAD^+ and 0.05 mmol/1 [^{35}S]hypotaurine (0.2 GBq/1).
          Since hypotaurine oxidation was partially inhibited in the
          glycine-NaOH buffer, the enzyme activities are expressed
          in relation to the respective maximal activity in both
          buffers at pH 9.0. Means ± S.E.M. (n=3) are shown.

detecting any hypotaurine oxidation in tissues and for the generally
very low enzyme activities found in vitro, in spite of the fast conversion of hypotaurine to taurine in vivo. The optimum pH for rat
liver cysteine dioxygenase is 8.5-9.5 (Sakakibara et al., 1976). It
is interesting that the pathway: cysteine – cysteine sulphinate –
hypotaurine – taurine, seems to be active only at pHs well above
the normal physiological ones. One might speculate whether or not
this circumstance has any regulatory significance in the organism,
keeping in mind the modulatory and inhibitory role of taurine in
electrically excitable tissues and the probable involvement of calcium ions in its action.

CONCLUSIONS

     In the present studies we have shown that the conversion of
hypotaurine to taurine in mouse tissues, particularly in liver and
brain, exhibits characteristic properties of an enzyme-catalyzed
reaction. It is pH and temperature-dependent, obeys Michaelis-
Menten kinetics and is enhanced by coenzymes, metal activators and

oxygen. The enzyme activity is unevenly distributed among subcellular fractions. The optimum incubation conditions for hypotaurine
oxidation found so far include pH 9.0, oxygenation of the reaction
mixture and the presence of 50 μmol/l Cu^{2+} ions and 50 mmol/l NAD^+.
The oxidation appears to be catalyzed by an oxidoreductase-like
enzyme or enzyme system, but we consider it too early to attempt a
definite classification of the enzyme at issue. Nor can we on the
basis of the present results infer whether or not the enzyme oxi-
dizing hypotaurine is specific for hypotaurine. The oxidation could
also be effected by some already known oxidoreductase as a side-
reaction.

ACKNOWLEDGEMENT

We are grateful to Ms. Pirkko Erkkilä for her expert technical
assistance.

REFERENCES

Agrawal, H. C., Davison, A. N., and Kaczmarek, L. K., 1971, Sub-
     cellular distribution of taurine and cysteinesulphinate
     decarboxylase in developing rat brain, Biochem. J., 122:759.
Cavallini, D., De Marco, C., Mondovi, B., and Stirpe, F., 1954,
     The biological oxidation of hypotaurine, Biochim. biophys.
     Acta (Amst.), 15:301.
Cavallini, D., Scandurra, R., Dupré, S., Santoro, L., and Barra, D.,
     1976, A new pathway of taurine biosynthesis, Physiol. Chem.
     Physics, 8:157.
Di Giorgio, R. M., Macaione, S., and De Luca, G., 1977, Subcellular
     distribution of hypotaurine oxidase activity in ox retina,
     Life Sci., 20:1657.
Di Giorgio, R. M., De Luca, G. and, Macaione, S., 1978, Hypotaurine
     oxidase activity in retina, Bull. mol. Biol. Med., 3:115.
Fiori, A., and Costa, M., 1969, Oxidation of hypotaurine by peroxide,
     Acta vitamin. (Milano), 23:204.
Fleischer, S., and Kervina, M., 1974, Subcellular fractionation of
     rat liver, in: "Methods in Enzymology", Vol. XXXI,
     S. Fleischer and L. Parker, eds., Academic Press, New York,
     p. 6.
Jacobsen, J.G., and Smith, L.H., Jr., 1968, Biochemistry and
     physiology of taurine and taurine derivatives, Physiol. Rev.,
     48:424.
Jacobsen, J. G., Thomas, L. L., and Smith, L. H., Jr., 1964, Proper-
     ties and distribution of mammalian L-cysteine sulfinate
     carboxy-lyases, Biochem. biophys. Acta (Amst.), 85:103.

Macaione, S., Tucci, G., and Di Giorgio, R. M., 1975, Taurine
    distribution in rat tissues during development, Ital. J.
    Biochem., 24:162.
Misra, C. H., and Olney, J. W., 1975, Cysteine oxidase in brain,
    Brain Res., 97:117.
Lowry, O. H., Rosebrough, N. J., Farr, A. L., and Randall, R. J.,
    1951, Protein measurement with the Folin phenol reagent,
    J. biol. Chem. 193:265.
Oja, S. S., Uusitalo, A. J., Vahvelainen, M.-L., and Piha, R. S.,
    1968, Changes in cerebral and hepatic amino acids in the rat
    and guinea pig during development, Brain Res., 11:655.
Oja, S. S., Karvonen, M.-L., and Lähdesmäki, P., 1973, Biosynthesis
    of taurine and enhancement of decarboxylation of cysteine
    sulphinate and glutamate by the electrical stimulation of rat
    brain slices, Brain Res., 55:173.
Pasantes-Morales, H., Loriette, C., and Chatagner, F., 1977, Regional
    and subcellular distribution of taurine-synthesizing enzymes
    in the rat central nervous system, Neurochem. Res., 2:671.
Peck, E. J., Jr., and Awapara, J., 1967, Formation of taurine and
    isethionic acid in rat brain, Biochim. biophys. Acta (Amst.),
    141:499.
Perry, T. L., and Hansen, S., 1973, Quantification of free amino
    compounds of rat brain: identification of hypotaurine,
    J. Neurochem., 21:1009.
Rassin, D. K., and Gaull, G. E., 1975, Subcellular distribution of
    enzymes of transmethylation and transsulphuration in rat
    brain, J. Neurochem. 24:969.
Rassin, D. K., Sturman, J. A., Hayes, K. C., and Gaull, G. E., 1978,
    Taurine deficiency in the kitten. Subcellular distribution
    of taurine and [^{35}S]taurine in brain, Neurochem. Res., 3:401.
Sakakibara, S., Yamaguchi, K., Hosokawa, Y., Kohashi, N., Ueda, I.,
    and Sakamoto, Y., 1976, Purification and some properties of
    rat liver cysteine oxidase (cysteine dioxygenase), Biochim.
    biophys. Acta (Amst.), 422:273.
Scandurra, R., Fiori, R., and Cannella, C., 1969, Preparation of
    ^{35}S-labeled hypotaurine, Ital. J. Biochem., 18:19.
Scandurra, R., Politi, L., Dupré, S., Moriggi, M., Barra, D., and
    Cavallini, D., 1977, Comparative biological production of
    taurine from free cysteine and from cysteine bound to
    phosphopantothenate, Bull. mol. Biol. Med., 2:172.
Sumizu, K., 1962, Oxidation of hypotaurine in rat liver, Biochim.
    biophys. Acta (Amst.), 63:210.
Whittaker, V. P., and Barker, L. A., 1972, The subcellular frac-
    tionation of brain tissue with special reference to the
    preparation of synaptosomes and their component organelles,
    in: "Methods of Neurochemistry", Vol. 2, R. Fried, ed.,
    Marcel Dekker, New York, p. 1.

SULFUR AMINO ACIDS IN THE REPRODUCTIVE TRACT OF MALE GUINEA PIGS, RATS, MICE AND SEVERAL MARINE ANIMALS[‡]

Charles D. Kochakian

Laboratory Experimental Endocrinology
University of Alabama Medical Center
Birmingham, Alabama 35294   USA

INTRODUCTION

Androgens stimulate growth[1] in a number of tissues other than the accessory sex organs of the male. The influence of castration and androgen on the free amino acids of several of the responsive tissues[2-6] was determined as an aid in the delineation of the anabolic action of androgens. On analysis of the seminal vesicles of the guinea pig[2] an unknown amino acid peak appeared just before taurine on slight adjustment of the pH of the elution buffer. The unknown compound was of particular interest because it nearly disappeared after castration and was restored to or near normal by androgen administration. The elution rate of this compound proved to be identical with that of hypotaurine, thus, suggesting an influence of androgens on the metabolism of sulfur amino acids. The studies, therefore, were extended to the reproductive tract of two other laboratory rodents[4-7] and five marine animals.

‡   These investigations were supported by grant AM11060 from the NIAMDD of the National Institute of Health USPHS.

‡   The guinea pig[2,3] and rat[4] data are presented with the approval of the Ala. J. Med. Sci. and those of the mouse[5] with the approval of the Amer. J. Physiol. Some of the data on the marine animals[8] is presented with the approval of Raven Press.

MATERIALS AND METHODS

Animals: The guinea pigs, rats, and mice were purchased from stock dealers. The lobsters and squid were gathered in the Cape Cod region by NEMSCO during July 1975 and the blue crab during July 1974. The tissues were removed, frozen on dry ice and transported to Birmingham. The flounder tissues were removed, frozen and shipped via air by Dr. D. R. Idler, Newfoundland in July 1973 and the octopus material was obtained by Dr. T. Mann.

Methods: The tissues were extracted by homogenization with cold 6% perchloric acid containing norleucine as an internal standard and analyzed on a Beckman/Spinco 120B Analyser.[2-6]

RESULTS*

Seminal Vesicles: Guinea Pig: The guinea pig seminal vesicles (Table 1) contained a high and almost equal concentration of both taurine and hypotaurine. The taurine was modestly increased during the first few days after castration but after 36 days of castration, when the organ was only about one-fourth normal size, the concentration of taurine was at the normal level. Hypotaurine, on the other hand, was greatly decreased six days after castration and had nearly disappeared after 35 days. The administration of testosterone for 21 days restored the concentration of the hypotaurine and the weight of the seminal vesicle to or nearly normal. The other sulfur amino acids occurred in low concentrations except glutathione (2.07 µmoles/ml) and were not significantly changed by castration or androgen treatment.

The guinea pig seminal vesicle fluid (Table 2) also had a high concentration of taurine but an even higher concentration of hypotaurine. Castration for 36 days resulted in a complete absence of fluid but after three and six days, the fluid content was still nearly normal and small changes were observed in both compounds. The other sulfur amino acids were in low concentration with glutathione non-detectable. Testosterone treatment for 21 days restored the fluid volume to the normal level with concomitant reappearance of normal concentrations of the sulfur compounds.

* Because of the limitation in space, the standard error of the means have been omitted. These values can be found in the original articles.[2-6]

Table 1: Changes in the Free Sulfur Amino Acids of the Guinea Pig Seminal Vesicles After Castration and Testosterone Treatment.

Compound	Normal	Castration Days				Testosterone Days	
	μmoles/g	3 %	6 %	36 %	(36) %	21 %	(21) %
Taurine	8.58	32**	62**	-12*	(8)*	-3	(-2)
Hypotaurine	9.74	36**	-45*	-77	(-72)*	-28	(-9)
Sem. Ves.,g	1.05	1.19	0.78	0.26	(0.24)	0.86	(0.75)

The tissues were extracted immediately after autopsy. There were six normal, five castrated for 3 and 6 days and 14 for 36 days. The values in parenthesis are for tissues stored at -20C for three to four months prior to extraction. There were 10 normal, 28 castrated for 36 days and 20 testosterone treated guinea pigs. The values are the means of five, seven and six analyses of the respective groups and each analysis represents the pool of two normal, four castrated and two testosterone treated guinea pigs except one of the treated groups represents the pool of ten guinea pigs. The differences are from those of the respective normal values. * p <0.01, ** p <.05 <.01. Glutathione (1.13) and cysteic (0.07) showed no consistent changes. Storage before extraction resulted in a nearly complete disappearance of glutathione no changes in taurine and hypotaurine, and varying increases in the other compounds.

Table 2: Changes in the Free Sulfur Amino Acids of Guinea Pig Seminal Vesicle Fluid After Castration and Testosterone Treatment.

Compound	Normal	Castration		Testosterone
	μmoles/g	3 days %	6 days %	21 days %
Taurine	8.69	-26*	-23	6
Hypotaurine	25.80	30*	-5	-35*
Methionine	0.08	90*	520*	22
Sem.Ves. Fluid, g	1.26	1.90	0.92	0.96

See Table 1 for details. Glutathione (0.0, Cysteine-sulfinic acid (0.08), Cysteic (tr), and Cystine/2 (0.04) showed no consistent changes.

Table 3: Changes in the Free Sulfur Amino Acids of the
Rat Seminal Vesicles After Castration and Testosterone
Propionate (μmoles/g).

Compounds	Normal (4)	Castrate (4)	TP (5)
Taurine	7.53	8.13	8.93
Hypotaurine	0.06	0.01*	0.06
Cysteinesulfinic and Cysteic	0.15	0.09*	0.17
Glutathione	0.05	0.93	0.34**
Methionine	0.06	0.06	0.10**
Cystine/2	tr	tr	0.02
Sem. Ves.mg. (No)	475(8)	115(13)	502(7)

The rats were two months old at castration and autopsy
was one month later. Fourteen days before autopsy testosterone propionate (TP) was implanted subcutaneously
as a 15 mg pellet. The amount absorbed was 2.9 mg. The
number of analyses are indicated in parentheses at the
head of each column. Each analysis of the castrated rats
represents the pool of two to four organs.* p <0.01,
** p <0.05.

Rat: The taurine concentration of the rat seminal
vesicles (Table 3) was similar to that of the guinea pig
but the hypotaurine concentration was very low. Castration
and testosterone treatment had no influence on the concentration of taurine in spite of major changes in the
weight of the organ. The low concentration of hypotaurine
and also that of cysteinesulfinic acid, however, were
markedly reduced by castration and restored to normal
by testosterone treatment. The seminal vesicle fluid of
both normal and testosterone propionate treated rats had
only a fraction (1.13) of the taurine of the tissue and
no detectable glutathione or hypotaurine. The concentration of hypotaurine in both the organ and the fluid
is in marked contrast with that in the guinea pig.

Mouse: The taurine concentration in the seminal
vesicles of the mouse (Table 4) was roughly five times
that in the guinea pig and rat and decreased to one half
with a 90% loss in tissue weight after castration. The
hypotaurine concentration was not as high as in the
guinea pig but much greater than in the rat and nearly
disappeared after castration with a restoration to
normal with testosterone propionate treatment.
Glutathione, methionine and cystine/2 occurred in low

concentrations but were increased after castration and restored to normal by testosterone propionate treatment. Separate analyses of the fluid from the seminal vesicles revealed essentially the same values as that of the organ with fluid except that glutathione was absent and hypotaurine was slightly higher (1.80 vs 1.15).

Table 4: Changes in the Free Sulfur Amino Acids of the Mouse Seminal Vesicles (with fluid) After Castration and Testosterone Propionate (μmoles/g).

Compounds	Normal (4)	Castrate (4)	TP (4)
Taurine	41.90	21.01*	38.10
Hypotaurine	2.08	0.13*	1.86
Cysteic	0.15	0.08	0.15
Cysteinesulfinic	0.06	0.07	0.05
Glutathione	0.26	1.03*	0.28
Methionine	0.04	0.16*	0.04
Cystine/2	0.03	0.11*	0.03
Sem.Ves., mg (No)	180(32)	17(67)	231(26)

The mice were two months old at castration; two weeks later testosterone propionate (TP) was implanted subcutaneously as a 15 mg pellet and the three groups of mice were killed after 11 days. TP absorbed was 2.0 mg. The number of analyses are indicated at the head of each column. Each analysis represented the pool of 6-10 normal, 12-20 castrated and 5-8 TP-treated mice.

Table 5: Changes in the Free Sulfur Amino Acids of the Guinea Pig Prostate After Castration and Testosterone.

Compounds	Normal	Castration Days				Testosterone Days	
	μmoles/g	3 %	6 %	36 %	(36) %	21 %	(21) %
Taurine	6.50	45*	45*	-1	(22)	-1	(-7)
Hypotaurine	5.70	-61*	-62*	-90*	(-85)*	-66	(-25)
Cysteine-sulfinic	0.08	-14	-31	-63*	(-27)**	-17	(-6)
Methionine	0.08	34**	130*	160*	(90)*	23	(5)
Cystine/2	0.07	-80*	-67*	-43	(38)	97	(57)
Prostate,gm	0.79	0.62	0.41	0.26	(0.26)	0.47	(0.51)

Prostate: Guinea Pig: Taurine and hypotaurine were
present at a high and nearly identical concentration
(Table 5) while cysteinesulfinic, methionine and cystine/2
were present at less than 0.1 μmole/g. Taurine exhibited
an increase after three and six days of castration and
normal values after 36 days of castration and 21 days
of testosterone. Hypotaurine and also cysteinesulfinic
and cystine/2, exhibited an immediate decrease after
castration with a restoration to near normal after 21
days of testosterone. Methionine exhibited an increase
after castration with a return to normal after testosterone treatment.

Rat: The taurine concentration in the rat prostate
(Table 6) was only about one half that in the guinea pig
and showed a 100% increase after one month of castration
and a restoration to normal after testosterone propionate.
Hypotaurine, on the other hand, was very low and almost
disappeared after castration with a partial restoration
after testosterone propionate. Interestingly, cysteinesulfinic plus cysteic acid and methionine were present
in relatively high concentrations which were markedly
decreased by castration and restored to normal by testosterone propionate.

Mouse: The prostate of the mouse (Table 7) contained
the highest concentration of taurine which was not altered
by either castration or androgen. Hypotaurine and cysteine
sulfinic plus cysteic acid were present in low concentration and were markedly decreased after castration.

Testis: The mouse exhibited a high concentration of
taurine (Table 8) but the guinea pig showed a relatively
low value and the rat an intermediate concentration which
was practically identical with that of two other
reports.[9,10] Hypotaurine was present but at a relatively
low concentration. The concentration of glutathione was
practically identical for the three animals and was
strikingly high. The other sulfur amino acids were
present but in concentrations less than 0.10 μmole/g.

The testis of immature[3] guinea pigs showed practically the same concentration of the different compounds
as that of the adult guinea pig except for glutathione
which was approximately 50% greater (6.31 μmoles/g).

Table 6: Changes in the Free Sulfur Amino Acids of the
Rat Prostate After Castration and Testosterone
Propionate (μmoles/g).

Compounds	Normal (8)	Castrate (4)	TP (8)
Taurine	3.20	6.40*	3.20
Hypotaurine	0.23	0.01*	0.08
Cysteinesulfinic and Cysteic	1.04	0.26*	1.06
Methionine	0.68	0.07*	0.73
Prostate,mg (No)	958(8)	139(13)	982(8)

Table 7: Changes in the Free Sulfur Amino Acids of the
Mouse Prostate After Castration and Testosterone
Propionate (μmoles/g).

Compounds	Normal (4)	Castrate (4)	TP (4)
Taurine	13.04	13.10	14.24
Hypotaurine	0.48	0.01*	0.57
Cysteinesulfinic and Cysteic	0.48	0.11*	0.44
Glutathione	0.92	0.61	0.65
Methionine	0.06	0.08	0.08
Cystine/2	---------- trace ---------		
Prostate,mg (No)	71(32)	19(68)	90(26)

Table 8: Concentration of Free Sulfur Amino Acids in the
Testes of Normal Adult Animals (μmoles/g).

Compounds	Guinea Pig (16)	Rat (5)	Mouse (5)
Taurine	0.86	2.66	6.88
Hypotaurine	0.59	0.12	0.21
Glutathione	4.29	4.67	4.60
Cystine/2	0.06	0.00	tr
Methionine	0.03	0.05	0.05
Cysteic	0.02	0.04	0.05
Cysteinesulfinic	0.01	0.09	0.08
Organ Wgt.g (No)	4.25(12)	3.73(5)	0.23(25)

Glutathione of immature guinea pig 6.31 μmoles/g.

   Epididymes: Taurine was present in high concentration
in both the cauda and caput epididymes (Table 9). The
cauda contained approximately twice that of the caput in
the rat and mouse. A similar differential has been
reported by Shimazaki et al.[10] The caput of the guinea
pig was very small and was not analysed. The concentration
of hypotaurine, also, was high but did not exhibit a
significant difference between the caput and the cauda.
Glutathione was present in appreciable concentration but
much less than seen in the testis.

   The other sulfur amino acids were present in concentrations less than 0.1 μmole/g except cysteic acid
(0.10 to 0.18 μmole/g).

Table 9: The Concentration of Free Sulfur Amino Acids
in the Epididymes of Normal Adult Animals (μmoles/g).

Compounds	Guinea Pig(12)	Rat(8)		Mouse(7)	
	Cauda	Caput	Cauda	Caput	Cauda
Taurine	5.88	3.89	6.73	9.80	20.50
Hypotaurine	8.34	7.73	6.44	6.50	5.90
Glutathione	0.59	0.82	0.78	0.77	0.47
Cystine/2	0.03	0.05	0.03	tr	tr
Methionine	0.02	0.04	0.04	0.13	0.06
Cysteic	0.18	0.16	0.10	0.12	0.14
Cysteinesulfinic	0.05	0.08	0.10	0.08	0.05
Organ Wgt.mg (No)	923(16)	442(23)	358(25)	48(57)	31(42

Table 10: The Concentration of Free Sulfur Amino Acids
in the Testes of Adult Marine Animals (μmoles/g).

Compounds	Flounder(2)	Lobster(6)	Blue(10) Crab	Squid(10)
Taurine	22.12	21.90	39.30	146.00
Hypotaurine	1.89	0.51	1.39	0.09
Methionine	0.27	0.91	3.19	0.38
Cysteic	0.07	0.62	0.48	0.02
Cysteinesulfinic	--	0.12	0.31	0.04
Glutathione	--	0.75	0.30	--
Cystine/2	--	0.31	--	--
Organ Wgt., mg (No)	712(1)	804(6)	161(10)	1580(10)

MARINE ANIMALS

Testis: The four animals (Table 10) showed high concentrations of taurine, with an exceptionally high
value in the squid. The hypotaurine occurred in a much
lower concentration with considerable variation among
the animals. The other sulfur amino acids were present
in variable low concentrations with a high methionine
value in the blue crab testis. Glutathione, in contrast
to the rodents, was not detected or was at low concentration. This, however, may be due to loss in transportation in spite of the short period and the frozen
(dry ice) state of the testes.

Spermatophores: Taurine again was present in high
concentration with the value in the squid three to four
times that of the octopus (Table 11). Cysteic acid was
present in high concentration and the other compounds
were present in small but appreciable concentration
except that glutathione and cystine/2 were absent. The
fluid in the spermatophore sac[11] of the octopus was
separated by centrifugation and analysed separately. The
values for the individual compounds were slightly higher
in the fluid than in the spermatophores.

Accessory Sex Organs: Taurine was present in high
concentration (Table 12). The accessory sex organs of
the blue crab[12-14] were the only ones which were readily
dissected into the component parts. The squid again
contained an exceptionally high concentration. The concentration of taurine was distinctively different for
the three components. Hypotaurine was present at relatively low values except in the anterior lobe of the
blue crab. The other sulfur amino acids also were in
higher concentration in the anterior lobe of the blue
crab.

Table 11: The Concentration of Free Sulfur Amino Acids
in the Spermatophores of the Squid and Octopus (μmoles/g).

Compounds	Squid (10)	Octopus (2)	
	Sperm. & Sac	Sperm.	Fluid
Taurine	45.40	13.70	16.50
Hypotaurine	0.09	0.66	0.77
Methionine	0.29	0.28	0.34
Cysteic	3.50	3.74	5.35
Cysteinesulfinic	0.26	0.32	0.31
Organ Wgt.mg (No)	396(10)	9,500(1)	12,000(1)

Table 12: The Concentration of Free Sulfur Amino Acids
in the Accessory Sex Organs of Marine Animals (μmoles/g).

Compounds	Lobster(6)	Squid(10)	Blue Crab(10)		
			Ant.	Med.	Post.
Taurine	16.90	73.00	17.90	10.34	41.90
Hypotaurine	0.19	0.18	1.18	0.18	0.40
Methionine	0.35	0.55	2.48	0.52	0.45
Cysteic	0.27	0.30	1.02	0.27	0.10
Cysteinesulfinic	0.07	0.12	0.42	0.18	0.12
Glutathione	--	--	0.22	0.16	--
Organ Wgt.,mg	668	267	184	1047	1045

DISCUSSION AND SUMMARY

Taurine occurs in high concentration in all of the
sex organs of guinea pigs, rats, mice and was exceptionally high in the several marine animals. Castration
and testosterone did not change the concentration of
taurine in the organs of the rodents except the very
high value in the mouse seminal vesicle was reduced to
one half with a 90% loss in organ weight while that of
the rat prostate was increased. In both instances,
testosterone restored the concentration to normal with
a concomitant restoration of organ weight.

In contrast to taurine, the concentration of hypotaurine not only varied widely among the different
tissues but also was always markedly decreased by
castration and restored to or nearly normal by the
administration of testosterone. Cysteinesulfinic and
cysteic acids occurred in general in low concentration
but in several instances responded in the same manner
as hypotaurine to androgen administration. Glutathione
was detected in most tissues with exceptionally high
values in the mammalian testis in confirmation of the
high values reported earlier[9] for the rat.

The observation of van der Horst and Brand[15] that
the concentration of hypotaurine in the follicle of the
ewe is related to fertility suggests that the changes
in hypotaurine after castration and testosterone
administration may be related to fertility in the male.
The high concentration of hypotaurine in the epididymes
could be related to maturation of the spermatozoa.[16,17]
The wide variation in the concentration of hypotaurine
in the seminal vesicles and prostates of the three
mammals probably is an indication of a difference in the

role of these tissues in the individual animals. The
possible role of hypotaurine in fertility gains support
from the reported high concentration in the spermatozoa
and seminal plasma of the boar, bull and dog,[18] and the
seminal vesicles and epididymal fluid of the boar.[19] The
very high concentration of hypotaurine in the plasma of
the total ejaculate was greatly reduced in boars with
low sperm counts.[19] Taurine also showed a correlation
with sperm count. The fractional collection of the
ejaculate of normal boars showed a close correlation
of concentration of spermatozoa and hypotaurine.[18]
Hypotaurine has been reported to be contained in or
attached to the spermatozoa.[18] The epididymal plasma
contained high concentration of hypotaurine (44.94
μmoles/ml) while taurine was present at 9.3 μmoles/ml.
The accumulated evidence strongly supports a role for
hypotaurine in fertility but the mechanism remains to
be elucidated. Van der Horst and Brand[15] have suggested
an antioxidant function while Johnson et al[19] postulate
a possible osmoregulatory role[20] similar to that
described in marine vertebrates for taurine.[21]

REFERENCES

1.   C. D. Kochakian, Anabolic-Androgenic Steroids.
Springer-Verlag (1976).
2.   C. D. Kochakian and J. Marcais, The free amino acids,
hypotaurine and other ninhydrin reacting substances of
the prostate and seminal vesicles of the guinea pig:
Regulation by testosterone, Ala. J. Med. Sci. 11:240
(1974).
3.   C. D. Kochakian, Correlation of epididymal hypotaurine and amino acids with sexual development of the
guinea pig, Ala. J. Med. Sci. 15: 21 (1978).
4.   C. D. Kochakian, The free amino acids of the reproductive organs of the male rat: Regulation by androgen,
Ala. J. Med. Sci. 12: 336 (1975).
5.   C. D. Kochakian, Free amino acids of sex organs of
the mouse: Regulation by androgen, Amer. J. Physiol.
228: 1231 (1975).
6.   C. D. Kochakian, The free amino acids of the mouse
kidney: Effect of castration and androgen, Ala. J. Med.
Sci. 11: 333 (1975).
7.   C. D. Kochakian, Influence of testosterone on the
concentration of hypotaurine and taurine in the reproductive tract of the male guinea pig, rat and mouse,
in "Taurine", R. Huxtable and A. Barbeau (Eds), p 375
Raven Press, New York (1976).

8.   C. D. Kochakian, Taurine and related sulfur compounds
in the reproductive tract of marine animals, in "Taurine"
R. Huxtable and A. Barbeau (Eds), p 375, Raven Press,
New York (1976).
9.   I. K. Mushawar and R. E. Koeppe, Free amino acids
of testes, Biochem. J. 132: 353 (1973).
10. J. Shimazaki, H. Yamanaka, I. Taguchi and K. Shida,
Free amino acids in the caput and the cauda epididymis
of adult rats, Endocrinol. Japan 23: 149 (1976).
11. T. Mann, 5-Hydroxytryptamine in the spermatophoric
sac of the octopus, Nature 199: 1066 (1963).
12. L. E. Cronin, Anatomy and histology of the male
reproductive system of callinectes sapidus rathbun,
J. Morphol. 81: 209 (1947).
13. Bull. U. S. Bur. Fisheries, Natural History of
American Lobster, pp 288-366 (1909).
14. L. W. Williams, The anatomy of the common squid,
(Lolligo Pealli), Brill. Leiden (1909).
15. C.J.G. van der Horst and A. Brand, The occurrence
of hypotaurine in ovarian follicle walls in sheep and
other animals, Zuchthyg. 6: 97 (1971).
16. T. D. Glover and L. Nicander, Some aspects of
structure and function in the mammalian epididymis,
J. Reprod. Fertil. (Suppl) 13: 39 (1971).
17. J. A. Blaquier, M.S. Cameo and M. H. Burgos, The
role of androgens in the maturation of epididymal
spermatozoa in the guinea pig, Endocrinol. 90: 839 (1972).
18. C.J.G. van der Horst and H.J.G. Grooten, The
occurrence of hypotaurine and other sulfur containing
amino acids in seminal plasma and spermatozoa of boar,
bull and dog, Biochem. Biophys. Acta 117: 495 (1966).
19. L. A. Johnson, V. G. Pursell, R. J. Gerrits and
C. H. Thomas, Free amino acid composition of porcine
seminal, epididymal and seminal vesicle fluids, J.
Anim. Sci. 34: 430 (1972).
20. J. G. Jacobsen and L. H. Smith, Jr., Biochemistry
and physiology of taurine and taurine derivatives,
Physiol. Rev. 48: 422 (1968).
21. A.P.M. Lockwood, The osomergulation of crustacea,
Biol. Rev. 37: 257 (1962).

HYPOTAURINE IN THE REPRODUCTIVE TRACT

C.J.G.van der Horst

Laboratory Biochemistry of Reproduction
Oudwijk 11, 3581 TE Utrecht
The Netherlands

INTRODUCTION

Hypotaurine was first detected in boar semen (Horst and Grooten, 1966). The free amino acid composition of boar seminal plasma was examined by paper electrophoresis at pH 4.5 and at pH 1.9 followed by ascending chromatography in the solvent system butanol-acetic acid-water (12:3:5). The amino acids were detected with ninhydrin. Leucine, phenylalanine, valine, alanine, glycine, serine and glutamic acid were present in readily detectable amounts; basic amino acids were only found in very small quantities. The most striking phenomenon, however, was a very large spot of an amino acid, which after electrophoresis at pH 4.5 appeared in the neutral fraction approximately in the same position as glycine, but, after electrophoresis at pH 1.9, the spot was found far to the left of glycine and of glutamic acid. This showed that the unknown amino acid would contain, besides the amino group, an acidic group stronger than the carboxyl group as e.g. a sulfonic acid group. Micro reactions according to Feigl (1960) established the presence of sulfur containing groups. Also thanks to the report of the chromatographic behaviour of eight sulfur containing amino acids, all related to cysteine, given by Italian investigators (De Marco et al., 1965) we were able to show that the unknown amino acid was hypotaurine. Later on the presence of hypotaurine in boar semen and in semen of other species of animals has been confirmed by other investigators (among others Johnson, 1972 and Kochakian, 1973).

DETERMINATION OF THE AMINO ACIDS

The quantities of leucine, valine, alanine, glycine, glutamic acid and of hypotaurine were determined by a semi-quantitative

method (accuracy about 10 %). After spraying the papers with
ninhydrin and copper nitrate solution, the red spots were eluted
with 3 ml methanol and then the extinctions were measured at 525 nm.
The present amounts could be calculated using standard curves.
Hypotaurine could also be titrated with 0.004 N $KMnO_4$ solution after
elution from the papers with water (Grooten, 1967).

SEMEN AND THE MALE GENITAL TRACT

Up to then glutamic acid has always been considered as the main
amino acid in boar semen and indeed, also according to our determinations the concentration of glutamic acid in boar seminal plasma
is about twice as high as that of hypotaurine (the average values
of 15 samples were 4.8 and 2.2 mg % respectively). Inside the
spermatozoa, however, hypotaurine was the most important amino acid;
it dominated by far.

Then experiments were carried out to establish where hypotaurine
is formed. For this purpose testis and epididymis were homogenized
with saline, after which the amino acids were investigated in the
same way as mentioned above (Grooten, 1967). It appeared that
hypotaurine was present in hardly detectable amounts in the testis
and in that part of the caput epididymidis adjacent to the testis;
in the rest of the epididymis and particularly in corpus and cauda
considerable larger quantities were found. In Figure 1 the quantities
of hypotaurine and some other amino acids are given relative to
alanine; the aim of these experiments was solely to investigate
where and to what extent hypotaurine is formed. If it is taken that

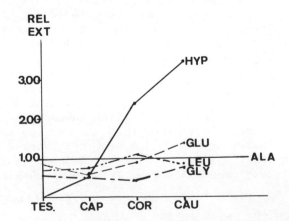

Fig. 1.   The occurrence of hypotaurine relative to ala, glu, leu
          and gly in the testicle and in the caput, corpus and cauda
          epididymidis of the boar.

the extinction of alanine spots on the chromatograms is equal to 1, then the relative extinction of hypotaurine in the testis appeared to be approximately 0.02, whereas in the corpus and cauda it varied from 2.0 to 6.0. Extinctions of the other amino acids do not exhibit this increase with the exception of that of glutamic acid. The increase in the latter is, however, smaller than in the case of hypotaurine.

In order to determine whether the production of hypotaurine is dependent on the presence of spermatozoa, an experiment was carried out with young boars, starting at the age of four weeks. Figure 2 shows that in the 4 weeks old boar extremely little hypotaurine was present in the cauda epididymidis (the relative extinction was 0.04) but the extinction then increased steadily in the ages up to about 16 weeks. In the weeks which followed, when spermatozoa began to be visible, the relative hypotaurine extinction remained about equal to that of glutamic acid but after the age of about 24 weeks more hypotaurine was formed. From this age onwards the hypotaurine extinction was always higher than that of glutamic acid. These experiments, therefore, make clear that the formation of hypotaurine occurs at all ages, independent of the presence of spermatozoa. The formation of hypotaurine might be influenced by testosterone (Grooten, 1967, Kochakian, 1973).

Finally an attempt was made to obtain spermatozoa from caput epididymidis and from corpus and cauda. After homogenization with water the amino acids were investigated in the known way. The relative extinctions are given in Figure 3 for 2 boars. In the caput the glutamic acid content is higher than that of hypotaurine but in the cauda the reverse is the case. It is apparent that in boar the hypotaurine content in spermatozoa rises strongly during their

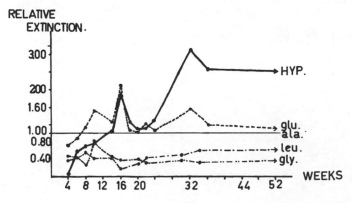

Fig. 2.   The occurrence of hypotaurine relative to ala, glu, leu and gly in the cauda epididymidis of boars from the age of 4 weeks onwards.

passage through the epididymis and this must be an important aspect
of the maturation process of boar spermatozoa.

Hypotaurine is also found in the epididymis of bull, stallion
and dog. The amounts, however, are considerable smaller than in boar,
particularly in the dog, where more cysteine is found. The relative
extinction of cysteine in testis is in boar as well as in bull and
stallion much higher than in cauda epididymidis. This leads to the
supposition that hypotaurine is formed from cysteine.

Hypotaurine does not occur in the semen of all species of
animals. We ourselves could show it in boar, bull, stallion and
ram semen in well measurable amounts, but we could not detect it in
turkey semen, in cock semen and also not in human semen. When hypotaurine was not found, cysteine appeared to be present. In turkey
semen cysteine occurs exclusively inside the spermatozoa (Litjens
and Horst, 1973).

THE FEMALE GENITAL TRACT

The Fallopian tube and uterus form the biochemical environment
in which fertilization and early embryonic development occur. We
compared amongst other things the free amino acid content in the

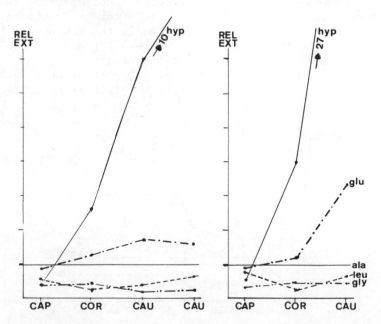

Fig. 3.   The occurrence of hypotaurine relative to ala, glu, leu
          and gly in boar spermatozoa prepared from caput, corpus
          and cauda epididymidis.

Fallopian tube of ewes during various stages of the oestrous cycle
(Horst and Brand, 1969). The day of oestrus was established by means
of a vasectomized ram twice daily. Immediately after slaughter the
tubes were removed and after homogenization with water the amino
acids were examined in the same way as mentioned for semen. A
striking feature was the strong resemblance between the amino acid
pattern on the chromatograms made with the Fallopian tube and that
with normal ram seminal plasma. The occurrence of hypotaurine in the
Fallopian tube is surprising; it is the only amino acid the concentration of which is influenced by the oestrous cycle. From Figure 4
it can be seen that the greatest amounts of hypotaurine (maximum
concentration was 22 mg %) are found during the first and the last
days of the cycle. Ova enter the uterus until about 80 to 90 hours
after the onset of oestrus in the ewe. Passage of the ovum into the
Fallopian tube therefore coincides with a period of maximum concentration of hypotaurine. Hypotaurine may serve as an antioxidant; it
can easily be set free. We rinsed out the Fallopian tube with water
on day 12 and on day 0 of the cycle. The first solution contained

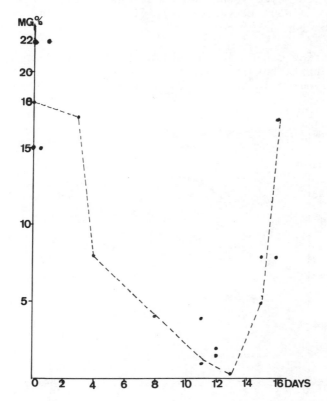

Fig. 4.   Hypotaurine content (expressed as mg/100 g wet tissue)
          in the Fallopian tube of the ewe at various stages of the
          oestrous cycle.

0.6 mcg and the second solution 24 mcg hypotaurine.

Small amounts of hypotaurine were found now and then in the uterus, mostly in the second part of the cycle, but the number of findings is too low to draw any conclusion, though it points to the possibility that it might play a role in early embryonic development.

In the pig (Horst and Kuiper, 1972) hypotaurine was found in the Fallopian tube from about D-0 to D-5 in rather large amounts and further in smaller amounts up to D-8. While taurine was found during the whole cycle, hypotaurine did not appear again before D-17 of the cycle and the amounts increased to D-21, the last day or the first day of the cycle. In the uterus hypotaurine was found in decreasing amounts from about D-14 to D-17 and at the same time taurine and cysteic acid were found in increasing amounts; it seems that hypotaurine is actually being oxidized. These findings show that both in the ewe and in the pig hypotaurine is found in the uterus at the same part of the cycle (the length of the cycles of ewe and sow are about 16 and 21 days respectively), viz. at the time when the corpora lutea begin to regress.

Hypotaurine could also be shown in the Fallopian tube of cow, horse and rabbit, but it could not be detected in the rat, turkey and hen. Curiously in these species of animals hypotaurine was also

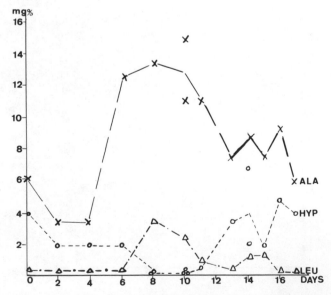

Fig. 5.   The contents of hypotaurine, alanine and of leucine (expressed as mg/100 g wet tissue) in the ovarian follicle walls of the ewe at various stages of the oestrous cycle.

not found in the semen. There seems to be an evident resemblance between the amino acid composition of semen and of the Fallopian tube in the same species of animal.

After the detection of the occurrence of hypotaurine in the Fallopian tube and its regulation during the oestrous cycle in sheep, an experiment was started for the presence of hypotaurine in different parts of sheep ovary during the cycle (Horst and Brand, 1971). Hypotaurine could never be detected in sheep follicular fluid, but it did occur in the follicle wall. The concentration varied from 0 - 7 mg %. In Figure 5 the course of the concentration of alanine, leucine and of hypotaurine are given. It is evident that also in the follicle wall hypotaurine is exceptional among the other amino acids investigated. In midcycle when the concentration of the normal amino acids reach their highest values (only 2 are given in Figure 5), hypotaurine could hardly be detected and the greatest amounts were found during the last days of the cycle and on the day of oestrus. In the corpora lutea hypotaurine could be shown, but the concentrations were very low; they varied from 0 - 0.1 mg %. Also in the remaining parts of the ovaries hypotaurine was found only in traces and particularly when follicles were still present. It can be concluded, therefore, that in sheep ovary hypotaurine is exclusively formed in the follicle wall and that the concentration is regulated by the oestrous cycle. The values are lowest when the progesterone level in the corpora lutea is high, the same as in the Fallopian tube. Hypotaurine was also found in sheep fimbriae round the day of oestrus.

Summarising it is evident that in sheep female tract hypotaurine is present in the follicle wall chiefly during the last days of the oestrous cycle and on the day of oestrus. Before ovulation the egg lies at one side of the ovarian follicle embedded in a solid mass of follicular cells called the cumulus oöphorus. The egg has two distinct membranes from which one, the zona pellucida is penetrated by egg and granulosa cell microvilli which may provide biochemical communication between the oocyte and the follicle (Cole and Cupps, 1977). In this way hypotaurine might play a role in the protection of the egg. After ovulation the egg comes in contact with the ciliated cells lining the fimbriae and there it finds hypotaurine again. Then it passes into the Fallopian tube where on those days hypotaurine has its greatest concentration. The spermatozoa which penetrate in the Fallopian tube most probably contain still their hypotaurine; it seems therefore that hypotaurine will be important for the fertilization process, probably as an anti-oxidant.

POSSIBLE INFLUENCE OF HYPOTAURINE ON FERTILIZATION

The idea that hypotaurine will influence fertilization is supported by findings made with turkey semen. As already mentioned

turkey spermatozoa contain cysteine instead of hypotaurine and
normally cysteine does not occur in seminal plasma. When leakage of
cysteine out of the spermatozoa had taken place fertilization did
not occur. Many investigators looked in vain for a suitable diluent
for turkey semen, but after addition of cysteine in order to prevent
leakage of cysteine out of the spermatozoa a diluent was found with
which very satisfying fertility results were obtained (Litjens,
1973, Litjens and Horst, 1973). The first A.I. centre for turkeys
could be set up. We had the impression that in boars with a low
fertility rate relatively more hypotaurine was present in the
seminal plasma and smaller amounts inside the spermatozoa.

Formation of hypotaurine in the Fallopian tube is regulated
during the oestrous cycle, most probably by progesterone. It seems
possible, therefore, that during experiments with progesterone-
treated animals (as for instance carried out with rabbits by
Chang, 1969) the formation of hypotaurine is retarded or reduced.
Lack of hypotaurine might contribute to a change of the biochemical
environment of the genital tract in such a way, that development
or transport of eggs may be influenced unfavourably.

Then a small number of ewes became available which had been
treated for 13 days with intravaginal sponges containing 60 mg of
6α-methyl-17α-acetoxyprogesterone (MAP), applicated in the periods
D-11 to D-13 and D-2 to D-5 respectively. The animals were investigated immediately after removal of the sponge and 24, 48 and 72
hours later. The hypotaurine content in the Fallopian tube of
animals with sponge application in the period D-11 to D-13 was
strongly lowered and that after sponge application in the period
D-2 to D-5 was but little less than in the untreated ewes. Willemse
(1968) had shown that application of MAP for 13 days when started
in the period D-8 to D-14 resulted in a significant lowered fertility
rate with respect to application of MAP in the period D-1 to D-7.
This suggests the hypothesis that the observed decreased fertility
rate after sponge application in the period D-8 to D-14 might be
due partly to the strongly lessened hypotaurine concentrations
(Horst and Brand, 1971).

Other experiments showed that there is also a cyclical change
in carbohydrate metabolism in the uterus of cow, pig and dog
(Horst, 1973, Horst and Brand, 1974, Zikken, 1978, Horst and Vogel,
1979). It would be interesting to know the influence of treatments
with progesterone- and oestradiol- derivatives or with other
specialities on the biochemical environment both in Fallopian tube
and in uterus. A similar investigation might lead to a new field
of contraceptive intervention.

REFERENCES

Chang, M. C., 1969, Fertilization, Transportation and Degeneration
    of Eggs in Pseudo-pregnant or Progesterone-treated Rabbit,
    Endocrinology, 84:356
Cole, H. H. and Cupps, P. T., 1977, Reproduction in Domestic Animals,
    Academic Press, New York.
De Marco, C., Mosti, R. and Cavallini, D., 1965, Column chromatogra-
    phy of some sulfur containing amino acids. J. Chromatog.
    18:492.
Feigl, A., 1960, Spot tests in organic analyses, Elsevier, Amsterdam.
Grooten, H. J. G., 1967, An investigation into the occurrence of
    amino acids, chiefly of hypotaurine, and of keto acids in
    semen and reproductive tract of the boar, Thesis, Utrecht.
Horst, C. J. G. van der and Grooten, H. J. G., 1966, The occurrence
    of hypotaurine and other sulfur-containing amino acids in
    seminal plasma and spermatozoa of boar, bull and dog,
    Biochim. Biophys. Acta, 117:495.
Horst, C. J. G. van der and Brand, A., 1969, Occurrence of hypotau-
    rine and inositol in the reproductive tract of the ewe and
    its regulation by pregnenolone and progesterone, Nature,
    223:67.
Horst, C. J. G. van der and Brand, A., 1971, Preliminary investi-
    gation into the effect of administration of MAP on the for-
    mation of oestradiol-17β, hypotaurine and inositol in sheep,
    Tijdschr. Diergeneesk., 96:1291.
Horst, C. J. G. van der and Brand, A., 1971, The occurrence of
    hypotaurine in ovarian follicle walls in sheep and in the
    Fallopian tube of sheep and other animals, Zuchthyg., 6:97.
Horst, C. J. G. van der and Kuiper, C. J., 1972, Investigation into
    the occurrence of some steroids, amino acids and carbohydrates
    on specific days of the oestrous cycle of the pig. Neth. J.
    vet. Sci., 5:35.
Horst, C. J. G. van der, 1973, Glucose metabolism in the female tract
    and the inhibitory hormone dependent effect of glucuronic
    acid by linkage to carboglutelin, Cytobios, 8:15.
Horst, C. J. G. van der and Brand, A., 1974, Carbohydrates in the
    lumen of the cow during the oestrous cycle, Congress of the
    Sterility and Fertility, London.
Horst, C. J. G. van der and Vogel, F., 1979, Cyclical changes in the
    carbohydrate composition of dog endometrium, Cytobios, in
    press.
Johnson, L. A., Pursell, V. G., Gerrits, R. J. and Thomas, C. H.,
    1972, Free amino acid composition of porcine seminal, epidi-
    dymal and seminal vesicle fluids, J. Animal Sci., 34:430.
Kochakian, C. D., 1973, Hypotaurine : Regulation of production in
    seminal vesicles and prostate of Guinea-pig by testosterone,
    Nature, 241:202.

Litjens, J. B. and Horst, C. J. G. van der, 1973, Artificial inse-
    mination in turkeys and investigation into a diluent of
    turkey semen suitable for field use, Neth. J. vet. Sci., 5:
    148.
Litjens, J. B., 1973, An investigation into the occurrence of car-
    bohydrates and amino acids in oviduct and semen of turkey
    and into an efficient diluent for semen, Thesis, Utrecht.
Willemse, A. H., 1968, Relation between the day of the oestrous
    cycle at the time of the intravaginal application of a MAP
    impregnated sponge and synchronization rate and conception
    rate in Texel sheep, Vle Cong. Intern. Reprod. Anim. Insem.
    Artif., Paris, 11:1539.
Zikken, A., 1978, Some clinical and biochemical aspects of post-
    parturient cows, Thesis, Utrecht.

TAURINE IN HUMAN NUTRITION AND DEVELOPMENT

Gerald E. Gaull[1,2] and David K. Rassin[1,3]

[1]Department of Human Development and Nutrition
New York State Institute for Basic Research in Mental
Retardation, Staten Island, New York 10314
and
Departments of Pediatrics[2] and Pharmacology[3]
Mount Sinai School of Medicine of the City University
of New York, New York, New York

Despite intensive studies by a number of investigators the functions of taurine, and, in particular, the reason for the large amounts found in most mammalian tissues remain unknown. Taurine has been associated with many different biological functions (Jacobsen and Smith, 1968; Huxtable and Barbeau, 1976; Barbeau and Huxtable, 1978) but only its role in conjugating with bile acids is well established. We have concentrated upon possible functions of taurine during development. If taurine is of importance to the brain it is during this vulnerable period that a failure to supply adequate amounts of this compound might have the most long-lasting effects. Thus, the study of the transfer of taurine to the fetus and taurine nutrition in the neonate is of considerable potential importance.

The high concentration of taurine in brain during fetal development of the rat, monkey, rabbit and man at a time when the presumed synthetic pathway, via cysteinesulfinic acid decarboxylase, has little measurable activity implies that a dietary source of taurine is necessary (Sturman et al., 1977b, 1978a; Gaull and Rassin, 1979). In studies designed to investigate the effects on preterm infants of feeding formulas containing different protein quantity and quality and comparing these infants with infants fed pooled human milk, an interesting pattern of plasma and urine amino acid concentrations was observed. The amino acid concentrations in plasma and urine generally were either increased or unchanged when compared with those infants fed pooled human milk (Rassin et al., 1977a,b). The striking

exception was taurine (Gaull et al., 1977). The concentration of
taurine in the urine of infants fed casein-predominant formulas was
lower from the first week of study than that of infants fed pooled
human milk. The plasma taurine decreased steadily and by the fourth
week of study was significantly lower in the plasma of infants fed
formulas than it was in infants fed the pooled human milk. Preliminary studies in term infants fhow generally similar results (Rassin
et al., 1979). Furthermore, as we measure it, the concentration of
taurine in plasma and urine of breast-fed term infants is considerably higher than that of pre-term infants fed pooled human milk at a
fixed volume. The reason for this difference is not as yet clear.
In effect, however, the differences in plasma and urine taurine concentration between the infants fed formulas and those fed human milk
are greater in term infants than they are in preterm infants. The
small concentrations of taurine in plasma of infants fed artificial
formulas has been documented by others but not commented on (Dickinson
et al., 1970).

The finding of a dietary requirement for taurine in the human
infant is consistent with the negligible activity of cysteinesulfinic
acid decarboxylase present both in fetal and in mature human liver
(Gaull et al., 1977).

This evidence for the relative inability of man to synthesize
taurine and the pattern of deficiency of taurine in the plasma and
urine led us to an examination of the milk of various species, as
well as of some currently available commercial formulas. Taurine is
a major constituent of the free amino acid pool of milk in a number
of species (Rassin et al,, 1978). Human milk contains a considerable
amount of taurine whereas bovine milk, from which most of the formulas are prepared, contains only minimal amounts of taurine (Gaull
et al., 1977). The lack of taurine in the casein-predominant formulas in this study of preterm infant feeding reflected the lack of
taurine in the bovine milk from which the formulas were prepared.
The three major commercial infant formulas available in the United
States contained only small amounts of taurine, and soy-based formulas contained none at all. For most species studied, taurine is a
major constituent of the free amino acid pool of milk, and it has a
greater concentration early in lactation.

These clinical findings led us to examine the transfer of
taurine from the mother to the infant using the rat as a model. The
concentration of taurine in rat milk is considerable for the first
few days after birth. At this time it is the ninhydrin positive compound present in highest concentration. The concentration of taurine
in milk decreases rapidly thereafter, and by a week after birth it
has reached an approximate constant value (Sturman et al., 1977a).
The total amount of taurine transferred to the pup via the milk decreases initially but increases again by the first week after birth

because there is an increase in the quantity of milk ingested. We studied the transfer, via milk to brain and liver of the pups of [^{35}S]taurine injected into the mother 6-8 hours after birth. Labelled taurine from the mother was secreted in the milk and accumu- lated in the liver and brains of the pups. We calculated that approximately 5 micromoles of taurine is supplied to each pup during the first five days, and a minimum of 0.5 micromoles ended up in the brain during this period (Sturman et al., 1977a). Thus even in the developing rat, which has a greater capacity for biosynthesis of taurine than most other species, dietary taurine is a significant source of tissue taurine, especially that of brain. It seems likely, therefore, that there is a dietary requirement for taurine in the rapidly growing human infant. Indeed, even in adult man only 1% of an oral load of L-cystine is recovered as increased urinary excre- tion of taurine, giving further evidence for a relatively limited ability to synthesize taurine. The limited ability of adult man to convert dietary cystine to taurine is in striking contrast with the considerable ability of the rat to convert dietary cysteine to taurine (Swan et al., 1964; Sturman and Cohen, 1971).

To date no adverse clinical signs attributed to taurine defi- ciency have been identified in human infants fed commercial formulas, although no systematic investigations have been made. There are a number of lines of evidence, however, which suggest that such inves- tigations should be made.

Cats and kittens fed a synthetic diet containing partially- purified casein as the source of protein become taurine deficient and develop retinal degeneration. This degeneration, which even- tually results in blindness, can be prevented or reversed by supple- menting these diets with taurine but not by supplementation via ostensible taurine precursors, methionine, cysteine or inorganic sulfate (Hayes et al., 1975a,b; Schmidt et al., 1976).

Retina is especially dependent on taurine and actively resists taurine depletion. When the taurine concentration of most organs has decreased 10-fold that in retina has decreased less than 50%. Retinal degeneration appears to begin when the taurine concentration has been reduced by more than 50%. Prior to structural changes in the retina, changes in the electroretinogram can be detected and can be reversed by taurine supplementation (Schmidt et al, 1976). In addition, the retina of taurine-deficient kittens takes up labelled taurine more avidly than retina from control kittens (Sturman et al., 1978). Taken together these findings suggest that taurine has a role in the struc- tural integrity of the retina, especially in the photoreceptor cells, in addition to its putative role as a neurotransmitter or a neuro- modulator in this system (Pasantes-Morales et al., 1972; Mandel et al., 1976).

It is likely that the effects of feeding a taurine-deficient diet to a human infant may be considerably more moderate than those observed in the kitten for a number of reasons: The human infant conjugates bile acids solely with taurine at birth but later develops the ability to conjugate with glycine (Encrantz and Sjovall, 1959; Sjovall, 1960; Poley et al., 1964; Challacombe et al., 1975). The cat conjugates bile acids with taurine only (Rabin et al., 1976). It seems possible that the ability alternatively to conjugate bile acids with glycine serves to spare taurine in man. Infants fed human milk remain predominantly taurine conjugators of bile acids, whereas those fed taurine-deficient formulas become predominantly glycine conjugators of bile acids (Brueton et al., 1978; Watkins et al., 1979). Those human infants fed the currently available synthetic formulas often have taurine-containing foods added to the diet within a few weeks or months of birth, and the time scale for induction of retinal degeneration in the cat is of the order of months. Furthermore, the concentration of taurine in retina of most species, although high, is not as great as in the retina of the cat. This high concentration of taurine may render the cat retina especially vulnerable.

The findings that formula-fed infants tend to become predominantly glycine conjugators rather than taurine conjugators of bile acids suggest that the deficiency of taurine may not be in the best interests of the infant. The taurine conjugates may be better detergents for fat absorption in the gut because they are not precipitated at the acidic pH frequently present in the proximal small intestine (Hoffman and Small, 1967). Since taurine conjugates are fully ionized at body pH, passive absorption does not occur. A higher intraluminal concentration of bile acids can be maintained, thus facilitating the solubilizing activity of these conjugated bile acids. Taurine conjugation thus decreases passive absorption in the gut. Therefore, bile acids conjugated with taurine are more efficient than those conjugated with glycine (Schersten, 1970).

Taurine deficiency can be created in man under other circumstances. In infants fed total parenteral nutrition, in which the amino acid solution does not contain taurine, the concentration of taurine in the plasma and urine decreases (Rigo and Senterre, 1978). In a beagle pup model of total parenteral nutrition, the liver pool of taurine decreases more rapidly than the plasma pool, whereas the bile taurocholate pool is much slower to reflect taurine deficiency (Malloy et al., in preparation). In short-term experiments in adult man (up to 9 days), the concentration of taurocholate reflected the concentration of taurine available in the liver and the diet rather than that in plasma and muscle, indicating that all taurine pools in man are not in equilibrium (Hardison 1978).

The plasma concentration of cholesterol in breast-fed infants is remarkably higher than that in infants fed synthetic formulas. It

has been suggested that the human infant may have a dietary requirement for preformed cholesterol (Fomon 1974). The recent work of Nervi
and Dietschy (1978) clearly establishes the importance of conjugated
bile acids in the control of hepatic cholesterol synthesis and therefore in the control of the total body cholesterol pool. It is possible also that taurine conjugated bile acids in man are more efficient
in abeting the absorption of dietary cholesterol. The interaction of
taurine and cholesterol may be important and it is apparently species
specific. Taurine reduces serum, liver and aortic cholesterol in the
rat, an animal that primarily conjugates bile acids with taurine. It
does not reduce serum, liver and aortic cholesterol content in the
rabbit, an animal that primarily conjugates bile acids with glycine
(Hermann 1959). It is of some interest, therefore, that animals which
conjugate bile acids with taurine only are very resistant to arteriosclerosis, whereas animals which conjugate with glycine only are remarkably susceptible to arteriosclerosis. Animals such as man which
can conjugate in either way show intermediate susceptibility to arteriosclerosis, and this susceptibility seems to be related to diet.
Recent work of Stephan and Hayes (1979) in the cebus monkey suggests
that taurine may facilitate high density lipoprotein-cholesterol removal. They suggest that it does this by enhancing incorporation of
cholesterol into bile acids. The concept of the long-term assessment
of the relationship of taurine nutrition to later disease is one which
is of particular interest. Certainly the relationship of taurine to
the absorption of lipids, sterols and fat-soluble vitamins in human
infants is an area of considerable importance.

Two recent papers also suggest that dietary taurine may interact
with genetic factors in the production of disease. Taurine supplementation of the diet may reduce the development of hypertension in
strains of mice genetically susceptible to hypertension. Taurine is
apparently not involved in the normal regulation of blood pressure;
however, in the genetically susceptible rat the lack of dietary
taurine interacts with genetic factors and results in hypertension
(Nara et al., 1978). Another example of the interaction of genetic
susceptibility and dietary taurine is found in a strain of mice
susceptible to gall stones (Fujihara et al., 1978). The strain is
genetically prone to cholelithiasis when given an atherogenic diet
containing cholesterol and cholic acid. The formation of gall stones
in these animals is reduced by including 2% taurine in the drinking
water. The interaction of taurine with epinephrine at the beta receptors both in the heart (Chubb and Huxtable, 1978) and in the pineal
gland (Wheler et al., 1979) suggest that it may well have considerable
importance in cardiac and other β-receptor function. Taurine is known
to accumulate in higher concentration in the ventricles of heart
failure both in man and in animals (Huxtable and Bressler, 1974).
There is also recent evidence for taurine binding to proteins in
ventricular sarcolemma (Kulakowski et al., 1978).

There is evidence that taurine is useful in determining current nutritional status in children. In a recent investigation of plasma amino acids in malnourished children five groups were studied: Normal (Group 1), well-nourished but suffering from an infection (Group 2), well-nourished but having fasted for 2-4 days due to infection (Group 3), moderate protein-calorie malnutrition (Group 4) and severe protein-calorie malnutrition (Group 5). The diets for the control groups were not fully described; nevertheless, plasma taurine concentrations were far lower in the malnourished individuals Group 4 and 5, than they were in Groups 1-3. These differences were more dramatic than those observed for a number of the classical indicies of malnutrition based on various ratios of other plasma amino acids (Ghisolfi et al., 1978). It is the conclusion of these authors that plasma taurine may be the best discriminant between the normal or acutely fasting and the malnourished state in children.

Finally, there is interesting new evidence suggesting a biological role of fundamental nutritional importance of taurine in the spiny lobster. The antennular system of this lobster contains specific taurine receptors which are insensitive to alpha amino acids, which are common constituents of seawater (Fuzzessery et al., 1978). These antennular taurine receptors may have specifically evolved to allow this animal to identify food sources. For the lobster, food usually contains taurine as one of the five most abundant amino acids, against a background of other amino acids present in the lobster's environment. The use of taurine to identify food may be the earliest important nutritional function of this compound.

REFERENCES

Barbeau, A., and Huxtable, R. J., 1978, "Taurine and Neurological Disorders," Raven Press, New York.
Brueton, M. J., Berger, H. M., Brown, G. A., Ablitt, L., Iyangkaren, N., and Wharton, B. A., 1978, Duodenal bile acid conjugation patterns and dietary sulphur amino acids in the newborn, Gut 19:95.
Challacombe, D. N., Edkins, S., and Brown, G. A., 1975, Duodenal bile acids in infancy, Arch. Dis. Child. 50:837.
Chubb, J., and Huxtable, R., 1978, Isoproterenol-stimulated taurine influx in the perfused rat heart, Europ. J. Pharmacol. 48:369.
Dickinson, J. C., Rosenblum, H., and Hamilton, P. B., 1970, Ion exchange chromatography of the free amino acids in the plasma of infants under 2,500 gm at birth. Pediatrics 45:606.
Encrantz, J. D., and Sjovall, J., 1959, On the bile acids in duodenal contents of infants and children. Clin. Chim. Acta 4:793.
Fomon, S. J., 1974, "Infant Nutrition" (2nd ed.) pp. 172-174, Saunders Company, Philadelphia, Pa.

Fujihara, E., Daneta, S., and Oshima, T., 1978, Strain difference in
     mouse cholelithiasis and the effect of taurine on the gall-
     stone formation in C$_{57}$BL/C mice, Biochem. Med. 19:211.

Fuzzessery, Z. M., Carr, W. E. S., and Ache, B. W., 1978, Antennular
     chemosensitivity in the spiny lobster, Panuliris argus:
     Studies of taurine sensitive receptors, Biol. Bull. 154:226.

Gaull, G. E., and Rassin, D. K., 1979, Taurine and brain development:
     Human and animal correlates, in "Developmental Neurobiology,
     IBRO Symposium No. 5" E. Meisami and M. A. B. Brazier, eds.,
     Raven Press, New York, in press.

Gaull, G. E., Rassin, D. K., Raiha, N. C. R., and Heinonen, K., 1977,
     Milk protein quantity and quality in low-birth-weight infants.
     3. Effects on sulfur amino acids in plasma and urine,
     J. Pediatr. 90:348.

Ghisolfi, J., Charlet, P., Ser, N., Salvayre, R., Thouvenot, J. P.,
     and Duole, C., 1978, Plasma free amino acids in normal children
     and in patients with protein-calorie malnutrition: Fasting and
     infection, Pediat. Res. 12:912.

Hardison, W. G. M., 1978, Hepatic taurine concentration and dietary
     taurine as regulators of bile acid conjugation with taurine,
     Gastroent. 75:71.

Hayes, K. C., Carey, R. E., and Schmidt, S. Y., 1975a, Retinal degen-
     eration associated with taurine deficiency in the cat, Science
     188:949.

Hayes, K. C., Rabin, A. R., and Berson, E. L., 1975b, An ultrastruc-
     tural study of nutritionally induced and reversed retinal de-
     generation in cats, Am. J. Pathol. 78:505.

Herrmann, R. G., 1959, Effect of taurine, glycine and β-sitosterols on
     serum and tissue cholesterol in the rat and rabbit, Circ. Res.
     7:224.

Hofmann, A. F., and Small, D. M., 1967, Detergent properties of bile
     salts: Correlation with physiological function. Ann. Rev. Med.
     18:333.

Huxtable, R., and Barbeau, A., (eds.) 1976, "Taurine" Raven Press, New
     York.

Huxtable, R., and Bressler, R., 1974, Taurine concentrations in con-
     gestive heart failure, Science 184:1187.

Jacobsen, J. G., and Smith, L. H., Jr., 1968, Biochemistry and physi-
     ology of taurine and taurine derivatives, Physiol. Rev. 48:424.

Kulakowski, E. C., Maturo, J., and Schaffner, S. W., 1978, The iden-
     tification of taurine receptors from rat heart sarcolemma,
     Biochem. Biophys. Res. Comm. 80:936.

Mandel, P., Pasantes-Morales, H., and Urban, P. F., 1976, Taurine, a
     putative transmitter in retina, in "Transmitters in the Visual
     Process," S. L. Bonting, ed., Pergamon Press, New York.

Nara, Y., Yamori, Y., and Lovenberg, W., 1978, Effect of dietary
     taurine on blood pressure in spontaneously hypertensive rats,
     Biochem. Pharmacol. 27:2689.

Nervi, F. O., and Dietschy, J. M., 1978, The mechanisms and the interrelationship between bile acid and chylomicron-mediated regulation of hepatic cholesterol synthesis in the liver of the rat. J. Clin. Invest. 61:895.

Pasantes-Morales, H., Klethi, J., Urban, P. F., and Mandel, P., 1972, The physiological role of taurine in retina: Uptake and effect on electroretinogram (ERG), Physiol. Chem. Phys. 4:339.

Poley, J. R., Dower, J. C., Owen, C. A., and Stickler, G. B., 1964, Bile acids in infants and children, J. Lab. Clin. Med. 63:838.

Rabin, B., Nicolosi, R. J., and Hayes, K. C., 1976, Dietary influence on bile acid conjugation in the cat, J. Nutr. 106:1241.

Rassin, D. K., Gaull, G. E., Heinonen, K., and Raiha, N. C. R., 1977a, Milk protein quantity and quality in low-birth-weight infants. 2. Effects on selected aliphatic amino acids in plasma and urine, Pediatrics 59:407.

Rassin, D. K., Gaull, G. E., Raiha, N. C. R., and Heinonen, K., 1977b, Milk protein quantity and quality in low-birth-weight infants. 4. Effects on tyrosine and phenylalanine in plasma and urine, J. Pediatr. 90:356.

Rassin, D. K., Sturman, J. A., and Gaull, G. E., 1978, Taurine and other free amino acids in milk of man and other mammals, Earl Hum. Develop., 2:1.

Rassin, D. K., Jarvenpaa, A.-L., Raiha, N. C. R., and Gaull, G. E., 1979, Breast feeding versus formula feeding in full-term infants: Effects on taurine and cholesterol, Pediat. Res. 13: 406.

Rigo, J., and Senterre, J., 1977, Is taurine essential for the neonate? Biol. Neonate 32:73.

Schersten, T., 1979, Bile acid conjugation, in "Metabolic Conjugation and Metabolic Hydrolysis," W. H. Fishman, ed., Academic Press, New York.

Schmidt, S. Y., Berson, E. L., and Hayes, K. C., 1976, Retinal degeneration in cats fed casein. I. Taurine deficiency, Invest. Ophthalmol. 15:47.

Sjovall, J., 1960, Bile acids in man under normal and pathological conditions, Clin. Chim. Acta 5:33.

Stephan, Z. F., and Hayes, K. C., 1979, Dietary taurine and hepatic clearance of HDL vs. LDL in Cebus monkeys, Fed. Proc. 38:386.

Sturman, J. A., and Cohen, P. A., 1971, Cystine metabolism in vitamin B_6 deficiency, Evidence of multiple taurine pools. Biochem. Med. 5:245.

Sturman, J. A., Rassin, D. K., and Gaull, G. E., 1977a, Taurine in developing rat brain: Transfer of [^{35}S]taurine to pups via the milk, Pediatr. Res. 11:28.

Sturman, J. A., Rassin, D. K., and Gaull, G. E., 1977b, Taurine in development, Life Sci. 21:1

Sturman, J. A., Rassin, D. K., and Gaull, G. E., 1978a, Taurine in the development of the central nervous system, in "Taurine and Neurological Disorders," A. Barbeau and R. J. Huxtable, eds.,

Raven Press, New York.

Sturman, J. A., Rassin, D. K., Hayes, K. C., and Gaull, G. E., 1978b, Taurine deficiency in the kitten: Exchange and turnover of [35S]taurine in brain, retina and other tissues, J. Nutr., 108:1462.

Swan, P., Wentworth, J., and Linksweiler, H., 1964, Vitamin B_6 depletion in man: Urinary taurine and sulfate excretion and nitrogen balance, J. Nutr. 84:220.

Watkins, J. B., Jarvenpaa, A.-L., Raiha, N., Szczepanik Van-Leween, P., Klein, P. D., Rassin, D.K., and Gaull, G., 1979, Regulation of bile acid pool size: Role of taurine conjugates, Pediat. Res. 13:410.

Wheler, G. H. T., Weller, J. L., and Klein, C., 1979, Taurine: Stimulation of pineal N-acetyltransferase activity and melatonin production via a beta-adrenergic mechanism, Brain Res., 166:65.

THE EFFECT OF TOTAL PARENTERAL NUTRITION ON TAURINE

Michael H. Malloy[1,2], Gerald E. Gaull[1], William C. Heird[2], and David K. Rassin[1]

Department of Human Development and Nutrition[1]
New York State Institute for Basic Research in Mental
Retardation, Staten Island, New York 10314
and
Department of Pediatrics[2], Columbia University College of
Physicians and Surgeons, New York, New York 10032

Nutrition during early growth and development has received considerable attention because of the observed relationship between malnutrition in young animals and subsequent abnormal brain development (Dobbing et al., 1971; Winick and Rosso, 1969). Therefore, attempts have been made to provide nutritionally complete regimens to human infants as early in life as possible. Very premature infants and infants with gastrointestinal abnormalities, however, are sometimes unable to achieve an adequate nutritional intake by the oral route. Total parenteral nutrition using a central venous infusion of a hypertonic glucose solution, a free amino acid solution, a fat emulsion, and minerals and vitamins together has provided an alternative means of nutrition for these infants (Heird and Winters, 1975).

Total parenteral nutrition (TPN) in human infants optimally should provide all of the nutrients essential for growth and development. The requirement for taurine during infancy, however, is uncertain. It is not surprising, therefore, to find that taurine is absent from amino acid solutions used for TPN. Regardless of whether or not taurine is essential for growth and development during infancy, it is apparent that taurine pools, both in infants and in adults, are supplied predominantly by exogenous sources (Gaull et al., 1977). Plasma taurine concentrations have been observed to decrease in infants, both those receiving enteral and those receiving parenteral diets deficient in taurine (Gaull et al., 1977; Rigo and Senterre, 1977). The requirement for an exogenous source of taurine is attributed to the low

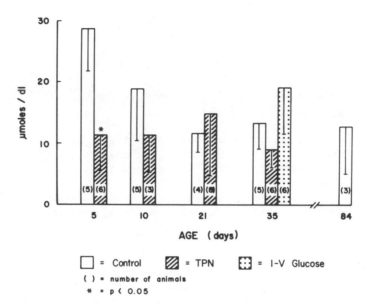

Figure 1:   Controls were nursed until the age indicated. Experimental
            animals received TPN as follows: 5 day old animal, TPN
            from 0-5 days; 10 day = 0-10 days; 21 day = 10-21 days;
            35 day = 21-35 days (for both TPN and I-V glucose).

activity, both in infants and in adults, of hepatic cysteinesulfinic
acid decarboxylase (CSAD), the enzyme in man probably responsible for
catalyzing taurine synthesis from cysteine (Gaull et al., 1977).

     The functional significance of decreases in plasma taurine is
uncertain. As the only known biochemical function of taurine is conjugation with bile acids (Encrantz and Sjovall, 1959), measurement of
taurine pools in liver and bile could be helpful in assessing the
potential functional effects of a taurine-deficient diet. Measurement
of the cerebral pool of taurine might also be helpful in assessing the
availability of taurine for its putative function as a neurotrans-

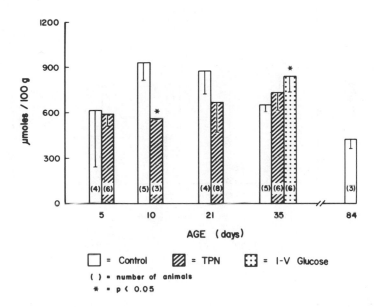

Figure 2: See legend to figure 1.

mitter or neuromodulator. It is difficult in human infants, however, to measure the liver and cerebral taurine pools. These tissues were, therefore, investigated in beagle puppies given a taurine-deficient parenteral diet.

Total parenteral nutrition was administered to beagle puppies through a central venous catheter. The solution provided 180-200 kcal/kg/day, 6 g free amino acids/kg/day, minerals and vitamins. Periods of TPN were carried out in the puppies from 0-4, 0-10, 10-21, and from 21-35 days of age. A second parenteral feeding regimen consisted of 60 kcal/kg/day of glucose from 21-35 days of age. Nursed beagle puppies were used as age-matched controls.

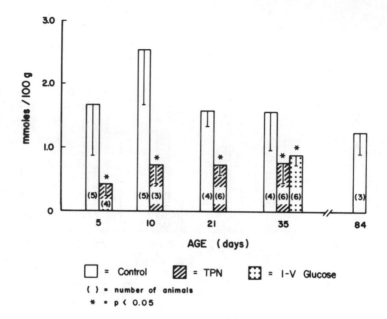

Figure 3:   See legend to figure 1.

As has been described for other species (Sturman et al., 1977), the developmental pattern of most of the taurine pools in the normally fed beagle puppy is one of decreasing concentrations with increasing age.   Although hepatic and cerebral taurine concentrations in the beagle appeared to increase during the first 10 days of life they later followed the general pattern of decreasing concentrations with age as observed in the plasma (Figs. 1, 2 & 3).   The concentration of taurine-conjugated bile salts in gall bladder bile increased with age, the only instance in which such a pattern of increase was observed in the beagle puppy (Fig. 4).   Because the dog conjugates bile acids predominantly with taurine (Haslewood, 1967), this development pattern might be expected.   The beagle puppy, unlike man or the cat develops substantial hepatic CSAD activity (Fig. 5).   At 4 days of age the beagle liver CSAD activity of 4 $nmCO_2$ produced/mg protein/hour is far

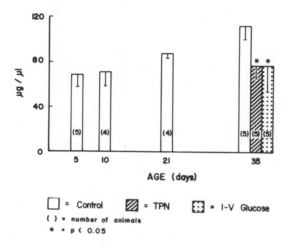

Figure 4:   See legend to figure 1.

above the activity (0.3) reported in adult man (Gaull et al., 1977) and is equal to the activity (4) reported in mature cats (Knopf et al., 1978). At 12 weeks of age the beagle has attained a hepatic CSAD activity of approximately one-fourth that reported for mature rats which are known to have a large capacity to synthesize taurine.

     Taurine-deficient diets fed to human infants for 3-4 weeks (Gaull et al., 1977) and to kittens for 10 weeks (Sturman et al., 1978) have been reported to cause a decrease in plasma taurine concentrations. Because of the limited tissues available for assaying taurine in human infants, it is difficult to determine how accurately plasma taurine concentrations reflect the effect of a taurine-deficient diet on tissue taurine concentrations. In the taurine-deficient kitten, plasma was second to liver only as the most adversely affected taurine pool (Sturman et al., 1978). The affect of a relatively short period of a

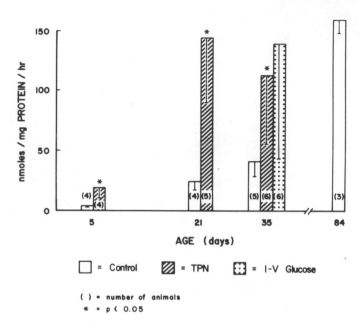

Figure 5: See legend to figure 1.

taurine-deficient diet on taurine pools is observed in plasma and
tissues of the beagle puppies given total parenteral nutrition or
only IV glucose. The youngest animals that received TPN developed
plasma taurine concentrations significantly less than those observed
in plasma of nursed puppies (Fig. 1) and the 0-10 days of age TPN
puppies developed cerebrum taurine concentrations significantly less
than cerebrum of control animals (Fig. 2). In contrast to the singu-
lar episodes of deficiencies of plasma and cerebral taurine pools in
the parenterally-nourished puppies a significant decrease of taurine
in liver was observed after every period of TPN studied (Fig. 3).

    Bile acid conjugation has been suggested to be a function of the
availability of taurine and glycine in the liver (Hardison 1978). In
the only period of TPN in which bile salts were measured, i.e. 21-35

days of age, the concentration of taurine-conjugated bile salts in
gall bladder bile of puppies receiving parenteral nutrition was
significantly less than that of controls (Fig. 4). The decreased
concentrations of taurine-conjugated bile salts in the animals on
taurine-deficient diets may well be a reflection of the limited
taurine available in the liver. Alternatively, because of the absence of an enteral stimulus the decreased taurine-conjugated bile
salt concentration may simply reflect a lack of expansion of the
taurine-conjugated bile acid pool that occurs between 21 and 35 days
of age in the control puppies. Thus, the taurine-conjugated bile salt
concentration at the end of the 21-35 day period of TPN (77.1 ± 12 μg/
ul, $\bar{x}$ ± S.D.) is comparable to the concentration of taurine-conjugated
bile salts observed in the control animals at 21 days of age (87.1 ±
4.1 μg/ul, $\bar{x}$ ± S.D.) (Fig. 4).

In the taurine-deficient kitten CSAD activity does not increase
to supplement taurine pools with endogenously synthesized taurine
(Knopf et al., 1978). The dog, however, does appear to respond to a
taurine-deficient diet with increased endogenous synthesis. Hepatic
cysteinesulfinic acid decarboxylase activity was increased during
every period of TPN examined (Fig. 5). There did appear to be a developmental limitation of the hepatic CSAD activity as the youngest
animals, 0-4 days of age, did not increase their enzyme activity to
the extent observed in the 10-21 day and 21-35 day TPN animals
(Fig. 5). In the puppies that received only IV glucose from 21-35
days of age the mean hepatic CSAD activity increased; however, due to
the large variation in activity the increase was not significant.

There is growing evidence that taurine is essential for normal
development in certain species. For example, in the taurine-deficient
kitten retinal degeneration has been observed (Hayes et al., 1975).
Identification of the taurine pools which are affected earliest by a
taurine-deficient diet is important in understanding what functional
biochemical role taurine may play during development. Although the
only accepted physiological function of taurine at present is the conjugation with bile acids (Encrantz and Sjorvall, 1959) taurine may
play a part in central nervous system development (Gaull and Rassin,
1979). The present finding in the cerebrum of the taurine-deficient
puppy and those in retina of the taurine deficient cat (Sturman et al.,
1978) indicate a relative sparing of taurine pools, compared with
those of the liver. These findings give further evidence that the
various tissue taurine pools are not in equilibrium.

In summary, total parenteral nutrition in the beagle puppy provides a useful tool in which further to examine the consequences of
taurine-deficient diets during development. These data suggest that
the hepatic taurine pool is the pool most rapidly depleted during
taurine deficiency. Hepatic CSAD activity in the beagle is increased
in response to a taurine-deficient diet, but this response is not

sufficient to maintain the concentration of taurine in the liver.

REFERENCES

Dobbing, J., Hopewell, J. W., and Lynch, A., 1971, Vulnerability of developing brain. VII. Permanent deficit of neurons in cerebral and cerebellar cortex following early mild undernutrition, Exp. Neurol., 32:439.

Encrantz, J. C., and Sjovall, J., 1959, On the bile acids in duodenal contents of infants and children, Clin. Chim. Acta, 4:793.

Gaull, G. E., and Rassin, D. K., 1979, Taurine and brain development: Human and animal correlates, in: "Developmental Biology, IBRO Symposium No. 5," E. Meisemi and M. A. B. Brazier, eds., Raven Press, New York, in press.

Gaull, G. E., Rassin, D. K., Raiha, N. C. R., and Heinonen, K., 1977, Milk protein quantity and quality in low-birth-weight infants. III. Effects on sulfur amino acids in plasma and urine, J. Pediatr., 90:348.

Hardison, W. G. M., 1978, Hepatic taurine concentration and dietary taurine as regulators of bile acid conjugation with taurine, Gastroenterology, 75:71.

Haslewood, G. A. D., 1967, Bile salts, Richard Clay, The Chaucer Press, Ltd., Great Britain.

Hayes, K. C., Carey, R. E., and Schmidt, S. Y., 1975, Retinal degeneration associated with taurine deficiency in the cat, Science, 188:949.

Heird, W. C., and Winters, R. W., 1975, Total parenteral nutrition, J. Pediatr., 86:2.

Knopf, K., Sturman, J. A., Armstrong, M., and Hayes, 1978, Taurine: An essential nutrient for the cat, J. Nutr. 108:773.

Rigo, J., and Senterre, J., 1977, Is taurine essential for the neonates? Biol. Neonate, 32:73.

Sturman, J. A., Rassin, D. K., and Gaull, G. E., 1977, Taurine in development, Life Sci. 21:1.

Sturman, J. A., Rassin, D. K., Hayes, K. C., and Gaull, G. E., 1978, Taurine deficiency in the kitten: Exchange and turnover of [^{35}S]taurine in brain, retina, and other tissues, J. Nutr., 108:1462.

Winick, M., and Rosso, P., 1969, Effects of severe early malnutrition on cellular growth of human brain, Pediat. Res., 3:181.

TAURINE AND ISETHIONATE IN SQUID NERVE

Francis C.G. Hoskin and Patrick K. Noonan

Department of Biology
Illinois Institute of Technology
Chicago, Illinois   60616

INTRODUCTION

The concentration of isethionate (2-hydroxyethanesulfonate) in the squid giant axon is about 150 mM[1-3]; taurine (2-aminoethanesulfonate) has been reported as about 75 mM, and according to our own measurements may be higher. Thus this single giant cell is 0.2 to 0.3 molar in these two seemingly closely related organosulfur compounds. While taurine is widely distributed, isethionate appears to be limited to cephalopod nerve. Other organs of cephalopods, and organs of other animals, including even the giant axon of the annelid Myxicola contain little or no isethionate[3-5]. Although we had reported a few micromoles isethionate per gram fresh tissue in other organs and species, that was near the limit of our method of analysis[3]; the GLC assay since developed and applied to this question seems to confirm the sharp limitation to cephalopod nerve[6].

Until recently it has generally been assumed that isethionate is synthesized from cysteine with the -C-C-S- backbone passing intact through taurine[7,8]. Now it appears this is not so for squid[3,9], and even if isethionate were present in other species it would be unlikely to have been formed via taurine[10]. These two negative statements still would not add up to a positive identification of the pathway by which isethionate is synthesized. Recently evidence for one possible pathway has begun to emerge from the laboratory of the Chairman of this meeting[11], and for another from our own laboratory. A part of our work was presented two years ago[12], and was subsequently published elsewhere. That, and the most recent progress is presented here.

In turning to squid nerve our rationale is that such a high level of isethionate must reflect a potentially high rate of synthe-

253

sis and a high level of enzyme activities. On the assumption that
isethionate constitutes a substantial part of the ionic pool it might
be speculated that its synthesis is subject to ionic perturbations
demonstrable at the cellular level using the intact squid giant axon,
and at the subcellular level using homogenates and eventually purified enzymes. For several years we have been concerned with an enzyme that hydrolyzes the organophosphorus cholinesterase inhibitor
DFP (diisopropylphosphorofluoridate)[13],[14], and have noted a concom-
itance of the squid-type-DFPase as we term it[5], and isethionate.
Another enzyme, rhodanese, until recently regarded, like DFPase, as
having no physiological role, is also found in squid nerve[15],[16],
although it must be noted that rhodanese is widely distributed. Near
the end of this presentation we will speculate about a possible role
for so-called DFPase, and for rhodanese, in isethionate synthesis.

The squid giant axon, essentially a single cell surrounded by a
Schwann cell and basement layer and some connective tissue, is typically about 0.7 mm across and can be dissected to a length of 70 or 80
mm. In the simplest case it can be laid across stimulating and recording electrodes to monitor the intactness of the plasma membrane.
For penetration studies the axoplasm can be extruded and measured
for, say, radioactivity, if that was the design of the experiment.
For metabolic studies, after an incubation period we have precipitated protein from the axoplasm, performed other operations such as reduction in volume, and separated reactants and products by paper and
column chromatography[3],[15].

SULFUR TO ISETHIONATE IS NOT VIA TAURINE[3],[9],[15]

Cysteine-U-^{14}C is taken up by the squid giant axon and metabolized to ^{14}C-cysteinesulfinate, ^{14}C-cysteate, ^{14}C-hypotaurine, [14]C-
taurine, and [14]CO$_2$, but not to ^{14}C-isethionate. When ^{35}S-cysteine is
taken up, ^{35}S-isethionate is formed. ^{35}S-Sulfide diffuses into the
squid axon, probably as unionized H$_2$[35]S, and is also converted to
[35]S-isethionate. These results are shown in Tables 1 and 2. While
the amount of sulfide taken up by the squid axon is much less than
the amount of cysteine taken up, the amount of sulfide converted to
isethionate as a fraction of the uptake is much greater than the amount of sulfur from cysteine considered in the same way. This suggests that the sulfur of cysteine is metabolized to isethionate via
sulfide, or that both may pass through another common intermediate.
It has occurred that this might be thiosulfate, but the failure of
thiosulfate to be taken up by the squid axon, as seen in Table 1,
precludes experimental testing at the cellular level. Again, it is
noteworthy that none of the carbons of cysteine are converted to
isethionate. Now the question arises as to the source of the carbons
of isethionate. This points to the need for a method of assaying
[14]C and ^{35}S in a double-labelled sample, a measurement not normally
possible due to the virtually identical decay energies of the two
isotopes. We now report the development of such a measurement, its

Table 1.   Uptake of Possible Isethionate
Precursors by Squid Giant Axons

Compound[a]	Penetration[b], % of Equivalent Distribution
^{35}S-Sulfide	43
^{35}S-Sulfate	0.15
Outer-^{35}S-Thiosulfate	0.59
^{14}C- or ^{35}S-Cysteine[c]	200
^{14}C-Pyruvate[c]	24

[a]External conc., 10^{-5} M; time, 1 hr.

[b]Average of all work from this laboratory.

[c]Corrected for ^{14}C metabolized to $^{14}CO_2$.

Table 2.   Formation of Isethionate from Sulfur Sources
by Squid Giant Axon[a]

Source	^{35}S-Sulfide[b]	^{35}S-Cysteine[b]	^{14}C-Cysteine
	picomoles		
Expected at equivalent distribution	500	500	100
Found after 5 hrs.	175	955	200
Isethionate formed	28	16	0

[a]All values based on a "standard" 10 mg axon.

[b]External conc., 5×10^{-5} M.

application to the question of the carbon source of isethionate in
squid nerve, and its possible application to other studies of sulfur
metabolism, including the metabolism of taurine.

^{14}C-^{35}S MEASUREMENT

All radioactive measurements were made in a Nuclear Chicago
Unicap 300 scintillation counter. However, there are three distinct

methods of sample preparation and counting. One is the introduction of radioactive sample into the appropriate scintillant; this is termed "conventional". The second is the complete oxidation of radioactive sample in the modified Van Slyke apparatus[9,17] and the exclusive trapping of $^{14}CO_2$ in a scintillant containing an organic base, for example New England Nuclear Corporation's Oxyfluor. Although the actual counting is by the usual scintillation means, the entire procedure of sample oxidation, CO_2 trapping, and counting is termed "Van Slyke". The third method is the complete oxidation of radioactive sample by a modified Hoskin-Brande procedure[3] and the exclusive retention of $Ba^{35}SO_4$ on a 0.22 μm Millipore filter. Again, although the actual counting of the filter pad is by the usual scintillation means, the entire procedure of sample oxidation, precipitation, filtration, and counting is termed "Hoskin-Brande". However, the organic-base-containing scintillants are incompatible with Millipore filters, and thus for the Hoskin-Brande procedure New England Nuclear Corporation' Aquasol was used. In summary, samples have been counted by conventional, Van Slyke, and Hoskin-Brande procedures with the measurement of ^{14}C in the presence of ^{35}S, and the measurement of ^{35}S in the presence of ^{14}C, the respective goals. To do this two series of $^{14}C-^{35}S$ "double labelled" samples of cystine were formed, one having a uniformly high level of ^{14}C and amounts of ^{35}S varying from zero upward, and the other having this order reversed. Aliquots of the variously singly labelled and "double labelled" cystines were counted by the conventional, Van Slyke, and Hoskin-Brande procedures.

Van Slyke Procedure[9,17]

The modified apparatus is shown in Fig. 1, and the procedure is recounted here so as to note the modifications. A 0.1 ml aliquot of the sample to be assayed and 2-3 mg of a carrier carbon source, e.g., non-radioactive cystine, were placed in the bottom of a combustion tube and evaporated to dryness under nitrogen. The dry oxidants, $K_2Cr_2O_7$ and KIO_3, were placed in a 1 ml beaker and lowered into the bottom of the combustion tube. Introduction of liquid reagent, combustion, collection of volatile product, expulsion of air, and so on, were performed exactly as originally described[17]. The final step in the transfer of $^{14}CO_2$ into scintillant was modified from that previously described[9]. The modification is evident in Fig. 1; 5 ml of Oxyfluor was placed in a 10 ml pear-shaped flask which was then submerged in liquid nitrogen up to the level of the scintillant. The rest of the transfer was carried out as has been described[9,17]. The 5 ml Oxyfluor now containing $^{14}CO_2$ was transferred, with rinsing, into a scintillation vial for counting.

Hoskin-Brande Procedure[3]

A 0.1 ml aliquot of the sample to be assayed and 1.4-1.5 mg carrier Na_2SO_4 were evaporated to dryness under nitrogen in a combustion

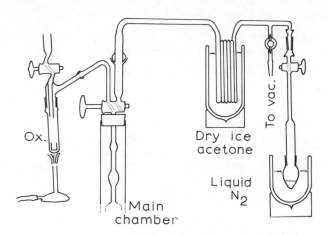

Fig. 1 Modified Van Slyke apparatus for ^{14}C assay in the presence
        of ^{35}S. The dropping funnel and oxidation vessel are not
        used simultaneously with the gas-transfer line; the entire
        apparatus is shown to illustrate consecutive operations of
        sample oxidation and ^{14}CO$_2$ trapping, ^{14}CO$_2$ release on acidi-
        fication of the main chamber, and transfer of ^{14}CO$_2$ to Oxy-
        fluor in a pear-shaped flask submerged in liquid nitrogen.

tube, and 0.25 ml of each of fuming nitric and conc. perchloric acids
were added. The tube was heated for 3 hrs at 140-160°, evacuated on
a water aspirator at 100° for 10 min, and cooled; 1.2 ml 2 M NaOH was
added and the tube was vortexed; 2.0 ml 0.02 M BaCl$_2$ was added and
the tube was thoroughly vortexed and ice-chilled for 30 min or longer. The tube was again vortexed and the mildly acidic cold suspension of Ba^{35}SO$_4$ was collected on a 0.22 μm Millipore filter. The
filter pad was washed with 3.5 ml ice-cold water and transferred to
10 ml Aquasol in a scintillation vial. With occasional shaking over
the next 10-20 min the filter pad dissolved and the small amount of
BaSO$_4$ was so finely dispersed as to be almost invisible. Millipore
filters cause a significant difference in Aquasol response. Hence a
filter was added to each Aquasol-containing background and conventional vial and was, of course, already present in Hoskin-Brande
vials. This gave statistically constant external standard ratios.

The results have shown that the Van Slyke method can measure

Table 3. Assay of Either ^{14}C or ^{35}S in the
Presence of the Other Isotope[a]

Composition of Sample[b], Conventional Counting		Van Slyke Assay	Hoskin-Brande Assay
^{14}C	^{35}S	^{14}C	^{35}S
6746 ± 124	0	6679 ± 171	2 ± 1
0	8230 ± 34	2 ± 2	6419 ± 322[c]

[a]All values are counts per min, average of 4
determinations, ± std. dev.

[b]Intermediate samples yielded correspondingly
intermediate values; extremes shown here to
emphasize quantitative and exclusive nature of
assay.

[c]Filtrates contained 1926 ± 181 counts per min;
thus 78% of radioactivity identified as $Ba^{35}SO_4$;
101% accounted for.

95 ± 3% (std. dev.) of ^{14}C. Even more important, there is no contamination of the ^{14}C counts with ^{35}S. This is shown in Table 3. The
Hoskin-Brande method for the measurement of ^{35}S is less quantitative
than the ^{14}C assay. On the average the method can determine 75 ± 11%
(std. dev.) of ^{35}S. The filtrates from the Hoskin-Brande method account for 30 ± 13% (std. dev.) of the ^{35}S originally present. Thus
essentially all of the ^{35}S is accounted for. As with the Van Slyke
assay, it is important to note that in the Hoskin-Brande assay there
is no contamination of the ^{35}S counts with ^{14}C. This is also shown
in Table 3.

ISETHIONATE BIOSYNTHESIS

Glucose-U-^{14}C and $Na_2{}^{35}S$, both at 10^{-4} M, were incubated with
homogenized frozen squid optic ganglia in 10^{-3} M HEPES (N-2-hydroxy-
ethylpiperazine-N'-2-ethanesulfonate) buffer, pH 7.5, for 5 hrs at
30°. Thereafter protein was precipitated, and from the soluble part,
unused glucose and sulfide were separated from the sulfate-plus-
isethionate peak on an Aminex A-4 (BioRad) column[15], sulfate and carrier sulfate were precipitated as $BaSO_4$, and hexose phosphates were
hydrolyzed by heating in 1 M NaOH. These procedures were intersper-

Table 4. Biosynthesis of $^{14}C-^{35}S$ Double Labelled
Isethionate by Squid Nerve Tissue

Operation	Radioactivity		$^{14}C/^{35}S$
	Counts/min	%	
Incubation mixture[a]	29,769,000	100	
Glucose-U-^{14}C	19,584,000	66	1.94
Na$_2$$^{35}S$	10,185,000	34	
Isethionate isolated	23,180	0.078	
Van Slyke for ^{14}C	12,100	0.041	1.46
Hoskin-Brande for ^{35}S	8,360	0.028	
Incubation mixture[a]	15,115,000	100	
Glucose-U-^{14}C	6,754,000	45	0.82
Na$_2$$^{35}S$	8,361,000	55	
Isethionate isolated	62,550	0.41	
Van Slyke for ^{14}C	9,180	0.061	0.20
Hoskin-Brande for ^{35}S	45,450	0.30	

[a]Substrates are each present at 10^{-4} M, 80 mg homogenized squid optic ganglion, 10^{-3} M HEPES buffer to
give a final volume of 1.0 ml, pH 7.5, 30°, 5 hrs.

sed with additional Aminex A-4 purifications, sulfate precipitations,
AG50(H) (BioRad) removal of cations, and finally a paper chromatographic isolation of a radioactive product migrating with authentic
1,2-^{14}C-isethionate, and not with $^{35}SO_4^=$. This material was assayed
for total radioactivity by conventional counting, and for ^{14}C and
^{35}S by the Van Slyke and Hoskin-Brande procedures, respectively.
The results of the two experiments performed to date are given in
Table 4. All samples constituting one complete experiment were
counted in one 24-hr period, thus correcting for ^{35}S radioactive
decay. The different radioactive compositions of the two incubation
mixtures are due only to matters of laboratory convenience, and do
not reflect changes in chemical composition. So far as an illustra-
tion of the $^{14}C-^{35}S$ counting method, it can be seen that 88-89% of

the radioactivity found in isethionate is accounted for. If the ^{14}C
determination is quantitative, then the ^{35}S determinations are 77%
and 86%, both within the 75 $\pm$ 11% (std. dev.) that the Hoskin–Brande
method is capable of determining.

Turning now to the metabolic significance of the results, it may
appear that the isethionate yield is very small, 0.1–0.4% without regard to which isotope is considered. It should be emphasized that
these values do not represent isethionate yield. In the course of
purifying the isethionate there have probably been large losses of
all of the metabolic products. Indeed, the first Aminex A–4 separation suggested that about 4% of the total starting radioactivity
became isethionate-plus-sulfate, and the first BaSO$_4$ precipitation
suggested that about half of this was isethionate. This is similar
to the yields we reported[15] for isethionate as a percentage of substrate taken up by intact squid giant axons. But results on freshly
dissected intact single cells (as the giant axon is for the purpose
of considering intracellular concentrations) cannot be compared with
results obtained from homogenates of frozen tissues.

While the yields of ^{14}C and ^{35}S in isethionate differ rather
markedly between the two experiments, in neither case has there been
a mere trickle of radioactivity into the isethionate from one or the
other isotopic source. In fact, the proportions of ^{14}C and ^{35}S in
the isethionate are not unreasonably different from the proportions
in the incubation mixture. Thus in the one experiment the starting
^{14}C/^{35}S ratio is 1.94 and the product ratio is 1.46; in the other
the same two ratios are 0.82 and 0.20, respectively. On the whole,
these results suggest a relatively direct pathway from glucose to a
2-carbon fragment, and the transfer of sulfide or a sulfide intermediate to this 2-carbon acceptor. A possibly analogous transfer is
already known, namely that of the outer (sulfane) sulfur of thiosulfate to cyanide to produce thiocyanate, catalyzed by the enzyme
rhodanese[18,19], mentioned earlier. Recently, evidence has been
presented that a physiological role of rhodanese is the incorporation
of the outer sulfur of thiosulfate into the protein structure of
certain iron-sulfur proteins[20,21]. Might it be that another physiological role for rhodanese is the formation of the sulfur–carbon
bond of isethionate? This question and the question of a possible
role for so-called squid-type-DFPase are posed in Fig. 2 . The
right-hand side is mainly fact, many of the details of which are
the subject of other papers in this Symposium. The lack of a pathway from taurine to isethionate, and the initial and final parts of
a pathway from cysteine via sulfide, and from glucose, to isethionate
appear to be true for squid nerve. A possible role for rhodanese,
the existence of an acid anhydride S∿X sulfinic intermediate, and
especially a role for squid-type-DFPase on that pathway are all
highly speculative. Clearly more work remains to be done. One of
the messages of this presentation is that taurine and isethionate

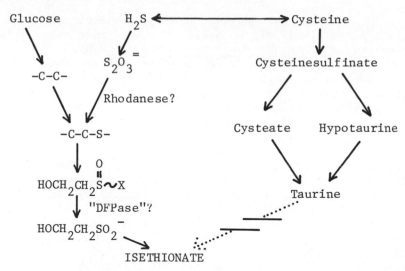

Fig. 2 Pathways to account for the observations that sulfide and
the sulfur of cysteine are incorporated into isethionate,
that the carbons of cysteine are not, and that there appears
to be a fairly direct route from glucose to isethionate.
The involvement of thiosulfate, of an S~X compound, of a
sulfinate precursor of isethionate (isethiinate), and of
rhodanese and DFPase are speculative.

are only indirectly related as contrasted to the usual view. Yet
while isethionate is not synthesized from taurine, both isethionate
and taurine are synthesized from sulfide[22]. The methods reported
here will now be applied to problems of the metabolism of taurine,
a compound which seems to be of more physiological interest than
isethionate.

ACKNOWLEDGMENTS

     This research was supported by NIH Grants NS-09090 and
ES-02116. A large part of the work was performed at the Marine
Biological Laboratory, Woods Hole, Mass.

REFERENCES

1. B.A. Koechlin, On the chemical composition of the axoplasm of
squid giant nerve fibers with particular reference to its
nerve pattern, J. Biophys. Biochem. Cytol. 1:511 (1955).

2. G.G.J. Deffner and R.E. Hafter, Chemical investigations of the
    giant nerve fibers of the squid. IV. Acid-base balance in
    axoplasm, Biochim. Biophys. Acta 42:200 (1960).

3. F.C.G. Hoskin and M. Brande, An improved sulfur assay applied
    to a problem of isethionate metabolism in squid axon and
    other nerves, J. Neurochem. 20:1317 (1973).

4. F.C.G. Hoskin, Squid nerve type DFPase: a consideration of mol-
    ecular structures, in: "Jerusalem Symposium on Molecular and
    Quantum Pharmacology," E.D. Bergman and B. Pullman, eds.,
    D. Reidel, Dordrecht, Holland (1974).

5. J.M. Garden, S.K. Hause, F.C.G. Hoskin, and A.H. Roush, Compar-
    ison of DFP-hydrolyzing enzyme purified from head ganglion
    and hepatopancreas of squid (Loligo pealei) by means of iso-
    electric focusing, Comp. Biochem. Physiol. 52C:95 (1975).

6. M.A. Remtulla, D.A. Applegarth, D.G. Clark, and I.H. Williams,
    Analysis of isethionic acid in mammalian tissues, Life Sci.
    20:2029 (1977).

7. M.K. Gaitonde, Sulfur amino acids, in: "Handbook of Neurochem-
    istry," A. Lajtha, ed., Plenum Press, New York (1970).

8. P.L. McGeer, J.C. Eccles, and E.G. McGeer, "Molecular Neurobiol-
    ogy of the Mammalian Brain," Plenum Press, New York (1978).

9. F.C.G. Hoskin, M.L. Pollock, and R.D. Prusch, An improved
    method for the measurement of $^{14}CO_2$ applied to a problem of
    cysteine metabolism in squid nerve, J. Neurochem. 25:445
    (1975).

10. J.H. Fellman, E.S. Roth, and T.S. Fujita, Taurine is not meta-
    bolized to isethionate in mammalian tissue, in: "Taurine and
    Neurological Disorders," A. Barbeau and R.J. Huxtable, eds.,
    Raven Press, New York (1978).

11. D. Cavallini, S. Duprè, G. Federici, S. Solinas, G. Ricci, A.
    Antonucci, G. Spoto, and M. Materese, Isethionic acid as a
    taurine co-metabolite, in: "Taurine and Neurological Disord-
    ers," A. Barbeau and R.J. Huxtable, eds., Raven Press, New
    York (1978).

12. F.C.G. Hoskin and E.R. Kordik, Hygrogen sulfide as a precursor
    of isethionate synthesis in squid axon, presented at: "Taurine
    in Neurological Disorders," R.J. Huxtable and A. Barbeau,
    organizers, Tucson, Arizona (1977).

13. F.C.G. Hoskin and R.L. Long, Purification of a DFP-hydrolyzing
    enzyme from squid head ganglion, Arch. Biochem. Biophys.
    150:548 (1972).

14. F.C.G. Hoskin, Distribution of diisopropylphosphorofluoridate-
    hydrolyzing enzyme between sheath and axoplasm of squid
    giant axon, J. Neurochem. 26:1043 (1976).

15. F.C.G. Hoskin and E.R. Kordik, Hydrogen sulfide as a precursor
    for the synthesis of isethionate in the squid giant axon,
    Arch. Biochem. Biophys. 180:583 (1977).

16. H. Schievelbein, R. Baumeister, and R. Vogel, Comparative in-
    vestigations on the activity of thiosulfate transferase,
    Naturwissenschaften 56:416 (1969).

17. D.D. Van Slyke, R. Steele, and J. Plazin, Determination of total carbon and its radioactivity, J. Biol. Chem. 192:769 (1951).
18. K. Lang, Die Rhodanbildung in Thierkörper, Biochem. Z. 259:243 (1933).
19. J. Westley, Rhodanese, Advan. Enzymol. Relat. Areas Mol. Biol. 39:327 (1973).
20. A. FinazziAgro, C. Cannella, M.T. Graziani, and D. Cavallini, Possible role for rhodanese. Formation of "labile" sulfur from thiosulfate, FEBS Lett. 16:172 (1971).
21. T. Taniguchi and T. Kimura, Role of 3-mercaptopyruvate sulfur-transferase in the formation of the iron-sulfur chromophore of adrenal ferredoxin, Biochim. Biophys. Acta 364:284 (1974).
22. F.C.G. Hoskin and R.D. Prusch, unpublished observations; see also Fig. 1 of reference 15.

TAURINE BIOSYNTHESIS IN OX RETINA

Salvatore Macaione, Rosa Maria Di Giorgio and Grazia
De Luca
Department of Biochemistry,University of Messina,Italy
Giuseppe Nisticò
Department of Pharmacology, University of Messina

DISTRIBUTION AND ROLE OF TAURINE IN RETINA

Taurine is present in high concentrations in the retina (1-7) where it constitutes 40-50% of the total aminoacid pool (4, 6).

In every animal species so far studied, although substantial levels of taurine are present in all retinal layers, this amino acid has been found to be predominantly concentrated at photo-receptor level (8-10). In particular detailed analytical studies on the retina layers of different vertebrate species (8) have shown that the highest concentration of taurine occurs at the level of the outer nuclear layers, 95% of which as volume is represented by the photoreceptors and only 5% by Müller glial cells according to morphometric studies (11). In mice and rats (4, 5) a sharp increase in taurine occurs when the receptors start to differentiate their inner and outer segments, in contrast to other neuronal tissues in which the concentration of taurine decreases during development (12). A significant fall in taurine concentration is found in mice and rats affected by retinal degeneration whereas other amino acids are unaffected (2, 3). On the other hand in the mouse retinal lesions by sodium monoglutamate, which produce a deficit in ama-crine and ganglion cells leaving unaffected photoreceptors, result in a relative increase in taurine level for the surviving retina (3). In addition cat photoreceptors degenerate when these animals are kept on taurine-free diet and only taurine and no other amino acid is able to antagonize such degeneration (13-15).

Although the precise role of taurine in retina is not completely known there is evidence that it can play an inhibitory transmitter role (16, 17) since it is able to affect potassium and calcium fluxes at the ionic membrane level (18-20). At the light of these findings it is easily interpreted the functional inhibitory role of taurine in experiments in which from perfused chick and rat retina preloaded with 35S-taurine light flashes enhance endogenous and labeled taurine release (21, 22). Taurine is the most potent amino acid able to reversibly abolish electroretinogram b wave (23-25) as well as to depress several neuronal populations from cerebral cortex, subcortical areas and spinal cord (26-29). In addition a specific uptake system has been described to occur in rat retina (30, 31) as well as in retinal pigment epithelium (32) and in slices and synaptosome preparations from various areas of the brain (33, 34).

TAURINE BIOSYNTHESIS IN RETINA

Several metabolic pathways can lead to taurine biosynthesis (35). The most important route in mammalian tissues, including brain and retina, has been shown to originate by pathway I (methionine $\longrightarrow$ cysteine $\longrightarrow$ cysteine sulphinate $\longrightarrow$ hypotaurine $\longrightarrow$ taurine), while pathway II (methionine $\longrightarrow$ cysteine $\longrightarrow$ cysteine sulphinate $\longrightarrow$ cysteate $\longrightarrow$ taurine) would have some significance. Other suggested alternative pathways from cysteamine (36-38) or from sulphate (39, 40) would be only minimally or not at all operative.

There are wide differences amongst species in the capacity to synthesize taurine. The chick embryo and the chick can convert inorganic sulphate to taurine (41-43). The adult rat is able to convert inorganic sulphate to taurine whereas this conversion cannot be demonstrated to occur in the cat (44, 45). However until recently it was a matter of discussion whether in retina taurine is synthesized in situ either it is taken up from the blood stream through the pigment epithelium (32).

In a previous work we have demonstrated that cysteine is oxidized to cysteine sulphinate by cysteine dioxygenase in rat retina (46). In the retina the final reaction product is almost exclusively cysteine sulphinate, but in brain a considerable portion of it is further oxidized to cysteate (47).

Cysteine sulphinate decarboxylase activity has been detected in retina of several species (48, 49). In chick retina this enzymatic activity was found associated with the crude nuclear fraction for approximately 18% and with the crude mitochondrial fraction for about 27% whereas about 55% was found in the soluble fraction (49.

The oxidation of hypotaurine to taurine is poorly demonstrated. However recently we have been able to show the presence of hypotaurine oxidase activity in retina from several species (50, 51).

In the present chapter we will report the subcellular distribution pattern of taurine and enzymatic activities involved in taurine biosynthesis in ox retina and in particular in the subcellular nuclear fraction which contains besides nuclei, the outer and inner segments of photoreceptors as well as pinched-off nerve-endings of the photoreceptors.

TAURINE DISTRIBUTION IN PRIMARY AND SECONDARY FRACTIONS FROM OX RETINA

Taurine is the predominant free amino acid in ox retina (48) where it is present at a concentration of 20-22 µmoles/g w. w. tissue and of 16-18 µmoles/100 mg protein. Such concentrations are higher in comparison to those reported in chick and rat retina by Pasantes-Morales et al. (6) and lower than that found in rat retina (4). In addition in the pigment epithelium of ox retina taurine is present at a concentration of 1.8-2.2 µmoles/100 mg protein thus the retina/pigment epithelium ratio is about 8.5 : 1.

Data showing taurine distribution in primary and secondary subcellular fractions are given in Table I.

Table I. Taurine distribution pattern in primary and secondary subcellular fractions from ox retina

Fraction	Taurine (µmoles/g w. w.)	Percentage
Homogenate	22.5 ± 0.42	100
Nuclear	3.2 ± 0.25	14.2
Mitochondrial	1.2 ± 0.02	5.3
Microsomal	0.4 ± 0.03	1.7
Soluble	17.9 ± 0.46	79.5
Recovery		100.7
Mitochondrial	1.25 ± 0.02	100
Myelin	0.14 ± 0.01	11.2
Synaptosomes	0.48 ± 0.12	38.5
Mitochondria	0.57 ± 0.10	45.6
Recovery		95.3

Values are the mean of six experiments ± SEM. Crude nuclear and mitochondrial fractions were obtained according to Papa et al. (52). The secondary subcellular fractions and taurine were obtained and assayed respectively according to Macaione et al. (48).

It is evident that the distribution pattern is similar to that of free amino acids (53); in fact approximately 80% of taurine is present in soluble fraction whereas the remainder is associated to the particulate fractions from which is released after adding Triton X-100. In the secondary fractions separated from crude mitochondria and examined by electron microscopy taurine was found to be associated (about 40%) with the synaptosomal fraction (Table I). This is in agreement with similar findings in the rat retina (54).

CYSTEINE DIOXYGENASE ACTIVITY

In comparison to taurine distribution, cysteine dioxygenase activity was found to be present predominantly in particulate fractions whereas only about 39% was in the soluble fraction (Table II).

Table II. Percentage subcellular distribution of cysteine dioxygenase activity in ox retina

Fraction	Percentage	Fraction	Percentage
Homogenate	100	Mitochondrial	100
Nuclear	26.1	Myelin	9.0
Mitochondrial	28.0	Synaptosomes	37.5
Microsomal	16.3	Mitochondria	58.3
Soluble	39.1		
Recovery	109.6		104.8

Cysteine dioxygenase activity was assayed in presence of Fe^{++} and NAD^+ according to Yamaguchi et al. (55).

In the secondary subcellular fractions approximately 58% of the total activity recovered from the mitochondrial fraction was contained in purified mitochondrial fraction. The remainder activity was found in synaptosomes, in which the enzymic activity was present for about 37%, and in myelin (Table II).

Cysteine dioxygenase activity was assayed in presence of only cysteine or with cysteine, Fe^{++} and NAD^+. In all the primary and secondary subcellular fractions Fe^{++} and NAD^+ were able to increase the cysteine sulphinate production (Table III). This activating effect was different in the various subcellular fractions studied, the maximum being obtained in the soluble and synaptosomal fractions and the lowest in the nuclear and purified mitochondrial fractions.

In all the subfractions NAD^+ and $NADP^+$ showed a more marked activating effect in comparison to reduced forms (S. Macaione et al. unpublished data). When the specific cysteine dioxygenase activity was compared to activity referred per g w. w. in the synaptosomes there was a specific activity 3 fold higher than that of mitochondrial fraction.

In 55, 000 x g supernatant of rat retina, enzymic activity is ten

fold higher than in that one of ox retina (47 nmoles in rat retina
and 4.37 nmoles in ox retina). This quantitative discrepancy between

Table III. Effects of Fe^{++} and NAD^+ on cysteine dioxygenase activity in primary and secondary subcellular fractions from
ox retina

Fraction	Cysteine dioxygenase activity (nmoles CSA/g w. w./ 60 min)		Activation
	no Fe^{++} no NAD^+	+ Fe^{++} 0.125 mM + NAD^+ 2 mM	
Homogenate	142.6 ± 7.2	217.6 ± 11.9	52.6
Nuclear	44.1 ± 2.3	56.9 ± 4.0	29.2
Mitochondrial	37.2 ± 1.6	61.0 ± 2.9	63.8
Microsomal	21.3 ± 0.3	35.5 ± 0.8	66.9
Soluble	35.1 ± 3.6	85.1 ± 4.5	142.3
Myelin	2.7 ± 1.0	5.5 ± 0.9	103.7
Synaptosomes	10.0 ± 1.2	22.9 ± 1.2	128.2
Mitochondria	24.0 ± 1.0	35.6 ± 2.0	47.9

Values are the mean of 8 experiments ± SEM.

ox and rat retina may be due only to difference in species utilized
as the enzymic assays were carried out in the same way.

CYSTEINE SULPHINATE DECARBOXYLASE ACTIVITY

This activity is mainly located in particulate components of the
primary subcellular fractions. About 35% of the enzyme activity is
present in the crude mitochondrial fraction. Further subfractionation has shown that half of the cysteine sulphinate decarboxylase
activity, present in this fraction, is localized in synaptosomes
(Table IV).

Cysteine sulphinate and cysteate are obviously decarboxylated
by the same enzyme. The ratio between the decarboxylation of cysteine sulphinate and of cysteate was constant throughout all the
steps of our fractionation procedure. This ratio is approx. 6.6 and
is similar to that reported in rat liver (56).

HYPOTAURINE OXIDASE ACTIVITY

The discovery of the decarboxylation of cysteine sulphinate to
hypotaurine, which is then oxidized to taurine, led to the proposi-
tion that this pathway represents the main biosynthetic route for
taurine in animal tissues (57, 58). The finding of consistent amounts
of hypotaurine in organs and body fluids after injection or feeding

of cysteine (59-61) strongly supported the occurrence of this pathway in animals, nevertheless the possible oxidation of hypotaurine
to taurine has been very poorly demonstrated. In adult rats no measurable oxidation in vitro by brain tissue is found but some oxidation may occur in immature brain tissue (62).

Table IV. Cysteine sulphinate decarboxylase activity in primary
and secondary subcellular fractions from ox retina

Fraction	CSA decarboxylase activity (μmoles CO_2/g w. w. /h)	Percentage
Homogenate	111.36 $\pm$ 4.50	100
Nuclear	20.25 $\pm$ 1.25	18.1
Mitochondrial	39.25 $\pm$ 0.90	35.3
Microsomal	14.25 $\pm$ 0.75	12.7
Soluble	33.75 $\pm$ 0.93	30.3
Recovery		96.6
Mitochondrial	39.25 $\pm$ 0.90	100
Myelin	1.14 $\pm$ 0.21	2.9
Synaptosomes	17.48 $\pm$ 0.70	44.5
Mitochondria	19.38 $\pm$ 0.58	49.3
Recovery		96.7

Values are the mean of 15 experiments $\pm$ SEM. Cysteine sulphinate
decarboxylase activity was assayed according to Macaione et al. (48

Table V. Hypotaurine oxidase activity in primary and secondary
subcellular fractions from ox retina

Fraction	Hypotaurine oxidase activity (nmoles taurine/g w. w. /120 min)	Percentage
Homogenate	3,500 $\pm$ 61	100
Nuclear	203 $\pm$ 20	5.8
Mitochondrial	455 $\pm$ 18	13.0
Microsomal	61 $\pm$ 12	1.7
Soluble	2,620 $\pm$ 48	74.8
Recovery		95.5
Mitochondrial	455 $\pm$ 18	100
Myelin	14 $\pm$ 20	3.1
Synaptosomes	254 $\pm$ 11	55.8
Mitochondria	207 $\pm$ 10	45.5
Recovery		104.4

Values are the mean of six experiments $\pm$ SEM. Enzymic activity
assayed according to Oja et al. (62) was linear as function of time
(up to 90 min) and of protein content (up to 2.5 mg).

The results on the distribution of hypotaurine oxidase activity in subcellular fractions of ox retina is given in Table V. About 78% of enzyme activity was found in the soluble fraction while about 22% was associated with particulate components. In the synaptosomal fraction was present about 55% of enzyme activity, recovered from crude mitochondria.

TAURINE BIOSYNTHESIS IN PHOTORECEPTOR CELLS

The presence of the enzymic activities involved in taurine synthesis in the nuclear fraction is indicative of their location at level of photoreceptors.

Table VI. Cysteine dioxygenase activity in nuclear (N_1) subcellular fractions from ox retina handly homogenized

Fraction	Predominant constituents	Cysteine dioxygenase activity		
		total activity	recovery %	specific activity
N_1		50.72	–	7.05
N_1C	outer segments	absent	–	absent
N_1D	inner and outer	4.83	6.4	3.52
N_1	segments			
N_1E	inner segments + some pedicle synaptosomes	22.71	30.1	7.34
N_1F	nuclei + synaptosomes + inner segments	16.96	22.4	12.21
N_1P	nuclei + inner segments + some synaptosomes	2.39	3.2	3.06
N_1G	pedicle synaptosomes + some mitochondria	17.26	22.8	144.24
N_1H	synaptosomes + nuclei	11.40	15.1	101.80
N_1I	nuclei	absent	–	absent

Values are the mean of 4 experiments. Total activities are nmoles CSA/g retina/60 min. Specific activities are nmoles CSA/mg protein/60 min. Nuclear subcellular fractions were obtained according to Marshall et al. (63).

The ox nuclear pellet obtained from retinas homogenized mechanically contains, as shown by EM, heterogeneous populations of nuclei, aggregation of mitochondria derived from photoreceptor inner segments, fragmented outer segments and isolated photoreceptor terminals. Thus the enzymic activities reported above in crude nuclear fractions are only indicative of their occurrence as the photoreceptor level without specifying the precise site of their localization in photoreceptors.

Therefore we have investigated the effects of more gentle procedures according to Marshall et al. (63) in order to avoid excessive disruption and to isolate synaptic pedicles and inner segments of photoreceptor cells. Crude nuclear pellets obtained by the above procedure were further fractionated by density gradient centrifugation and as marker of metabolic pathway for taurine biosynthesis the cysteine dioxygenase activity was followed. On the basis of the gradient fractions observed by EM we were able to obtain a purified fraction (N1C, see Table VI), which sedimented at the interface 1.0-1.2M of sucrose density, constituted by outer segments. This fraction was found devoid of cysteine dioxygenase activity. This finding is in agreement with the lack of cysteine sulphinate decarboxylase activity in chick outer segments isolated by a different procedure (64). In contrast the highest cysteine dioxygenase total recovered activity occurred in fractions where synaptosomes and mitochondria are the predominant constituents, whereas cysteine dioxygenase specific activity was maximal in fractions N1G and N1H, sedimented at the interface 1.8-2.0 and 2.0-2.2M respectively, which contain almost exclusively pedicle synaptosomes (Table VI). These results are in agreement with our previous experiments which have demonstrated that in mitochondrial fraction from rat retina during development the highest cysteine dioxygenase activity was obtained at the time of photoreceptor maturation (46).

CONCLUSIONS

In ox retina, similarly to other vertebrate retinas studied, the highest content in taurine was found at the photoreceptor level (unpublished data). The existence of the enzymic activities leading to taurine biosynthesis in photoreceptors exclude that taurine uptake from plasma is the only mechanism as suggested for some animal species, responsible for taurine accumulation at this level.

The results obtained in subcellular nuclear fractions, using two different sucrose gradients, show that cysteine dioxygenase is localized predominantly in inner segments and photoreceptor nerve endings without being present in the outer segments. This confirms data obtained for cysteine sulphinate decarboxylase activity in other species in which photoreceptor single constituents were fractionated by microdissection (64) or in experiments in which retinal lesions gave rise to photoreceptor loss (65).

Our methodology show that at the used sucrose gradients pedicle synaptosomes were shrunken and synaptic vesicles were clumping, thus suggesting that further studies with different gradients might enable to avoid these artifacts and better clarify the biochemical mechanisms and functional role of taurine in retina.

REFERENCES

1. R. Kubicek, and A. Dolenek, Taurine et acides aminés dans la rétine des animaux, J. Chromatogr., 1:266 (1958).
2. J. Brotherton, Studies on the metabolism of the rat retina with special reference to retinitis pigmentosa. II. Amino acid content as shown by chromatography, Ex. Eye Res., 1:246 (1962).
3. A. I. Cohen, M. McDaniel, and H. T. Orr, Absolute levels of some free amino acids in normal and biologically fractionated retinas, Invest. Ophtalmol., 12:686 (1973).
4. S. Macaione, P. Ruggeri, F. De Luca, and G. Tucci, Free amino acids in developing rat retina, J. Neurochem., 22:887 (1974).
5. H. T. Orr, A. I. Cohen, and J. A. Carter, The levels of free taurine, glutamate, glycine and γ-amino butyrric acid during the postnatal development of the normal and dystrophic retina of the mouse, Exp. Eye Res., 23:377 (1976).
6. H. Pasantes-Morales, J. Klethi, M. Ledig, and P. Mandel, Free amino acids of chicken and rat retina, Brain Res., 41:494 (1972).
7. K. Yamamoto, Y. Yoshitani, H. Fujiware, and K. Matsuura, Study on free amino acids in the retina, Acta Soc. Ophtalmol. Jpn., 74:1561 (1970).
8. H. T. Orr, A. I. Cohen, and O. H. Lowry, The distribution of taurine in the vertebrate retina, J. Neurochem., 26:609 (1976).
9. A. J. Kennedy, and M. J. Voaden, Free amino acids in the photoreceptor cells of the frog retina, J. Neurochem., 23:1093 (1974).
10. P. Keen, and R. A. Yates, Distribution of amino acids in subdivided rat retinae, Br. J. Pharmacol., 52:118 (1974).
11. K. E. Rasmussen, A morphometric study of the Müller cell cytoplasm in the rat retina, J. Ultrastruct. Res., 39:413 (1972).
12. J. A. Sturman, and G. E. Gaull, Taurine in the brain and liver of the developing human and monkey, J. Neurochem., 25:831 (1975).
13. E. L. Berson, K. C. Hayes, A. R. Rabin, S. Y. Schmidt, and G. Watson, Retinal degeneration in cats fed casein. II. Supplementation with methionine, cysteine or taurine, Invest. Ophthalmol., 15:52 (1976).
14. K. C. Hayes, R. E. Carey, and S. Y. Schmidt, Retinal degeneration associated with taurine deficiency in the cat, Science, 188:949 (1975).
15. K. C. Hayes, A. R. Rabin, and E. L. Berson, An ultrastructural study of nutritionally induced and reversed retinal degeneration in cats, Am. J. Pathol., 78:504 (1975).
16. D. R. Curtis, and G. A. R. Johnston, Amino acid transmitters in the mammalian central nervous system, Ergeb. Physiol., 69:97 (1974).
17. K. Krnjević, Chemical nature of synaptic transmission in vertebrates, Physiol. Rev., 54:418 (1974).
18. P. Dolara, A. Agresti, A. Giotti, and G. Pasquini, Effect of taurine on calcium kinetics of guinea pig heart, Eur. J. Pharmacol., 24:352 (1973).
19. R. Gruener, H. Bryant, D. Markovitz, R. Huxtable, and R. Bressler,

Ionic actions of taurine on nerve and muscle membranes:Electrophysiological studies, in:"Taurine", R. Huxtable and A. Barbeau eds., Raven Press, New York (1976).

20. J. D. Welty, M. J. McBroom, A. W. Appelt, M. B. Peterson, and W. O. Read, Effect of taurine on heart and brain electrolyte imbalances, in: "Taurine", R. Huxtable and A. Barbeau eds., Raven Press, New York (1976).

21. H. Pasantes-Morales, J. Klethi, P. F. Urban, and P. Mandel, The effect of electrical stimulation, light and amino acids on the efflux of ^{35}S-taurine from the retina of domestic fowl, Exp. Brain Res., 19:131 (1974).

22. A. Lopez Colomé, D. Erlij, and H. Pasantes-Morales, Different effects of calcium-flux-blocking agents on light and potassium stimulated release of taurine from retina, Brain Res., 113:527 (1976).

23. H. Pasantes-Morales, J. Klethi, P. F. Urban, and P. Mandel, The physiological role of taurine in retina:Uptake and effect on electroretinogram(ERG), Physiol. Chem. Phys., 4:339 (1972).

24. N. Bonaventure, N. Wioland, and P. Mandel, Antagonists of the putative inhibitory transmitter effects of taurine and GABA in the retina, Brain Res., 80:281 (1974).

25. P. F. Urban, H. Dreyfus, and P. Mandel, Influence of various amino acids on the bioelectrical response to light stimulation of a superfused frog retina, Life Sci., 18:473 (1976).

26. L. K. Kaczmarek, and W. R. Adey, Modification of the direct cortical response by taurine, Electroencephalogr. Clin. Neurophysiol., 39:292 (1975).

27. K. Krnjević, and E. Puil, Electrophysiological studies on actions of taurine, in:"Taurine", R. Huxtable and A. Barbeau eds., Raven Press, New York (1976).

28. H. L. Haas, and L. Hosli, The depression of brain-stem neurons by taurine and its interaction with strychnine and bicuculline, Brain Res., 52:399 (1973).

29. V. Sonnhof, P. Grafe, J. Krumnik, M. Linder, and L. Schindler, Inhibitory post-synaptic actions of taurine, GABA and other amino acids on motoneurons of the isolated frog spinal cord, Brain Res., 100:327 (1975).

30. M. S. Starr, and M. J. Voaden, The uptake, metabolism ans release of ^{14}C-taurine by rat retina in vitro, Vision Res., 12:1261 (1972).

31. M. J. Neal, D. G. Peacock, and R. D. White, Kinetic analysis of amino acid uptake by the rat retina in vitro, Br. J. Pharmac., 47:656 (1973).

32. N. Lake, J. Marshall, and M. J. Voaden, Studies on the uptake of taurine by the isolated neural retina and pigment epithelium of the frog, Biochem. Soc. Trans., 3:524 (1975).

33. G. G. S. Collins, The rates of synthesis, uptake and disappearance of ^{14}C-taurine in eight areas of the rat central nervous system, Brain Res., 76:447 (1974).

34. J. B. Lombardini, Regional and subcellular studies on taurine in the rat central nervous system, in: "Taurine", R. Huxtable and A. Barbeau eds., Raven Press, New York (1976).

35. J. G. Jacobsen, and L. H. Smith Jr, Biochemistry and physiology of taurine and taurine derivatives, Physiol. Rev., 48:424 (1968).

36. D. B. Hope, The persistence of taurine in the brains of pyrido-
xine-deficient rats, J. Neurochem., 1:364 (1957).
37. L. Eldjarn, The conversion of cystinamine to taurine in rat,
rabbit and man, J. Biol. Chem., 206:483 (1954).
38. D. Cavallini, R. Scandurra, S. Duprè, G. Federici, L. Santoro,
G. Ricci, and D. Barra, Alternative pathways of taurine biosynthesis, in: "Taurine", R. Huxtable and A. Barbeau eds.,
Raven Press, New York (1976).
39. W. G. Martin, N. L. Sass, L. Hill, S. Tarka, and R. C. Truex, The
synthesis of taurine from sulphate. IV. An alternative pathway for taurine synthesis by the rat, Proc. Soc. Exp. Biol.
Med., 141:632 (1972).
40. W. G. Martin, R. C. Truex, S. Tarka, W. Gorby, and L. Hill, The
synthesis of taurine from sulphate VI. Vitamin B6 deficiency
and taurine synthesis in the rat, Proc. Soc. Exp. Biol. Med.,
147:835 (1974).
41. I. P. Lowe, and E. Roberts, Incorporation of radioactive sulphate sulfur into taurine and other substances in the chick
embryo, J. Biol. Chem., 212:477 (1955).
42. L. J. Machlin, and P. B. Pearson, Studies on utilization of sulphate sulfur for growth of the chicken, Proc. Soc. Exp. Biol.
Med., 93:204 (1956).
43. L. J. Machlin, P. B. Pearson, and C. A. Denton, The utilization of
sulphate sulfur for the synthesis of taurine in the developing
chick embryo, J. Biol. Chem., 212:469 (1954).
44. W. G. Martin, R. C. Truex, S. Tarka, L. Hill, and W. Gorby, The
synthesis of taurine from sulphate. VIII. A constitutive enzyme in mammals, Proc. Soc. Exp. Biol. Med., 147:563 (1974).
45. K. C. Hayes, A review of the biological function of taurine, Nutr.
Rev., 34:161 (1976).
46. R. M. Di Giorgio, G. Tucci, and S. Macaione, Cysteine oxidase
activity in rat retina during development, Life Sci., 16:429
(1975).
47. C. H. Misra, and J. W. Olney, Cysteine oxidase in brain, Brain
Res., 97:117 (1975).
48. S. Macaione, G. Tucci, G. De Luca, and R. M. Di Giorgio, Subcellular distribution of taurine and cysteine sulphinate decarboxylase activity in ox retina, J. Neurochem., 27:1411
(1976).
49. H. Pasantes-Morales, R. Salceda, A. M. Lopez Calomè, and P.
Mandel, Cysteine sulphinate decarboxylase in chick and rat
retina during development, J. Neurochem., 27:1103 (1976).
50. R. M. Di Giorgio, S. Macaione, and G. De Luca, Subcellular distribution of hypotaurine oxidase activity in ox retina, Life
Sci., 20:1657 (1977).
51. R. M. Di Giorgio, G. De Luca, and S. Macaione, Hypotaurine oxidase activity in retina, Bull. Molec. Biol. Med., 3:115 (1978).
52. S. Papa, A. G. Secchi, N. E. Lofrumento, F. D'Ermo, and E. Quagliariello, Studi sulla fosforilazione ossidativa nei mitocon-
dri di retina, Italian J. Biochem., 14:175 (1975).
53. J. L. Mangan, and V. P. Whittaker, The distribution of free amino
acids in subcellular fractions of guinea-pig brain, Biochem.
J., 98:128 (1966).
54. S. Macaione, G. Tucci, and R. M. Di Giorgio, Taurine distribution

in rat tissues during development, Italian J. Biochem.,24:162 (1975).

55. K. Yamaguchi, S. Sakakibara, and I. Veda, Indution and activation of cysteine oxidase of rat liver, Biochim. Biophys. Acta, 237:502 (1971).

56. M. C. Guion-Rain, and F. Chatagner, Rat liver cysteine sulphinate decarboxylase :some observations about substrate specificity, Biochim. Biophys. Acta, 276:272 (1972).

57. A. N. Davison, Amino acid decarboxylase in rat brain and liver, Biochim. Biophys. Acta, 19:66 (1956).

58. E. Peck, and J. Awapara, Formation of taurine and isethionic acid in rat brain, Biochim. Biophys. Acta, 141:499 (1967).

59. J. Awapara, and W. J. Wingo, On the mechanism of the taurine formation from cysteine in the rat, J. Biol. Chem., 203:189 (1953).

60. D. Cavallini, and B. Mondovi, Sulla presenza di un nuovo metabolita nelle urine di ratto alimentato con L-cistina, Giornale di Biochimica, 1:170 (1951).

61. D. Cavallini, B. Mondovi, and C. De Marco, The isolation of pure hypotaurine from the urine of rats fed cysteine, J. Biol. Chem., 216:577 (1955).

62. S. S. Oja, M. L. Karvonen, and P. Lähdesmäki, Biosynthesis of taurine and enhancement of decarboxylation of cysteine sulphinate and glutamate by the electrical stimulation of rat brain slices, Brain Res., 55:173 (1973).

63. J. Marshall, P. A. Medford, and M. J. Voaden, Subcellular fractionation of the rabbit retina : the isolation of synaptic pedicles and inner segments of photoreceptor cells, Exp. Eye Res., 19:559 (1974).

64. R. L. Mathur, J. Klethi, M. Ledig, and P. Mandel, Cysteine sulphinate decarboxylase in the visual pathway of adult chicken, Life Sci, 18:75 (1976).

65. H. Pasantes-Morales, R. Salceda, and A. M. Lopez Colomé, Taurine in normal retina, in: "Taurine and Neurological Disorders", A. Barbeau and R. J. Huxtable eds., Raven Press, New York (1978).

THE REGULATION AND FUNCTION OF

TAURINE IN THE HEART AND OTHER ORGANS

Ryan J. Huxtable

Department of Pharmacology
University of Arizona
Health Sciences Center
Tucson, Arizona   85724

The regulation of taurine in the heart has proved difficult to elucidate. This is in part due to the difficulty of modifying taurine content experimentally, and in part due to the complexity of putative metabolic routes to taurine. Insight into the function of taurine has been retarded also by the lack of a suitable taurine antagonist. Knowledge of the regulation of taurine is becoming increasingly important, due to recent findings that dietary deficiency of taurine in the cat leads to retinal degeneration and blindness, and due to the possibility that taurine may be an essential amino acid in humans (Schmidt et al. 1976; Gaull et al. 1977). This is potentially alarming, in that the commercial milks that most babies are reared on in the developed nations contain no taurine. A disturbance in taurine regulation occurs in congestive heart failure, in which a doubling of taurine concentration is observed (Peterson et al. 1973; Huxtable and Bressler 1974). This observation is of interest, apart from its possible physiological significance, in that experimental increases in cardiac taurine concentrations have yet to be achieved in the absence of cardiac dysfunction.

The problem of regulation is complicated by the metabolic complexity of sulfur amino acids, and the wide variation in organ taurine concentrations between species. The major putative metabolic routes to taurine from cysteine are three: These involve the intermediacy respectively of cysteine sulfinic acid, cysteic acid, and cysteamine. The first two utilize the enzyme cysteine sulfinic acid decarboxylase (CSAD), and the latter the enzyme cysteamine dioxygenase (CD). The distribution of these enzymes differ both quantitatively and qualitatively in corresponding organs of various species. Other pathways of taurine biosynthesis have also been proposed. For

277

none of these pathways is there conclusive evidence that the path is of quantitative importance in taurine biosynthesis. In the cysteine sulfinic acid pathway, for example, a poor correlation is obtained between the distributions of CSAD and taurine. Inhibition of the enzyme does not affect tissue content of taurine. Further- more, it is possible that one of the isozymes of CSAD is identical to glutamic acid decarboxylase. If the cysteamine pathway is con- sidered, the biochemistry whereby cysteine is converted to cysteamine has yet to be worked out, although there have been a number of ingenious suggestions. Both CD and CSAD form hypotaurine, yet no enzymatic activity has been found that will convert hypotaurine to taurine, although this is a facile chemical step (Reviewed in Huxtable 1976, 1978).

It is probable that the relative contributions of diet and biosynthesis to taurine regulation varies from species to species. Taurine is almost absent from the plant kingdom, but is present in large amounts in animal tissues, particularly muscle. A carnivore, therefore, receives a large amount of taurine in its diet, whereas a herbivore receives none. Omnivores fall between these two ex- tremes. A herbivore must either make its own taurine, or derive it microbiologically from gut or rumen flora. A carnivore is not faced with this necessity, and, as the example of the cat shows, biosyn- thesis may be insufficient to maintain taurine balance.

In this article, I shall review the evidence from my laboratory as to the source of taurine in the heart, and its possible function. I will also describe a simple method of modifying taurine content, using a compound, guanidinoethyl sulfonate, which is a competitive antagonist of taurine transport, and which is also, in part, a physiological antagonist.

TAURINE BIOSYNTHESIS IN THE HEART

CSAD activity is absent from mammalian heart. We therefore examine the possibility that taurine is formed from cysteamine.

When rat or mouse heart homogenates were incubated with 2mM ^{35}S cysteamine, radioactive taurine was produced. The production of taurine was dependent on time, and the concentration of protein incubated. CD activity was assayed in heart homogenates and semi- purified fractions by measurement of the rate of production of hypotaurine. CD activity was found by these means, in homogenates of heart and lung of rat, guinea pig, mouse, and rabbit. Activity was dependent on length of incubation and protein concentration, and exhibited saturable kinetics in each case. Specific enzyme activity (per mg protein) could be increased five to ten-fold by partial purification: On centrifugation of heart homogenate at 1100xg, 85% of the enzyme activity remained in the supernatant. Further enrichment was achieved by the addition of ammonium sulfate to 50% of saturation. Precipitated protein from cow heart, for example, had cysteamine dioxygenase activity of 13.1 nMole/mg/hr, compared with 1.3 nMole/mg/hr in the initial homogenate. These results indicate that cardiac tissue contains CD activity. Small quantities of cysteamine were detected in rat heart and other tissues, indicating that the substrate for this activity was also available. The concentrations detected were low, being of the order of nMole/g tissue, and the recovery of cysteamine at this level is only of the order of 3%. However, the results demonstrate that cysteamine does occur endogenously in rat heart (Huxtable and Bressler 1976).

Cysteamine presumably arises from cysteine. We have found in the rat, that injection of ^{35}S-methionine or ^{3}H-cystine to the intact animal leads to the appearance of radiolabeled taurine and radiolabeled cysteamine in all tissues examined. We have shown therefore that the rat contains cysteamine, and has the ability to form this substance from cystine. Furthermore, a number of species, including the rat, can convert cysteamine to hypotaurine and taurine in the heart. These data, taken together, indicate that heart has the capability of biosynthesizing taurine. However, the quantitative significance of this in the maintenance of taurine balance in the heart has not been established. The movement of radiolabeled compounds along a pathway does not necessarily indicate that mass transfer of substrate occurs by the same route. Additionally, a number of serious problems remain with the CD pathway. Although cysteamine is the decarboxylated analog of cysteine, no decarboxylase activity has ever been found. A complex scheme has been proposed for the conversion of cysteine to cysteamine, involving the intermediacy of 4'-phosphopantothenoyl cysteine, 4'-phosphopantetheine and pantetheine, acting as carriers for the cysteinoyl residue while it is decarboxylated to a cysteamine residue. Dephosphorylation of 4'-phosphopantetheine involves an, as yet, unknown phosphatase. This pathway awaits experimental validation (Huxtable 1978). Furthermore, as discussed above, no enzymatic activity has been found for the conversion of hypotaurine to taurine. Although hypotaurine spontaneously converts to taurine, it appears unlikely that an important step in biosynthesis would be nonenzymatic.

TAURINE TRANSPORT IN THE HEART

The concentration gradient for taurine across the heart cell membrane is approximately 400:1. In the isolated, perfused rat heart, taurine is taken up by an active transport process, which is saturable at taurine concentrations of 200 mM or higher. The process exhibits a Km of 45 mM and a Vmax of 32 nMole/g dry weight/min. The transport site is specific for β amino acids, showing no affinity for α amino acids. In addition, as discussed in more detail below, the structure-activity requirements for transport are amazingly

precise. β-Alanine, the carboxylic analog of taurine, is transported, but β-aminoethane phosphonic acid is not. The latter is
isoelectronic with taurine, and differs only in being a dibasic
acid. Close structural analogues of taurine, such as 1,2-dimethyl
taurine and 2-methyl-cyclohexane sulfonic acid, show no affinity for
the transport site (Huxtable and Chubb 1977, Chubb and Huxtable
1978a, Azari et al. 1979).

BIOSYNTHESIS VERSUS TRANSPORT IN THE NORMAL HEART

It is apparent that the heart has the capability to biosynthesize
taurine, and also to take it up from the circulation. We have
initiated quantitative studies as to the relative contributions of
these two sources. Rats were maintained on a diet containing taurine
of a known specific activity. Rats were sacrificed at intervals,
and the specific activity of taurine determined in various tissues.
From this, the content of taurine that had been derived from dietary
sources could be calculated, as shown in Table 1. By the 15th day,
in all tissues except the liver, influx had accounted for 1/3 of
the total taurine. The remainder must either arise from biosynthesis, or be part of the taurine pool present in the tissue at the
beginning of the experiment. These data, at face value, appear to
be inconsistent with the results of experiments described below
using guanidinoethyl sulfonate, a transport inhibitor of taurine.
In these experiments, decreases in taurine content of up to 80% were
seen within two to three weeks. One interpretation of this apparent
discrepancy is that recently influxed taurine is not equilibrating
with the total taurine pool in the cell, but comprises a pool from
which a high percentage of the effluxed taurine is derived; in other
words, a fraction of the total cell taurine turns over faster than
the average. This is in keeping with previous findings of multiple
pools of taurine in rat organs, as deduced from kinetic data
(Sturman 1973).

These experiments are still continuing as are parallel experiments in which animals are being maintained on a diet containing
sulfur amino acids of a known specific activity, in order that the
contribution of biosynthesis to taurine balance may be measured
directly.

ADRENERGIC STRESS ON TRANSPORT AND BIOSYNTHESIS IN THE HEART

The heart is an autonomic organ in that it can beat and pump
blood in the absence of any neural or hormonal influence. However,
the heart is neurally and hormonally regulated so that the cardiac
output may be varied to suit the varying demands of the body. A
major modulator of the heart is the β-adrenergic system. Sympathetic
nerve stimulation induces the release of norepinephrine which binds

Table 1. Contribution of Dietary Taurine to Organ Levels

μmole taurine/g organ

	All sources[a]	Diet			
		4d	7d	10d	15d
Serum[c]	0.53 ± 0.06	0.33 ± 0.06	0.48 ± 0.09	0.68 ± 0.16	0.89 ± 0.02
Heart	30.51 ± 1.30	4.41 ± 0.38	8.44 ± 0.48	8.59 ± 0.51	10.26 ± 0.10
Lung	16.77 ± 0.16	5.97 ± 0.52	6.22 ± 0.47	4.81 ± 0.50	5.59 ± 0.05
Liver	12.24 ± 0.53	2.82 ± 0.37	2.40 ± 1.02	5.05 ± 0.64	2.96 ± 0.85
Kidney	12.13 ± 1.34	3.61 ± 0.50	3.87 ± 0.05	4.92 ± 0.59	3.77 ± 0.12
Muscle	18.50 ± 2.18	2.69 ± 0.58	4.24 ± 0.07	4.23 ± 0.61	6.13 ± 0.07
Spleen	12.49 ± 0.34	3.74 ± 0.36	4.69 ± 0.16	4.01 ± 0.01	4.67 ± 0.06
Intestine	17.14 ± 1.18	5.18 ± 0.07	4.70 ± 0.20	3.85 ± 0.23	3.87 ± 0.01
Cerebellum	4.90 ± 0.38	0.99 ± 0.04	1.44 ± 0.05	1.93 ± 0.75	1.43 ± 0.14
Midbrain	4.72 ± 0.32	0.81 ± 0.07	0.69 ± 0.55	1.16 ± 0.16	1.41 ± 0.06
Pons Medulla	1.98 ± 0.36	0.76 ± 0.01	1.02 ± 0.09	0.88 ± 0.16	1.25 ± 0.01
Cerebral hemispheres	9.38 ± 0.34	0.72 ± 0.30	1.38 ± 0.34	2.30 ± 0.13	2.34 ± 0.31

[a]Endogenous taurine concentration. [b]Taurine derived by influx after animal has been consuming [³H]taurine for indicated number of days. [c]Values are μmole/ml. Data are reported as means ± SEM for 2 animals per time.
Procedure: Male Sprague Dawley rats (initial weight 100g, range ± 20g) were maintained on a taurine-free diet (Bio-Mix 900, Bio-Serve, Frenchtown, New Jersey) from the first day of the experiment. They were allowed free access to water containing 5 mM [³H]taurine (24625 dpm/μmole). The average daily intake of taurine approximated that which rats receive in our animal facility when maintained on Wayne rat chow (1.6 μmole taurine/g) or Purina rat chow (4.2 μmole taurine/g). After the indicated number of days, animals were sacrificed and radioactivity and taurine content determined in the listed organs. From the measurement of radioactivity, the quantity of taurine arising in each organ from dietary intake was calculated.

to β-adrenergic receptors on the heart, inducing a sequence of changes resulting in increased force of contraction, increased heart rate, and increased cardiac output. β-Adrenergic stimulation thus causes the physiological changes known as the "fight or flight" response.

If the heart is stressed sufficiently, cardic hypertrophy occurs (that is to say, the mass of the heart muscle increases) leading eventually to fibrosis and congestive heart failure. There is a doubling in taurine content in the affected ventricle in congestive heart failure in both man and animals. Except for a slight altera- tion in methionine content, the content of no other amino acid is altered (Peterson et al. 1973, Huxtable and Bressler 1974).

We have investigated the effects of adrenergic stress on the transport and biosynthesis of taurine in the heart, in order to ascertain the origin of the increased taurine content in congestive failure.

(a) Biosynthesis

Isoproterenol is a potent synthetic β-adrenergic agonist. Cardiac hypertrophy was produced in rats by twice daily injections of isoproterenol (5mg/kg sc) for periods of up to ten days (Stanton et al. 1969). After 24 hours, a significant increase in heart weight occurs, and the weight increases smoothly for the duration of isoproterenol treatment. On discontinuation, weight reverts to the initial heart weight over an eight-day period. Hypertrophy is accompanied by a marked increase in the total taurine content of the heart (fig. 1). Taurine biosynthesis was determined in this exper- iment by two means. In the first, radiolabeled cysteine was injected IP and the animals sacrificed four hours later. The appearance of radiolabeled taurine in the heart was taken as a measure of the conversion rate. The other index of biosynthesis used was the measurement of CD activity in heart homogenates. The results of these measurements are shown in figure 2. Correlation between the two methods was good, both showing no change in biosynthetic rats at the time when total taurine content was increasing most rapidly. Conversely, when taurine content was falling during regression of hypertrophy, biosynthetic rates were highest (Chubb and Huxtable 1978 a,b,c).

(b) Transport

Under the same conditions of isoproterenol-induced hypertrophy described above, a rapid and marked stimulation in taurine influx occurs (fig. 1). More detailed studies of the time course of this stimulated influx reveals that it occurs within two hours of an

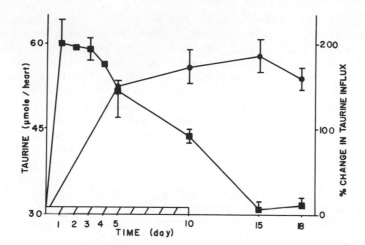

Fig. 1.   Cardiac taurine content and influx rate during
          isoproterenol-induced hypertrophy. Isoprotere-
          nol (5 mg/kg) injected s.c. every 12 hr for
          period indicated by hatched bar. Symbols:
          Squares indicate influx rate/heart as percent
          control; circles indicate μmole/heart. Each
          point is mean ± SEM of 5 rats.

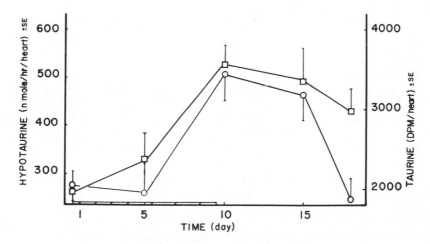

Fig. 2.   Biosynthesis of taurine during isoproterenol-
          induced hypertrophy. Conditions as fig. 1.
          Symbols: Squares indicate conversion of cys-
          tine to taurine per heart; circles indicate
          CD activity per heart (formation of hypo-
          taurine).

isoproterenol injection, before any hypertrophy has occured. Furthermore, the effect is more prolonged than the pharmacological duration
of action of isoproterenol, persisting for an excess of 36 hr, whereas
the action of isoproterenol on arterial blood pressure terminates
within 12 hr. A similar response to isoproterenol is seen in the
Langendorff perfused heart. At pharmacological concentrations of
isoproterenol, the influx of taurine (and other β-amino acids) is
stimulated, but α-amino acids are unaffected (Chubb and Huxtable
1978 a,b,c). The kinetics of this stimulated influx are interesting.
The Lineweaver-Burk plot reveals parallel lines for influx in the
presence and absence of isoproterenol (fig. 3). In other words,
stimulation is accompanied by an increase in both the Michaelis
constant, Km (indicating decreased affinity of taurine for the
transport site), and the maximum velocity of reaction (Vmax). This
type of kinetic behavior implies that there is an increased rate of
breakdown of the taurine:receptor complex into the interior of the
cell. We have suggested that a transport site phosphorylation
process is involved (Huxtable and Chubb 1977).

Sympathetic stimulation of the heart produces a cascade of
responses, any of which could be responsible for the observed increased influx of taurine. The one response we have found to be
essential, however, is increased cAMP concentration. Influx is
stimulated following perfusion of the isolated heart with either
dibutyryl cAMP (a transportable cAMP analog), glucagon (a hormone
that raises cAMP concentrations independently of the adrenergic
system), or theophylline (a phosphodiesterase inhibitor). Coper-
fusion of isoproterenol with propanolol - α β-adrenergic antagonist
- blocks the stimulation of influx, confirming the the response is
an adrenergic one (Chubb and Huxtable 1978c).

It does not appear that events in the adrenergic cascade past
the increase in cAMP concentrations are involved in the stimulation
of taurine influx. For example, increased calcium influx does not
appear to be a factor. If the increase in calcium flux produced by
isoproterenol is blocked by the calcium antagonist verapamil, the
increase in cAMP still occurs, as does increased taurine influx.

We have, therefore, found a very unusual modulation of the
influx of an amino acid. Stimulated uptake of taurine in the
presence of adrenergic activation is probably the reason for the
high levels found in congestive heart failure. Such a state is
produced as the end result of prolonged stress on the heart, in
which there has probably also been a prolonged or continuous activation of the β-adrenergic system. A possible physiological function
for this modulation is discussed below.

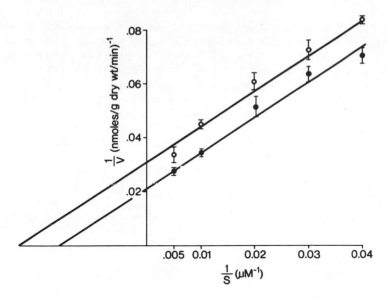

Fig. 3. Lineweaver-Burk plot of taurine transport in the
         isolated, perfused heart. Open symbols: control
         hearts (Vmax 32 nMole/g dry wt/min; Km 45µM).
         Filled symbols: hearts coperfused with 4×10^{-7}M
         isoproterenol (Vmax 46 nMole/g dry wt/min; Km
         62µM). Each point is mean ± SEM of 4 hearts.

(c) <u>Biosynthesis Versus Transport in the Stressed Heart</u>

As figures 1 and 2 illustrate, the increase in taurine content
in cardiac hypertrophy is accompanied by a marked stimulation of
taurine influx, with no alteration in biosynthesis. This indicates
a priori that, regardless of the role that biosynthesis plays in the
normal heart, during hypertrophy the increase in taurine is derived
from influx.

Dietary experiments also emphasize the importance of influx
relative to biosynthesis. Three groups of rats were compared: One
group ("high diet") was maintained on standard chow and drinking
water containing 100 mMole taurine: Another group ("normal diet")
was maintained on standard chow containing 3.9 µMole taurine/g; and
a third group ("low diet") was maintained on a taurine-free diet.
In these animals CD activity varied inversely with dietary taurine;
i.e. animals receiving the highest amount of taurine had the highest
CD activity in the heart. A partially purified fraction from the

hearts of rats on the high taurine diet has a Vmas of 49 nMole hypotaurine/mg protein/hr, compared to 16 from control animals and 11 from "low taurine" animals. However, taurine biosynthesis from cystine did not vary between the three for the heart, although in the kidney the high taurine diet resulted in the greatest rate of biosynthesis, and vice versa. Conversely, a good correlation was obtained between diet and taurine influx (table 2). Animals maintained on the low diet showed the highest rate of influx, and those on the high diet the lowest rate of influx. Response to isoproterenol-induced stress was exaggerated in the low diet animals, and diminished in the high diet animals (table 2).

All these experiments combine to indicate that, although the heart can biosynthesis taurine, the influx of taurine is more important in maintaining the cardiac cell content of this substance.

HYPOTHESES FOR THE FUNCTION OF TAURINE IN THE HEART

Does the stimulation in taurine influx as a result of adrenergic excitation serve any function? The editors have invited speculation, so I will take advantage of that to suggest two functions for the adrenergic modulation of taurine flux into the heart, one presynaptic and the other postsynaptic.

Taurine increases calcium flux into the heart cell, and increases the pool size of calcium available for contraction. There is recent evidence that these changes are achieved by an allosteric action of taurine, increasing the binding affinity of calcium for a membrane site (Chovan et al. 1979). I propose that the magnitude of the effect is regulated by the flux of taurine across the sarcolemmal membrane: thus, in terms of modifications in calcium movement, the total taurine content of the heart is less important than the membrane flux of taurine. β-Adrenergic receptor activation increases the taurine flux, which on this hypothesis would increase the affinity of calcium for sites on the inner aspect of the sarcolemma, favor increased calcium entry, and achieve a magnification of the adrenergic response.

I also suggest a presynaptic action of taurine, acting by the following sequence. In response to some as yet unknown stimulus, possibly stretch of the sarcolemmal membrane, the permeability of the cell membrane to taurine increases. As a result of the high concentration gradient from inside the cell to outside, there is a sudden rise in the rate of efflux of taurine. It has been shown that subcutaneous injections of taurine cause increased release of norepinephrine. Norepinephrine, in turn, binds to the adrenergic receptor causing an increase in intracellular cAMP content. The reuptake of taurine is thereby stimulated, and the concentration of taurine in the synaptic cleft falls, decreasing the release of

Table 2.   Effect of Dietary Taurine on the Transport and Biosynthesis of Taurine in the Heart.

	High[a]	Diet Control[b]	Low[c]
Cysteine influx	10280 ± 430	9940 ± 370	10920 ± 430
Taurine biosynthesis	2280 ± 860	2190 ± 370	2600 ± 330
Taurine influx	50410 ± 2700[e]	69900 ± 1500	85630 ± 5700[e]
Taurine influx after isoproterenol	87523 ± 4500[e]	159944 ± 3750	240360 ± 6000[e]

All values are dpm/heart, reported as means ± SEM of 4 animals.

[a]Male Wistar rats were given free access to 100 mM taurine solution and to Wayne rat chow (Chicago) containing 3.9 μMole taurine/g. The diet was maintained for 2 weeks.   [b]Protocol the same except drinking water contained no taurine.   [c]Protocol the same except casein-derived chow (taurine-free) was used.   [d]Four injections of isoproterenol (5 mg/kg) were given at 12 hr intervals, and influx determined 12 hr after last injection.   [e]p <.05 relative to control diet.   Methodology as for Chubb and Huxtable 1978a.

---

norepinephrine (Prous et al. 1978).

This hypothesis suggests that taurine is having a modulatory role on norepinephrine release, such that both the onset and offset of release is accelerated.   This is in keeping with the concept that taurine serves as a neuromodulator rather than as a neurotransmitter. Furthermore, in terms of the functionality of taurine, the rate of transport across the cell membrane is of greater importance than the absolute concentration of taurine within the cell.   Verification of this hypothesis would explain the wide variations in cardiac taurine content from species to species and would explain why it is possible to lower taurine concentrations in the cat and other species to about 10 or 20% of their original values without any marked pathology being observed (Knopf et al. 1978; Huxtable et al. 1979). As the normal intercellular:extracellular concentration ratio is about 400:1, a decrease to a ratio of 40-80:1 still permits a marked rate of efflux on the cell membrane becoming permeable to taurine. Indeed, in both the above referenced models, serum taurine concentrations also fall.

INHIBITORS OF TAURINE TRANSPORT

The structure-activity requirements of the taurine transporting system in the heart are quite precise.   Compounds we have found to

Table 3. Inhibitors of Taurine Influx

Taurine Influx Rate
(nmole/min/g dry weight)

Compound	Concentration (µM)	Control	Treated
Hypotaurine	150	10.6 ± 1.2	1.8 ± 1.0
β-Alanine	500	21.4 ± 2.4	3.6 ± 1.6
β-Guanidinoethyl	150	18.4 ± 1.2	10.9 ± 1.0
sulfonate	250	10.2 ± 0.2	4.3 ± 0.1
β-Guanidino-	150	13.2 ± 1.3	10.4 ± 0.3
propionate			

Values are given as means ± SD. Hearts per group 4-11 (average of 6). Hearts were perfused with 25µM ^{3}H-taurine, except for the β-alanine experiment, where 100µM was perfused. Inhibitors were coperfused at the concentrations given.

_____

be effective inhibitors in the Langendorff heart preparation are shown in table 3. The inhibitors hypotaurine and β-alanine both occur naturally in the rat, and both are straight chain β-amino acids, as is taurine. Hypotaurine is the sulfinic acid analog and β-alanine the carboxylic acid analog of taurine. Guanidinoethyl sulfonate and guanidinopropionate, the amidino derivatives of taurine and β-alanine respectively, are also transport inhibitors of taurine.

Active transport of β-alanine occurs in the heart; apparently by the same system that transports taurine. However, as shown in figure 4, there is a nonsaturable component to transport. The pKa of β-alanine is 3.6. Thus, at a pH of 7.2, the proportion of β-alanine with a non-ionized acid function is 125 times higher than for an equivalent concentration of taurine. Guanidinoethyl sulfonate is transported in place of taurine by a saturable process (fig. 5). The Lineweaver-Burk plot for transport is shown in figure 5. Over the concentration range 0-400µM, transport occurs by a one component system with an affinity (Km) of 153 µM, and a Vmax of 65 nmole/g dry weight/min.

Can these inhibitors be used to decrease organ taurine content in the intact animal? Hypotaurine is of little practical use as a transport inhibitor, as it is readily oxidized to taurine both in the perfusion medium, and after transport into the heart.

β-Alanine has potential usefulness as a transport inhibitor. However, it has physiological and pharmacological actions of its own, and there is a nonsaturable component to its uptake by the heart. Guanidinoethyl sulfonate, on the other hand, does not occur

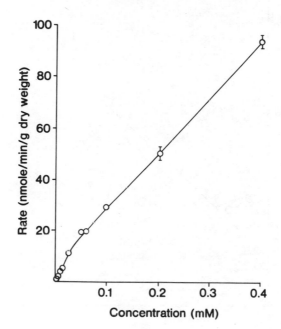

Fig. 4.   Transport of β-alanine in the isolated, per-
          fused heart. Each point is mean ± SEM of
          4-7 hearts.

in mammalian heart (Guidotti and Costagli, 1970). Furthermore, its
transport is saturable. We have, therefore, tested this compound
in vivo for its effects on tissue taurine concentrations.

DEPLETION OF TAURINE CONTENT BY GUANIDINOETHYL SULFONATE

     Guinea pigs, rats, and mice given free access to drinking water
containing 1% guanidinoethyl sulfonate show marked decreases in the
concentration of taurine in all tissues examined, regardless of the
taurine content of their chow. The data reported here are for rats
maintained on a taurine-free diet. The changes in taurine concentration of various organs following different durations of exposure
to guanidinoethyl sulfonate are shown in table 4. Within 9 days,
there is marked decrease in taurine levels in all areas outside of

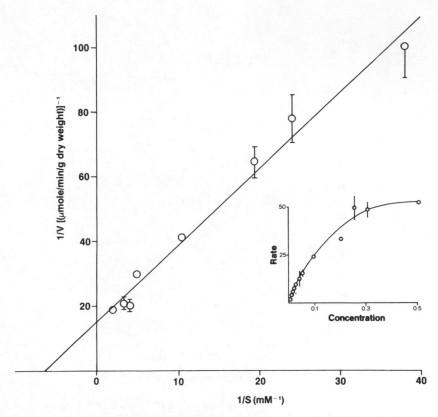

Fig. 5.   Lineweaver-Burk plot of guanidinoethyl sulfon-
          ate transport in the isolated, perfused heart.
          (Vmax 65 nMole/min/g dry wt; Km 153μM).

the brain. In the brain, depletions range from 52% in the frontal
cortex, to 30% in the cerebellum and inferior colliculus. The
spinal cord shows the smallest decrease, at 14%. Depletion of
taurine content is accompanied by accumulation of guanidinoethyl
sulfonate. Concentrations of free amino acids apart from taurine
are unaltered in the brain or the heart (other organs were not
examined).

Despite these dramatic modifications in taurine content, the
only gross behavioral abnormality noted was hyperactivity. However,
no specific behavioral tests were applied. Intracerebral injections
of 9.2 μmole taurine into seizure susceptible rats cause marked
attenuation of seizure severity, in keeping with the anticonvulsant
actions of taurine shown in a number of laboratories (Laird and
Huxtable 1976). Similar injections of guanidinoethyl sulfonate
lead to intense seizure activity within five seconds, before the
injection had been completed. The animal suffered continuous

Table 4. Effect of Guanidinoethyl Sulfonate
and Taurine Free Diet on Tissue Taurine Content

Days of Treatment	Percent of Control Concentrations						
	2	4	6	9	20	30	40
Serum	72±12	48±3	27±2	23±3	42±8	38±3	53±6
Heart	82±4	74±8	40±3	31±4	24±2	24±2	30±2
Liver				28±5	39±6		
Lung				25±2	24±1		
Kidney				36±5	45±6		
Spleen				22±1	31±4		
Muscle	96±0	70±6	63±5	39±2	24±0		
Intestine				30±3	40±8		
Cerebellum	89±2	78±6	69±4	70±7	45±10	43±1	45±6
Hypothalamus				63±2	51±3	40±1	51±6
CH				56±8	41±2	50±1	47±3
Midbrain				49±5	43±0	50±1	47±3
Medulla-pons				57±5	34±5	26±2	42±5
Frontal cortex				48±2	37±3	53±1	63±3
IC	88±3	82±5	68±4	70±7	47±7	39±1	53±4
Spinal cord				84±4	55±5	47±7	52±6

Data are mean percentages of control values for corresponding period
± SEM. 3-4 Animals used per point. Male Sprague-Dawley rats (initial weight 125g) were maintained on water containing 1% guanidinoethyl sulfonate and on taurine-free chow (Biomix 900).
CH: Cerebral hemispheres   IC: Inferior colliculus

clonic seizures, and died that night. A seizure resistant animal,
injected in the same manner, started seizuring within five minutes,
and died 48 hr later (Laird, unpublished data). This suggests that
guanidinoethyl sulfonate is not only a transport antagonist, but at
least in the brain, may also be a physiological antagonist to
taurine.

A number of conclusions follow from these experiments:
(i) A marked decreased in cardiac taurine content may be achieved
within a few days of animals receiving drinking water containing
1% guanidinoethyl sulfonate: (ii) Cardiac taurine content cannot
be decreased below 20-30% of normal, regardless of the length of
time guanidinoethyl sulfonate is administered. This irreducible
component of the taurine content may represent the proportion of
taurine made available by biosynthesis rather than influx; and
(iii) Taurine levels are modified without effects on other free
amino acid levels.

CONCLUSIONS

(1) Cysteamine dioxygenase activity is present in the heart, and the heart has the capability to biosynthesize taurine. The heart can also take up taurine from the circulation.

(2) In the stressed heart, transport is more important than biosynthesis in the regulation of cardiac taurine content.

(3) The transport of taurine into the heart is stimulated by adrenergic stimulation of the heart. It is proposed that this mechanism accounts for the large increase in taurine observed in hearts in congestive failure.

(4) A hypothesis is proposed for the function of taurine in the heart, and for the importance of a transport process in these functions. It is proposed that taurine has both a pre- and postsynaptic action.

(5) β-Alanine, hypotaurine, guanidinoethyl sulfonate and guanidinopropionate are inhibitors of taurine in the Langendorff perfused rat heart.

(6) Guanidinoethyl sulfonate is transported into the rat heart by a one component saturable system. The transport of β-alanine into the heart has both a saturable and a non-saturable component.

(7) Marked depletion of tissue taurine concentrations occur within days when rats and other small mammals are maintained on drinking water containing 1% guanidinoethyl sulfonate.

(8) Depletion in taurine content occurs without alteration in the concentrations of other free amino acids.

(9) Guanidinoethyl sulfonate administration is proposed as a simple and effective method for the experimental manipulation of taurine concentration.

ACKNOWLEDGEMENTS

This work was supported by USPHS grants HL19394, HL20087, and NS14405.

J. Azari, J. Bahl, and R. Huxtable, Guanidinoethyl sulfonate and other inhibitors of the taurine transporting system in the heart, Proc. Western Pharm., Soc. 22: in press (1979).

J. P. Chovan, E. C. Kulakowski, B. W. Benson, and S. W. Schaffer, Taurine enhance of calcium binding to rat heart sarcolemma, Biochim. Biophys. Acta, 551:129 (1979).

J. Chubb and R. Huxtable, The effects of isoproterenol on taurine concentration in the rat heart, Eur. J. Pharmacol., 48:357 (1978a).

J. Chubb and R. J. Huxtable, Transport and biosynthesis of taurine in the stressed heart, in: "Taurine and Neurological Disorders," A. Barbeau and R. Huxtable, eds., New York (1978b).

J. Chubb and R. Huxtable, Isoproterenol-stimulated taurine influx
     in the perfused rat heart, Eur. J. Pharmacol., 48:369 (1978c).
G. E. Gaull, D. K. Rassin, N. C. R. Räihä, and K. Heinonen, Milk
     protein quantity and quality in low-birth-weight infants,
     J. Pediatrics, 90:348 (1977).
A. Guidotti and P. F. Costagli, Occurrence of guanidotaurine in
     mammals: variation of urinary and tissue concentration after
     guanidotaurine administration, Pharmacol. Res. Comm., 2:341
     (1970).
R. Huxtable, Metabolism and function of taurine in the heart, in:
     "Taurine," R. Huxtable and A. Barbeau, eds., Raven Press,
     New York (1976).
R. J. Huxtable, Regulation of taurine in the heart, in: "Taurine
     and Neurological Disorders," A. Barbeau and R. Huxtable, eds.,
     Raven Press, New York (1978).
R. J. Huxtable and R. Bressler, Taurine concentrations in congestive
     heart failure, Science, 184:1187 (1974).
R. Huxtable and R. Bressler, The metabolism of cysteamine to taurine,
     in: "Taurine," R. Huxtable and A. Barbeau, eds., Raven Press,
     New York (1976).
R. Huxtable and J. Chubb, Adrenergic stimulation of taurine trans-
     port by the heart, Science, 198:409 (1977).
R. J. Huxtable, H. E. Laird and S. Lippincott, Rapid depletion of
     tissue taurine content by guanidinoethyl sulfonate, in: "Actions
     of Taurine on Excitable Tissues," S. L. Baskin, S. W. Schaffer,
     and J. Kocsis eds., Spectrum Press, New York (1979).
K. Knopf, J. A. Sturman, M. Armstrong, and K. C. Hayes, Taurine: an
     essential nutrient for the cat, J. Nutr., 108:773 (1978).
H. Laird and R. Huxtable, Effects of taurine on audiogenic seizure
     response in rats, in: "Taurine," R. Huxtable and A. Barbeau,
     eds., Raven Press, New York, (1976).
M. B. Peterson, R. J. Mead, and J. D. Welty, Free amino acids in
     congestive heart failure, J. Mol. Cell Cardiol., 5:139 (1973).
J. G. Prous, A. Carlsson, and M. A. Gomez, The effect of taurine on
     motor behavior, body temperature and monoamine metabolism in
     rat brain, N-S. Arch. Pharmacol., 304:95 (1978).
S. Y. Schmidt, E. L. Berson, and K. C. Hayes, Retinal degeneration
     in cats fed casein.1. taurine deficiency, Invest. Ophthalmol.,
     15:47 (1976). .
H. C. Stanton, G. Brenner, and E. Mayfield, Studies on isopro-
     terenol-induced cardiomegaly in rats, Am. Heart J., 77:72
     (1969).
J. Sturman, Taurine pool sizes in the rat: effects of vitamin B-6
     deficiency and high taurine diet, J. Nutr., 103:1566 (1973).

EFFECTS OF ISCHEMIA ON TAURINE LEVELS

J.B. Lombardini

Department of Pharmacology and Therapeutics
Texas Tech University School of Medicine
Lubbock, Texas 79430

INTRODUCTION

The attempts to correlate the high levels of taurine which
are present in cardiac tissue with a physiological function have
generated much speculation. However, success has thus far been
quite fleeting as unfortunately a mechanism(s) of action for taurine in the myocardium has eluded all investigations and the role
of taurine remains an unknown entity in the heart. The only observation concerning taurine in cardiac tissue that is non-controver-
sial is that it constitutes approximately 50% of the free amino
acid pool of the heart (Jacobsen and Smith, 1968).
    In this chapter some recent data on the effects of ischemia
produced by the sympathomimetic agents, DL-isoproterenol and methoxamine, on taurine levels in rat heart, blood and liver will be
discussed. Also previously published experiments from our laboratory concerning the ischemic effects of ligation of a coronary
vessel in the dog (Crass and Lombardini, 1977) and anoxic perfusion
of isolated rat hearts (Crass and Lombardini, 1978) on cardiac taurine levels will be reviewed.

METHODS

    The circumflex branch of the left main coronary artery 2 cm
distal to the bifurcation was ligated in adult mongrel dogs to produce ischemia. Cardiac tissue samples were obtained for taurine
analyses four hours after ligation as described by Crass and Lombardini (1977).
    Cellular hypoxia was produced in Sprague-Dawley rat hearts by
perfusing with medium equilibrated with 95% N_2 - 5% CO_2. Taurine
levels were measured in ventricular tissue after either 5 minutes

of a nonrecirculated washout or 30 minutes of a recirculated perfusion (Crass and Lombardini, 1978).

Heart and liver tissues were homogenized in 2 or 3 volumes of
2% perchloric acid. Blood was deproteinized with an equal volume
of 5% perchloric acid. Tissues were analyzed for taurine by either
an enzymatic double isotope derivative assay or the amino acid analyzer technique (Lombardini, 1975). Quantitative determination of
other individual amino acids was performed on a Beckman 121 automatic amino acid analyzer. (Glutamine and asparagine eluted as a
single peak under the conditions employed.)

RESULTS

Coronary Artery Occlusion (Dog)

The effects of ligation of the circumflex branch of the main
coronary artery are shown in Fig. 1. The left ventricle demonstrated the greatest (p < 0.001) decrease in taurine content (55.6

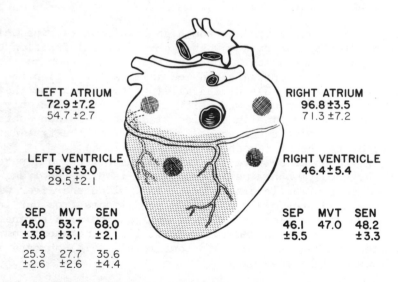

*Figure 1. Posterior surface of the canine heart. Schematic dia-
gram of area of myocardial ischemia (hatched area) after
ligation of circumflex branch of the left main coronary
artery. Taurine levels of cardiac regions (cross-hatched
circles) obtained from hearts which were non-ligated (controls) are shown in bold type; taurine levels from ischemic heart regions are shown in light type. SEP = subepi-
cardium; MVT = midventricle; SEN = subendocardium. Each
value represents the mean ± S.E. of samples from 4 or 5
animals. (Data from Crass and Lombardini, 1977.)*

± 3.0 to 29.5 ± 2.1 μmoles • g dry wt^{-1}). A transmural gradient was observed in the control left ventricle with the subendocardial layer containing the highest concentration of taurine (68.0 ± 2.1 μmoles). However, the gradient was not apparent in the ischemic left ventricle thus indicating a greater loss of taurine in the subendocardial region (Δ 32.4 μmoles) then in either the subepicardial (Δ 19.6) or midventricular regions (Δ 26.0). No gradient was observed in the control right ventricle.

The atria, regions considered to be "non-ischemic", in this preparation also decreased in taurine content. The loss of taurine was greater in the right atria (96.8 ± 3.5 to 71.3 ± 7.2) ($p < 0.02$) than in the left atria (72.9 ± 7.2 to 54.7 ± 2.7) ($p > 0.10$).

Anoxic Perfusion (Rat)

The ventricular taurine content of rat hearts perfused for 5 minutes in a nonrecirculating system with medium equilibrated with 95% N_2 - 5% CO_2 was 146.4 ± 3.7 μmoles • g dry wt^{-1} as compared to a control (95% O_2 - 5% CO_2) value of 154.4 ± 2.6 μmoles ($p > 0.1$). Additional 30 minutes of perfusion with oxygen-deficient medium significantly reduced ($p < 0.02$) the taurine content of the ventricles to 132.2 ± 3.6 μmoles. Perfusion with oxygenated medium (95% O_2 - 5% CO_2) did not change the taurine content (159.6 ± 2.7). Taurine was quantitatively recovered in the perfusate. Moreover, the release of taurine into the perfusate increased with time during the 30 minute sampling period. (Published data of Crass and Lombardini, 1978.)

Sympathomimetic Agents (Rat)

Taurine content of control rat ventricles was shown to be 97.6 ± 2.9 μmoles • g dry wt^{-1} (Fig. 2). Administration of cardiotoxic doses (80 mg/kg) of DL-isoproterenol (ISO) decreased the taurine content in rat ventricles by 29.6% (68.7 ± 2.6 μmoles) after 7 hours and by 42.0% (56.6 ± 8.4 μmoles) after 24 hours. The concentration of taurine in the blood increased from 0.226 ± 0.016 to 0.442 ± 0.033 μmoles • ml^{-1} seven hours after ISO administration and returned to control values (0.256 ± 0.023 μmoles) after 24 hours. Liver taurine content did not change 7 hours after administration of ISO although after 24 hours taurine levels decreased but not significantly ($p > 0.05$).

The total amino acid pattern of ventricular tissue is shown in Table 1. Administration of ISO (7 hours) altered the levels of both acidic protein amino acids, glutamate and aspartate. Aspartate levels increased by 40.3% while glutamate levels decreased by 13.4%. The ventricular content of both of these acidic amino acids reverted to control values after 24 hours. Of the neutral amino acids, threonine decreased as did glutamine plus asparagine (combined analyses). Glutamine-asparagine content decreased by 17.4 μmoles • g dry wt^{-1}, a loss of 37.3%. Alanine increased (12.59 ± 0.25 to 17.27 ± 0.78

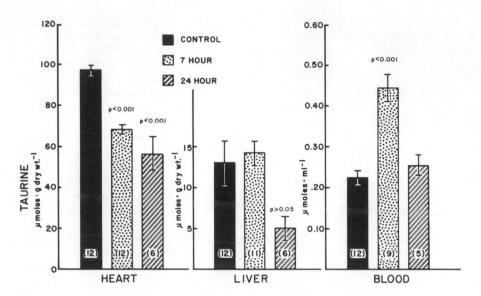

*Figure 2.   Effect of DL-isoproterenol on taurine levels in rat ven-
tricles, liver and blood.   Animals were killed seven or
24 hours after administration of 80 mg/kg of DL-isoproterenol or saline.   Number of animals per group is shown
in parentheses.   (Mean ± S.E.)*

μmoles).   Two ninhydrin positive compounds, phosphoethanolamine and
ethanolamine increased significantly in the rat ventricles.   Tissue
levels of the non-protein amino acid, citrulline, significantly decreased only 24 hours after ISO administration while ornithine levels
increased.   The basic protein amino acids did not change after treatment with ISO.

The total amino acid pattern of blood was also analyzed (Table
2).   The concentrations of most of the essential amino acids (threonine, valine, isoleucine, leucine, and arginine) decreased significantly after administration of DL-isoproterenol.   In addition, the
concentrations of serine, alanine, tyrosine and ornithine were reduced.   Changes in amino acid concentrations varied from 33% to 63%.

A summary of the effects of DL-isoproterenol on all ninhydrin
positive substances (NPS) in the rat ventricle is shown in Table 3.
In the control animals taurine (102.5 ± 2.4 μmoles) constituted 41.8%
of all amino acids (245 ± 3.3 μmoles) and 34.5% of the total NPS
(297.0 ± 7.9 μmoles).   Seven hours after ISO administration 36.5% of
the total amino acid (198.4 ± 7.2 μmoles) and 29.0% of the total NPS
pools (249.9 ± 7.4 μmoles) were taurine (72.4 ± 3.8 μmoles).   Twenty-
four hours after ISO administration the quantity of taurine was reduced to 28.8% of the total amino acids (196.7 ± 4.4 μmoles) and
23.7% of the total NPS (238.8 ± 6.3 μmoles).   The total amino acid
content of the rat ventricles (not including taurine) decreased by

Table 1. *Effects of DL-isoproterenol on the content of amino acids and other ninhydrin positive substances in rat ventricular tissue.*

	ISOPROTERENOL		
	Control	7 Hours	24 Hours
	(μmoles · g dry tissue^{-1})		
Protein Amino Acids			
Acidic amino acids			
Aspartic acid	8.13 ± 0.40	11.41 ± 0.29[a]	8.20 ± 0.30
Glutamic acid	33.92 ± 0.45	29.38 ± 1.75[d]	40.79 ± 2.92
Neutral amino acids			
Threonine	4.23 ± 0.28	3.47 ± 0.06[d]	3.60 ± 0.17
Serine	7.73 ± 0.71	6.71 ± 0.24	6.16 ± 0.38
Glutamine & Asparagine	46.46 ± 5.00	29.11 ± 2.50[c]	32.80 ± 3.70
Glycine	5.34 ± 0.55	5.95 ± 0.24	5.83 ± 0.24
Alanine	12.59 ± 0.25	17.27 ± 0.78[a]	16.62 ± 0.75[a]
Valine	2.05 ± 0.17	2.00 ± 0.25	2.06 ± 0.25
Methionine	0.63 ± 0.07	0.71 ± 0.07	0.69 ± 0.04
Isoleucine	1.44 ± 0.13	1.53 ± 0.10	1.46 ± 0.03
Leucine	3.12 ± 0.24	3.00 ± 0.21	3.21 ± 0.11
Tyrosine	1.32 ± 0.07	1.17 ± 0.07	1.31 ± 0.08
Phenylalanine	1.44 ± 0.12	1.51 ± 0.08	1.53 ± 0.09
Basic amino acids			
Lysine	6.88 ± 0.75	6.54 ± 0.62	9.35 ± 0.53[c]
Histidine	1.94 ± 0.16	2.07 ± 0.32	1.24 ± 0.09[d]
Arginine	3.44 ± 0.36	2.52 ± 0.24	3.24 ± 0.11
Non-Protein Amino Acids **Plus Other Ninhydrin Positive Compounds**			
Phosphoserine	0.59 ± 0.09	0.36 ± 0.04	0.50 ± 0.05
Phosphoethanolamine	1.47 ± 0.13	2.66 ± 0.27[b]	3.76 ± 0.24[a]
Urea	48.60 ± 6.60	46.11 ± 4.30	36.55 ± 2.72
Citrulline	0.95 ± 0.10	0.97 ± 0.14	0.56 ± 0.03[b]
Ornithine	0.25 ± 0.05	0.16 ± 0.04	0.44 ± 0.06[d]
Ethanolamine	1.54 ± 0.34	2.73 ± 0.20[c]	1.75 ± 0.22
Taurine	102.60 ± 2.40	72.40 ± 3.80[a]	56.60 ± 8.30[a]

Each value represents the mean ± standard error of 5 animals per group.

[a]equals $p < 0.001$ relative to control values.

[b]equals $p < 0.01$ relative to control values.

[c]equals $p < 0.02$ relative to control values.

[d]equals $p < 0.05$ relative to control values.

Table 2.   Effects of DL-isoproterenol on the content of amino acids
           and other ninhydrin positive substances in rat blood.

	ISOPROTERENOL	
	Control	7 Hours
	(μmoles $\cdot$ ml^{-1})	
Protein Amino Acids		
Acidic amino acids		
Aspartic acid	0.060 ± 0.006	0.049 ± 0.004
Glutamic acid	0.307 ± 0.035	0.327 ± 0.047
Neutral Amino Acids		
Threonine	0.316 ± 0.026	0.128 ± 0.012[a]
Serine	0.293 ± 0.026	0.156 ± 0.017[b]
Glutamine & Asparagine	1.193 ± 0.150	1.429 ± 0.184
Glycine	0.356 ± 0.046	0.304 ± 0.022
Alanine	0.535 ± 0.042	0.356 ± 0.031[b]
Valine	0.189 ± 0.011	0.075 ± 0.008[a]
Isoleucine	0.094 ± 0.005	0.043 ± 0.004[a]
Leucine	0.172 ± 0.010	0.077 ± 0.007[a]
Tyrosine	0.110 ± 0.008	0.041 ± 0.004[a]
Phenylalanine	0.065 ± 0.003	0.064 ± 0.007
Basic amino acids		
Lysine	0.727 ± 0.085	0.645 ± 0.045
Histidine	0.079 ± 0.005	0.082 ± 0.008
Arginine	0.297 ± 0.022	0.160 ± 0.011[a]
Non-Protein Amino Acids		
Plus Other Ninhydrin Positive Compounds		
Phosphoserine	0.146 ± 0.015	0.145 ± 0.012
Phosphoethanolamine	0.048 ± 0.007	0.043 ± 0.006
Urea	6.110 ± 0.454	6.766 ± 0.593
Citrulline	0.073 ± 0.007	0.055 ± 0.004
Ornithine	0.080 ± 0.013	0.033 ± 0.003[b]
Ethanolamine	0.023 ± 0.003	0.018 ± 0.004
Taurine	0.263 ± 0.021	0.479 ± 0.041[a]

Each value represents the mean ± standard error of 5 animals per group.

[a]equals $p < 0.001$ relative to control values.

[b]equals $p < 0.01$ relative to control values.

11.8% seven hours after ISO administration and returned to the control value after 24 hours.

The taurine content of rat ventricular tissue was also significantly reduced by methoxamine, an α-adrenergic agonist; however, both liver and blood levels increased (Fig. 3). The administration

Table 3. Content of ninhydrin positive substances (NPS) in rat ventricular tissue (μmoles · g dry wt^{-1}).

	DL-ISOPROTERENOL		
	Control	7 Hours	24 Hours
Taurine	102.5 ± 2.4	72.4 ± 3.8	56.6 ± 8.3
Amino acids	142.9 ± 3.7	126.0 ± 6.3	140.1 ± 7.0
Other ninhydrin positive substances[a]	51.6 ± 6.4	51.5 ± 4.2	42.1 ± 2.7
Total amino acids plus taurine	245.4 ± 3.3	198.4 ± 7.2	196.7 ± 4.4
Total ninhydrin positive substances	297.0 ± 7.9	249.9 ± 7.4	238.8 ± 6.3

Each value represents the results from five animals (Mean ± S.E.).

[a]Includes urea, phosphoethanolamine and ethanolamine.

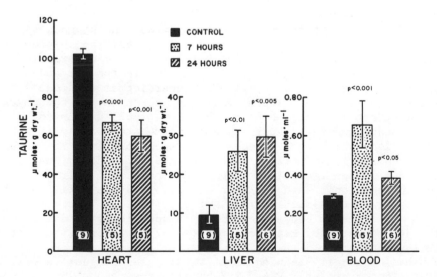

Figure 3. Effect of methoxamine on taurine levels in rat ventricles, liver and blood. Animals were killed seven or 24 hours after administration of 20 mg/kg of methoxamine. Control animals received saline. The number of animals in each group is shown in parentheses. (Mean ± S.E.)

of methoxamine (20 mg/kg) decreased the ventricular taurine content from a control value of 101.8 ± 2.6 µmoles • g dry wt^{-1} to 66.3 ± 3.8 µmoles after seven hours and to 59.8 ± 7.8 µmoles after 24 hours. The taurine content of liver increased from a control value of 9.58 ± 2.43 to 26.1 ± 5.3 µmoles after seven hours and to 29.75 ± 5.16 µmoles after 24 hours. However, as was observed after DL-isoproterenol administration, blood taurine concentrations first increased from 0.288 ± 0.011 µmoles • ml^{-1} to 0.652 ± 0.128 µmoles (seven hours) and then returned to a value of 0.380 ± 0.033 µmoles (24 hours).

DISCUSSION

     A regional heterogeneity in taurine content in the dog heart has been observed by Kocsis et al. (1976) and Crass and Lombardini (1977). It has also been reported that taurine can alter electrophysiological parameters of cardiac tissue, specifically that taurine is an antiarrhythmic agent (Read and Welty, 1963; Welty and Read, 1963; Novelli et al., 1969; Guidotti and Giotti, 1970; Chazov et al., 1974) effective against toxic doses of epinephrine, digoxin and digitoxin (Read and Welty, 1963; Welty and Read, 1963; Novelli et al., 1969) and strophanthin-K (Chazov et al., 1974). Read and Welty (1963) demonstrated that the intravenous administration of taurine terminated acute digoxin-induced arrhythmias and returned chronic digitoxin-induced premature ventricular contractions to a normal sinus rhythm. Chazov et al. (1974) suggested that taurine is perhaps necessary for the normal function of cardiac tissue by regulating the excitability of the myocardium and by modifying membrane permeability to potassium ions. In opposition, Hinton et al. (1975) reported that taurine was ineffective in counteracting ventricular arrhythmias induced by deslanoside in the cat and was even shown to intensify the abnormal electrocardiogram by producing ventricular fibrillation. In suggesting a mechanism for the physiological effects of taurine Read and Welty (1963) and Chazov et al. (1974) considered that taurine was converted to isethionic acid, a strong anion, which was subsequently responsible for the intracellular accumulation of either potassium and/or calcium ions. However, it has been recently communicated that taurine is not metabolized to isethionic acid (Fellman et al., 1978; Cavallini et al., 1978) and thus this mechanism is not possible.

     A relationship between cardiac pathophysiology and taurine was first demonstrated by Peterson et al. (1973) who observed that taurine levels increased three-fold in the right ventricle of the dog which had surgically-induced right ventricular congestive failure. Huxtable and Bressler (1974a, 1974b) also observed a similar phenomenon, that is, an increase in taurine content, in human patients who died of left ventricular congestive failure. Moreover, Huxtable and Bressler (1974a, 1974b) also observed that hypertensive rats and humans also had elevated levels of taurine in cardiac tissue. On the other hand, a _loss_ of taurine from the myocardium has been reported by Crass and Lombardini (1977, 1978) in ischemic or oxygen-deficient cardiac tissue.

The left ventricle of the dog released 47% (26.1 μmoles ·g dry wt^{-1}) of its taurine content after ligation of the circumflex branch of the left main coronary artery. Furthermore, it was demonstrated that a specific region of the left ventricle, the subendocardium, lost the greatest quantity of taurine (48%). These observations suggested that a loss of tissue taurine might be related to conduction disturbances which result in electrical abnormalities since this region is anatomically rich in Purkinje fibers (Cardwell and Abramson, 1931) and is most susceptable to ischemic damage (Moir, 1972). It is well known that the subendocardium is metabolically more active than the midventricular or subepicardial regions in experimental ischemia due to the underperfusion of this region. Thus it has also been reported (Griggs et al., 1972) that ischemia produced greater biochemical changes, such as fluctuations in pyruvate, lactate, ATP and phosphocreatine levels, than in any other region.

It has been previously reported by Guidotti et al. (1971) that guinea-pig auricles lost 15% of their taurine content when incubated for one hour in oxygenated Tyrode solution. In our laboratory (Crass and Lombardini, 1978) loss of taurine (17%) from rat ventricular tissue was observed only after hearts were perfused with anoxic medium. In this latter study the release of taurine was time dependent; moreover, taurine was quantitatively recovered in the perfusate. Thus the loss of intracellular taurine which was observed in this experiment as was also demonstrated in the experiments involving the ischemic dog myocardium may share in the mechanism responsible for the arrhythmias which are commonly encountered in acute myocardial ischemia.

A third procedure for studying the release of taurine from the myocardium, that is, administration of cardiotoxic doses of sympathomimetic agents, has been tested in our laboratory. DL-Isoproterenol in large doses produces infarct-like myocardial necrosis (Kutsuma, 1972; Rona et al., 1973). In addition DL-isoproterenol lowers coronary artery perfusion pressure which results in myocardial ischemia (Handforth, 1962; Daniell et al., 1967). Rats were administered 80 mg/kg of DL-isoproterenol and killed after either seven hours or 24 hours. Cardiac tissue lost 30% of its taurine content after seven hours and 45% after 24 hours whereas liver taurine levels did not change [at 24 hours there is a trend for the content of taurine to decrease (not significant; p > 0.05)]. Blood levels of taurine increased at seven hours and then returned to control values after 24 hours. When the pattern of total ninhydrin positive substances was analyzed (seven hours post DL-isoproterenol administration) it appeared quite remarkable that only eight compounds changed in concentration. Aspartic acid, alanine, phosphoethanolamine, and ethanolamine increased in content while glutamic acid, threonine, glutamine-asparagine and taurine decreased. Furthermore, 24 hours after DL-isoproterenol administration five (aspartic acid, glutamic acid, threonine, glutamine-asparagine, and ethanolamine)

of the eight compounds returned to control levels. In terms of
total quantitative importance it should be noted that glutamine-
asparagine decreased by 17.4 µmoles • g dry wt^{-1}, an amount second
only to taurine loss. One explanation for the relatively constant
levels of the ninhydrin positive compounds may be that degradation
of protein and other macromolecules occurred during this time period
(seven hours) which in turn supplied the individual pools. However,
taurine levels were depleted with no immediate source for restoration of pool size. Alternatively, loss of components from the damaged myocardium may be quite specific and taurine depletion which
is quantitatively the most significant is involved in a yet undefined process.

     Blood levels of many of the ninhydrin positive substances were
also affected by DL-isoproterenol. However, taurine was the only
substance that increased while many of the essential amino acids decreased. Utilization of the essential amino acids for protein synthesis in the damaged myocardium might account for their depletion
in the circulation. The increase in taurine concentration was most
likely a result of taurine loss from cardiac tissue.

     Administration of methoxamine, an α-agonist and vasoconstrictor,
to rodents also produced a decrease in cardiac tissue and an increase
in blood levels of taurine. On the other hand, taurine levels in the
liver increased approximately 2.5- to 3-fold. While it is known that
liver is capable of sequestering exogenous taurine, it can be calculated that the loss of taurine from the heart was quantitatively
not sufficient to account for the increase in the liver. Thus the
source of the increased taurine content of the liver is not known.
Whether methoxamine stimulated *de novo* taurine synthesis in the
liver or caused a loss of taurine from tissues other than the heart
(such as skeletal muscle) which in turn was taken up by the liver
requires further experimentation.

SUMMARY

     The effects of ischemia on taurine levels were measured in
three types of animal models: ligation of the circumflex artery in
the dog; perfusion of rat hearts with oxygen-deficient buffer; and
administration of toxic doses of DL-isoproterenol and methoxamine.
In all experiments taurine levels decreased in cardiac tissue. In
the model which utilized sympathomimetic agents an increase in blood
taurine levels was observed. The total amino acid pattern was measured in ventricular and blood tissues after DL-isoproterenol administration. In ventricular tissue the levels of most of the amino
acids remained constant thus attesting to the specificity of taurine
loss. On the other hand, in blood most of the essential amino acids
decreased seven hours after DL-isoproterenol administration and returned to control values after 24 hours.

ACKNOWLEDGMENT

   This study was supported in part by a grant from the Texas Af-
filiate of the American Heart Association.
   The skillful technical assistance of Mrs. S. Paulette Decker
is gratefully acknowledged. Mr. Harvey O. Olney, Department of Biochemistry, is thanked for his help in performing the amino acid
analyses. I am also grateful to Ms. Cindy Frisbie and Dr. Roger R.
Markwald, Department of Anatomy, for their time and effort in help-
ing to draw Figure 1.

REFERENCES

Cardwell, J.C. and Abramson, D.I. (1931). The atrioventricular con-
     duction system of the beef heart. *Am. J. Anat. 49*: 167-192.
Cavallini, D., Dupre, S., Federici, G., Solinas, S., Ricci, G., An-
     tonucci, A., Spoto, G. and Matarese, M. (1978). Isethionic
     acid as a taurine co-metabolite. In: *Taurine and Neurologi-
     cal Disorders*, A. Barbeau and R.J. Huxtable, eds., Raven Press,
     New York, pp. 29-34.
Chazov, E.I., Malchikova, L.S., Lipina, N.V., Asafov, G.B. and
     Smirnov, V.N. (1974). Taurine and electrical activity of the
     heart. *Circ. Res. 34 and 35 (Suppl. III)*: 11-20.
Crass, M.F.,III and Lombardini, J.B. (1977). Loss of cardiac muscle
     taurine after acute left ventricular ischemia. *Life Sciences
     21*: 951-958.
Crass, M.F.,III and Lombardini, J.B. (1978). Release of tissue
     taurine from the oxygen-deficient perfused rat heart. *Proc.
     Soc. Exp. Biol. Med. 157*: 486-488.
Daniell, H.B., Bagwell, E.E. and Walton, R.P. (1967). Limitation
     of myocardial function by reduced coronary blood flow during
     isoproterenol action. *Circ. Res. 21*: 85-98.
Fellman, J.H., Roth, E.S. and Fujita, T.S. (1978). Taurine is not
     metabolized to isethionate in mammalian tissue. In: *Taurine
     and Neurological Disorders*, A. Barbeau and R.J. Huxtable, eds.
     Raven Press, New York, pp. 19-24.
Griggs, D.M., Jr., Tchokoev, V.V. and Chen, C.C. (1972). Transmural
     differences in ventricular tissue substrate levels due to coro-
     nary constriction. *Am. J. Physiol. 222*: 705-709.
Guidotti, A., Badiani, G. and Giotti, A. (1971). Potentiation by
     taurine of inotropic effect of strophanthin-K on guinea-pig
     isolated auricles. *Pharmacol. Res. Comm. 3*: 29-38.
Guidotti, A. and Giotti, A. (1970). Taurina e sistema cardiovasco-
     lare. *Recent. Progr. Med. (Roma) 49*: 61-97.
Handforth, C.P. (1962). Isoproterenol-induced myocardial infarc-
     tion in animals. *Arch Path. 73*: 161-165.
Hinton, J.R., Souza, J.D. and Gillis, R.A. (1975). Deleterious ef-
     fects of taurine in cats with digitalis-induced arrhythmias.
     European J. Pharmacol. 33: 383-387.

Huxtable, R. and Bressler, R. (1974a). Taurine concentrations in congestive heart failure. *Science 184*: 1187–1188.

Huxtable, R. and Bressler, R. (1974b). Elevation of taurine in human congestive heart failure. *Life Sciences 14*: 1353–1359.

Jacobsen, J.G. and Smith, L.H., Jr. (1968). Biochemistry and physiology of taurine and taurine derivatives. *Physiol. Rev. 48*: 424–511.

Kocsis, J.J., Kostos, V.J. and Baskin, S.I. (1976). Taurine levels in the heart tissues of various species. In: *Taurine*, R. Huxtable and A. Barbeau, eds., Raven Press, New York, pp. 145–153.

Kutsuna, F. (1972). Electron microscopic studies on isoproterenol-induced myocardial lesions in rats. *Jap. Heart J. 13*: 168–175.

Lombardini, J.B. (1975). An enzymatic derivative double isotope assay for measuring tissue levels of taurine. *J. Pharmacol. Exp. Ther. 193*: 301–308.

Moir, T.W. (1972). Subendocardial distribution of coronary blood flow and the effect of antianginal drugs. *Circ. Res. 30*: 621–627.

Novelli, G.P., Ariano, M. and Francini, R. (1969). Un nuova medicamento per la prevenzione delle aritmie: la taurina. *Minerva Anest. 35*: 1241–1250.

Peterson, M.B., Mead, R.J. and Welty, J.D. (1973). Free amino acids in congestive heart failure. *J. Mol. Cell. Cardiol. 5*: 139–147.

Read, W.O. and Welty, J.D. (1963). Effect of taurine on epinephrine and digoxin induced irregularities of the dog heart. *J Pharmacol. Exp. Ther. 139*: 283–289.

Rona, G., Boutet, M., Huttner, I. and Peters, H. ((1973). Pathogenesis of isoproterenol-induced myocardial alterations: functional and morphogical correlates. In: *Recent Advances in Studies on Cardiac Structure and Metabolism*, N.S. Dhalla, ed., University Park Press, Baltimore, vol. 3, pp. 507–525.

Welty, J.D. and Read, W.O. (1963). Studies on the function of taurine in the heart. *Proc. S.D. Acad. Sci. 42*: 157–163.

TAURINE EFFECTS ON CALCIUM TRANSPORT IN NERVOUS TISSUE

H. Pasantes-Morales and A. Gamboa

Centro de Investigaciones en Fisiología Celular, Univer sidad Nacional Autónoma de México, México 20, D.F., México

Is a Neurotransmitter Role the Main Function of Taurine in the Central Nervous System?

In the last years there have been increasing interest in the role of taurine in the central nervous system (CNS), because of its possible association with the mechanisms regulating nervous excitability[1]. After the finding of the generalized anticonvulsant action of taurine[2-4], as well as of its effects on human epilepsies [5,6], many investigations have been undertaken in an attempt to elucidate the intimate mechanism through which taurine is involved in the control of nervous excitation. Taurine, when applied iontophoretically, has depressant effects on neuronal activity[7,8] and has been proposed to be acting as an inhibitory neurotransmitter[9,10]; actually, it appears to possess many of the properties expected for such a role. However, a rigorous analysis of the criteria a substance must fulfill in order to be identified as a transmitter must be made when evaluating such a role for taurine in the CNS. Particular emphasis should be made on the specificity of taurine actions in order to exclude the possibility that some of its effects were mediated through receptors for neurotransmitter amino acids with which taurine has structural analogies, like GABA or glycine. It is also of extreme interest to localize the precise sites at which taurine is being released or taken up, or at which it is modifying membrane excitability. Finally, the search of specific postsynaptic receptors and of their localization and distribution is essential in clarifying the question of a neurotransmitter role for taurine. In this chapter, the evidence relating taurine to a neurotransmitter function will be briefly reviewed. An alternate role as a modulator of calcium gradients in nervous tissue will be discussed in the light of our recent findings on taurine effects on calcium transport.

Ocurrence and distribution. Taurine is unevenly distributed among the different regions of the CNS. The highest levels are found in cerebral cortex, olfactory bulb, hippocampus, striatum and cerebellum, and the lowest in spinal cord, pons medulla and hypothalamus[11-13]. Until now, no correlation has been demonstrated between taurine distribution and a functional parameter relating it with a neurotransmitter role. Studies on taurine distribution in subcellular fractions isolated from different regions of the CNS have not demonstrated a preponderant location of taurine in nerve terminals[12]. The distribution pattern of taurine in discrete areas of the CNS has been studied only in cerebellum, spinal cord and thalamus[14,15]. In cerebellum, taurine appears to be concentrated at the molecular layer, although it is present at significant amounts in all the other cerebellar layers[15]. In spinal cord and thalamus, an homogeneous distribution of taurine was found[14]. This result contrasts with observations in similar studies on GABA and glycine, which show an uneven distribution in the spinal cord and in some brain regions, with their highest concentration localized at sites where there is a high density of neurons sensitive to their physiological action[14,16-18]. Therefore, this microdistribution studies do not favor a role for taurine as neuro-transmitter, at least in the aforementioned regions. Obviously, systematic studies on the microdistribution of taurine are needed to ascertain if the amino acid is found particularly concentrated in definite areas, which might suggest the existence of specific taurinergic pathways. However, the results obtained until now, showing a rather homogeneous distribution of taurine, suggest its involvement in a more generalized function, not restricted to specific synapses.

Electrophysiological effects. Taurine actions on neuronal activity has not been extensively studied in nervous preparations. Most of the investigations showing taurine effects on nerve cells have been designed to test the effects of other amino acids and taurine has been used mainly with comparative purposes. There are only very few systematic studies on the effect of taurine using optimum conditions for taurine iontophoretic applications[8]. This makes difficult to compare strictly the effects of taurine with those of other neuroactive amino acids. In any case, it appears that in the spinal cord and brain stem taurine shows inhibitory effects on neuronal activity with a potency roughly comparable to that of GABA or glycine[7,19-22]. In contrast, in cortical neurons iontophoretically applied taurine is consistently less effective than GABA, although its effect is probably stronger than that of glycine[23]. The inhibitory effect of taurine is associated to an increase in membrane conductance which leads to hyperpolarization. An increased permeability to Cl^- or K^+ appears to be responsible for the membrane changes observed. Taurine, like GABA or glycine depolarizes some presynaptic fibers in frog dorsal roots[24,25]. In these preparations, the depolarizing action of taurine is weaker than that of GABA. A change in Cl^- permeability has been also associated to the depolarizing effect of taurine. It is noteworthy that the effects of taurine in both membrane potential and membrane permeability to Cl^- or K^+ have been observed in lobster axons, in which postsynaptic receptors are obviously absent[26].

Taurine effects on spinal cord and brain stem are blocked by strychnine, an antagonist apparently specific for glycine post-synaptic receptors, but are unaffected by the GABA antagonists picrotoxin and bicuculline[21,22]. Since microdistribution studies do not favor a role for taurine as a neurotransmitter in the spinal cord, the above observations suggest that taurine effects in this area might be mediated through glycine receptors. In cortical areas, as well as in the cerebellum, taurine depressant effect is blocked by both glycine and GABA antagonists[23]. Taurine depolarizing effects on frog dorsal roots are also antagonized by strychine and by picrotoxin[24,25]. These observations might be interpreted as an indication that taurine is able to combine with both GABA and glycine receptors in these preparations. However, this assumption implies that GABA receptors would be different in spinal cord and in brain, which has not been demonstrated.

The study of receptors for neurotransmitters has been greatly facilitated by the technique of radiolabeled ligand binding to membrane preparations developed by Snyder, Enna and their colleagues[27]. This technique differentiates binding to receptor sites from binding to neuronal uptake sites by its independence on sodium and temperature, as well as by the specific antagonism exerted by compounds known to block the neurophysiological neurotransmitter action on neurons. The distribution of post-synaptic receptor sites for neuroactive amino acids, studied by this technique, correlates well with the localization of other parameters relating these amino acids to a function as neuro-transmitters; GABA receptor binding has been demonstrated in spinal cord, cerebral cortex, cerebellum and retina[27-29]; glycine binding has been mainly observed in the spinal cord. Following the procedure of Snyder et al.[27] we have investigated the binding of taurine to membranes showing the properties of postsynaptic receptor binding; in several regions of the CNS, retina, spinal cord and cerebral cortex, we were unable to demonstrate a taurine binding exhibiting these properties, under conditions in which GABA or glycine binding is easily detected. If we consider that this technique is effectively a marker of the presence and distribution of postsynaptic receptors, it may be concluded that taurine receptors are extremely scarce in the CNS. However, also using this technique, we have found a widely distributed binding to uptake sites; this result, which agrees with the demonstration of a high affinity taurine transport in most regions of the CNS, suggests a widespread involvement of taurine in nervous function but probably not as a neurotransmitter.

   Uptake. The presence of a sodium-dependent, high affinity
uptake mechanism for a given substance, occurring particularly in
subcellular fractions enriched in synaptosomes, has been considered
as a support for a transmitter function of this substance. It has
been suggested that such a selective mechanism is responsible for
the termination of transmitter action and for the restoration of
the presynaptic pool. The existence of this high affinity transport system is critical in postulating a neurotransmitter role for
taurine, in view of the virtual absence of known pathways of
taurine metabolic degradation. In nervous tissue, both non
saturable and saturable taurine transport system have been
described. The saturable mechanism can be resolved into two
different components: a low affinity transport system and a high
affinity transport system[31]. The high affinity transport system
has been demonstrated in tissue slices as well as in synaptosomes
from various regions of the CNS[13,32,33]. It is highly specific
for taurine, it is sodium and energy dependent and it exhibits K_m
values in the range of 11-60 μM. Studies on the distribution of
the high affinity transport system in nervous regions, which in
some cases have contributed to localize the CNS areas where a
substance is acting as neurotransmitter, have not given such
information for taurine. The high affinity taurine uptake has been
found in practically all nervous regions, except probably in the
cerebellum[12]. For glycine, which is presumably a neurotransmitter
in the spinal cord but not in cerebral cortex, both the high
affinity and the low affinity transport systems are present in the
spinal cord, whereas only the low affinity transport system is
observed in cerebral cortex[34]. Therefore, the widespread
localization of the high affinity transport system for taurine
would indicate a widely distributed function as neurotransmitter,
which seems not to be supported by electrophysiological
investigations, microdistribution studies or localization of postsynaptic receptors. However, the presence of this high affinity
transport system, which is very specific for the amino acid, is
again suggestive of its involvement in a general aspect of nervous
function.

   Release. The effect of experimental stimulus which simulate
physiological depolarization of nerve terminals on the release of
taurine has been extensively examined. In pioneer experiments,
Jasper and Koyama[35] demonstrated a taurine release from the exposed
cat cerebral cortex in vivo, after electrical stimulation of the
midbrain reticular formation. An increase in the resting release
of endogenous taurine from the rat visual cortex in vivo was
demonstrated under electrical stimulation or by topical application
of depolarizing KCl concentrations[36,13]. In vitro, an increase in
the efflux of labeled taurine, recently taken up by the tissue, has
been reported to occur in rat cerebral cortex[37,38]. However, in a

comparative study of the release of ^{35}S-taurine, ^{3}H-GABA and
^{14}C-glycine, from eight regions of the rat CNS, taurine release was
not stimulated by depolarizing KCl concentrations added to an
isotonic medium, under conditions which caused a significant
increase on GABA and glycine release[39]. The stimulated release of
neurotransmitters is considered to require calcium ions; yet, the
induced release of taurine is not clearly calcium-dependent. In
synaptosomes isolated from rat brain, K^+ stimulated release of
^{35}S-taurine is not reduced but enhanced, in the absence of
calcium[40]. These results might suggest that the taurine released
upon depolarizing stimulus originates from glial cells, from where
amino acid release appears to be a calcium-independent process,
rather than from the nerve terminals. This would be particularly
true when exogenous, labeled taurine is used, since it is known
that it may be accumulated in glial cells as well as in the nerve
endings. Particular emphasis should be placed in future studies
of taurine release on the identification of the subcellular sites
from which taurine is released, in order to ascertain if taurine
efflux conforms the stimulus-secretion coupling mechanism.

If we now return to the question initially posed in this
chapter: is taurine a neurotransmitter?, it emerges clearly that
much more work is necessary before an adequate answer could be
given. However, if we compare taurine responses with those of
amino acids for which a role as neurotransmitter seems well
established, like GABA or glycine, it seems that if taurine is
actually acting as a synaptic transmitter, the extent of its
involvement in such a role is rather restricted. Its neurophysiological action is weak, its release upon depolarization is
controversial and probably calcium-independent and, most impor-
tantly, taurine postsynaptic receptors appear to be very scarce.
Therefore, it may be considered that the main role of taurine in
the CNS is not that of a neurotransmitter, although such an action
in restricted areas cannot be ruled out.
Taurine is not an exclusive constituent of nervous tissue; it
is also present at very high levels in contractile and secretory
tissues. Since, except for the synthesis of taurocholic acid,
taurine does not participate in metabolic reactions, its occurrence
in animal tissues at substantial amounts should be related with a
specific function which remains still unknown. If we consider
that most taurine in nervous system is not related with a function
as a neurotransmitter, the possibility that a single role is played
by taurine in nervous, contractile or secretory tissues may be
raised. Evidently, a multiplicity of taurine roles may also be
envisaged, but we rather favour the hypothesis of a single action,
probably at a basic mechanism underlying contractile, secretory
and nervous function. A common denominator in all these functions
is a modification of calcium gradients related to the excitation
state of the tissue and a subsequent redistribution of calcium in
order to return towards the resting intracellular levels. Taurine

Table 1. Factors Affecting ^{45}Ca Transport by Rat Brain Synaptosomes

Treatment	^{45}Ca uptake (n moles/mg protein)	
Control: 37°C, 1 mM ATP	1.11 ± 0.04	(10)
4°C, 1 mM ATP	0.22 ± 0.01	(4)
Without ATP	1.12 ± 0.19	(6)
Ouabain (100 μM)	1.07 ± 0.09	(4)
KCl (70 mM)	1.02 ± 0.11	(4)
Ruthenium red (100 μM)	0.99 ± 0.28	(4)
Low Sodium medium	1.88 ± 0.09	(3)

Synaptosomes were obtained from rat brain according to the
procedure of Hajós et al[42]. Synaptosomes were resuspended in a
Krebs-bicarbonate medium (118 mM NaCl; 4.7 mM KCl; 1.17 mM $MgSO_4$;
1.2 mM KH_2PO_4; 25 mM $NaHCO_3$; 5.6 mM glucose), pH 7.4, and aliquots
of the suspension containing 0.5-0.7 mg protein were incubated
during 5 min. The incubation mixture contained 1 mM ATP, 2 μCi of
$^{45}CaCl_2$ and 2.5 mM unlabeled $CaCl_2$ in a final volume of 1 ml. At
the end of the incubation, aliquots of 0.3 ml were withdrawn and
centrifuged in a Beckman microfugue; the sediment was solubilized
with NCS (tissue solubilizer, Amersham), and radioactivity
measured after addition of Tritosol. Synaptosomes were exposed
to ouabain, ruthenium red, high KCl concentration and low sodium
medium from the beginning of the incubation. In experiments with
ruthenium red fractions were preincubated for 10 min with the dye.
In the low sodium medium condition, isoosmolar choline chloride
replaced NaCl, and synaptosomes were preincubated for 10 min in a
Ca-free Na-containing medium in the presence of 10^{-3} M ouabain and
then incubated in the low sodium medium[47]. Values are means ±
S.E.M. of the number of experiments indicated in parenthesis.

has been implicated in the modulation of calcium fluxes in heart[41],
and the possibility of a similar role in nervous tissue may be
considered. In order to explore this possibility, we have studied
the effect of taurine on calcium transport into brain synaptosomes.

Calcium Transport into Brain Synaptosomes.

Taurine effects on ^{45}Ca transport were studied in a
synaptosomal fraction isolated from the rat brain by the procedure
described by Hajós et al[42]. ^{45}Ca uptake by synaptosomes was
measured in a Krebs-bicarbonate medium, since this medium simulates
very closely the ionic composition of the cerebrospinal fluid.
When incubated in this medium at 37°C synaptosomes rapidly
accumulate $^{45}CaCl_2$, reaching 1.0-1.3 nmoles/mg protein after 3 min
of incubation. ^{45}Ca accumulation by the synaptosomes, under these
conditions, is not increased by depolarizing concentrations of
KCl, or by ouabain (Table 1); it is not sensitive to ruthenium red

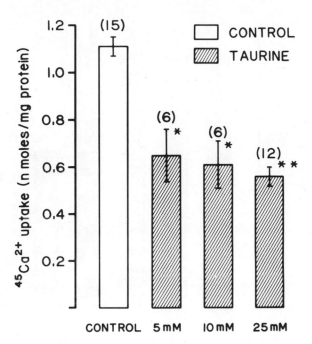

Fig. 1.   The effect of taurine on ^{45}Ca accumulation in rat brain
           synaptosomes. Synaptosomes (0.5-0.7 mg protein) were
           incubated in Krebs-bicarbonate medium at 37°C, in the
           presence of ^{45}CaCl$_2$ (2 μCi/ml). Taurine at the
           concentrations indicated was present throughout the
           incubation period. Results are means ± S.E.M. of the
           number of experiments indicated in parenthesis.
           *Significantly different from controls at P∠0.001.

and it is not stimulated by ATP, and is highly reduced by
incubation at low temperature (Table 1). When the sodium gradient
is reversed by preincubating synaptosomes in a sodium containing
medium in the presence of ouabain, and then incubating them in a
sodium-free medium[43], ^{45}Ca accumulation is greatly increased
(Table 1). These properties of the calcium transport suggest that
the ^{45}Ca accumulation observed under these conditions is being
carried out through the mechanism of Na-Ca exchange. This calcium
transport mechanism has been described in a variety of tissues,
including squid and crab axons,[44] many types of muscle[45], secretory
tissues[46], retinal photoreceptors[47] and rat isolated nerve

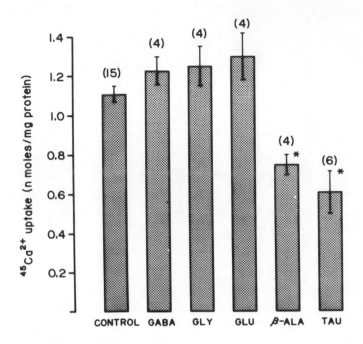

Fig. 2.   The effect of various amino acids on $^{45}CaCl_2$
accumulation in rat brain synaptosomes. Assay
conditions were the same as described in Fig. 1. All
the amino acids, including taurine were present at 10 mM
concentration. Results are means ± S.E.M. of the number
of experiments indicated in parenthesis. *Significantly
different from controls at P∠0.001.

endings[48]. The Na-Ca exchange process seems to be also capable of
a Ca-Ca exchange and appears to be different from the mechanism
which causes calcium uptake during depolarization.

Taurine Effects on Calcium Transport in Brain Synaptosomes

Taurine at 2.5-25 mM concentrations produces a marked decrease
in the accumulation of ^{45}Ca by the synaptosomes (Fig.1). Taurine-
induced decrease varied from 35% to 55% with respect to the
control values. The effect of taurine is rapid; it is observed at
the shortest time we measured, 1 min. Taurine effect is specific;
a similar decrease in ^{45}Ca transport is not observed in the
presence of other neuroactive amino acids, like GABA or glycine;

only β-alanine, a very close structural analogue of taurine, produces a similar decrease on ^{45}Ca accumulation. Taurine seems not to affect the release of previously incorporated ^{45}Ca nor calcium binding. The effect of taurine on ^{45}Ca accumulation by synaptosomes requires the presence of bicarbonate in the incubation medium; when TRIS, HEPES or phosphate replaces bicarbonate for maintaining the pH of the medium, the effects of taurine are no more apparent. The ionic composition of the incubation medium modifies taurine action on ^{45}Ca accumulation: in a Krebs bicarbonate medium containing 118 mM NaCl, which mimics the composition of the extracellular fluid, taurine reduces ^{45}Ca accumulation; when NaCl is replaced by equiosmolar concentrations of KCl, which simulate the intracellular medium, taurine increases ^{45}Ca transport.

The Na Ca exchange mechanism, which appears to be affected by taurine has been implicated in calcium transport across the nerve terminals. Nerve excitation produces a change in the calcium permeability of the nerve ending membrane; exogenous calcium then easily enters the terminal because of the large, inwardly directed calcium electrochemical gradient. The subsequent rise in the intracellular calcium in the terminal triggers the release of neuro transmitters. A very rapid return to resting calcium levels occurs, probably due to a process of calcium sequestration carried out by intraterminal organelles (mitochondrial and non mitochondrial) and this causes the rapid termination of the phasic period of transmitter release. The final extrusion of calcium out of the terminal, for restoring the normal, resting intracellular calcium levels, occurs probably by a plasma membrane mechanism and the Na-Ca exchange process seems to be directly implicated in this long term regulation of the intraterminal calcium concentration[43].

Since taurine is present in very low amounts in the extracellular space and it is mainly concentrated intracellulary, the effect of exogenous taurine observed in the present study may represent a pharmacological action rather than a physiological effect. However, the different action of taurine observed in media which reflect the intracellular or the extracellular ionic environment, increasing calcium transport in intracellular ionic conditions and decreasing it in extracellular ionic conditions, might suggest a physiological action of taurine in the mechanism of calcium extrusion. The presence of taurine in particularly high amounts in contractile or secretory tissues, where the redistribution of calcium plays a critical role in their normal function, suggests that this proposed action for taurine in the regulation of the intracellular calcium levels may not be restricted to nervous tissue but might represent a major role for taurine in excitable and secretory tissues.

Acknowledgements. We are grateful to Dr. R. Tapia for his comments on the chapter. We also thank Miss Virginia Godínez for typing

the manuscript.   This work was supported in part by grants 1 R01
EY 025 40-01 from the National Eye Institute and 1621 from Consejo
Nacional de Ciencia y Tecnología (México).

References

1. A. Barbeau and R. Huxtable eds. "Taurine and Neurological
     Disorders", Raven Press, New York (1978).
2. N. M. Van Gelder, Antagonism by taurine of cobalt induced
     epilepsy in cat and mouse, Brain Res. 47: 157 (1972).
3. K. Izumi, H. Igisu, and T. Fukuda, Supression of seizures by
     taurine-specific or nonspecific?, Brain Res. 76:171 (1974).
4. R. Mutani, L. Bergamini, R. Fariello, and M. Delsedime, Effects
     of taurine in chronic experimental epilepsy, Brain Res. 70:
     170 (1974).
5. L. Bergamini, R. Mutani, M. Delsedime, and L. Durelli, First
     clinical experience on the antiepileptic action of taurine,
     Eur. Neurol. 11:261 (1974).
6. A. Barbeau and J. Donaldson, Zinc, taurine and epilepsy, Arch.
     Neurol. 52: 1 (1974).
7. D. R. Curtis and J.C. Watkins, The excitation and depression
     of spinal by structurally related amino acids, J. Neurochem.
     6:117 (1960).
8. K. Krnjević and E. Puil, Electrophysiological studies on
     actions of taurine, in "Taurine", R. Huxtable and A.
     Barbeau, eds., pp. 179, Raven Press, New York (1976).
9. A.N. Davison and L.K. Kaczmarek, Taurine - A possible neuro-
     transmitter, Nature, 234: 107 (1971).
10. P. Mandel and H. Pasantes-Morales, Taurine: a putative neuro-
     transmitter, in Advances in Biochemical Psychopharmacology,
     E. Costa ed. pp. 141, Raven Press, New York (1976).
11. R. P. Shank and M.H. Aprison, The metabolism in vivo of
     glycine and serine in eight areas of the rat nervous system,
     J. Neurochem. 17:1461 (1970).
12. J. B. Lombardini, Regional and subcellular studies on taurine
     in the rat central nervous system, in "Taurine", R.
     Huxtable and A. Barbeau eds., pp. 311, Raven Press, New
     York (1976).
13. G. G.S. Collins, The rate of synthesis, uptake and disap-
     pearance of ^{14}C-taurine in eight areas of the rat central
     nervous system, Brain Res. 76:447 (1974).
14. Y. Yoneda and K. Kuriyama, A comparison of microdistribution
     of taurine and cysteine sulphinate decarboxylase activity
     with those of GABA and L-glutamate decarboxylase activity
     in rat spinal cord and thalamus. J. Neurochem. 30: 821
     (1978).
15. N. S. Nadi, W.J. McBride and M.H. Aprison, Distribution of
     several amino acids in regions of the cerebellum of the rat,
     J. Neurochem. 28:453 (1977).
16. Y. Okada, Distribution of GABA and GAD activity in the layers

of superior colliculus of the rabbit, in "GABA in Nervous System Function" E. Roberts, T.N. Chase and D.B. Tower eds. pp. 229 Raven Press, New York (1976).

17. K. Kuriyama and H. Kimura, Distribution and possible functional roles of GABA in retina, lower auditory pathway and hypothalamus, in "GABA in Nervous System Function" E. Roberts, T.N. Chase and D.B. Tower eds. pp. 203 Raven Press, New York (1976).

18. M. H. Aprison and R. Werman, The distribution of glycine in cat spinal cord and roots, Life Sci. 4, 2075 (1965).

19. D. R. Curtis, L. Hösli and G.A.R. Johnston, A pharmacological study of the depression of spinal neurons by glycine and related amino acids, Exp. Brain Res. 6:1 (1968).

20. D. R. Curtis and A.K. Tebécis, Bicuculline and thalamic inhibition, Exp. Brain Res. 16:210 (1972).

21. D. R. Curtis, A.W. Duggan, D. Felix and G.A.R. Johnston, Bicuculine an antagonist of GABA and synaptic inhibition of the spinal cord of the cat, Brain Res. 33: 57 (1971).

22. D. R. Curtis, L. Hösli, and G.A.R. Johnston, A pharmacological study of the depression of spinal neurons by glycine and related amino acids. Exp. Brain Res. 6:1 (1968)

23. D. R. Curtis, A.W. Duggan, D. Felix, G.A.R. Johnston and H. Mc Lennan, Antagonism between bicuculline and GABA in the cat brain, Brain Res. 33: 57 (1971).

24. J. L. Barker, R.A. Nicoll, and A. Padjen, Studies on convulsants in the isolated frog spinal cord. I. Antagonism of amino acid responses, J. Physiol. 245:521 (1975).

25. J. L. Barker, R.A. Nicoll, and A. Padjen, Studies on convulsants in the isolated frog spinal cord. II. Effects on root potentials, J. Physiol. 245:537 (1975).

26. R. Gruener and H.J. Bryant, Excitability modulation by taurine. Action on axon membrane permeabilities, J. Pharmacol. Exp. Ther. 194:514 (1975).

27. S. R. Zukin, A.B. Young, and S.H. Snyder, Gamma-aminobutyric acid binding to receptor sites in the rat central nervous system. Proc. Nat. Acad. Sci 71:4802 (1974).

28. A. B. Young and S.H. Snyder, Strychnine binding in rat spinal cord membranes associated with the synaptic glycine receptor: cooperativity of glycine interactions, Mol. Pharmacol. 10: 790 (1974).

29. S. J. Enna and S.H. Snyder, Properties of gamma-aminobutyric acid (GABA) receptor binding in rat brain synaptic membrane fractions, Brain Res. 100: 81 (1975).

30. S. J. Enna and S.H. Snyder, Gamma aminobutyric acid (GABA) receptor binding in mammalian retina, Brain Res. 115:174 (1976).

31. P. Lahdesmaki and S.S. Oja, On the mechanism of taurine transport at brain cell membranes, J. Neurochem. 20:1411 (1973).

32. R. E. Hruska, R.J. Huxtable, R. Bressler, and H.I. Yamamura,

Sodium-dependent, high affinity transport of taurine into rat brain synaptosomes, Proc. West. Pharmacol. Soc. 19:152 (1976).

33. W. Sieghart and M. Karobath, Evidence for specific synaptosomal localization of exogenous accumulated taurine, J. Neurochem. 23:911 (1974).

34. W. J. Logan and S.H. Snyder, High affinity uptake systems for glycine, glutamic acid and aspartic acid in synaptosomes of rat central nervous system, Brain Res. 42:413 (1972).

35. H. H. Jasper and I. Koyama, Rate of release of amino acids from the cerebral cortex in the cat as affected by brain stem and thalamic stimulation, Can. J. Physiol. Pharmacol. 47:889 (1969).

36. R. M. Clark and G.G.S. Collins, The release of endogenous amino acids from the mammalian visual cortex, J. Physiol. (Lond.) 246:16 (1975).

37. L. K. Kackzmarek and A.N. Davison, Uptake and release of taurine from rat brain slices, J. Neurochem. 19:2355 (1972).

38. F. Orrego, R. Miranda and C. Saldate, Electrically induced release of labelled taurine, α - and β -alanine, glycine glutamate and other amino acids from rat neocortical slices in vitro. Neuroscience, 1:325 (1976).

39. A.M. López-Colomé, R. Tapia, R. Salceda, and H. Pasantes-Morales, K$^+$-stimulated release of labeled γ -aminobutyrate, glycine and taurine in slices of several regions of rat central nervous system, Neuroscience 3:1069 (1978).

40. W. Sieghart and K. Heckl, Potassium evoked release of taurine from synaptosomal fractions of rat cerebral cortex, Brain Res. 116:538 (1976).

41. P. Dolara, A. Agresti, A. Giotti and G. Pasquini, Effect of taurine on calcium kinetics of guinea pig heart, Eur. J. Pharmacol. 24:352 (1973).

42. F. Hajós, An improved method for the preparation of synaptosomal fractions in high purity, Brain Res. 93:485 (1975).

43. M. P. Blaustein and C.S. Oborn, The influence of sodium on calcium fluxes in pinched-off nerve terminals in vitro, J. Physiol. 247:657 (1975).

44. P. F. Baker, Transport and metabolism of calcium ions in nerve, Progr. Biophys. Mol. Biol. 24: 117 (1972).

45. M. P. Blaustein, The interrelationship between sodium and calcium fluxes across cell membranes, Rev. Physiol. Biochem. Pharmacol. 70:33 (1974).

46. T. J. Rink, The influence of sodium on calcium movements and catecholamine release in thin slices of bovine adrenal medulla, J. Physiol. 266:297 (1977).

47. P. P.M. Schnetkamp, F.J.M. Daemen, and S.L. Bonting, Biochemical aspects of the visual process, XXXVI. Calcium accumulation in cattle rod outer segments: evidence for a calcium-sodium exchange carrier in the rod sac membrane, Biochim. Biophys. Acta 468: 259 (1977).

Taurine in Retinas of Taurine Deficient Cats and RCS Rats

Susan Y. Schmidt

Berman-Gund Laboratory for the Study of
Retinal Degenerations
Harvard Medical School
243 Charles Street, Boston, Massachusetts 02114

Cats fed a taurine-free, casein diet develop retinal taurine deficiency and subsequently photoreceptor cell death. (Berson, Hayes, Rabin, Schmidt, and Watson, 1976; Schmidt, Berson and Hayes, 1976; Schmidt, Berson, Watson and Huang, 1977). Supplementation of this diet with methionine, cysteine, inorganic sulfate, vitamin B_6, or vitamin B_6 plus cysteine did not prevent development of retinal taurine deficiency and retinal malfunction. A synthetic amino acid diet devoid of casein and taurine also resulted in retinal taurine deficiency and retinal malfunction. Only taurine-containing diets (i.e., chow or casein plus taurine) preserved normal retinal taurine concentrations and electroretinogram (ERG) amplitudes. These findings have firmly established a role for exogenous taurine in maintaining normal retinal function in the cat.

In taurine-deficient cats the decreases in retinal taurine concentrations and reductions in ERG amplitudes have been closely correlated (Fig. 1). This correlation could be demonstrated prior to detectable cell death as measured by changes in retinal deoxyribonucleic acid (DNA) concentrations. Reductions in retinal taurine concentrations below 50% of normal were associated with the appearance of abnormal granularity in the area centralis and reduction in retinal DNA concentrations. At a time when retinal taurine concentrations were reduced 70-80% below normal, the ERG responses were small or non-detectable, and ultrastructural studies showed photoreceptor cell death (Hayes, Rabin and Berson, 1975).

In vitro studies (Schmidt, 1979) have shown that homogenates of

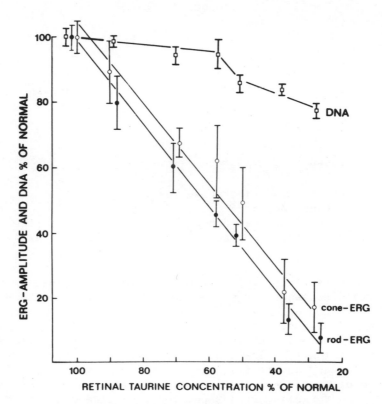

Figure 1. Peak-to-peak amplitudes and retinal DNA concentrations
related to retinal taurine concentration in 10 control and 38 taurine-
deficient cats. Controls were considered to have 100% retinal taurine
concentration, ERG amplitude, and retinal DNA concentration. Taurine-
deficient cats were divided into six groups of 4 to 8 according to
the amount of taurine in their retinas. For each group, average
amplitudes (mean ± SEM) for rod (black dots) and cone (circles)
responses and average DNA values (squares) are presented. The
coefficients of correlation for rod ERG amplitude and cone ERG
amplitude to retinal taurine concentration were 0.90 and 0.84
respectively. (Reproduced with permission from Schmidt, Berson,
Watson and Huang, Invest. Ophthalmol. 16:673–678, 1977).

cat livers fail to convert ^{35}S-cysteine and ^{35}S-cysteic acid to taurine. Under identical conditions homogenates of rat livers synthesized appreciable amounts of ^{35} S-taurine (0.01 - 0.03 nmoles/ mg protein/min).* The failure to detect measurable taurine synthesis in vitro in cat livers is consistent with previous findings of low levels of cysteine - decarboxylase activity in cat livers (Jacobsen et al 1964 and Hardison et al 1977). No taurine synthesis could be detected in retinal homogenates of either cats or rats (Schmidt, 1979). This suggests that in both species retinal taurine concen- trations are maintained by uptake of taurine from the plasma and not by synthesis in situ in the retina.

The uptake of taurine into the outer nuclear layer (where photoreceptor cell bodies and Müller cell processes are located) undoubtedly occurs at least in part via transport of taurine from the plasma across the pigment epithelium. The pigment epithelium has been shown to be an active site of taurine accumulation in vivo in rats and frogs (Young, 1968) and in tissue culture (Edwards, 1977). Schmidt, Berson, Watson and Huang (1977) have shown that 48-72 hours after injection of radioactive taurine of differing specific activi- ties the accumulation of labelled taurine by the retina could be correlated with plasma taurine concentrations; accumulation was higher in slightly taurine-deficient retinas than in normal retinas for a given plasma taurine concentration (Figure 2). Studies with tritiated taurine of high specific activity showed that uptake of taurine into the retina could be detected even when plasma levels of taurine were reduced to 10-20% of normal in slightly taurine-deficient cats; this could explain why retinal levels of taurine remained nearly normal as plasma concentrations of taurine were decreasing in these animals. In control cats, retinal taurine concentrations are mainted in vivo against a 400 fold gradient (taurine concentrations in retina and plasma were about 40mM and 0.1mM respectively). Moreover, in the slightly taurine-deficient cat the retina has a capacity to con- centrate taurine against a 1600 fold gradient (taurine concentrations in retina and plasma were about 32mM and .02mM respectively) (Schmidt, 1979).

---

*For studies of taurine synthesis from cysteic acid incubations were conducted according to MacDonnell and Greengard (1975) in the presence of 20mM ^{35} S-cysteic acid. For studies of taurine synthesis from ^{35}S-cysteine the incubations were done according to Misra and Olney (1975). Pyridoxal - 5' - phosphate (0.5mM) was present in all incub- ations. Taurine synthesis was evaluated by thin layer chromatography as previously described (Schmidt, Berson, Watson and Huang 1977).

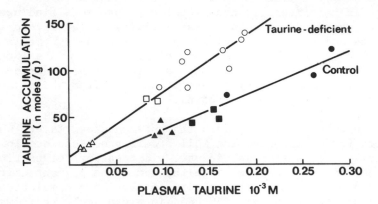

Figure 2. Taurine accumulation (nmoles/g wet weight) by the retina of control (▲,■,●) and slightly taurine-deficient cats (△,□, ○), related to plasma taurine concentration. The cats were inject-ed with either ^{3}H-taurine (▲,△), ^{35}S-taurine (■,□) or ^{14}C-taurine (●,○). The coefficients of correlation are 0.81 for taurine-deficient cats and 0.90 for the control group. (Reproduced with permission from Schmidt, Berson, Watson and Huang, Invest. Ophthalmol. 16:673-678, 1977).

Kinetic analysis of taurine uptake by isolated cat retinas (Figure 3) revealed that two mechanisms exist for uptake of taurine. One process designated as a high-affinity uptake mechanism has an apparent Michaelis constant (K_m) of 50μM and a maximal velocity (V_{max}) of 0.55 nmoles/mg dry wt/min (Schmidt 1979). The second process designated as a nonsaturable mechanism has a rate constant of 3.20 nmoles/mg. dry wt./min/mM taurine in the medium. In iso-lated retinas from slightly taurine-deficient cats the Km value was 29μM suggesting an increased affinity for taurine in these retinas. The high-affinity uptake mechanism, in contrast to the nonsaturable mechanism, could not be detected in isolated cat retinas after the photoreceptor cells had degenerated at late stages of taurine defi-ciency. Comparison of taurine uptake in normal and photoreceptorless retinas showed that the high affinity uptake mechanism is inhibited to a greater extent than the nonsaturable mechanism by reduced temperature, ouabain (100μM) and omission of glucose from the incubation medium (Table 1). At low concentrations (0.005 –

0.01mM) of taurine in the media, 60-70% of taurine uptake was due to the high affinity uptake mechanism and 30-40% to the nonsaturable mechanism. Autoradiograms of isolated retinas of control and slightly taurine-deficient cats incubated in dim light, in media containing low concentrations of ^{3}H-taurine showed that accumulations of ^{3}H-taurine was greatest over the photoreceptor cell layer (Schmidt and Szamier 1978, Lake, Marshall and Voaden 1978). Microdissection of the same incubated cat retinas (Schmidt 1979) showed that 60-70% of the retinal radioactivity was in the outer retina (Figure 4) where endogenous taurine is also known to be concentrated (Cohen et al 1973, Kean and Yates 1976, Kennedy et al 1977, Orr et al 1976, Schmidt et al 1976). Within the inner retina uptake has been observed in Müller cells, in some horizontal, and in some amacrine cells (Lake, Marshall and Voaden, 1978).

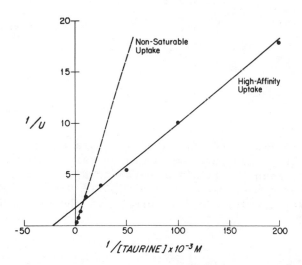

Figure 3. Representative double reciprocal plot of ^{3}H-taurine uptake (U) vs. taurine concentration in the medium in isolated normal control cat retinas. The high affinity uptake mechanism (solid line) was derived with least square analysis to best approximate the data points for the linear portion of the double reciprocal plot (between 0.005 - 0.10 mM taurine in the medium). At concentrations above 0.1 mM taurine in the medium, the data points for the nonsaturable mechanism approach zero (hatched line).

TABLE I

EFFECTS OF METABOLIC INHIBITORS ON THE HIGH-AFFINITY AND NONSATURABLE
MECHANISMS FOR TAURINE UPTAKE BY ISOLATED CAT RETINAS

CONDITIONS OF INCUBATION	NORMAL RETINAS	PHOTORECEPTORLESS RETINAS
37°C (control)	100 $\pm$ 8.2	100 $\pm$ 8.6
34°C	66 $\pm$ 3.2	77 $\pm$ 3.7
30°C	23 $\pm$ 6.7	54 $\pm$ 5.0
Ouabain (100 μM)	50 $\pm$ 3.5	69 $\pm$ 5.0
No Glucose	54 $\pm$ 4.1	76 $\pm$ 6.2

Sections of retinas were incubated in media containing ^{3}H taurine
(0.005mM). The values are expressed as percentages of control and
represent the mean $\pm$ S.E.M. for three to six separate incubations.
The uptake rates were respectively 0.059 $\pm$ 0.005 and 0.025$\pm$
0.0024 nmol/mg dry wt./min for normal and photoreceptorless retinas.

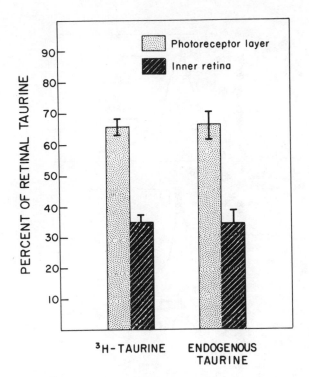

Figure 4. Distribution of labeled taurine taken up from the medium
and endogenous taurine in the microdissected photoreceptor cell layer
and inner retina. Isolated cat retinas were incubated for 10 minutes
in dim light in media containing 0.05 mM ^{3}H-taurine. The vertical
lines within the bars indicate $\pm$ S.E.M. for 6 different experiments.

In contrast to the taurine-deficient cat in which retinal deficiency precedes photoreceptor cell death, decreases in retinal taurine
concentrations in the RCS/p+ rat occur simultaneously with photoreceptor cell death after the third week of postnatal life. During
early postnatal life, the RCS rat retina develops comparably to that
of normal rats, and until about 21-23 postnatal days, the RCS rat
retina is similar to normal with respect to the thickness of the outer
nuclear layer, retinal DNA and taurine concentrations, and ERG amplitudes. After the 23rd postnatal day, the photoreceptor cells begin
to degenerate, and thereafter the reductions in retinal taurine
content (Figure 5) can be correlated with reduction in the thickness
of the outer nuclear layer, decrease in retinal DNA concentrations,
and a decline in ERG amplitudes.

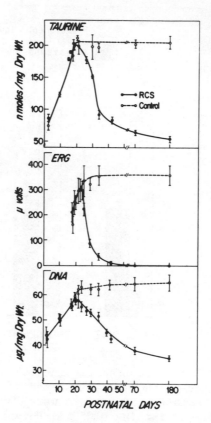

Figure 5. Retinal taurine concentrations, ERG amplitudes and retinal
DNA concentrations at various postnatal ages in normal Long-Evans
rats and pigmented RCS rats with hereditary retinal degeneration.
Each data point and vertical bar represents the mean ± S.E.M. for
measurements from 3-5 rats. (Reproduced with permission from
Schmidt and Berson, Exp. Eye Res. 27:191-198, 1978).

As discussed above for the cat retina, two processes exist for
the uptake of taurine in isolated rat retinas. The high affinity
uptake mechanism has an apparent K_m value of about 50μM (Schmidt and
Berson, 1978, Neal et al 1973) and a V_{max} of 0.3 /nmoles/mg/min.
The nonsaturable uptake mechanism has a rate constant of 0.63 nmoles/
mg dry wt/min/mM taurine in the medium (Schmidt and Berson 1978).
The high affinity uptake mechanism in the rat retina is inhibited

to a greater extent than the nonsaturable mechanism, by ouabain and reduced temperature. Retinas from normal rats studied at ages 22–180 days and from RCS rats studied at ages 22–45 days show both mechanisms at a time when photoreceptor cell function is detectable in vivo with the ERG. Retinas from 180–day–old photoreceptorless RCS rats retain the nonsaturable uptake mechanism at a time when no photoreceptor function can be detected in the ERG; this again demonstrates that the high affinity mechanism for taurine uptake depends on the presence of viable photoreceptor cells (Schmidt and Berson 1978). The high affinity uptake mechanism but not the nonsaturable mechanism is lost in the isolated normal rat retina when the photoreceptor cells are mechanically disrupted and is also lost when the isolated normal rat retina is maintained in room light for 20 minutes or in darkness for 60 minutes, conditions associated with a large efflux of taurine from the isolated retina and presumed photoreceptor cell death (Schmidt, 1978).

In addition to the high–affinity uptake mechanism for taurine, light–evoked fluxes of taurine have also been associated with the photoreceptor cells (Schmidt 1978). Onset and cessation of illumination have been shown to be associated with a prompt transient release followed by reuptake of taurine by isolated retinas incubated in media containing labelled taurine. Light–evoked taurine fluxes are also observed in the isolated retinas from normal rats and 30–day–old RCS rats but cannot be demonstrated in the photoreceptorless retinas from 180–day old RCS rats.

Studies with taurine–deficient cats have provided a new approach to the study of the cell biology of photoreceptor cells. The mechanism by which retinal taurine–deficiency leads to photoreceptor cell death remains to be defined. The close correlation of retinal taurine deficiency with peak–to–peak ERG amplitudes in taurine–deficient cats suggests that taurine deficiency may have some effect on the ionic fluxes of Na^+ and K^+ involved in the generation of the ERG. Since the generation of the ERG depends on hyperpolarization of photoreceptor cells and depolarization of Müller cells, it is possible that taurine deficiency has led to abnormal ionic concentrations Na^+ and K^+) in photoreceptor and Müller cells. This possibility is currently under investigation.

REFERENCES

Berson, E. L., Hayes, K. C., Rabin, A. R., Schmidt, S. Y., and Watson, G.: Retinal degeneration in cats fed casein: II. Supplementation with methionine, cysteine or taurine. Invest. Ophthalmol. 15:52–58, 1976.

Edwards, R.B.: Accumulation of taurine by cultured retinal pigment
    epithelium of the rat.   Invest. Ophthalmol. & Vis. Sci. 16:
    201-208, 1977.

Cohen, A. I., McDaniel, M. and Orr, H.: Absolute levels of some free
    amino acids in normal and biologically fractionated retinas.
    Invest. Ophthalmol. 12:686-93, 1973.

Hardison, W. G. M., Wood, C. A. and Proffitt, J. H.: Quantification
    of taurine synthesis in the intact rat and cat liver.   Proc.
    Soc. Exp. Biol. Med. 155:55-58, 1977.

Hayes, K. C., Rabin, A. R., and Berson, E. L: An ultrastructural
    study of nutritionally induced and reversed retinal degeneration
    in cats, Am. J. Pathol. 78:505-524, 1975.

Jacobsen, J. G., Thomas, L. L. and Smith, L. H., Jr.: Properties and
    distribution of mammalian L-cysteine sulfinate carbosylases.
    Biochem. Biophys. Acta 85: 103-116, 1964.

Keen, P., and Yates, R. A.: Distribution of amino acids in subdivided
    rat retinae. Br. J. Pharmacol. 52:118P, 1974.

Kennedy, A. J., Neal, M. J. and Lolley, R. N.: The distribution of
    amino acids within the rat retina. J. Neurochem. 29:157-159,
    1977.

Lake, N., Marshall, J. and Voaden, M. J.: High affinity uptake sites
    for taurine in the retina. Exp. Eye Res. 27:713-718, 1978.

MacDonnell, P. and Greengard, O.: The distribution of glutamate
    decarboxylase in rat tissues; isotopic vs fluorometic assays.
    J. Neurochem: 24:615-618, 1975.

Misra, C.H. and Olney, J.W. Cysteine oxidase in brain. Brain Res.
    97:117-126, 1975.

Neal, M. J., Peacock, D. G. and White, R. D.: Kinetic analysis of
    amino acid uptake by the rat retina in vitro. Br. J. Pharmacol.
    47:656-670, 1973.

Orr, H. T., Cohen, A. I., and Lowry, O. H.: The distribution of
    taurine in the vertebrate retina. J. Neurochem. 26:609-611,
    1976.

Schmidt, S. Y., Berson, E. L. and Hayes, K. C.: Retinal degeneration
    in cats fed casein: I. Taurine deficiency. Invest. Ophthalmol.
    15:47-52, 1976.

Schmidt, S. Y., Berson, E. L., Watson, G., and Huang, C.: Retinal
    degeneration in cats fed casein: III. Taurine deficiency and
    ERG amplitudes.   Invest. Ophthalmol. 16:673-678, 1977.

Schmidt, S. Y., and Berson, E. L.: Taurine uptake in isolated retinas
    of normal rats and rats with hereditary retinal degeneration.
    Exp. Eye Res. 27:191-198, 1978.

Schmidt, S. Y. and Szamier, R. B.: Taurine uptake by the photoreceptor
    cells in the isolated cat retina.   Abstract, Association for
    research in vision and ophthalmology, Sarastota, Florida, 1978.

Schmidt, S. Y.: Taurine fluxes in isolated cat and rat retinas:
    Effects of illumination. Exp. Eye Res. 26, 529-535, 1978.

Schmidt, S. Y. Unpublished observations. 1979.

Young, R. W.: The organization of vertebrate photoreceptor cells. In
    The Retina: Structure, function and clinical characteristics
    (eds. Straatsma, B., Allen, R., Hall, M. and Crescitelli, F.,
    pp. 177-209.   University of California Press, Los Angeles, 1969.

This work was supported by Research Grant EY01687 from the
National Eye Institute and by grants from the National
Retinitis Pigmentosa Foundation, Baltimore, Maryland and
the George Gund Foundation, Cleveland, Ohio, USA.

This article will also appear in the Proceedings of the
21st Annual A. N. Richards Symposium: "Actions of taurine
on excitable tissues".

THIALYSINE, THIAISOLEUCINE AND THIAPROLINE AS INHIBITORS IN PROTEIN SYNTHESIZING SYSTEMS

C.De Marco° and M.Di Girolamo[+]

Centro di Biologia Molecolare del CNR,Istituto di Chimica Biologica (°),Centro per lo studio degli Acidi Nucleici del CNR,Istituto di Fisiologia Generale ([+]), Università di Roma, Rome, Italy

Thialysine, thiaisoleucine and thiaproline are sulfur-containing analogs of natural aminoacids. Thialysine,or S-2-aminoethyl-cysteine,is a lysine analog having the 4 methylene group substituted by a sulfur atom; thiaisoleucine,or 2-amino-3-methylthio-butyric acid, is an isoleucine analog in which the 4 methylene group of the valerianic acid carbon chain is substituted by a sulfur atom.

$$H_2N.CH_2.CH_2.S.CH_2.CH(NH_2).COOH \qquad \text{thialysine}$$

$$CH_3.S.CH(CH_3).CH(NH_2).COOH \qquad \text{thiaisoleucine}$$

β-thiaproline   γ-thiaproline

Two thiaprolines will be considered: β-thiaproline, or thiazolidine-2-carboxylic acid, in which the β carbon atom of proline is substituted by a sulfur atom, and γ-thiaproline,or thiazolidine-4-carboxylic acid,in which is the γ carbon atom of proline that is substituted by a sulfur atom.

THIALYSINE

        Thialysine was first synthesized in 1954-55 inde-
pendently by Eldjarn(1) and by Cavallini et al.(2),who
also investigated its metabolism in vivo in the rat,
showing that it was excreted in the urine mainly as the
 α-N-acetyl derivative,in part as lanthionamine,its decarboxylation product,and in part as cystamine(3).
Thialysine was then tested by various investigators with
some lysine metabolizing enzymes. It was thus shown that
thialysine may be a substrate for bacterial lysine de-
carboxylase(4) and 2-oxo-glutarate aminotransferase(5),
for L-aminoacid oxidase from Mitylus edulis(6) or from
snake venom(7),for pea seedling diamineoxidase(8).
It was also shown that thialysine may be acetylated on
the terminal amino group by some microrganisms(9-11).
Thialysine may be cleaved non enzymatically by pyridoxal
phosphate through an α-β elimination giving cysteamine,
ammonia and pyruvate(12). It has to be recalled that
thialysine may be synthesized from serine and cysteamine
by yeast serine sulfhydrase(13).

        The activity of thialysine as lysine antimetabolite
was first demonstrated by Shiota et al.in lactic acid
bacteria(15).Further it was shown by Rabinowitz et al.in
1959 that thialysine acts as competitive inhibitor for
lysine incorporation into proteins in rat bone marrow
cells(16) and in rabbit reticulocytes(17).In 1965 using
purified E.coli lysil-tRNA synthetase,Kalousek and Rychlik showed that thialysine inhibits the lysine binding
to tRNA(18) and Stern and Mehler using labelled thialysine
demonstrated that it is a substrate for the enzyme,even
with a reduced affinity with respect to lysine,showing in
addition that it may be incorporated into proteins(19).
In 1977 Fabry et al. confirmed these observations and
showed poly(A)-directed formation of thialysine oligopeptides in E.coli protein synthesizing systems(20);the
incorporation of thialysine into peptidyl-tRNA was by one
order of magnitude lower than that of lysine.It has been
reported that in the presence of thialysine anomalous
proteins were synthesized in E.coli,characterized by a
higher rate of degradation(21).

        Recently we had the opportunity to test the acti-
vity of thialysine in protein synthesizing systems from
either E.coli,rat liver and rabbit reticulocytes,in a
study directed to investigate the activity of another
lysine analog,selenalysine(22). It was first studied the
activation of thialysine by E.coli and rat liver aminoacyl-tRNA synthetases,by assaying either the ATP-PPi

exchange or the effect of thialysine on the binding
of labelled lysine to tRNA.

In the ATP-PPi exchange with both E.coli and rat
liver synthetases the Km values obtained with thialysine
as substrate were similar to those of lysine,as shown in
Table I. On the contrary the V values for thialysine were
in both systems around forty per cent of that of lysine.
When lysine and thialysine were present together in 1:1
molar ratio,the obtained V value indicated that lysine
and thialysine are activated by the same enzyme.

TABLE I

Km and V values for lysine and thialysine in the ATP-PPi
exchange reaction.Experimental details in(22).V values
expressed as per cent of those obtained with lysine.

	Aminoacyl-tRNA synthetases from			
	E.coli		rat liver	
	Km	V	Km	V
Lysine	0.08	100	0.10	100
Thialysine	0.05	37	0.08	43
Lysine + thialysine		58		81

Studying the lysyl-tRNA synthesis it was shown that
thialysine inhibits the binding of lysine to tRNA: with
tRNAs and synthetases from E.coli the inhibiting effect
of thialysine was more evident than that observed with
tRNAs and synthetases from rat liver (Fig.1). The results
obtained using "mixed systems" containing synthetases
from E.coli and tRNAs from rat liver,or viceversa,sug-
gested that the higher inhibitory effect of thialysine
on the E.coli system is imputable to the higher ambiguity
of the E.coli synthetases,that is to their lower capacity
to differentiate lysine from thialysine. It was further
shown that the inhibition caused by thialysine may be reversed by lysine,and it was also excluded that the inhibition might be caused by inactivation of the lysyl-tRNA
synthetases.

Further the effects of thialysine on lysine or leucine incorporation into polypeptides were studied,using
polypeptide synthesizing systems from E.coli,rat liver or
rabbit reticulocytes. It was shown that in all three systems thialysine inhibits incorporation of labelled lysine into polypeptides,whereas it is without effect on leucine incorporation. The results obtained are reported in
Table II.

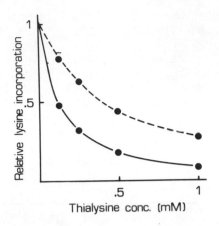

Fig.1. Inhibition of lysyl-tRNA synthesis by thialysine.
       Synthetases and tRNA from E.coli(full line) or
       from rat liver(Broken line).Relative lysine in-
       corporation = ratio of the amount of tRNA-bound
       ^{14}C-lysine in the presence of thialysine to the
       amount in its absence.Experimental details in(22).

TABLE   II

Per cent of inhibition of lysine or leucine incorporatio
into polypeptides by thialysine.Experimental details in
(22).

Substrate (0.04 mM)	Thialysine conc.	Polypeptide synthesizing system		
		E.coli	Rat liver	Rabbit reti culocytes
^{14}C-Lysine	1 mM	92	85	86
	5 mM	97	95	97
^{14}C-Leucine	1 mM	0	1	2
	5 mM	0	8	10

      Overall the results obtained clearly showed that
thialysine may be activated by aminoacyl-tRNA synthetase
may be transferred to tRNA^{lys},and may be incorporated
into polypeptides.In all these reactions thialysine acts
as a competitive inhibitor of lysine.Thus the previous
results obtained mainly with prokariotic cells were confirmed and extended for most aspects to eukariotic organisms.

THIAISOLEUCINE

Thiaisoleucine was first synthesized in 1965 by
McCord et al.(23) to test its possible antagonistic effects towards isoleucine in microrganisms. They showed
in fact that thiaisoleucine inhibits growth of E.coli
and other microrganisms,and that the toxicity of thiaisoleucine in E.coli was specifically and competitively
reversed by isoleucine. It was thus stated that thiaisoleucine is a specific and effective isoleucine antagonist
for the microrganisms studied. Szentirmai and Umbarger
confirmed these results and showed that thiaisoleucine
is a substrate for isoleucyl-tRNA synthetase(24) and
inhibits the transfer of labelled isoleucine to tRNA(25).
They showed moreover that thiaisoleucine inhibited the
deamination of threonine,and that it underwent transamination with 2-oxo-glutarate thus inhibiting isoleucine
transamination.

Further Treiber and Iaccarino(26) studying isoleucyl-tRNA synthetases partially purified from wild type
E.coli and from an isoleucine requiring mutant showed
that thiaisoleucine and another isoleucine analog,
0-methyl-threonine,inhibiting E.coli growth(27),were
substrates for the enzyme. It was also demonstrated in
E.coli crude extracts that thiaisoleucine and 0-methyl-
threonine inhibit threonine deaminase; this is the first
enzyme in isoleucine biosynthesis,and might thus be involved in the growth inhibitory effects caused by the two
isoleucine analogs(28).

In all the above studies thiaisoleucine was tested
with prokariotic organisms; it seemed interesting to study
its effects on protein synthesizing systems from rat liver
and rabbit reticulocytes. It was first investigated the
activation of thiaisoleucine by rat liver aminoacyl-tRNA
synthetases studying the aminoacid-dependent ATP-PPi
exchange reaction(29). It was shown that thiaisoleucine,
like 0-methyl-threonine,is a substrate for the enzyme.
The Km values obtained,reported in Table III,showed that
both analogs have a lower affinity than isoleucine for
the synthetase,but thiaisoleucine has a three-fold greater
affinity than 0-methyl-threonine. The V values were quite
similar for isoleucine and for both the analogs. From the
V values obtained when both isoleucine and thiaisoleucine
were present together in 1:1 molar ratio,or when isoleucine was incubated in the presence of equimolar amounts
of leucine,it was concluded that thiaisoleucine is activated by isoleucyl-tRNA synthetase.

TABLE   III

Km and V values for isoleucine,thiaisoleucine and
O-methyl-threonine in the ATP-PPi exchange reaction
catalyzed by rat liver aminoacyl-tRNA synthetase.
Experimental details in(29). V values expressed as per
cent of that obtained with isoleucine.

Substrates	Km (mM)	V
Isoleucine	0.15	100
Thiaisoleucine	1.12	91
O-methyl-threonine	3.50	97
Isoleucine + thiaisoleucine		93
Isoleucine + leucine		203

It was then tested if thiaisoleucine once activated
could be transferred to tRNA and this was done by assaying
the effect of thiaisoleucine on the binding of labelled
isoleucine to tRNA. Fig.2 shows that thiaisoleucine and
O-methyl-threonine inhibit isoleucyl-tRNA synthesis;
thiaisoleucine was ten times more effective than O-methyl
threonine.It was also demonstrated that the inhibition
was in both cases reversed by isoleucine.

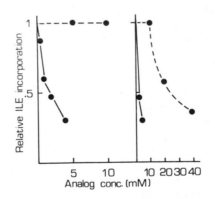

Fig.2. Inhibition of isoleucyl-tRNA synthesis by thiaiso
       leucine(full lines) and by O-methyl-threonine
       (broken lines).Relative ILE incorporation = ratio
       of the amount of tRNA-bound ^{3}H-isoleucine in the
       presence of the indicated concentrations of the
       analogs to the amount in their absence. Experi-
       mental details in (29).

Further the effect of thiaisoleucine on the incorporation of labelled leucine and isoleucine into protein
was tested. Protein synthesizing systems from either rat

liver and rabbit reticulocytes were used. The results
obtained,reported in Table IV,show that thiaisoleucine
inhibits the isoleucine incorporation and is more effective than O-methyl-threonine. Neither of the analogs
instead inhibited leucine incorporation. The inhibitory
effect of both analogs was completely reversed by increasing concentrations of isoleucine. The higher inhibitory
effect of thiaisoleucine has been correlated to its higher
capacity to compete with isoleucine for charging tRNA.

These results indicated that thiaisoleucine may be
incorporated in place of isoleucine into polypeptides,in
competition with isoleucine,and that its incorporation
does not impair the polypeptide synthesis. This was confirmed by the analysis of the polysomal pattern and of
the polysomal-bound radioactivity. Rabbit reticulocyte
lysates were incubated,in the presence of all the factors
necessary for protein synthesis,with labelled isoleucine
or leucine, and with or without thiaisoleucine. As shown
in Fig.3 the polysomal profile remained unchanged in the
presence or absence of thiaisoleucine,while the polysomal-bound radioactivity due to isoleucine was highly reduced in the presence of thiaisoleucine.On the other hand
the polysomal-bound radioactivity due to leucine was unaffected by the presence of thiaisoleucine. These results
showed that the incorporation of thiaisoleucine into the
polypeptide chain does not modify the ribosome run-off
nor the elongation rate of the growing polypeptide chain.

TABLE   IV

Per cent of inhibition of isoleucine and leucine incorporation into polypeptides by thiaisoleucine(T-ILE) and
O-methyl-threonine(O-MT).Experimental details in (29).

Substrate (0.04 mM)	Additions (mM)		Polypeptide synthesizing system from :	
			Rat liver	Rabbit reticulocytes
^{3}H-ILE	T-ILE	0.2	84	78
		1	93	89
	O-MT	2	65	19
		10	94	43
^{14}C-LEU	T-ILE	0.2	0	0
		1	0	4
	O-MT	2	0	0
		10	7	5

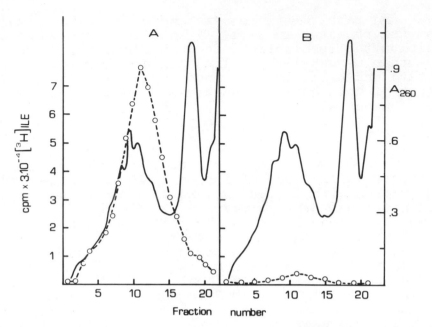

Fig.3. Polysomal profile (full lines,A_{260}) and polysomal bound radioactivity (broken lines) of reticulocyte lysates subjected to sucrose density gradients after incubation with ^{3}H-isoleucine in the absence (A) or in the presence (B) of 1 mM thiaisoleucine. Experimental details in (29).

Overall the results obtained show that all the enzymes involved in the various steps of isoleucine incorporation into proteins are effective also on thiaisoleucine. In other words the substitution of a carbon atom along the carbon chain of the substrate by a sulfur atom does not dramatically affect the substrate specificity of the enzymes. The comparative data obtained with O-methyl-threonine indicated that the substitution of a carbon atom by an oxygen atom along the carbon chain affects the substrate specificity of the enzymes more than does the substitution by a sulfur atom.

THIAPROLINES

It is well known that γ-thiaproline,or thiazolidine-4-carboxylic acid,acts as a competitive inhibitor of proline in protein synthesizing systems from rat liver (30,31) and in the activation of proline catalyzed by purified prolyl-tRNA synthetase from E.coli(32). In cell-free rat liver preparations γ-thiaproline is incorporated into ribosomal-bound proteins(31). It has been reported that γ-thiaproline inhibits cell division in growing cultures of E.coli(33),possibly by a toxic effect due to the production of abnormal proteins(34).

We had the opportunity to test the effects of γ-thiaproline on protein synthesizing systems in comparison with another proline analog,selenaproline,in which the sulfur atom of thiaproline is substituted by a selenium atom(35). The obtained results showed,in well agreement with the known data,that γ-thiaproline is a substrate for prolyl-tRNA synthetases from either E.coli or rat liver,even if with a much lower degree of affinity with respect to proline. It was also confirmed that γ-thiaproline may be incorporated in place of proline in the growing polypeptide chain,and it was demonstrated that once incorporated it affects the ribosome run-off and the chain elongation.

More recently we have studied the effects of β-thiaproline on protein synthesizing systems(36). β-Thiaproline,first synthesized by Fourneau et al.(37),has not been assayed in biological systems. We have obtained preliminary data indicating that it is a substrate for hog kidney D-aminoacid oxidase and for rat liver mitochondria proline oxidase,two enzymes active on γ-thiaproline.
It seemed interesting to test β-thiaproline,in comparison with γ-thiaproline,as a potential proline competitive inhibitor in protein synthesis.

It was demonstrated that aminoacyl-tRNA synthetases from E.coli or rat liver can activate β-thiaproline and transfer it to tRNA^{pro}. The Km values obtained in the ATP-PPi exchange reaction,reported in Table V,showed that both E.coli and rat liver synthetases have a greater affinity for β-thiaproline than for γ-thiaproline;in particular the affinity of the E.coli synthetase for β-thiaproline is very similar to that for proline,while that for γ-thiaproline is about hundred-fold lower.

The transfer of β-thiaproline to tRNA was demonstrated indirectly showing that it inhibits prolyl-tRNA synthesis in both E.coli and rat liver systems. In the rat liver system the inhibitory effect of β-thiaproline is more marked than that of γ-thiaproline (Fig.4).
In the E.coli system the inhibitory effect is more evident with respect to the rat liver,but there are no great differences between β- and γ-thiaproline. Controls made with labelled leucine showed that β-thiaproline,like γ-thiaproline,was without effect on leucyl-tRNA synthesis. It was thus demonstrated that β-thiaproline,like γ-thiaproline,bind to tRNA in competition with proline.

TABLE   V

Km and V values for proline, β-thiaproline(β-T) and
γ-thiaproline(γ-T) in the ATP-PPi exchange reaction.
Experimental details in(36). V values expressed as per
cent of those obtained with proline.

Substrates	Aminoacyl-tRNA synthetases from			
	E.coli		Rat liver	
	Km(mM)	V	Km(mM)	V
Proline	0.17	100	0.25	100
β-T	0.17	110	1.00	130
γ-T	14.00	160	8.00	24
Proline + β-T		103		108
Proline + γ-T		108		106

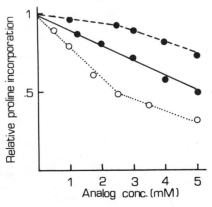

Fig. 4. Inhibition of prolyl-tRNA synthesis by β- and
γ-thiaproline.Full circles:synthetases and tRNAS
from rat liver;full line: β-thiaproline;broken
line: γ-thiaproline.Empty circles:synthetases and
tRNAs from E.coli; β- and γ-thiaproline gave the
same values.Relative proline incorporation = ratio
of the amount of tRNA-bound ^{14}C-proline in the pre
sence of the analogs to the amount in their absen
ce. Experimental details in (36).

The possibility that β-thiaproline could be incorporated into polypeptides in place of proline was then
investigated by studying its effect on the incorporation
of labelled proline or leucine.Protein synthesizing systems from E.coli,rat liver and rabbit reticulocytes wer

tested. It was shown that β-thiaproline,in all three systems,inhibits proline incorporation,and in this respect
is more active than γ-thiaproline. However,while in the
E.coli system β-thiaproline inhibits proline incorporation
but not leucine incorporation,in the rat liver and rabbit
reticulocytes β-thiaproline inhibits also leucine incorporation,even if at a minor extent than that of proline,
(Table VI). The inhibitory effect of β-thiaproline is
competitive with proline in all three systems,although
in the rabbit reticulocytes the inhibition observed at
the highest concentrations was not completely reversed
by proline.

TABLE   VI

Per cent of inhibition of ^{14}C-proline or ^{14}C-leucine incorporation into polypeptides by β-thiaproline(β-T) and
γ-thiaproline(γ-T).Polypeptide synthesizing systems from
E.coli,rat liver and rabbit reticulocytes. Experimental
details in (36).

	Substrate (0.04 mM)	β-T			γ-T			
		1	2	5	1	2	5	10
E.coli	Proline	27	57	83	17	38	60	82
	Leucine	0	0	0	0	0	4	21
Rat liver	Proline	16	27	79	0	0	0	21
	Leucine	0	3	48	0	0	0	15
Reticuloc.	Proline	36	46	79	0	0	10	27
	Leucine	27	42	77	0	0	5	16

The inhibition of leucine incorporation suggested
that in mammalian systems β-thiaproline once incorporated
inhibits polypeptide synthesis. This was checked by analyzing the polysomal patterns of reticulocyte lysates
incubated in the presence or absence of β-thiaproline.
It was shown that in the presence of β-thiaproline the
decrease in the total amount of polysomes and the increase in monosomes were not so marked as in the controls,
thus indicating that β-thiaproline inhibits ribosome
run-off.Also this effect was reversed by proline.

Further,studying the effect of puromycin on growing
polypeptide chains labelled with ^{14}C-leucine in the presence or absence of β-thiaproline,data were obtained indicating that β-thiaproline once incorporated into polypep-

tide chain impairs its further elongation,acting in this
respect like γ-thiaproline(35).

Overall the results obtained with β-thiaproline
showed that it acts as a competitive inhibitor of proline,but differs somewhat from the other proline analog,
 γ-thiaproline. We may distinguish between the E.coli
and the mammalian systems. In E.coli β-thiaproline has
a higher affinity than γ-thiaproline for the aminoacyl-
tRNA synthetases;on the other hand it is bound to tRNAᴾʳᵒ
at the same extent as γ-thiaproline. β-Thiaproline shows
a more evident inhibition on the proline incorporation
into polypeptides,and this may reflect a higher capacity
of β-thiaprolyl-tRNA with respect to γ-thiaprolyl-tRNA
to compete with prolyl-tRNA at ribosomal level.
In the mammalian systems the differences between the two
analogs are less evident as regards the affinity for
aminoacyl-tRNA synthetases and the capacity to bind tRNA
On the other hand β-thiaproline inhibits to a higher
extent the incorporation of both proline and leucine int
polypeptides,impairing further chain elongation once incorporated.

The results obtained with β- and γ-thiaproline indicate however that the presence of a sulfur atom either
in β or γ position does not substantially modify the act
vity of the two thiaprolines as proline competitive inhi
bitors.

In conclusion, the data here summarized on the effects of thialysine,thiaisoleucine and thiaprolines on
protein synthesizing systems show that all these thia-
analogs compete with the corresponding aminoacids in the
various steps of protein synthesis. The results of the
present studies add to those previously obtained with
other enzymatic systems in pointing out that the presenc
of a sulfur atom along the carbon chain does not greatly
impair the capacity of the sulfur analogs to be substrat
for enzymes acting on the corresponding aminoacids.
In protein synthesizing systems the thia-analogs,being
substrates for the different enzymes,obviously act as
competitive inhibitors. Some minor quantitative differences,like those observed between β- and γ-thiaproline,
suggest that slight structural differences may have some
influence on each single reaction.

R E F E R E N C E S

1 Eldjarn,L. Scand.Clin.Lab.Invest.6 Suppl.13,71(1954)
2 Cavallini,D.,De Marco,C.,Mondovì,B. & Azzone,G.F.
      Experientia, 11,61 (1955)

3 Cavallini,D.,Mondovì,B. & De Marco,C. Biochim.Biophys.
      Acta 18,122 (1955)
4 Work,E. Biochim.Biophys.Acta 62,173 (1962)
5 Soda,V.,Misono,H. & Yamamoto,T. Biochim.Biophys.Acta
      177,364 (1969)
6 Hope,D.B. Ciba Found.Study Gp. 19,88 (1965)
7 Cini,C.,Foppoli,C. & De Marco,C. Ital.J.Biochem.
      27,305 (1978)
8 Hermann,P. & Willhardt,L. 4 Int.Symp.Biochemie und
      Physiologie der Alkaloide,Halle(Saale)25-28 Juni
      1969 - Akademie-Verlag,Berlin (1972).
9 Hermann,P.,Asperger,O.,Kutschera,I. & Senkpiel,K.
      Vth FEBS Meeting,Prague (1968)
10 Soda,K.,Tanaka,H. & Yamamoto,T. Archiv.Biochem.
      Biophys. 130,610 (1969)
11 Tanaka,H. & Soda,K. J.Biol.Chem. 249,5285 (1974)
12 De Marco,C. Biochim.Biophys.Acta 85,62 (1964)
13 Hermann,P.,Müller,R.,Lemke,K. & Willhardt,L.
      VI Jahrversamml.Biochem.Ges.D.D.R. (1969)
14 Shiota,T.,Folk,J.E. & Tietze,F. Archiv.Biochem.
      Biophys. 77,372 (1958)
15 Shiota,T.,Mauraon,J. & Folk,J.E. Biochim.Biophys.
      Acta 53,360 (1961)
16 Rabinowitz,M. & Tuve,K. Proc.Soc.Expt.Biol.Med.
      100, 222 (1959
17 Rabinowitz,M. & Fisher,J.M. Biochem.Biophys.Res.
      Comm. 6, 449 (1961)
18 Kalousek,F. & Rychlik,I. Collection Czechoslov.
      Chem.Comm. 30, 3909 (1965)
19 Stern,R. & Mehler,A.H. Biochem.Z. 342, 400 (1965)
20 Fabry,M.,Hermann,P. & Rychlik,I. Collection Czecho-
      slov.Chem.Comm. 42, 1077 (1977)
21 Goldberg,A.L. Proc.Natl.Acd.Sci.U.S. 69, 422 (1972)
22 De Marco,C.,Busiello,V.Di Girolamo,M. & Cavallini,D.
      Biochim.Biophys.Acta 454, 298 (1976)
23 McCord,T.S.,Howell,D.C.,Tharp,D.L. & Davis,A.L.
      J.Med.Chem. 8, 290 (1965)
24 Szentirmai,A.,Szentirmai,M. & Umbarger,H.E.
      J.Bacteriol. 95, 1672 (1968)
25 Szentirmai,A. & Umbarger,H.E. J.Bacteriol.95,1666(1968)
26 Treiber,G. & Iaccarino,M. J.Bacteriol. 107,828(1971)
27 Smulson,M.E.,Rabinowitz,M. & Breitman,T.R.
      J.Bacteriol. 94, 1890 (1967)
28 Cervone,T. & Iaccarino,M. FEBS Lett. 26, 56 (1972)
29 Busiello,V.,Di Girolamo,M. & De Marco,C. Biochim.
      Biophys.Acta 561, 206 (1979)
30 Frazer,M.J. & Klass,D.B. Can.J.Biochem.41,2126 (1963)
31 Bekhor,I.J.,Mosheni,Z. & Bavetta,L.A. Proc.Soc.Exp.
      Biol.Med. 119, 765 (1965)
32 Papas,T.S. & Mehler,A.H. J.Biol.Chem. 245, 1588 (1970)

33 Unger,L. & De Moss,J.A.   J.Bacteriol. <u>91</u>, 1556 (1966)

34 Fowden,L.,Lewis,D. & Tristram,H. Adv.Enzym.<u>29</u>,89(1967)

35 De Marco,C.,Busiello,V.,Di Girolamo,M. & Cavallini,D.
      Biochim.Biophys.Acta <u>478</u>, 156 (1977)

36 Busiello,V.,Di Girolamo,M.,Cini,C. & De Marco,C.
      Biochim.Biophys.Acta (1979) in press.

37 Fourneau,P.,Efimovsky,O.,Gaignault,J.C.,Jacquier,R.
      & Le Ridant,C.   C.R.Acad.Sc.Paris <u>272</u> Serie C
      1515 (1971)

38 Coccia,R.,Blarzino,C.,Foppoli,C. & De Marco,C.
      Ital.J.Biochem. (1979) in press.

LANTHIONINE DECARBOXYLATION BY ANIMAL TISSUES

R.Scandurra,V.Consalvi,C.De Marco,
L.Politi and D.Cavallini

Institute of Biological Chemistry,
University of Rome,Rome,Italy

Taurine content in animal tissues,ranging from 1 to 38 μmoles/g of wet tissue(1),exceeds 5-200 fold the cellular concentration of each aminoacid.The metabolic pathways leading to taurine are summarized in fig.1.

The inorganic pathway,the enzymes of which have recently been isolated by Martin et al.(2),should be considered scarcely efficient in taurine production if it can be calculated that only one molecule of labeled taurine over a total of 40,000 has been found after administration of 20 μCi of labeled sulfate(3).In this route sulfate is firstly activated to adenosylphosphosulfate(APS), then to phosphoadenosylphosphosulfate(PAPS)which is transferred to α aminoacrylate(αAA)derived frcm serine.

The cysteinesulfinic acid(CSA)pathway has been till now considered the most important route for taurine production in animal tissues.However this claim needs some criticism for the following reasons:

a)cysteine oxygenase,the first enzyme of this pathway has been found only in rat liver(4).However rat heart and brain where cysteine oxygenase is absent,are able to convert cysteine into taurine in vitro(5,6);

b)it is highly improbable a transfer tc other tissues of CSA,being the latter a good substrate for the widely distributed aspartate aminotransferase(7);

c)in many tissues CSA decarboxylase shows a poor activity

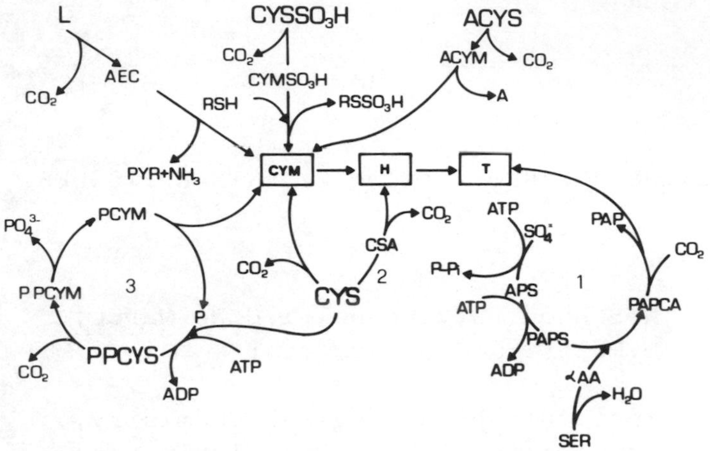

Fig.1-Metabolic pathways leading to taurine production.
1=The inorganic pathway.APS=adenosylphosphosulfate;
PAPS=phosphoadenosylphosphosulfate;SER=serine;αAA=
αaminoacrylate;PAPCA=phosphoadenosylphosphocysteic
acid;PAP=phosphoadenosylphosphate.CYM=cysteamine;
H=hypotaurine;T=taurine.2=The cysteinesulfinic acid
pathway.CYS=cysteine;CSA=cysteinesulfinic acid.
3=The cysteamine pathway.P=pantothenate;PPCYS=
phosphopantothenylcysteine;PPCYM=phosphopanthethe-
ine;PCYM=panthetheine;L=lanthionine;AEC=aminoethyl-
cysteine;PYR=pyruvate;$CYSSO_3H$=cysteinesulfonate;
$CYMSO_3H$=cysteaminesulfonate;ACYS=adenosylcysteine;
ACYM=adenosylcysteamine;A=adenosine.

and in others,as the heart,is lacking(5,8,9);
d)rats fed with a vitamin B_6 free diet decrease,without
extinguish,taurine production(10).
These observations allow to conclude that the CSA pathway should be never unique nor the most important metabolic route for taurine production.
The direct decarboxylation of cysteine to cysteamine has
never been observed.However when cysteine is bound to
phosphopantothenate it is decarboxylated by a specific

decarboxylase widely distributed in all animal tissues
(11)and purified from rat(12)and horse liver(13).The latter enzyme is not pyridoxal phosphate dependent but contains covalently bound pyruvate involved in the catalytic activity(14).Phosphopantetheine(PPCYM)obtained by
decarboxylation of phosphopantothenylcysteine(PPCYS) is
splitted in pantetheine(PCYM)and phosphate by a phosphatase and then,by pantetheinase(11)into pantothenate(P)
and cysteamine(CYM)which is oxidized to hypotaurine by
cysteamine oxygenase(15).The efficiency of this pathway,
the cysteamine pathway,has been evaluated by measuring
the production of either the amount of $^{14}CO_2$ or the sum
of hypotaurine and taurine incubating animal tissues
extracts with labeled phosphopantothenylcysteine and
omparing these results with those obtained using same
amounts of labeled cysteine in order to evaluate the efficiency of the CSA pathway(16).The results obtained indicate that when free cysteine is used,only in rat liver,
where cysteine oxygenase and CSA decarboxylase are present
the amount of hypotaurine plus taurine is higher.In all
other animal tissues the sum of hypotaurine plus taurine
derived from phosphopantothenylcysteine exceeds that from
free cysteine,demonstrating a preferential use of the
cysteamine pathway in most animal tissues.The efficiency
of the cysteamine pathway the enzymes of which are all
pyridoxal phosphate independent could then explain the
decrease,without extinction,of taurine production in a
rat fed a vitamin B_6 free diet.
However other routes should produce cysteamine since the
pathway described above is insufficient to explain the
high content of taurine in animal tissues.
Three other possible routes for taurine production,via
cysteamine,are indicated in fig.1.Adenosylcysteine(ACYS)
could be decarboxylated to adenosylcysteamine(ACYM)which
could be splitted in adenosine(A)and cysteamine(CYM).
Cysteinesulfonate($CYSSO_3H$)decarboxylated to cysteamine-
sulfonate($CYMSO_3H$)could react with a thiol(RSH)to give
cysteamine and a sulfonate($RSSO_3H$.
The third route stems from lanthionine(L)which has been
occasionally found in biological materials(17)where could
be generated either by condensation of cysteine with serine or of cysteine with aminoacrylate obtained by a

number of different mechanisms(18,19).

The one side decarboxylation of lanthionine could yield aminoethylcysteine(AEC)which is known as a precursor of taurine when injected into rats(20).

^{14}C-lanthionine,obtained by reaction of ^{14}C-cysteine with β-chloroalanine as reported previously(21),when incubated with animal tissues produces $^{14}CO_2$,demonstrating the presence of a decarboxylating activity in all animal tissues the most efficient being rat and beef kidney(21). The identification of the reaction products(aminoethylcysteine,cysteamine,hypoyaurine and taurine)has been performed either by paper or ion exchange chromatography, the latter allowed their quantitative estimation(21). These results suggested a purification of the enzyme. Beef kidney cortices were homogenized in 10 mM Tris-Cl buffer pH 8,centrifuged at 12,000 x g for 20 minutes.

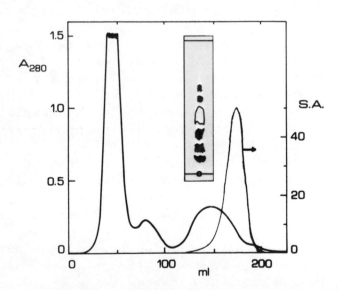

Fig.2-Elution profile from a DEAE cellulose column(cm
       2 x 30).Elution buffer 10 mM Tris-Cl,pH 8;500 mg
       proteins.Specific activity=μmoles$^{14}CO_2$/hr/g.
       Insert=chromatographic pattern of active fraction
       on cellulose thin layer developped in butanol-
       pyridine-acetic acid-water(45:30:9:36)and sprayed
       with ninhydrin.

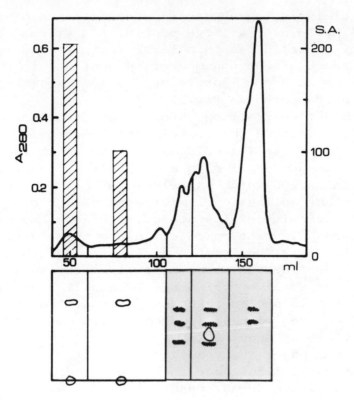

Fig.3-Elution profile from a Sephadex G 25 superfine
column(cm1 x 200)in ammonium bicarbonate 50 mM.
10 mg of protein.Specific activity=μmoles $^{14}CO_2$/
h/g.Fractions were collected as depicted in the
figure and tested on cellulose thin layer as
described in fig.2.

The supernatant was additioned with ammonium sulfate
till a 55% saturation,the precipitate,collected by
centrifugation,was dissolved in 10 mM Tris-Cl buffer pH
8 made 0.1 mM in pyridoxal-5'-phosphate and dialyzed
overnight against the same buffer.Clarified by centrifugation,the solution was applied to a DEAE(DE-23)cellulo-
se column equilibrated and eluted with 10 mM Tris-Cl buffer pH 8.The lanthionine decarboxylating activity was
found in the tail of the second eluted peak,as reported
in Fig.2.Concentrating by pressure dialysis the protein
solution containing the activity,this was not retained

by the dialysis bag,suggesting the presence of either
a small protein or a large peptide.Analytical chroma--
tography cn cellulose thin layer for peptides revealed
the presence of one non migrating spot due to a large
peptide or to a small proteine and some well migrating
spots ,as reported in Fig.2.
The DEAE active fraction gaves a purification factor of
about 70 fold over the centrifuged homogenate.This frac-
tionwas then lyophilized,dissolved in 50 mM ammonium
bicarbonate and applied to a Sephadex G25 superfine
column(cm1 x 200)which was eluted with the same solvent
at 5.7 ml/hr collecting 1 ml fractions.The elution profile,the activity of the pooled fractions and their
chromatographic pattern on cellulose thin layer is reported in Fig.3.The fractions with the decarboxylating
activity on lanthionine are the first two were large
peptides or small proteins are located.The specific activity of this pooled fraction is about 300 fold over
that of the centrifuged homogenate(0,5 μmoles$^{14}CO_2$/h/g,
with a yield of about 50%.
The optimum pH of the reaction is 8.5,the reaction is
linear for the first hour then stops demcnstrating a possible inhibition by the reaction products.Addition of
pyridoxal-5'-phosphate increases about 3 times the velocity of the reaction.The apparent K_m for lanthionine is
about 1.2 mM.

DISCUSSION

The experimental results reported demonstrate that
in beef kidney cortices is present a small protein operating a one-side decarboxylation of lanthionine with
the production of aminoethylcysteine which is furtherly
degraded into cysteamine,hypotaurine and taurine as
previously reported(21).At the present time cannot be
claimed that this decarboxylase is specific toward lan-
thionine:its activity on lanthionine could be an ancil-
lary activity of another decarboxylase.However this activity on lanthionine demonstrate the presence in animal
tissues of a new route which,via cysteamine,contributes
to taurine pocl.These results further   confirm that cysteamine pathway appears tc be the most important route

for taurine production.

REFERENCES

1   Jacobsen,J.G.and Smith,L.H.jr.Physiol.Rev.48,424(1968)
2   Gorby,W.G.and Martin,W.G.Proc.Soc.Exptl.Biol.Med.148,
    344(1975).
3   Martin,W.G.,Sass,N.L.,Hill,L.,Tarka,S.and Truen,R.
    Proc.Soc.Exptl.Biol.Mrd.141,632(1972).
4   Sakakibara,S.,Yamaguchi,K.,Hosokawa,Y.,Kohashi,N.,
    Ueda,I.and Sakamoto,Y.Biochim.Biophys.Acta 422,273,
    (1976).
5   Huxtable,R.and Bressler,R."The metabolism of cystea-
    mine to taurine"in"Taurine"Huxtable,R.and Barbeau,A.
    ed.Raven Press,New York,(1976).
6   Read,W.O.and Welty,J.D.J.Biol.Chem.237,1521(1962).
7   Leinweber,F.J.and Monty,K.J.Anal.Biochem.4,252(1962).
8   Jacobsen,J.C.,Thomas,L.T.and Smith,L.H.Biochim.Bioph.
    Acta 85,103(1964).
9   Huxtable,R."Metabolism and function of taurine in the
    heart"in"Taurine"Huxtable,R.and Barbeau,A.ed.Raven
    Press,New York(1976).
10 Yamaguchi,K.,Shigehisa,S.,Sakakibara,S.,Hosokawa,Y.
    and Ueda,I.Biochim.Biophys.Acta 381,1(1975).
11 Cavallini,D.,Scandurra,R.,Duprè,S.,Federici,G.,Santo-
    ro,L.,Ricci,G.and Barra,D."Alternative pathways of
    taurine biosynthesis"in"Taurine"Huxtable,R.and Barbeau
    A.ed.Raven Press,New York(1976).
12 Abiko,Y.J.Biochem(Tokyo)61,300(1967).
13 Scandurra,R.,Barboni,E.,Granat,F.,Pensa,B.and Costa,M.
    Eur.J.Biochem.49,1(1974).
14 Scandurra,R.,Moriggi,M.,Consalvi,V.and Politi,L.,
    Methods in Enzymol.62,245(1979).
15 Cavallini,D.,De Marco,C.,Scandurra,R.,Duprè,S.and
    Graziani,M.T.J.Biol.Chem.241,3189(1966).
16 Scandurra,R.,Politi,L.,Duprè,S.,Moriggi,M.,Barra,D.
    and Cavallini,D.Bull.Mol.Biol.and Med.2,172(1977).
17 Rao,D.R.,Ennor,A.H.and Thorpe,B.Bioch.Biophys.Res.
    Comm.22,163(1966).
18 Chapeville,F.and Fromageot,P.Bioch.Biophys.Acta,49,
    328(1961).
19 Tolosa,E.A.,Chepurnova,N.K.,Khomutov,R.M.and Severin,

E.S.Biochim.Biophys.Acta 171,369(1969).

20 Cavallini,D.,Mondovì,B.,De Marco,C.Biochim.Biophys.
   Acta 18,122(1955).

21 Scandurra,R.,Consalvi,V.,De Marco,C.,Politi,L.and
   Cavallini,D.Life Sciences,24,1925(1979).

SULFUR CONTAINING NUPHAR ALKALOIDS

Jerzy T.Wróbel, Halszka Bielawska
Agnieszka Iwanow, and Joanna Ruszkowska
Department of Chemistry,
University of Warsaw,
Poland

The chemistry of this group of Nuphar Alkaloids was not well known till now since the system has been shown not to be very reactive to most of common reagents, which usually are used for transformations and degradation of an alkaloid system [1].

In general sulfur containing alkaloids are resistant to heating and some oxidation agents, do not undergo quaternization and desulfuration in respect to exclusive removal of sulfur atom.

This behavior could be explained by very complex structure from both chemical and stereochemical point of view. All chemically active sites in the molecule, are either sterically hindered like nitrogen atoms by furan rings or compete in all reaction which involve electrophiles like sulfur and nitrogen. Therefore the results of any of chemical transformations were not clear cut and usually resulted in complex mixtures formed as a result of number of reactions which occur.

In alkaloids which posses some additional functional groups like sulfinyl or hydroxy groups, the chemical activity is more specific. We are now able to report on some chemical reaction which in proper conditions could be carried on both sulfides and sulfoxides types of alkaloids.

I s o m e r i z a t i o n   o f   s u l f o x i d e s. The naturaly occuring sulfoxides to which the β configuration of the oxygen has been assigned (cis to the C6-C7 bond) [2], could be isomerized to the α ones in the presence of hydrochloric acid.

The reaction is fast and results in a mixture of isomeric sulfoxides in the ratio 2,0 : 1 (β : ∝). This observation is in analogy with that of Mislow [3], who studied this isomerization for a number of aliphatic sulfoxides and proposed the mechanism which includes formation of an intermediate with the structure of sulfur dichloride type.

$$R_2SO + 2HCl \rightleftharpoons R_2SCl_2 + H_2O$$

/I/   /II/

It is reasonable to explain the isomerization of alkaloid sulfoxides in the same terms.

P u m m e r e r   r e a c t i o n. The hydrogen atoms in the sulfinylmethyline group in the sulfoxides of three isomeric sulfur alkaloids: neothiobinupharidine, thiobinupharidine, and thionuphlutine are acidic enough for Pummerer rearrangement.

Opposite to many other examples studied earlier [4],
in case of alkaloid sulfoxides in question the reaction
does not show high stereospecificity and both possible
isomers are formed. They could be separated and
assignments of the configurations to both isomers
could be given by interpretation of NMR spectra since
in each pair of isomers obtained from neothiobinupharidine, thiobinupharidine, and thionuphlutine only one
has the acetoxy group very close to the furan protons
which causes different values for the observed
chemical shifts in isomeric acetoxyderivatives.

α'-Acetoxyderivatives could be hydrolized in
alkaline solution to corresponding hydroxy derivatives
with retention of configuration. β -isomers are
resistant in the same conditions of hydrolysis and
under more vigorous conditions reaction seems to be
more complex and hydroxy derivatives are not obtained.

The hydroxyl group could be exchanged to chlorine atom by means of thionylchloride with the retention of configuration [5].

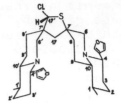

E l e c t r o p h i l i c   s u b s t i t u t i o n of methylenic protons in sulfides and sulfoxides. Alkaloidal β-sulfoxides could be chlorinated in α-position by means of sulfurylchloride in the presence of pyridine.
The α-sulfoxides in the same conditions suffer mostly decomposition.

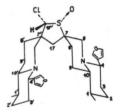

This phenomenon could be explained in terms of commonly accepted reaction mechanism[6] which includes either extraction of proton followed by electrophilic attack of Cl[+], or attack of Cl[+] on sulfur in the first stage followed by loss of proton and rearrangement resulting in α-chlorosulfoxide.

$$R-\overset{O}{\underset{}{S}}-CH_2R' + B \rightleftharpoons R-\overset{O}{\underset{}{S}}-\overset{\ominus}{C}HR' + BH^{\oplus} \xrightarrow{Cl_2}$$

$$R-\overset{O}{\underset{}{S}}-CHClR' + Cl^{\ominus}$$

$$R-\overset{O}{\underset{}{S}}-CH_2R' \xrightarrow{Cl_2} R-\overset{O}{\underset{Cl}{\overset{\oplus}{S}}}-CH_2R' \xrightarrow{B} R-\overset{O}{\underset{Cl}{S}}-\overset{\ominus}{C}HR' \rightarrow$$

$$R-\overset{O}{\underset{}{S}}-CHClR'$$

Since the product obtained had the configuration cis- in respect to the sulfinyl group and the chlorine atom, the reaction course could be rationalized in terms of mechanism proposed by S.Iriuchijima and G.Tsuchihashi for number of sulfoxides:

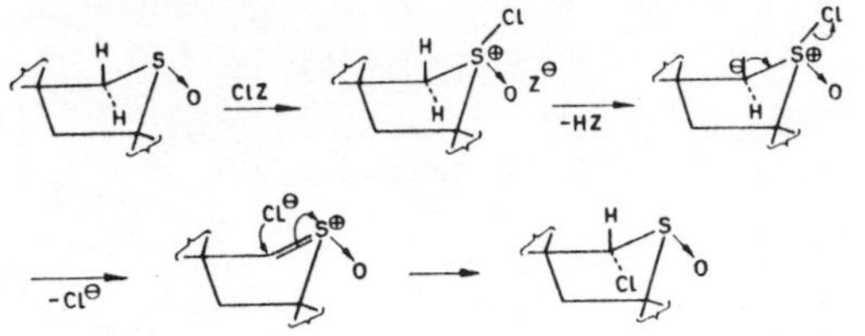

Crucial step is the trans attack of Cl⁻ in respect
to free electron pair on sulfur atom. It seems to be
reasonable to explain this stereochemistry of chlorina-
tion in terms of sterical hindrance caused by furan,
which enable the attack by chlorine only from the
opposite side to the furan ring. This is possible in
β -sulfoxides only. It has been shown that the sulfides
do not react with big variety of chlorinating agents
due possibly to the low acidity of hydrogen atoms in
the   position to sulfur atom.

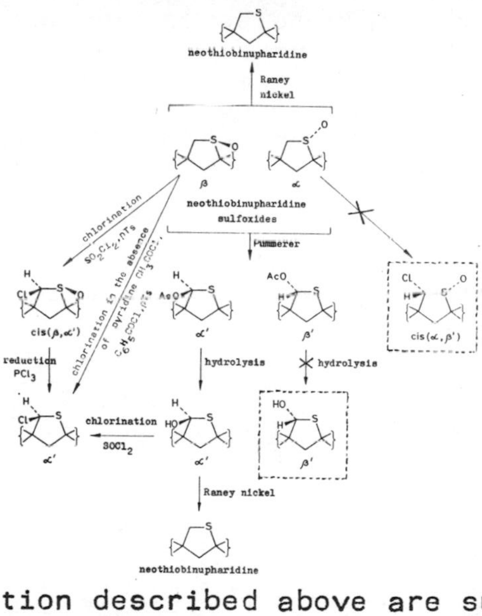

The reaction described above are summarized in the
scheme on the example of neothiobinupharidine.
     O x i d a t i o n   s t u d i e s. Furyl substituents
of the NTBN molecule are less susceptible for oxidation
in comparison with sulfur atom. Oxidation of NTBN by
h y d r o g e n   p e r o x i d e in acetic acid yields
the mixture of two isomeric NTBN S-oxides and NTBN-
sulfone. Furan rings of NTBN are affected only by
peracids in vigorous conditions or in milder ones after
prolonged reaction time.
     Oxidation of NTBN by means of o z o n e in acidic
medium in -15°C yields the mixture of products in which
both furan rings and sulfur atom are affected.
Spectroscopic analysis let us to assign to three
isolated products the following structures:

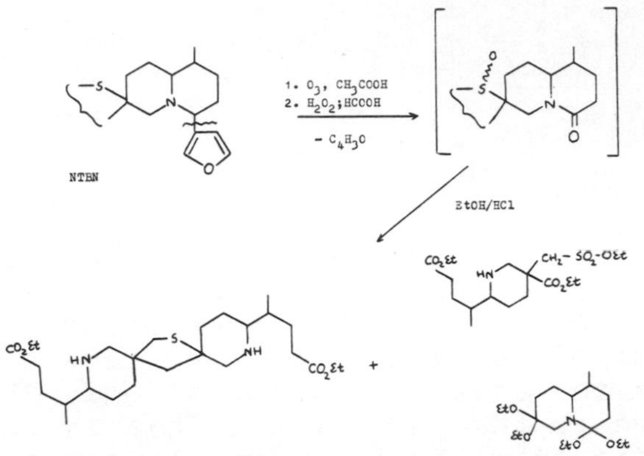

Ozonolysis causes usually in other furan containing natural compounds the loss of C_3H_3O unit from the molecule and a carboxylic group is formed. It can be assumed that in the NTBN molecule the positively charged nitrogen atom withdraws the electrons from the furan ring so the C4 - C11 bond bears some olefinic character. This bond could be attacked by ozone molecule in the subsequent reaction step and quinolizidone derivatives are formed. X-ray analysis of NTBN bromohydride shows that C4 - C11 bond is really shortened. In the next reaction step δ -lactam rings could be cleaved by acid in ethanol and corresponding esters are formed. The C-15 compounds are presumably formed via sulfenic acid intermediate.

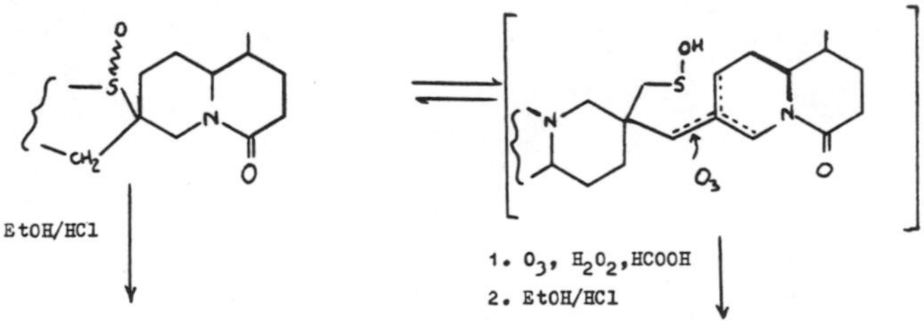

The results of the NTBN ozonolysis are comparable with that of ipomeamarone. The NTBN and ipomeamarone molecules are formally similar, having an electro-negative heteroatoms in the β -position to the furan ring.

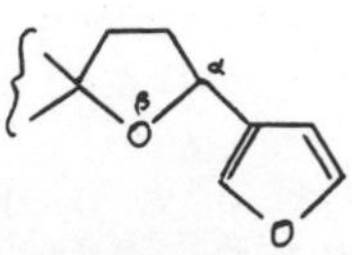

ipomeamarone 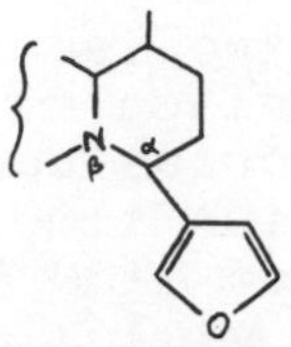　　　　　　　　　　neothiobinupharidine

Oxidation of NTBN by means of lead
t e t r a a c e t a t e in the ionic reaction conditions
does not take place. As the C-15 alkaloid - deoxy-
nupharidine- is also unreactive, the steric reasons
can be excluded. It seems probable that the positively
charged nitrogen atom situated in   -position to the
furan ring deactivates it with reagard to the
electrophilic attack of Pb(OAc)$_3^+$.

R e d u c t i o n   s t u d i e s . C-15 alkaloids
are reduced catalytically to the mixture of tetra- and
heksahydroderivatives. In the contrary, reductive
degradation of NTBN is unspecific and the reduction
yields are low. Depending on the reduction conditions
applied, the unchanged base is recovered or the mixtures
of hydrocarbons or lactonic products are obtained.
No reduction occurs on Adam's catalyst.

The sulfur atom of the alkaloid molecule is bound
strongly on the platinium surface, poisoning possibly
the catalyst. This presumption is strongly supported
by the fact of unexpected formation of NTBN β -S-oxide
during the reaction with equimolar quantity of PtO$_2$ in
acetic acid medium and under the increased hydrogen
pressure.

A l d e r - R i c k e r t   d e g r a d a t i o n,
NTBN yields two adducts with diaethyl acethylenecarbo-
xylate in the molar ratio 1:1 and 1:2 respectively.
So formed 3,6-dihydro-3,6-oxophtalic acid derivatives
yield diaethyl 3,4-furanedicarboxylate after an usuall
reaction sequence. Although the nitrogen-containing
reaction product polimerized in reaction conditions so
far applied.

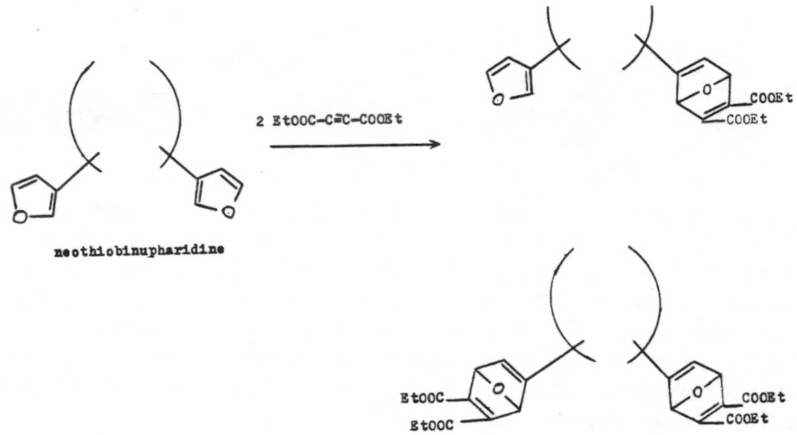

neothiobinupharidine

The sodium borohydride and borodeuteride reduction
of mono- and di- carbinolamines of sulfoxide type of
structure leads to similar results which were observed
in case of mono- and dihydroxyderivatives of alkaloid
sulfides [7]. The stereochemistry of the reduction in the
position C6 is not stereospecific, whereas reduction
in the position C6' introduces deuterium atom in the
axial configuration.

References

1. J.T.Wróbel in the Manske "The Alkaloids"
   Academic Press, N.Y. 1977, Vol.16, p.181
2. J.T.Wróbel, J.Ruszkowska, H.Bielawska
   Pol.J.Chem., 39 (1979)
   R.T.LaLonde, C.F.Wong, A.I-M.Tsai, J.T.Wróbel
   J.Ruszkowska, K.Kabzińska, T.I.Martin, D.B.MacLean,
   Can.J.Chem. 54, 3860 (1976)
3. K.Mislow, T.Simons, J.T.Melillo, A.L.Ternay,
   J.Amer.Chem.Soc., 86, 1452 (1964) ; 85, 2329 (1963)
4. M.Kise, S.Oae, Bull.Chem.Soc.Japan, 43, 1426 (1970)
5. D.J.Cram, J.Amer.Chem.Soc., 75, 332 (1953) ;
   N.Kitawaga, M.Nojima, N.Tokura, J.Chem.Soc., 22,
   2369 (1975)
6. G.Tsuchihashi, K.Ogura, S.Iriuchijima, S.Tomisawa,
   Synthesis, 89 (1971)
7. T.I.Martin, D.B MacLean, J.T.Wróbel, A.Iwanow,
   W.Starzec, Can.J.Chem., 52, 2705 (1974)

BIOLOGICALLY ACTIVE 1,2-DITHIOLANE DERIVATIVES

FROM MANGROVE PLANTS AND RELATED COMPOUNDS

Atsushi Kato and Yohei Hashimoto

Kobe Women's College of Pharmacy
Motoyama-Kitamachi, Higashinada-ku
Kobe, Hyogo 658, Japan

INTRODUCTION

The tropical plants of the family Rhizophoraceae may be generally devided into mangrove species and inland species. It has been recognized recently that either group of the plants contain organic sulfur compounds. Brugine (4)[1] and Gerrardine (5)[2,3] are 1,2-dithiolane compounds which were isolated from *Bruguiera sexangla* (mangrove species) and *Cassipourea gerrardii* (inland species), respectively, by J. W. Loder et al. We isolated two new 1,2-dithiolane compounds named Brugierol[4,5] and Isobrugierol from a mangrove species, *Bruguiera conjugata*, belonging to the Rhizophoraceae. There are few naturally occurring 1,2-dithiolane compounds to date. The main compounds are shown in Fig. 1. It is an interesting group of compounds as most of them have high biological activities.

Isolation and Structure of Brugierols

The extract of stem and bark at *Bruguiera conjugata* with chloroform was chromatographed on silica gel column to obtain rhombic crystals, mp. 84° of 6 and oily substance of 7.

An empirical formula $C_3H_6O_2S_2$ has been deduced for both compounds from elemental analysis and judging from the results of the UV, IR, NMR and MS spectra analyse, both compounds were apparently either of the cis- and trans-isomers of 4-hydroxy-1,2-dithiolane-1-oxide.

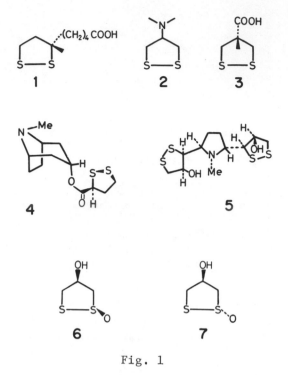

Fig. 1

The UV, IR and MS data were as follows; $6-(\lambda_{max}^{MeOH}$ 252 nm,

ν_{max} 1040–1060 cm^{-1}), $7-(\lambda_{max}^{MeOH}$ 248nm, ν_{max} 1030, 1090 cm^{-1}),

and mass spectra in which they gave parent peak at m/e 138 corresponding to the molecular ion with isotopic ion peaks at M+1 and
M+2 in accord with the formula. NMR data were summarized in Table
1. in which each isomer showed two sets of AB parts of ABX spectra
for a pair of methylenes, a multiplet for a methine and a doublet
exchangeable with D_2O for a hydroxyl, 6 at δ5.42 (J = 10 Hz) and
7 at δ5.43 (J = 5.3 Hz). The hydroxy group could be assumed to be
quasi-axial in a five membered ring, so the isomers should differ
in the conformation of sulfinyl oxygen. It was concluded that the
sulfinyl oxygen is assigned quasi-axial, i.e. *cis* to hydroxyl, in
Brugierol and quasi-equatorial, i.e. *trans* to hydroxyl, in Isobrugierol by the three following methods.

(1) The NMR measurements of acetylated brugierol and isobrugierol with the addition of a shift reagent Eu(dpm)₃. The
pseudo contact (P.S) value of each proton are in good accord with
the proposed structures. (Table 1. and Fig. 2)

(2) The dilution method using the IR spectra that was measured on CS_2-solutions of Brugierol or Isobrugierol, ranging in concentration from 1 to 0.003M. The IR spectrum showed the absorption peak at 3450 cm^{-1} that is assignable to the intramolecular bond between the quasi-axial hydroxyl and sulfinyl oxygen in Brugierol. This fact elucidated that Brugierol is a *cis* type.

Table 1. Chemicals Shifts (ppm) and Coupling Constants (Hz) with or without Eu (dpm)$_3$

COMPOUNDS			HA	HB	HX	HA$'$	HB$'$	JAB	JA$'$B$'$	JXA	JXB	JXA$'$	JXB$'$
6		δ	2.92 dd	4.05 dd	5.42 m	3.59 dd	4.09 dd	13.0	11.0	5.0	3.5	4.3	5.0
Ac–6	{	δ	3.79 dd	3.42 dd	5.97 m	3.99 dd	3.71 dd	14.0	12.0	5.0	3.0	4.0	5.0
		PS*	3.59 dd	1.82 dd	2.67 m	2.51 dd	1.52 dd	14.0	12.0	5.0	3.0	4.0	5.0
7		δ	3.54 m	3.54 m	5.33 m	3.85 dd	3.44 dd	–	11.0	5.0	5.0	5.0	6.0
Ac–7	{	δ	3.65 d	3.65 d	6.12 m	3.56 dd	4.07 dd	–	12.0	5.0	5.0	4.5	5.0
		PS*	2.18 dd	3.49 dd	2.44 m	1.68 dd	2.30 dd	14.0	12.0	5.0	5.0	4.0	4.8

* Down field proton shift induced with the addition of 0.33 equiv. Eu (dpm)$_3$.

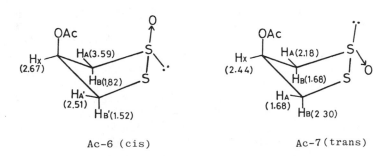

Ac–6 (cis)   Ac–7 (trans)

Fig. 2. PS. values in parentheses

(3) These two methods almost verified the conformation of these two compounds. However, there is no analytical data on the crystal structure of 1,2-dithiolane compounds having exocyclic sulfinyl groups as these compounds do. Thus, we conducted an X-ray structure analysis on the N-ethylcarbomate of brugierol and iso-brugierol and confirmed unambigucusly their conformation as well as obtaining various stereochemical informations on thier molecular structure.

The crystal data of N-ethylcarbamate of 6 were as follows: triclinic; space group P_1, a = 9.974(2), b = 10.098(2), c = 5.015(1) Å, z = 2, Dx = 1.437 g cm^{-3}, $\mu(\lambda$ = 0.7107 Å) = 5.37 cm^{-1}. Intensity data were collected on a Rigaku manual and automatic four-circle diffractometer with graphite-monochromated Mo Kα radiation for N-ethylcarbamate of 6. A total of 1625 independent reflections with 2θ less than 50° were measured in the θ-2θ scanning mode. The structure was solved by the block-diagonal least-squares method. Initially, all the approximate positions of atoms except the hydrogen atoms in the molecule were obtained by several cycles of diagonal least squares calculations and Fourier synthesis. Next, the positions of all the hydrogen atoms were found from the difference maps which were refined with anisotropic thermal parameters by block-diagonal least-squares method. Finally, isotropic thermal parameters for all the hydrogen atoms were refined and all the atomic coordinates and thermal factors were converged in a R factor of 4.2%.

From the X-ray analytical data, the various stereochemical informations due to exocyclic sulfinyl group was obtained, as a example, the torosion angle C_5SS/SSC_3 = 13.1Å is related to the bond length S(1) - S(2) = 2.086Å, one reason for which is that oxygen atoms of S = O radicals play a role similar to the lone pair electrone of the sulfur atom. Namely, we found that they had a repulsion action similar to that of lone pairs.

Survey of the above conformational analysis indicate that the orientations of sulfinyl oxygen and hydroxyl group (as acceptor and deceptor on hydrogen bond) of the compounds (cis- and trans-isomers) are significant with regard to stabilization of conformations, and are more dominant than another stabilization factors. Therefore, the four possible conformations (A - D) of five member ring compound having exocyclic sulfinyl and hydrogen group are stable in alphabetical order, A > B > C > D. (Fig. 4)

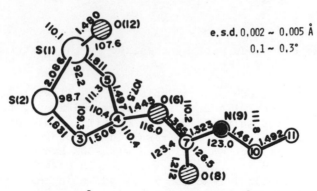

Fig. 3-1. Bond lengths (Å and valency angles (°) of cis-isomer (N-ethylcarbamate of 6).

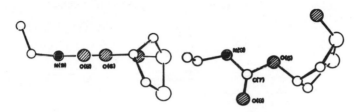

Horizontal and vertical projection to average plane O(6)–C(7)–O(8)–N(9).

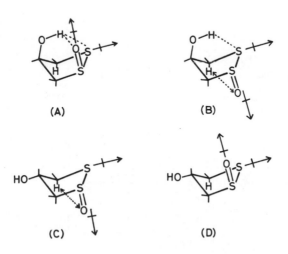

Fig. 4. Stability ---- A>B>C>D

Syntheses

The two compounds were synthesized by the next two novel methods.

The first method[6] of synthesis was to allow, 1,3-dimercapto-2-propanol to react with methyl (carboxy-sulfamoyl) triethyl ammonium hydroxy inner salt $(CH_3O_2CN^-SO_2N^+Et_3)$ in chloroform to afford 4-hydroxy-1,2-dithiolane. The resulting chloroform solution of 4-hydroxy-1,2-dithiolane was oxidized with 30% hydrogen peroxide to give Brugierol and Isobrugierol (in a ratio of 4 to 1). The speculation of the chemical reaction is as shown in the Fig. where N-carboxysulfamate of the dithiol is made as an intermediate by the reagent, which attacks the proton of the thiol from one side and leads to a dehydrogenated ring closure. This is the first time that this reagent was used for dehydrating sec- and tert-alcohols and was applied to the synthesis of 1,2-dithiolanes.

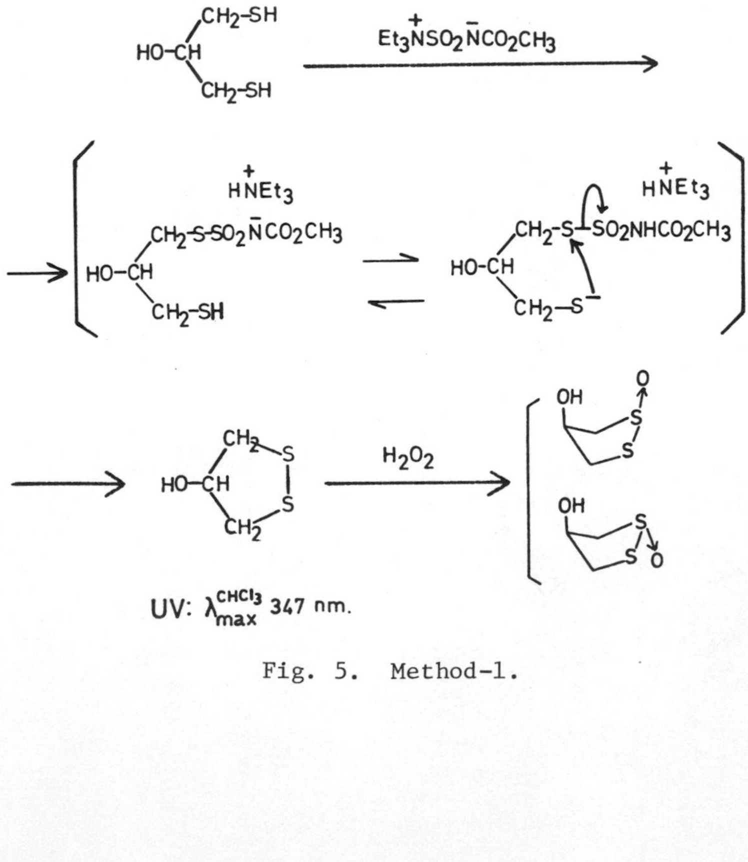

UV: $\lambda_{max}^{CHCl_3}$ 347 nm.

Fig. 5. Method-1.

The second method[7] of synthesis was to allow sodium tetrasulfide to react with glycerol-α,α'-dichlorohydrin in a water-
ethanol-chloroform solvent mixture, and introduce three sulfur
atoms at once to form v-trithiane ring. In order to synthesize
the brugierols from the v-trithiane, it was allowed to act with
excess m-chloroperbenzoic acid in chloroform. By doing so,
Brugierol and Isobrugierol were obtained in a ratio of 5 to 1.
A possible mechanism of this oxidative desulfurization is shown
in as Fig. 6, namely, the central sulfur atom in the v-trithiane
ring is oxidized to form sulfone, in which an intramolecular
nucleophillic rearrangement takes place to give the 1,2-dithiolane
ring and SO₂. The intermediate, 5-hydroxy-1,2,3-trithiane in this
reaction is a new v-trithiane compound. And, as we will mention
later, it is a compound of importance as the base of insecticidal
active derivatives. As there was no report on the crystal structure of v-trithiane, the stereochemical structure was investigated
by X-ray analysis.[7]

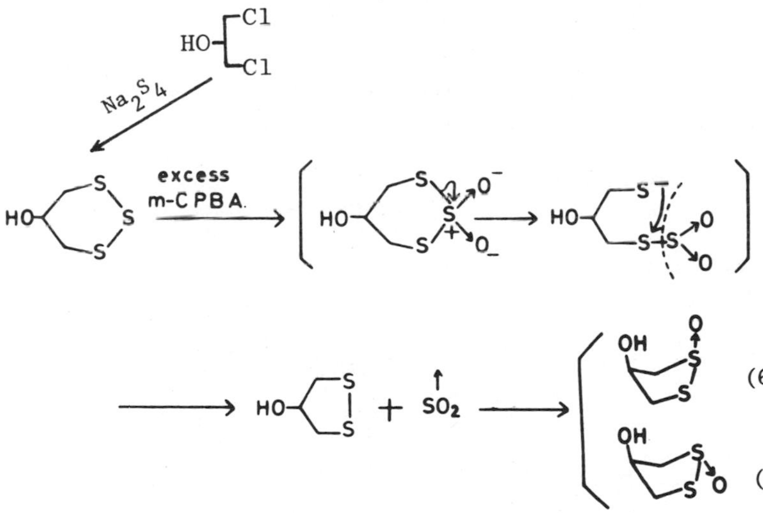

Fig. 6. Method-2.

As the crystals of (1) were unsuitable for X-ray work, the structure analysis was carried out on the crystals of 1,2,3-trithian-5-yl N-methylcarbamate(2) to elucidate the stereochemistry of the v-trithiane ring.

The single crystals were obtained from an acetone solution as colorless transparent plates. Crystal data: orthorhombic; space group $P2_1 2_1 2_1$; $a = 9.089(7)$, $b = 21.187(17)$, $c = 4.851(4)$ Å; $Z = 4$; $Dx = 1.502$ g cm^{-3}. A specimen, approximately 0.27 x 0.29 x 0.35 mm^3 was used for X-ray data collection on an automated four-circle diffractometer with graphite-monochromated Mo Kα radiation ($\lambda = 0.7107$ Å). A total of 1287 independent reflections with 2θ less than 50° were measured in the ω-2θ scanning mode, and of these 1116 reflections with $|Fo| > 3\sigma(F)$ were used for the structure determination.

The structure was solved by an application of the symbolic addition method and refined by the block-diagonal least-squares method. Anisotropic thermal vibrations were assumed for the non-hydrogen atoms. All the hydrogen atoms were clearly found from a difference Fourier map and their positional and isotropic thermal parameters were refined. The final conventional R index was 0.037.

The various stereochemical informations were supplied by the investigation of the X-ray analysis. For example, the molecule has an approximate mirror plane containing S(2), C(5), and non-hydrogen atoms of planar side-chain. The v-trithiane ring, having almost strict m(C$_s$) symmetry, assumes a chair form. The side chain at C(5) is equatorial. The interplanar angle of 64° between the planes S(1)-S(2)-S(3) and S(1)-C(6) ... C(4)-S(3) is slightly larger than that of 61° between the latter plane and the plane C(4)-C(5)-C(6).

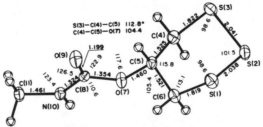

A perspective view of the molecular structure of 1,2,3-trithiane-5-yl N-methylcarbamate.

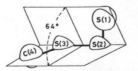

Torosion angle between three neighboring of four atoms
[S(1), S(2), S(3), C(4) or C(6), S(1), S(2), S(3)].

Fig. 7

Biological Activity

Antibacterial activities of Brugierol, as measured by MIC, against the 5 bacteria, hyochi baccilus, dysentery bacillus, typhoid bacillus, cholera bacillus and pneumobacillus were all 100 µg/ml. We attempted to obtain a different biological activity by modifying the chemical structure of this compound. Namely, we synthesized the 14 different derivatives such as carbamates, phosphates and N,N-dialkylates of dithiolane or trithiane, and found several strong biologically active compounds in these.

Table 2. The Insecticidal Screening Test
(Mortality % and Effect index ++, —)

Compounds			Pests (method, ppm.)				
No.	Structure		H.M. (Dipping, 50ppm)	S.B.P Rice Seedling, 200ppm	R.S.B Rice Seedling, 200ppm	T.S.M Spray, 500ppm	C.R.M Spray, 500ppm
37	R = cis	RHNCO-	0	0	—	6.5	41.7
37'	CH$_3$ trans		0	0	—	—	—
12	cis		0	11.1	—	-4.2	-9.5
12'	C$_2$H$_5$ trans		10	0	—	2.5	9.2
38	CH$_3$	RHNCO-	90	57.1	—	29.0	1.6
39	C$_2$H$_5$		80	23.1	—	2.5	49.6
36	CH$_3$	RHNCO-	100	100	—	1.1	27.1
40	C$_2$H$_5$		100	20	—	37.8	75.2
41	n-C$_3$H$_7$		100	30	—	32.4	-28.4
42	C$_6$H$_5$		100	20	—	29.7	7.3
47-H.Oxal	CH$_3$		100	20	++	4.7	2.9
-HCl			100	35	++	22.4	7.4
49	C$_2$H$_5$	RO$_2$PO-	90 (10ppm)	0	—	35.8	10.8
33	H	RO-	0 (10ppm)	8.3	—	12.9	17.5
34	CH$_3$CO		35 (10ppm)	0	—	1.0	-10.8
Comparatives							
Fenitrothion			100		++		
Bassa				100			
Carbaryl				100			
Nereistoxin				100	++		
Cartap·HCl					++		
Kelthane						100	100

H.M. (House mosquito), S.B.P. (Small Brown Planthopper), R.S.B. (Rice Stem Borer), T.S.M. (Two-Spotted Spider Mite), C.R.M. (Citrus Red Mite)

In the insecticidal test of these compounds, nine important pests including the house mosquito, and rice stem borer were used as pests. Table 2 shows the results of the primary screening test against 5 pests which can be commented on as follows. (1) compounds 36, 40, 41, 42 and 47 have strong insecticidal activity against the house mosquito. (2) against the small brown planthopper only 36 had an effect similar to that of the comparatives. (3) only compound 47 was effective against the rice stem borer, and (4) compounds 37, 39 and 40 had ome insecticidal activity against mites. On the compounds that had an effect in the first screening, a second screening involving different doses, exchanged pests, and various methods was conducted to give results of Table 3, 4.

Table 3. The Insecticidal Screening Test (Mortality %)

Compd. No.	pests (method, ppm, mg)			
	G.R.L Stem. dipping 500 ppm	T.C. Feed dipping 500 ppm	D.M. Leaf dipping 100 ppm	G.C. Contact 0.5 mg/mg
36	90	0	0	80
40	50	0	0	40
41	0	0	0	0
42	0	0	0	0
47-H.Oxal.	55	82.7	100	100
47-HCl	75	56.7	100	100
Comparatives				
Bassa	100			
Carbaryl	100			
Nereistoxin	100	55.6	100	100
Fenitrothion				100

G.R.L. (Green Rice Leafhopper), T.C. (Tobacco Cutworm), D.M. (Diamondback moth), G.C. (German Cockroach)

Table 4. The Insecticidal Activities to House Mosquito
(Mortality %*, LC-50)

Compd. No. \ ppm	3.125	6.25	12.5	25	50	LC-50
36	20	95	100	100	100	Ca. 4.0
40	15	70	100	100	100	4.5
41	10	90	100	100	100	4.5
42	30	40	80	85	100	6.0
Fenitrothion	100	100	—	—	—	0.015

Compd. No. \ ppm	0.063	0.125	0.25	0.5	1	LC-50
47-H.Oxal.	20	40	90	100	100	0.13
47-HCl	30	55	95	100	100	0.11
Nereistoxin	25	40	95	100	100	0.13

* control: 0%

47-H.Oxal
-HCl $\quad CH_3 \quad$

Table 5-1. The Insecticidal Actives to Rice Stem Borer
(Topical) Mortality %

Compd. \ μg/g	0.59	0.89	1.33	2	LD-50 (μg/g)
47-H.Oxal	35	55	90	100	0.78
47-HCl	40	55	90	100	0.74
Nereistoxin	40	50	95	95	0.78

Table 5-2. The Insecticidal Effect Against Rice Stem Borer
inside Rice Plants (pot test) Survival Rate %

Compd. \ Method Test. No.	Stem, Leaf spray (500 ppm)		Addition to Pot Water (2mg/stump.)	
	1	2	1	2
47-H.Oxal	0	0	13.3	16.7
47-HCl	—	0	—	8.1
Nereistoxin	0	1.1	0	0.6
Cartap.-HCl	—	0	—	0
Control	50.9	60.8	48.2	60.8

Summing up the activity from the results of the above 5 tables, the strong order is Fenitrothion $\gg$ Nereistoxin = 47 $\gg$ 36 $\geq$ 40 = 41 > 42 $\gg$ 38 = 39 > 12.

For the relationship between the insecticidal activity and the chemical structure on the carbamates, four activity factors are mentioned; (1) is the orbital resonance effect, (2) the electronic effect of substituents, (3) hydrogen bond effect, and (4) N-alkyl substituent groups and insecticidal activity.

On the first factor, if it is supposed that the carbamates act on acetyl choline esterase of the insect body as a cyclic form, dithilane ring as well as trithiane ring having a puckered configuration would receive stereostructural restriction when forming a complex with the anionic site of acetyl choline esterase. However, if we take into consideration the orbital resonance such illogicality should be dissolved. Dithiolane and trithiane rings have the plane electron ring by throughing conjugates (on atomic bond) which resembles the phenyl group in quality, because the sulfur atoms on the ring would be bound together with a π-bond due to 3d-3p orbital resonance and a π-bond with sulfur atom and neighboring carbon is also due to overlapping of the 3d-2p orbital. (Fig. 8)

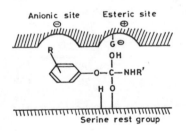

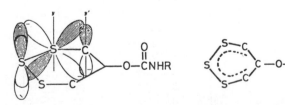

3d-3p or 3d-2p orbital.   Throughing conjugation

Fig. 8

Concerning the effect (2), as an orbital resonance occurs on the ring of dithiolane or trithiane, the electromeric effect on insecticide activity may be great, depending on the type of sub-stituents introduced to the ring. For example when a substituent with an electron-attracting effect is introduced, the rate of hydrolysis becomes greater and the insecticidal activity is gener-ally lowered. In fact, one reason that monosulfoxides 37, 37', 12 and 12' did not show any activity whereas non-sulfoxide compounds, 38 and 39 did in this test, would be due to the electron substi-tuent attracting effect.

(3) On the effect of the hydrogen bond, it has already been reported that a special hydrogen bond between the proton deceptor of acetylcholine E and various protons acceptors introduced into benzene ring of phenyl-N-alkylcarbamates is related to the acethyl-choline E inhibition (by Nishibe). In the carbamate group having a dithiolane or trithiane ring instead of a benzene ring, un-paired electrons of the sulfur atoms become proton acceptors and make hydrogen bonds possible, even if no special substituent as proton acceptor is available on the ring. (Fig. 9) The change of the N-alkyl group in (4) and the insecticidal activity is the same as those existing carbamate insecticides which are well known.

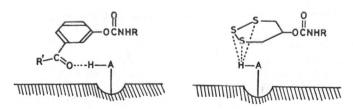

Fig. 9. Hydrogen bond effect.

As the compound 47(5-N,N-dimethylamino-1,2,3-trithiane) was
found to have the same strong insecticidal effect as Nereistoxin[8]
especially against rice stem borers, the ganglionic blocking
activity by the compound 47 in adult male American Cockroach was
investigated to reveal the insecticidal action of the compound.
To investigate the mechanism of this activity, the cercal-abdominal
nerve of the male American Cockroach was used to measure the potential of direct electrical stimulations. From the measurements of
the excitatory postsynaptic potential it is clear that 47 is bound
to the acetylcholine receptor of the postsynaptic membrane. However, if a strong direct electrical stimulus is given at this stage,
a large amount of acetylcholine is emitted. The binding of 47 to
the postsynaptic membrane is released, the acetylcholine is again
bound to the synaptic membrane, and depolarization and the excitatory postsynaptic potential is considered to occur a second time.
Thus, it is considered that 47 as well as Nereistoxin has a competitive blocking action against the postsynaptic membrane.

REFERENCES

1. J. W. Loder and G. B. Rursell, TROPINE 1,2-DITHIOLANE-3-
    CARBOXYLATE, A NEW ALKALOID FROM *BRUGUIERA SEXANGULA*,
    Tetrahedron Letters, 51: 6327 (1966).
2. W. G. Wright and F. L. Warren, Rhizophoraceae Alkaloids,
    Part I, Four Sulphur-containing Bases from Cassipourea spp.,
    J. chem, soc (C), 283 (1967).
3. ibid., 284 (1967).
    Rhizophoraceae Alkaloids, Part II, Gerrardine.
4. A. Kato and M. Numata, BRUGIEROL AND ISOBRUGIEROL, TRANS- AND
    CIS-4-HYDROXY-1,2-DITHIOLANE-OXIDE, *Tetrahedron Letters*, 3:
    203 (1972).
5. A. Kato, A NEW NATURALLY OCCURRING 1,2-DITHIOLANE FROM
    BRUGUIRA CYLINDRICA, *phytochemistry*, 15: 220 (1976).
6. A. Kato and T. Okutani, SYNTHESIS OF BRUGIEROL AND ISO-
    BRUGIEROL, TRANS- AND CIS-4-HYDROXY-1,2-DITHIOLANE-1-
    OXIDE, *Tetrahedron Letters*, 29: 2959 (1972).
7. A. Kato, Y. Hashimoto, Isoji Otsuka and Kazumi Nakatsu, REAC-
    TION AND STEREOCHEMISTRY OF 1,2,3-TRITHIANE RING, *Chemistry
    Letters*, 1219 (1978).
8. M. Sakai, Studies on the Insecticidal Action of Nereistoxin,
    4-N,N-dimethylamino-1,2-dithiolane, *BOTYU-KAGAKU*, 32: 21
    (1967).

BIOCHEMICAL AND CLINICAL EFFECTS OF PENICILLAMINE AND OTHER THIOLS USED IN THE TREATMENT OF RHEUMATOID ARTHRITIS

Egil Jellum and Eimar Munthe

Institute of Clinical Biochemistry
Rikshospitalet, Oslo and Oslo Sanitetsforening
University Rheumatism Hospital
Oslo, Norway

INTRODUCTION

Several years ago it was shown that rheumatoid factor molecules were split in vitro by the thiol penicillamine[1]. This was the rationale behind the original attempts to use penicillamine in the treatment of rheumatoid arthritis[2]. Although later studies have clearly shown that this simple mechanism does not operate in vivo, it was found that penicillamine was an antirheumatic drug[3] which offers an alternative to gold compounds which, despite their high toxicity, have been used in the treatment of rheumatoid arthritis for over 50 years. In clinical laboratories and rheumatological centres there has consequently been an increasing interest in the biological effects of thiols in general and of penicillamine in particular. Recently another thiol, thiopyridoxin[4,5] has been shown to posses antirheumatic properties. From the numerous scientific articles published during the last years, it appears that the thiol group may play important roles in the unlocking of the pathogenesis of rheumatoid arthritis.

In the present paper a short overview of the present situation will be given together with some new experimental data obtained in the authors laboratories.

CLINICAL EFFECTS OF D-PENICILLAMINE

The drug is normally administered in a low daily peroral dose (250-750 mg) and about 2/3 of the patients with rheumatoid arthritis respond to this treatment after 2-3 months. It is a tendency to use as low maintenance doses as possible. The destructions and deformities of the joints are arrested, and the drug protects against cartilage and bone erosions. Thus, penicillamine is a true antirheumatic drug and not only an anti-inflammatory agent (giving mostly symptomatic relief). Like gold-compounds, penicillamine unfortunately also has several side-effects, of which the most serious ones are damage of the kidneys and of the bone marrow.

MECHANISMS OF ACTION - BIOLOGICAL EFFECTS OF D-PENICILLAMINE

No good animal or other experimental models are available to test the anti-rheumatic effect of drugs. Therefore, although a number of biological effects of penicillamine are known it has not been possible to single out one particular mechanism responsible for its beneficiary effect in rheumatoid arthritis. One is consequently up against similar type of problems now as one has been for the last 25 years in understanding the mechanisms of protective action of thiols against ionizing radiation.

In the following the biological effects of penicillamine-like thiols are summarized.

Chelating properties

The thiols can act as chelators and bind heavy metal ions through mercaptide formation. The chelation may also involve the amino group. Copper-chelation by penicillamine, changes in copper metabolism and altered concentrations of coppercontaining proteins like ceruloplasmin and altered activity of the copper-dependent superoxide dismutase (an intracellular enzyme for removal of toxic superoxide radicals) may be of importance. Chelation of copper may also increase the prostaglandin E/F ratio, which is assumed to have an anti-inflammatory effect.

Thiol-disulphide interchanges

Exchanges between penicillamine and extra/intra-cellular physiological thiols and disulphides including proteins are known to take place. Penicillamine, with its SH-group surrounded by the bulky methyl-groups, is somewhat unique in this context, having only little tendency to form its own disulphide and a high tendency to form mixed disulphides[6]. Inside the cells, the SH/SS exchanges may have considerable metabolic consequences, affecting many enzymes, mitochondrial functions, nuclear functions, etc.[6]. Another effect (see fig. 1) is that penicillamine may liberate the protein-bound part of glutathione (normally about 20%), thereby increasing the intracellular glutathione concentration. This in turn might stimulate the glutathione peroxidase thus affording increased protection of the cells and membranes against toxic endoperoxides.

As mentioned above penicillamine may split the disulphide bonds of IgM in vitro, but in vivo the concentration of the thiol is far too low. Disappearance of rheumatoid factor after successful treatment is probably due to effects at the cellular level.

Interaction with aldehydes

Aldehyde groups are present in all tissues, for example in cellular membranes, in collagen fibers, as intermediates in glycolysis and in vitamin B_6. The thiols form hemithioacetal with the aldehyde groups, followed by ring closure to yield thiazolidines if the drug also contains an aminogroup. This type of reaction has been much discussed in connection with rheumatoid arthritis, as penicillamine may interfere with collagen-cross-linking.

Effects on immunity and immuno-competent cells.

Penicillamin reduces the activity of hemolytic complement, and has a wide range of in vitro effects on T and B lymphocytes which may be inhibitory as well as stimulatory. When lymph node cells are stimulated with con-A, penicillamine is enhancing the modulating effect of macrophages on the reaction, indicating influence on macrophage function[7]. The drug also influences the secondary lesion of adjuvant arthritis. In general, the many effects of penicillamine on the immune system probably are some of the most important

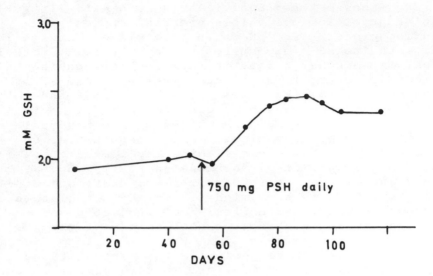

Fig. 1.   Effect of penicillamine treatment on the
           glutathione content of erythrocytes in a
           patient with rheumatoid arthritis. An enzy-
           matic method using glutathione reductase and
           NADPH was used for the specific determination
           of GSH in washed red blood cells. (Jellum and
           Munthe, unpublished results).

reactions to consider in understanding the mechanism of
action of this drug (see ref. 8).

OTHER ANTIRHEUMATIC THIOLS

     The success of D-penicillamine as an antirheumatic
drug have prompted the search for other thiols with
improved risk/benefit relationship. Recently 5-thio-
pyridoxine[4,5] was shown to have therapeutical value in
rheumatoid arthritis. This thiol caused no increased
urinary copper excretion, had no antipyridoxin effect,
had no effect on skin collagen and did not enter into
SH/SS exchange with the amino acid cystine[5]. Thus, the
redox potential and reactivity of thiopyridoxine appears
to be somewhat different from that of D-penicillamine.

     In our laboratory we[9] have for some time been

interested in the anion part of the gold salts used in treatment of rheumatoid arthritis. It turns out that all gold compounds in common use are complexes of gold with different thiols[8]: aurothiomalate (Myocrisin), aurothiosulphate (Sanochrysine, Crisalbine), gold thioglucose (Solganol) and Auranofin (gold alkyl-phosphine bound to tetracetyl-thio-glucopyranoside). Using double isotope-labelled aurothiomalate (Au^{198}-C^{14}-thiomalate) and experimental mice we have recently shown[10] that soon after intramuscular administration of the drug, the chemical forms present _in vivo_ are protein bound gold and free thiomalate which is partly taken up by the tissues and membranes and partly excreted. Also after administration of aurothiomalate to man thiomalate is released in the free form as determined by mass spectrometry[10]. Similar exchanges and transfer of gold to proteins with a concomitant liberation of the thiol carriers are likely to take place also upon admini-stration of Solganol and the other gold compounds. Thus, an important question to consider is whether the supposedly inert thiol moieties of these drugs may have biological and clinical effects. From _in vitro_ studies it has now been shown that thiomalate and thioglucose both have effects in biological model systems[11]. Moreover, preliminary clinical studies now under way in our hospital, seem to indicate that thiomalate (sodium salt) is well tolerated upon peroral admini-stration to patients, and that this thiol may have a beneficiary effect against rheumatoid arthritis. A controlled clinical trial remains to be done, however.

CONCLUSION

The last years have witnessed a growing interest in the treatment of rheumatoid arthritis with D-penicill-amine. The mechanism of action of this drug appears to be linked mainly to its thiol group, and seems to involve many different biological, immunological and cellular reactions. Another thiol (thiopyridoxin) also has antirheumatic effects, and it may prove that the thiol part of gold drugs, e.g. thiomalate, also possess antirheumatic properties. It is likely that thiol compounds in the future will play important roles in the treatment and understanding of rheumatoid arthritis.

REFERENCES

1. R. Heimer and M. Federico, Depolymerization of the 19S antibodies and 22S rheumatoid factor, _Clin._

Chim. Acta 25:41 (1958).

2. I. A. Jaffe, The treatment of rheumatoid arthritis
    and necrotizing vasculitis with penicillamine,
    Arthr. & Rheum. 13:436 (1970).

3. Multi-Centre Trial Group, Controlled trial of
    D(-)penicillamine in severe rheumatoid arthritis,
    Lancet 1:275 (1973).

4. J. P. Camus, Scand. J. Rheumatology (in press).

5. I. A. Jaffe, Scand. J. Rheumatology (in press).

6. E. Jellum and S. Skrede, Biological aspects of
    thiol-disulphide reactions during treatment with
    penicillamine, in "Penicillamine research in
    rheumatic disease", E. Munthe, ed., Fabritius/MSD,
    Oslo, (1977) p. 68.

7. L. Binderup, E. Bramm and E. Arrigoni-Martelli,
    D-penicillamine and macrophages. Modulation of
    lymphocyte transformation by concanavalin A (in
    press).

8. E. Munthe and E. Jellum, Mode of action of gold
    compounds and D-penicillamine, in "Proceedings
    from the European Colloquium on the mode of action
    of non-steroidal anti-inflammatory drugs", Cannes,
    France (1979) (in press).

9. E. Jellum and E. Munthe, Is the mechanism of action
    during treatment of rheumatoid arthritis with
    penicillamine and gold thiomalate the same? in
    "Perspectives in Inflammation. Future trends and
    developments", D. A. Willoughby, J. P. Giroud and
    G. P. Velo, eds., MTP-Press Ltd, Lanchester,
    England (1977) p. 575.

10. E. Jellum, E. Munthe, G. Guldal and J. Aaseth,
    Fate of the gold and the thiomalate part after
    intramuscular administration of aurothiomalate
    (Myocrisin) to mice, Ann. Rheum. Dis. (in press).

11. E. Arrigoni-Martelli, E. Bramm and L. Binderup,
    D-penicillamine-like activity of thiols, in EBRA
    symposium on anti-inflammatory and anti-rheumatic
    drug, Paris, (in press).

CYANOEPITHIOALKANES: SOME CHEMICAL AND TOXICOLOGICAL STUDIES

Jurg Luthy* and Michael Benn**

*Institut fur Toxikologie der Eidgenössischen
Technischen Hochschule und der Universität
Zürich, CH-8603 Schwerzenbach, Switzerland, and
**Department of Chemistry, The University, Calgary,
Alberta T2N 1N4, Canada

Abstract. The syntheses of (±)-1-cyano-2,3-epithiopro-
pane (*3*), (±)-1-cyano-3,4-epithiobutane (*2*), and (±)-1-
cyano-4,5-epithiopentane (*4*) are described, as are the
formation of *3* from allylglucosinolate, and its aglucone,
by a *Crambe abyssinica* preparation. Short term tests of *2*-
1 for mutagenicity and carcinogenicity gave weakly positive results.

INTRODUCTION

At Jabłonna, in the last symposium of this series one of us
reviewed the catabolism of glucosinolates.[1] To recapitulate, all
known glucosinolates can apparently undergo "normal" enzyme-induced
catabolism to yield, besides sulfate ion and D-glucose, isothiocyanates or nitriles (in which latter case the thioglucosidic sulfur
atom appears among the products as the element). However, a few
sometimes undergo other, "abnormal", modes of catabolism in which
the aglucone unit is transformed into thiocyanates, or cyanoepithioalkanes. Figure 1 portrays these alternatives for the specific
case of the allylglucosinolate ion.

In contrast with the reasonably well established pathways involved in "normal" metabolism, the reactions leading to the "abnormal" catabolites are very largely matters of conjecture, and pose
some interesting questions.

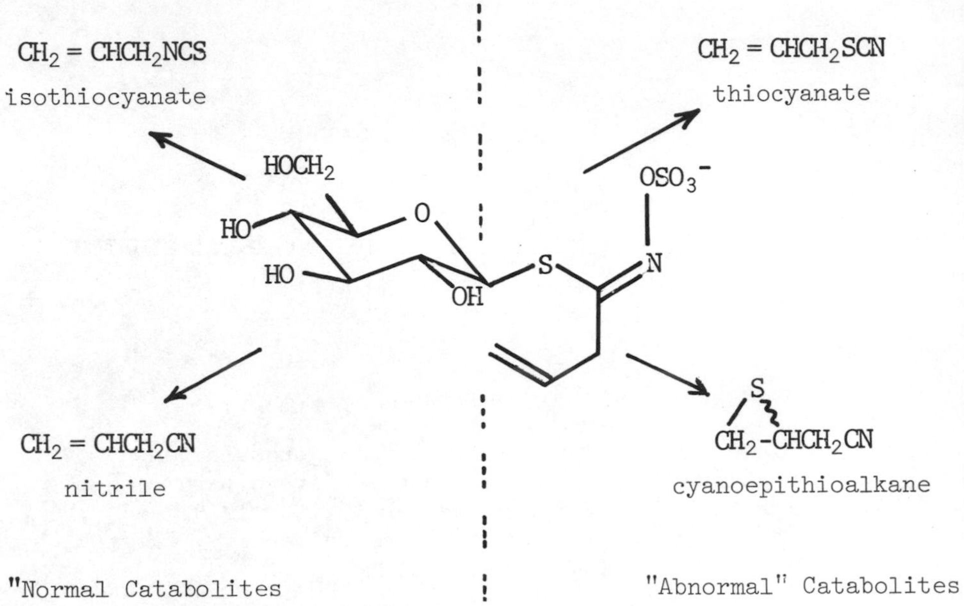

$CH_2 = CHCH_2NCS$

isothiocyanate

$CH_2 = CHCH_2SCN$

thiocyanate

$CH_2 = CHCH_2CN$

nitrile

CH_2-CHCH_2CN

cyanoepithioalkane

"Normal Catabolites

"Abnormal" Catabolites

Fig.1.   Alternative enzyme-induced transformations of the allyl-
         glucosinolate ion. D-Glucose and sulfate ion are also
         formed.

The formation of the 1-cyanoepithioalkanes is a particularly
intriguing process, and first drew our attention to these compounds.
However, the cyanoepithioalkanes were also of interest to us for
another reason: as potential environmental toxins. Thus, while
the toxicity of the "normal" catabolites, and thiocyanates, is
well recognised (especially the ability of the isothiocyanates to
yield goitrogens) little appeared to be known[2] about the toxicological properties of the cyanoepithioalkanes. To us it appeared
possible that, as a consequence of the reactivity of the epithio-
system, they might function as biological alkylating agents, and
should be suspect accordingly as mutagens, and carcinogens. These
sinister possibilities seemed to be urgent, and impelled the study
which we now present.

MATERIALS: THE SYNTHESIS, AND PROPERTIES OF SOME CYANOEPITHIOAL-
KANES.

The epithiocyanoalkanes reported to date are listed in Table
1. All were detected as autolysis products i.e. they were formed
by aqueous suspensions of disrupted plant-tissues from endogenous

Table 1. 1-Cyanoepithioalkanes identified
as glucosinolate catabolites

2S,3R and 2S,3S-1-Cyano-2-hydroxy-3,4-epithiobutane (S-1)[3]

2R,3R and 2R,3S-1-Cyano-2-hydroxy-3,4-epithiobutane (R-1)[4]

3R and 3S-1-Cyano-3,4-epithiobutane (2)[5]

1-Cyano-2,3-epithiopropane (3)[6,7,8]

1-Cyano-4,5-epithiopentane (4)[6,7,8]

$$CH_2 - CHCH(CH_2)_n CN$$
$$\diagdown S \diagup \quad X$$

S-1	X=OH(S),	n=1
R-1	X=OH(R),	n=1
2	X=H,	n=1
3	X=H,	n=0
4	X=H,	n=2

---

glucosinolates. Of them, S-1,[3] R-1,[4] and 2[5] were isolated, and
their structures established from spectroscopic studies, supplemented in the first case by some chemistry. As the discoverers
pointed out, the lack of stereospecificity in the formation of the
epithio functionality is striking. The formation of 3, and 4, was
first inferred by Cole,[6,7] during a study of the volatile autolysis
products of a number of fresh crucifers. Her identification of
these compounds was largely, and very reasonably, based upon similarities in their mass spectra (MS) with that reported[5] for 2. Subsequently Daxenbichler et al. also reported[8] the formation of 3
and 4 among the autolysis products of fresh cabbage leaves, again
on the basis of MS. Presumably, like 2, these preparations of 3,
and 4 are racemic.

None of these cyanoepithioalkanes had been synthesised. Since
we needed substantial amounts of material for our toxicological
investigations for which, moreover, we wanted to have them isoto-
pically labelled, we decided to work out a synthetic route to the
three simplest, 2-4.

Their synthesis was in fact readily achieved, in a straight-
forward way, as shown in Figure 2.

The only necessary cautionary comments concern the lability
of the epithio compounds: they appear to be very sensitive to excess aqueous base, so it is necessary to use care in the final step
of the synthesis, adding the carbonate solution slowly, in small

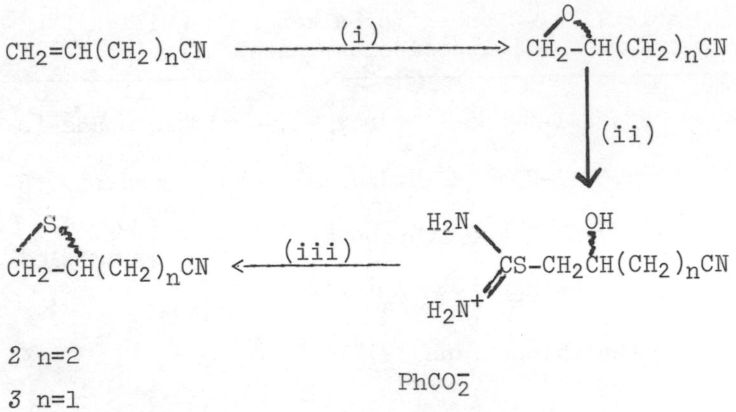

$$CH_2=CH(CH_2)_nCN \xrightarrow{\text{(i)}} \overset{O}{CH_2-CH(CH_2)_nCN}$$

(ii)

$$\overset{S}{CH_2-CH(CH_2)_nCN} \xleftarrow{\text{(iii)}} \underset{H_2N^+}{\overset{H_2N}{>}}CS-CH_2\overset{OH}{CH(CH_2)_nCN}$$

2 n=2

3 n=1 $PhCO_2^-$

4 n=3

(i) m-Chloroperoxybenzoic acid;[9] (ii) thiourea, and benzoic acid;[10] and (iii) aq. sodium carbonate.[10]

Fig. 2. The synthesis of three simple cyanoepithioalkanes.

portions, and extracting the reaction mixture from time to time during the addition in order to remove the products. Thus prepared, 2-4 could be purified by rapid distillation under reduced pressure. Slow distillation resulted in polymerisation and this usually also occurred rapidly if samples were allowed to stand at room temperature, in air; however, they could be kept for several months without decomposition if stored at -20°.

The distilled products were homogenous by gas liquid chromatographic analysis (GLC). They were completely characterised by elemental analysis, as well as infra-red (IR), MS and [1]H- and [13]C- nuclear magnetic resonance (NMR) spectroscopy: all the results being in accord with the structures 2-4. The ^{13}C-NMR data are collected in Table 2. The ^{1}H-NMR data for 2, and the MS of 2-4 agreed with the values reported for the natural products.[5,6]

BIOSYNTHESIS

With authentic 3 at hand we were able to confirm Cole's claim [6,7] that this was a natural catabolite of allylglucosinolate. Addition of potassium allylglucosinolate to a *Crambe abyssinica* Hochst ex R.E. Fries (var. Prophet) seed flour preparation gave excellent yields of 3 which was isolated by preparative GLC, and shown to have the same IR and ^{1}H-NMR spectra as the synthetic product.

Table 2. ^{13}C-NMR data* for compounds 2-4

Compounds	C-1	C-2	C-3	C-4	C-5
2	118.9	17.0 (137)	32.2 (135)	33.6 (172)	25.5 (173)	– –
3	116.6	24.6 (139)	28.5 (175)	24.6 (172)	– –	– –
4	119.6	16.8 (135)	25.1 (134)	35.2 (129)	34.4 (169)	25.4 (170)

* For solutions in CDCl$_3$, δ values p.p.m. from internal TMS (J_{CH} in Hz).

Small yields of 3 were also obtained when the aglucone 5 was used as a substrate. In these experiments the potassium silver salt 6[11] was added to the *Crambe* seed flour suspension, followed by potassium iodide, to generate 5 in situ.[11] The major product under these conditions was allylisothiocyanate, and this was formed essentially exclusively in the absence of the *Crambe* flour, in ac-cord with expectation,[11] given the pH, 6-7, of the reaction mixture. In another control experiment it was also shown that the yields of 3 could not be attributed to endogenous allyglucosinolate in the *Crambe* seed flour.

These results suggest that 3 (and more generally the other cy-anoepithioalkanes as well) is derived from the sinolate aglucone, like the "normal" isothiocyanate and nitrile catabolites (Fig. 3). Tookey,[12] and Cole[13] have shown that the enzyme systems responsible for producing the cyanoepithio compounds can be separated into thio-glucosidase and epithio specifier protein (ESP) fractions. ESP, itself incapable of cleaving alkenylglucosinolates, has to be added to the thioglucosidase in order to be able to make the cyanoepithio catabolites. It will be interesting to see if ESP alone can cata-lyse the formation of 3 from the aglucone, or whether both protein fractions are still required.

It has been shown[14] that elemental sulfur can react with some unsaturated sesquiterpene hydrocarbons under photo-chemical condi-tions at room temperature to yield the corresponding alkene epi-sulfides. Some of these episulfides occur as natural products in hops which have been sprayed with flowers of sulfur.[14] Episulfides

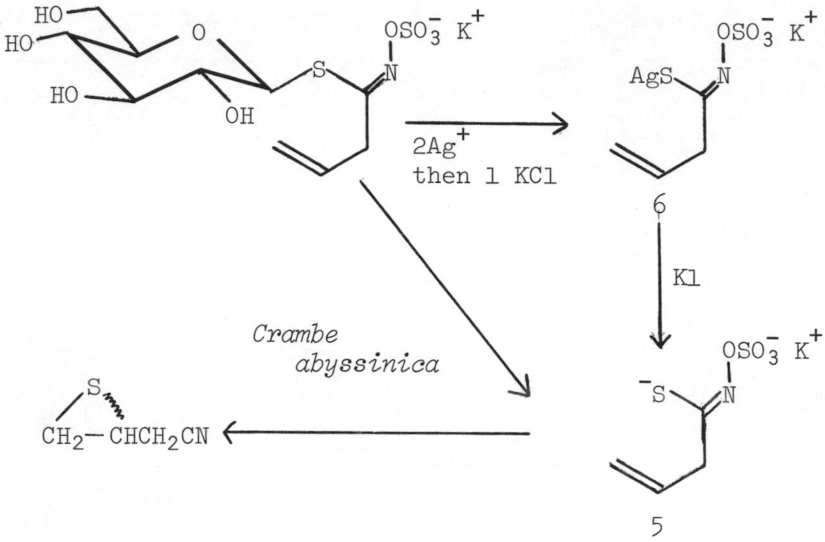

Fig. 3.   The formation of l-cyano-2,3-epithiopropane from the allyl
          glucosinolate.

are also known to be formed in the thermal reaction of sulfur with
alkenes.[15]   However, we are inclined to favour some process for the
formation of the 1-cyanoepithioalkanes which constructs them from
the 'sinolate aglucone by a non-stereospecific, intra-molecular reaction.   Various possibilities for such a reaction can be visualised,
and we are pursuing experimental tests in an attempt to decide the
matter.

TOXICOLOGY

     As stated in the introduction, the prime concern of this study
was with the toxicology of the cyanoepithioalkanes.   We had taken
note of the established toxicities of some simple episulfides[16], in
particular the report[17] that thiirane was a weak carcinogen.

     The acute toxicities of 2-4 in mice appear to be low, ca. 100-
500 mg Kg^{-1}, for compounds administered intraperitoneally or orally
as 5% suspensions in olive oil.   This is of the same order as found
earlier[2] for the compounds S-1 and R-1.

     In the Ames test[18] (with or without metabolism by extracts of
rat livers induced for microsomal mixed-function oxygenases with
methylcholanthrene, or Arochlor) 2-4 were inactive at low-medium

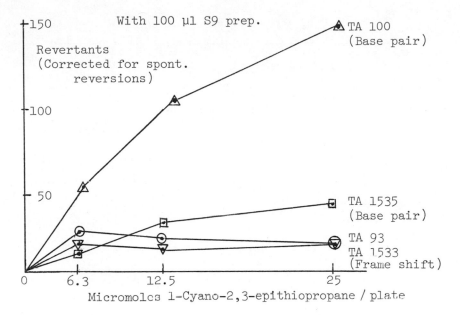

Fig. 4.   Ames Test results for *3*.

dose levels. Repetition of these tests with *3*, using the sensitive TA-100 strain of *Salmonella typhimurium* revealed[19] weak mutagenicity at rather high dose levels (see Fig. 4).

In the granuloma pouch assay[20] *3* showed no mutagenic activity. Interestingly it revealed a pronounced cytostatic effect in this test.

Finally, we decided to measure the covalent binding of *3* to DNA <u>in vivo</u>, i.e. determine its covalent binding index (CBI).[21] For this purpose we prepared ^{35}S- labelled *3*, by the route given in Fig. 2, employing ^{35}S- thiourea. For orally administered material there was a rapid excretion of the bulk of the radioactivity in the urine, as an acidic organic conjugate, but both tissue protein and DNA became labelled. Table 3 summarises these results[22] with the assumption that the radioactivity incorporated into the DNA represents covalently bound derivatives of *3*. To decide whether *3* is a weak carcinogen (as suggested by these results), or not, requires a classical long-term study with mammals.

It seems very probable that the other 1-cyanoepithioalkanes will behave similarly.

Table 3. <u>In vivo</u> covalent binding of *3* to DNA.

CBI of *3* for	liver DNA	0.9
	colon DNA	0.5
	stomach DNA	3.9

cf.[21] CBIs for liver DNA of NN-dimethylnitrosamine, 5600; benzo[a]pyrene, 10, benzene, 1.7; and saccharin, 0.005.

ACKNOWLEDGEMENTS

     We thank the Natural Sciences & Engineering Research Council of Canada, and the E.T.H. Zürich for the financial support of this work.

REFERENCES

1. M. H. Benn, Glucosinolates, <u>Pure</u>. <u>Appl</u>. <u>Chem</u>., 49:197 (1977).
2. C. H. VanEtten, M. E. Daxenbichler, and I. A. Wolff, Natural Glucosinolates(thioglucosides) in foods and feeds, <u>J</u>. <u>Agr</u>. <u>Food</u> <u>Chem</u>. 17:483 (1969).
3. M. E. Daxenbichler, C. H. VanEtten, and I. A. Wolff, Diastereomeric episulfides from <u>epi</u>-progoitrin upon autolysis of *Crambe* seed meal, <u>Phytochem</u>., 7:989 (1968).
4. M. E. Daxenbichler, C. H. VanEtten, W. H. Tallent, and I. A. Wolff, Rapeseed meal autolysis. Formation of diastereomeric (2R)-1-cyano-2-hydroxy-3,4-epithiobutanes from progoitrin, <u>Canad</u>. <u>J</u>. <u>Chem</u>., 45:1971 (1967).
5. J. T. O. Kirk and C. G. Macdonald, 1-Cyano-3,4-epithiobutane: a major product of glucosinolate hydrolysis in seeds from certain varieties of *Brassica campestris*, <u>Phytochem</u>., 13:2611 (1974).
6. R. A. Cole, 1-Cyanoepithioalkanes: major products of alkenyl-glucosinolate hydrolysis in certain Cruciferae, <u>Phytochem</u>., 14:2293 (1975).
7. R. A. Cole, Isothiocyanates, nitriles and thiocyanates as products of autolysis of glucosinolates in Cruciferae, <u>Phytochem</u>., 15:759 (1976).
8. M. E. Daxenbichler, C. H. VanEtten, and G. F. Spencer, Glucosinolates and derived products in cruciferous vegetables. Identification of organic nitriles from cabbage, <u>J</u>. <u>Agric</u>. <u>Food</u> <u>Chem</u>., 25:121 (1977).

9. H. K. Hall, jr., E. P. Blanchard, jr., S. C. Cherkovsky, S. B. Sieja, and W. A. Sheppard, Synthesis and polymerisation of 1-bicyclobutanecarbonitriles, J. Amer. Chem. Soc., 93:110 (1971).

10. F. G. Bordwell and H. M. Anderson, The reaction of epoxides with thiourea, J. Amer. Chem. Soc., 75:4959 (1953).

11. H. E. Miller, "The Aglucone of Sinigrin", M.A. Thesis, Rice University, Houston, Texas (1965).

12. H. E. Tookey, *Crambe* thioglucoside glucohydrolase (EC 3.2.3.1): separation of a protein required for epithiobutane formation, Canad. J. Biochem., 51:1654 (1973).

13. R. A. Cole, Epithiospecifier protein in turnip and changes in products of autolysis during ontogeny, Phytochem., 17:1563 (1978).

14. F. R. Sharpe and T. L. Peppard, Formation of sesquiterpene episulphides under mild conditions and their occurrence in the essential oil of hops, Chem. and Ind. (London), 664 (1977).

15. L. Bateman and C. G. Moore, Reactions of sulfur with olefins, in "Organic Sulfur Compounds", N. Kharasch, ed., Pergamon Press, Oxford (1961).

16. H. E. Christensen and T. T. Luginbyhl (eds.), "Toxic Substances List", U.S. Dept. of Health, Education and Welfare, P.H.S., U.S. Government Printing Office, Washington, D.C. 20402 (1973).

17. H. Druckrey, H. Kruse, R. Preussmann, S. Ivankovic, and Ch. Landschütz, Cancerogene alkylierende substanzen III. Alkyl-halogenide, -sulfate, sulfonate und ringgespannte hetero-cyclen, Zeit. fur Krebsforsch., 74:241 (1970).

18. B. N. Ames, J. McCann, and E. Yamasaki, Methods for detecting carcinogens and mutagens with the salmonella/mammalian-microsome mutagenicity test, Mutation Res., 31:347 (1975).

19. U. Friederich, E.T.H., Zürich, unpublished results.

20. P. Maier, P. Mauser and G. Zbinden, Granuloma pouch assay I. Introduction of oubain resistance by MNNG in vivo, Mutation Res., 54:159 (1978).

21. W. K. Lutz and Ch. Schlatter, Saccharin does not bind to DNA of liver or bladder in rats, Chem.-Biol. Interactions, 19:253 (1977).

22. A. von Däniken, E.T.H., unpublished results from a thesis in preparation.

CYSTEINE AND GLUTATHIONE IN MAMMALIAN PIGMENTATION[+]

Giuseppe Prota

Istituto di Chimica Organica e Biologica

Università di Napoli, Via Mezzocannone 16

During the past decade progress in the chemistry of melanins have led to several new and important discoveries which have largely changed the accepted concepts of melanogenesis,thus providing an improved background to look into the biochemical aspects of the pigmentary system of man and other mammals. It is not intended here to review these advances in any details but rather to highlight some aspects of melanin pigmentation with special reference to the critical role played by cysteine and glutathione in the control of the biosynthetic activity of the melanocyte. Insight into this mechanism follows primarily from the recognition that melanin pigmentation of mammalian hair, skin, or eye results from pigments of different structure and composition, which can be conveniently subdivided in three groups. These are (i) the dark nitrogenous melanins or eumelanins[1,2] accounting for black to brown colours, (ii) the nitrogen-sulphur containing phaeomelanins[3,4] which provide the lighter ones with a wide range from yellow to reddish-brown, and (iii) the trichochromes[5,6] closely related to phaeomelanins but found only in certain types of yellow or reddish hair as well as in the urine of patients with pigmented melanoma metastases[7,8].

This classification must be looked upon as provisional, however, as it is likely that further work in the field will extend

---

[+] This research was support in part by grants from C.N.R. (Rome).

the actual groups of melanin pigments. Thus, at present all black
and brown eumelanins are regarded as a fairly homogenous group of
virtually-insoluble irregular polymers consisting mainly of 5,6-
-dihydroxyindole units at various oxidative levels,[1,2] although
some are soluble in alkali, while others contain sulphur in various
percentages ranging from 1% as much as 5%.[9] On the other hand, in
the course of our studies on the pigments from reddish-hair, we
isolated some unusual types of phaeomelanins which, on analysis
showed little or no sulphur, and on degradation behaved similarly
to the black eumelanins. While the general nature of these pigments remains to be established, it is obvious that the identity
of an eumelanin or a phaeomelanin cannot be established simply on
the basis of colour and solubility properties, as these may well
be misleading. In fact, no satisfactory tests are available for
the identification of these pigments and even modern methods of
spectroscopic and X-ray crystallographic analysis are of little
diagnostic value. The main reason, apart from the generally untrac-
table nature of the material, lies in the complex chemical structures of both eumelanins and phaeomelanins, which, in spite of our
improved present knowledge, are still little understood.

Chemical investigation of the trichochromes which are much
smaller molecules has not been so hampered and several of these have bee
identified; indeed, they are the only group of "melanin" pigments
of well-defined structure, characterized by an unique pH-dependent
1,4-benzothiazine chromophore. Typical examples are the yellow

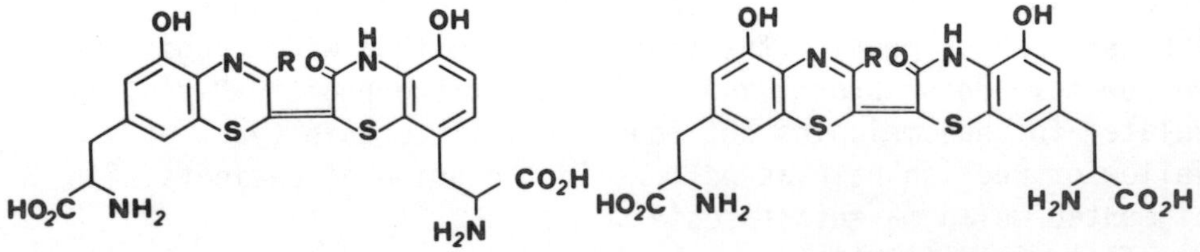

1 : R = CO₂H
3 : R = H

2 : R = CO₂H
4 : R = H

isomeric trichochromes B and C, identified as 1 and 2, respectively. They were first isolated from hen feathers of certain

strains, for instance New Hampshire or Rhode Island, and were later found in red human hair, in which they occur in very small amounts. Typically, on brief heating in acid solution both tricho-chrome B and C undergo selective decarboxylation of the acid group at position 3 to give new pigments (3 and 4, respectively) which are yellow-orange in alkaline solution and red-purple in acid.This characteristic behaviour is useful for analysis as extraction of hair with boiling mineral acids gives directly the decarboxylated pigments which can be readily purified by paper chromatography and identified by their pH-dependent spectra.

Despite their evident differences in molecular size and general properties, it is now well established that trichochromes, complex phaeomelanins, and eumelanins, are biogenetically related arising from a common metabolic pathway[10] (Fig.1) in which dopa-

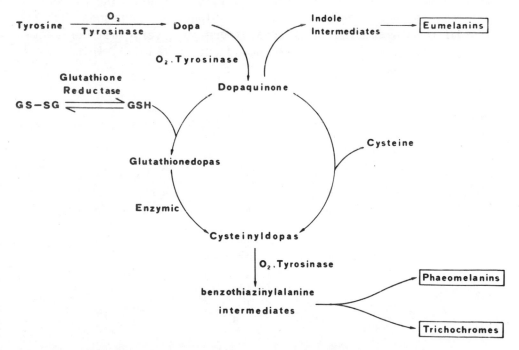

Fig. 1.   Suggested pathways of reactions leading to formation of the basic types of melanin pigments.[6,10]

quinone is the key intermediate. This is a most reactive o-quinone
(actually non-isolable) which is formed in melanocytes from
tyrosine by the action of copper(II) containing enzyme tyrosinase
consisting usually of as many as three molecular subforms ($T1,T_2$,
and T_3). Although the composition and activity of this enzymic
system may vary to a certain extent depending upon the type of
melanocytes[11] its catalytic role in each of the pathways remains
primarily connected with the conversion of tyrosine to dopaquinone,
the subsequent steps of melanogenesis requiring probably no further
specific enzymic assistence.

In eumelanin-forming melanocytes the subsequent fate of dopaquinone is essentially determined by its own chemical reactivity
which, as shown by in vitro experiments, leads ultimately to a
dark insoluble polymer by an oxidative process involving indole
intermediates, notably, dopachrome and indole-5,6-quinone. In
phaeomelanin-forming melanocytes, an alternative pathway exists
which leads to the production of the sulphur-containing pigments
instead, through the intervention of sulphydryl compounds such as
cysteine. This combines with dopaquinone in a very fast non-enzymic
reaction to give the cysteinyldopa adducts[12] (Fig.2), which by

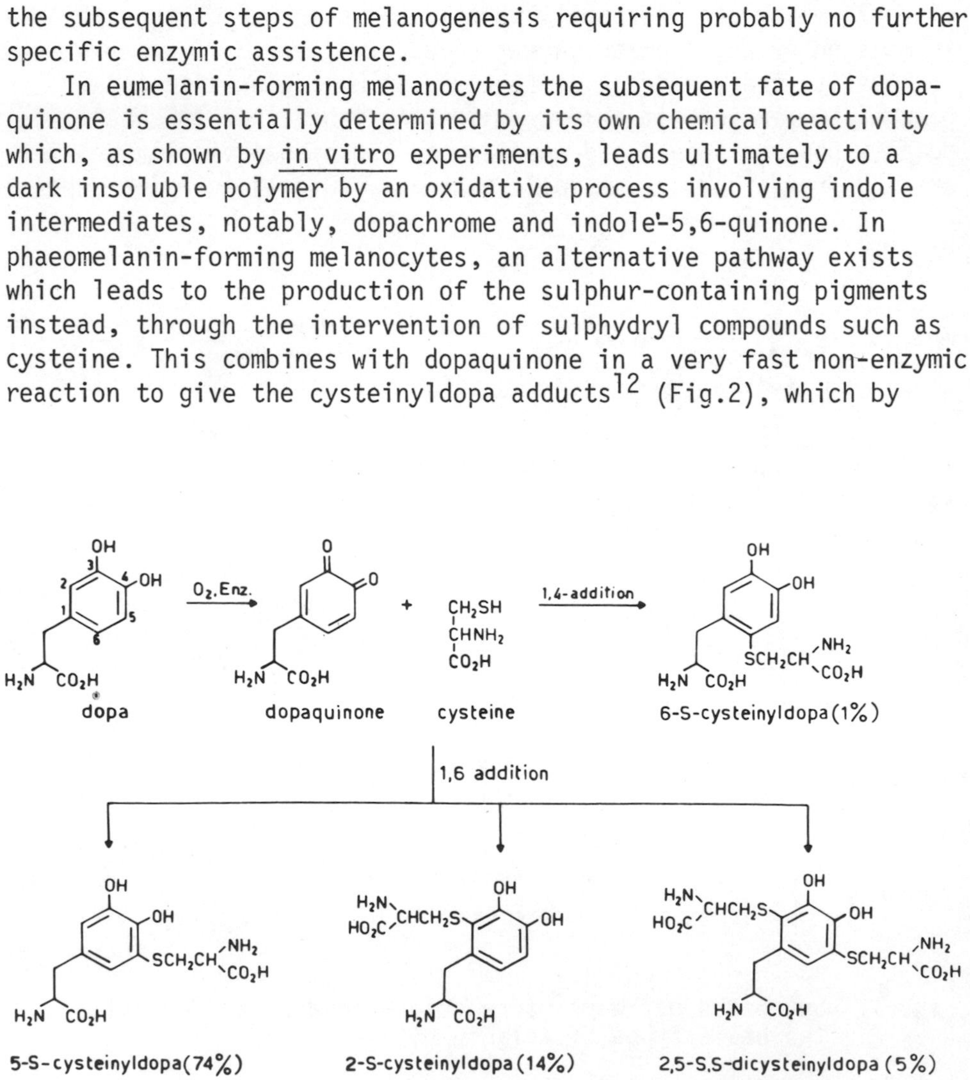

Fig. 2.   Structural formulae and yields of the cysteinyldopa adducts
          formed in the first stage of the co-oxidation of dopa and
          cysteine in the presence of tyrosinase.[12]

further oxidation are converted into phaeomelanins and/or trichochromes, presumably via 1,4-benzothiazine intermediates[13,14]
derived by cyclization of their cysteinyl residues. Interestingly,
5-S,cysteinyldopa[15] and related metabolites[16] are found in human
urines in varying amounts (from few micrograms to several milli-
grams) which are of clinical significance for the characterization
of normal and pathological melanocytes[17].

At present, little is known about the biologic conditions
which determine the ratio of formation of phaeomelanins and trichochromes in different mammalian species. But it is noteworthy that,
while in model experiments the cysteinyldopas are smoothly converted
into phaeomelanin-like pigments by the action of tyrosinase in the
presence of dopa as co-factor, attempts to convert them into trichochromes under similar conditions have been so far unsuccessful.
Thus, it seems that the later stages in the biosynthesis of trichochromes require certain specific biochemical conditions which, however, remain elusive.

Turning now to the switch mechanism leading to either eumelanin
or phaeomelanin biosynthesis we have seen that there is much
chemical evidence suggesting that this must be primarily linked
with the regulation of the sulphydryl content of the melanocytes
favouring the formation of the cysteinyldopas. Consistent with
this view are the now classical studies by Cleffman[18] that in
tissue cultures phaeomelanin forming melanocytes from agouti mouse
follicles produce only black pigment but revert to phaeomelanin
synthesis when sufficient glutathione or cysteine are added to the
nutrient medium. Whereas the effect of cysteine on the biosynthetic
activity of melanocytes is in line with the formation of the cysteinyldopas as in Fig.2, the participation of glutathione in the
phaeomelanin pathway is rather intriguing as it involves the
release of the bound cysteinyl residue after the coupling of the
thiol group to dopaquinone. Experimental evidence supporting such
a mechanism is provided by the occurrence of both 5-S-cysteinyl-
dopa, and 5-S-glutathiondopa[19] in melanoma tissues which seem to
contain also the hydrolytic enzymes (γ-glutamyl transferase and a
peptidase) capable of converting the latter into the former. Still,
it is not clear to what extent this alternative route to cysteinyldopas is relevant for pigment synthesis as it depends to whether
hydrolysis of glutathiondopa, occurs within or outside melanocytes.

In fact, in addition to serving as a possible substrate for
the synthesis of cysteinyldopa, glutathione could play a regulatory
role in pigment cell metabolism by converting part of the dopaqui-

none formed in the melanocytes into glutathiondopa which as such is unable to give rise to phaeomelanins unless transformed to cysteinyldopa, by enzymic hydrolysis. Hence, in the absence of this latter step, the formation of glutathiondopa would only have the effect of sidetracking dopaquinone without production of either phaeomelanin or eumelanin. Such a regulatory role of glutathione on pigment synthesis, mistakinkly reported[20] usually as inibitory effect on tyrosinase, is supported by early observations of Halprin and Ohkawara[21] that GSH levels are lower in black skin than in white skin and that in normal subjects hyperpigmentation following UV irradiation is preceded by a marked fall in the GSH levels with a corresponding rise in those of oxidized glutathione. Since this latter is unable to react with dopaquinone, the conversion might result in a more effective environment for pigment synthesis. In both cases, the observed changes in thiol content are associated with a reduced activity of the enzyme glutathione reductase that in combination with $NADPH_2$ effects the conversion GS-SG to GSH.

All this biochemical information points to a general role of the glutathione system in the metabolic activity of both eumelanic and phaeomelanic melanocytes. However, whether glutathione is also specifically involved in the biosynthesis of the cysteinyldopas required for phaeomelanin formation remains to be determined, and the answer obtained will unquestionably contribute to our understanding of the synthetizing activity of mammalian melanocytes.

REFERENCES

1. R. A. Nicolaus, "Melanins", Hermann, Paris (1968).
2. R. A. Nicolaus, "Melanins", in "Metodicum Chimicum", Vol.II, Part 3, E. Korte, ed., p.190, Academic Press, New York and George Thieme Publishers, Maruzen Co., Ltd. Tokyo (1978).
3. E. Fattorusso, L. Minale,and G. Sodano, Feomelanine da nuove fonti naturali, Gazz.Chim.Ital., 100:452(1970), and references therein.
4. R. H. Thomson, The pigments of reddish hair and feathers,Angew. Chem.Internat.Edit., 13:305 (1974).
5. G. Prota, Structure and biogenesis of phaeomelanins, in "Pigmentation, its genesis and biologic control", V. Riley, ed., p. 615, Appleton-Century-Crafts, New York (1972).
6. G. Prota and R.H. Thomson, Melanin pigmentation in mammals, Endeavour, 35:32 (1976).
7. G. Prota, H. Rorsman, A-M. Rosengren, and E. Rosengren, Occurrence of Trichochromes in the urine of a melanoma patient,

Experientia, 32:1122 (1976).

8. G. Agrup, C. Lindbladh, G. Prota, H. Rorsman, A-M. Rosengren, and E. Rosengren, Trichochromes in the urine of melanoma patients, J.Invest.Dermatol., 70:90 (1978).

9. G. Prota, H. Rorsman, A-M. Rosengren, and E. Rosengren, Phaeomelanic pigments from a human melanoma. Experientia, 32:1970 (1976).

10. G. Prota and A.G. Searle, Biochemical sites of gene action for melanogenesis in mammals, Ann.Genet.Sel.anim., 10:1 (1978).

11. K. Jimbow, W. D. Quevedo, T. B. Fitzpatrick, and G. Szabo, Some aspects of melanin biology: 1950-1975, J.Invest.Dermatol., 67:72 (1976).

12. S. Ito and G. Prota, A facile one-step synthesis of cysteinyl-dopas using mushroom tyrosinase, Experientia, 33:1118 (1977), and references therein.

13. S. Crescenzi, G. Misuraca, E. Novellino, and G. Prota, Reazioni modello per la biosintesi dei pigmenti feomelanici. Chimica Ind.(Milano), 57:392 (1975).

14. F. Chioccara, E. Novellino, G. Prota, and G. Sodano, A trimeric aldolization products of 1,4-benzothiazine, J.C.S.Chem.Comm., 50 (1977).

15. G. Agrup, B. Falck, K. Fyge, H. Rorsman, A-M. Rosengren, and E. Rosengren, Excretion of 5-S-cysteinyldopa in the urine of healthy subjects, Acta Dermatovener (Stockholm), 55:7 (1975).

16. G. Prota, H. Rorsman, A-M. Rosengren, and E. Rosengren,Isolation of 2-S-cysteinyldopa and 2,5-S,S-dicysteinyldopa from the urine of patients with melanoma, Experientia, 33:720 (1977).

17. G. Agrup, P. Agrup, T. Andersson, B. Falck, J. A. Hausson, S. Jacobson, H. Rorsman, A-M. Rosengren, and E. Rosengren, Urinary excretion of 5-S-cysteinyldopa in patients with primary melanoma or melanoma metastasis, Acta Dermatovener (Stockholm), 55:337 (1975).

18. G. Cleffmann, Function-specific changes in the metabolism of agouti pigment cells, Exp. Cell. Res., 35:590 (1964).

19. G. Agrup, B. Falck, B.M. Kennedy, H. Rorsman, A.-M. Rosengren, and E. Rosengren, Formation of cysteinyldopa from glutathionedopa in melanoma, Acta Dermatovener, (Stockholm), 55:1 (1975).

20. P. C. Jocelyn,"Biochemistry of the SH group", p 273, Academic Press, London-New York (1972).

21. K. M. Halprin and A. Ohkawara, Human pigmentation: the role of glutathione, in "Advances in biology of shin", W. Montagna and F. Hu, eds., p. 241, Pergamon Press, New York, 1967.

SULFUR COMPOUNDS IN MUSTELIDS

Kenneth K. Andersen and David T. Bernstein

Department of Chemistry
University of New Hampshire
Durham, N.H. 03824

INTRODUCTION

Secondary chemicals, especially alkaloids and terpenes isolated from plants, have interested chemists for many decades, yet only comparatively recently have their important roles in ecology been realized. They are no longer incorrectly considered to be just metabolic by-products.[1] Investigations of secondary compounds produced by mammals have a long history, but are not numerous in comparison to the studies of substances isolated from plants or even insects. Early studies on mammalian secretions were stimulated primarily by commercial concern with perfumes, whereas current research is motivated mostly by the great interest in chemical ecology, in particular by the influence these secretions acting as chemical signals have on animal behavior. Recent advances in the methodology of isolation and identification of small quantities of low molecular weight volatile compounds coincide with this growth of interest and have resulted in a dramatic increase in the number of mammalian secondary chemicals which have been identified. For the most part, mammalian secondary compounds are volatile and received by olfaction.[2] Since the odors of low molecular weight divalent sulfur compounds are usually very intense and distinct to humans, and apparently to other mammals as well, it is not surprising that such compounds, many examples of which are found in plants, are also represented in mammalian secretions.[3,4] Compound 1 was found in the vaginal secretion of the golden hamster[5], 2 in the anal scent gland of the striped hyena[6], and 3 and 4 in red fox urine.[7]

399

$$CH_3SSCH_3 \qquad\qquad CH_3\overset{\overset{\text{O O}}{\|\ \|}}{C}CH_2CH_2SCH_3$$

$$1 \qquad\qquad\qquad\qquad 2$$

$$CH_2\!\!=\!\!\overset{\overset{\displaystyle CH_3}{|}}{C}CH_2CH_2SCH_3 \qquad\qquad C_6H_5CH_2CH_2SCH_3$$

$$3 \qquad\qquad\qquad\qquad 4$$

We investigated the contents of the anal sac of the North American striped skunk (Mephitis mephitis) and the North American mink (Mustela vison), both members of the weasel family (mustelidae). Some of this work has been published.[8] In North America there are numerous members of this diverse and interesting family of animals ranging in size from the one meter, thirty kilogram wolverine (Gulo gulo) to the twenty centimeter, sixty gram weasel (Mustela rixosa) and including the otter (Lutra canadensis), the ermine (Mustela erminea), the martin (Martes americana), the fisher (Martes pennanti), and the badger (Taxidea taxus). Many other examples are found through-out the world.

Skunks are without doubt the most well-known and common musteline mammals in North America. They comprise three genera: Spilogale, Conepatus, and Mephitis with the most common species being the striped skunk (Mephitis mephitis) which ranges from southern Canada through most of the United States to northern Mexico. The hooded skunk (Mephitis macroura), a southern relative of the striped skunk, and the smaller spotted skunk (Spilogale putorius), found chiefly in the southwestern United States, are less common.[9]

All mustelids have anal glands and are said to be capable of spraying the gland secretions. For skunks, this spray is the animal's primary means of defense and their anal sacs are well developed for this purpose. Other mustelids such as the mink seem to use these secretions for marking rather than defense, and their anal glands contain relatively smaller amounts of vile-smelling fluids. The striped skunk[8,10], South American zorrino (Conepatus suffocans)[11], Phillipine teledu (Mydaus marchei Huet or Suillotaxus marchei)[12], European polecat (Mustela putorius)[13], American mink[13], and the ermine[14] are the only mustelids whose anal sac secretions have been examined chemically.

Beckmann[12] studied the scent of the Phillipine teledu and reported in 1896 that it contained 1-butanethiol. Aldrich reported in the same year that striped skunk scent contained "one of the butyl mercaptans" and, most likely 3-methylbutanethiol.[10] With the passage of time, the actual facts were mixed up and most North American chemists were erroneously taught that skunk scent owed its odor to 1-butanethiol. Elemental analysis of steam distilled zorrino scent

led Fester and Bertuzzi to conclude that it was a mixture of
2-butene-1-thiol and its disulfide, but their elemental analyses were
flawed by large differences between their experimental data and the
theoretical values. An account of these early studies has been
written.[15]

The large amount of physiological and nutritional work reported
on mink does not include mention of the anal sac. It was not until
1976 that Sokolov reported the isolation of several volatile
aliphatic acids from the anal sac secretion.[16] Butyric acid was the
most abundant, but acetic, propionic, 2-methylpropanoic, 3-methyl-
butanoic, 2-methylbutanoic, and a small amount of pentanoic acids
were also obtained. Schildknecht et al. published a communication
reporting the volatile sulfur compounds in the anal sac secretions
of the mink.[13] Crump identified the major malodorous compound in
the anal gland secretion of the male ermine or stoat (Mustela
erminea).[14] These results will be discussed subsequently together
with our own.

THE STRIPED SKUNK

In our initial study we investigated the major volatile organic
constituents of the scent of the striped skunk.[8] The scent was
collected by a commercial skunk breeder from live young male and
female skunks during surgical removal of their anal scent glands.
It consisted of two phases – an aqueous phase, which was not investigated further, and an oily organic phase which was responsible for
the scent's repulsive odor. Distillation of the latter with a bath
temperature kept at 40° or less (0.2 mm) gave the volatile portion
which was fractionated by gas-liquid chromatography. A viscous,
yellow-orange oil of much less odor remained in the distilling flask;
it constituted about one-third of the total organic layer.

The three major components were identified as trans-2-buten-1-
thiol (5), 3-methylbutanethiol (6), and trans-2-butenyl methyl
disulfide (7) by comparison of their proton nmr and ir spectra to
those obtained from authentic samples. Thioether derivatives prepared from 5 and 6 by treatment with 1-chloro-2,4-dinitrobenzene
proved identical to authentic samples.

$$
\begin{array}{ccc}
\underset{CH_3}{\overset{H}{>}}C=C\underset{H}{\overset{CH_2SH}{<}} & \underset{CH_3}{\overset{CH_3}{>}}CHCH_2CH_2SH & \underset{CH_3}{\overset{H}{>}}C=C\underset{H}{\overset{CH_2SSCH_3}{<}} \\
5 & 6 & 7
\end{array}
$$

More recently, we have investigated the minor volatile components in the scent using glc-mass spectrometry.[17] In this way we

were able to identify several trace components which avoided detection and/or isolation in our earlier efforts. Initially, we used a 6-ft x 1/4-in glass column packed with 15% Apiezon-M on Chromosorb-P 60/80 at 75° or 120° for our preparative work. Comparison of isothermal retention times of 5, 6, and 7 with authentic samples were also made using 12-ft x 1/4-in Teflon-lined aluminum columns packed with 20% SE-52 and 10% diisodecylphthalate both on Chromosorb-P 60/80 at temperatures between 46° and 70°. Subsequently, a 12-ft x 1/4-in column packed with 20% Carbowax 20M on Chromosorb-W 60/80 was used on the volatile oil from five different skunks - two males and three females (one lactating). Resolution of the main components was improved, but column bleed prohibited the use of this column with a mass spectrometer. Finally, a 10-ft x 1/8-in stainless steel column packed with 5% Carbowax 1500 on DMCS-treated Chromosorb-W 120/140 and temperature programmed from 50° to 120° at 2.5°/min proved satisfactory in conjunction with a modified Hitachi RMU-6 computer-assisted mass spectrometer for separation and identification of some of the minor scent components. Even though resolution was improved, some overlapping of peaks persisted. A minimum of twenty-two compounds were seen as resolved peaks or shoulders on the chromatograms. No particular differences were seen in the chromatograms of oils obtained from the five skunks.

A mass spectrum was recorded every three seconds during the glc-ms run giving a total of about four-hundred spectra. An ionization intensity plot was obtained for each m/e value. The total ionization plot for the glc-ms run was also recorded. This plot resembled the glc trace, since the amount of material entering the mass spectrometer from the gas chromatograph paralleled the amount undergoing ionization and detection. By overlaying the ionization intensity (for given m/e values) versus spectrum number plots on the total ionization energy versus spectrum number plot, it was possible to assign the m/e values to compounds giving particular peaks in the glc trace. Such comparisons were also useful in separating data in a given spectrum resulting from two compounds; i.e., incomplete resolution in the glc. In particular, 3-methylbutanethiol (6) eluted with 2-buten-1-thiol (5) and crotyl methyl disulfide (7) eluted with crotyl propyl sulfide (11).

Of the twenty-two or more compounds indicated by the glc recordings, nine were identified. Three were identified in our initial work[8] and six more by the glc-ms analysis. 3-Methylbutanethiol (6) was identified by comparison to an authentic mass spectrum. Also, its spectrum did not correspond to the spectra of isomers. 1-Butanethiol (8) was identified in a similar way. Its presence in skunk scent is less than one-percent of the total and any earlier reports of its presence must still be considered erroneous. The remaining five compounds, 9 to 13, were identified by analysis of their fragmentation patterns. Their structural formulas numbered in the relative elution order are shown below.

Except for 5, 6, and 7, none of the other compounds comprised more than one percent or so of the total volatile material.

$$CH_3CH_2CH_2CH_2SH$$

8

$$(CH_3)_2CHCH_2CH_2SCHCH{=}CHCH_3$$

9

$$(CH_3CH{=}CHCH_2)_2S$$

10

$$CH_3CH{=}CHCH_2{-}S{-}CH_3CH_2CH_2$$

11

$$(CH_3)_2CHCH_2CH_2S{-}CH_3CH{=}CHCH_2S$$

12

$$CH_3CH{=}CHCH_2S{-}CH_3CH{=}CHCH_2S$$

13

THE NORTH AMERICAN MINK

Mink anal sac secretions were removed from freshly killed and skinned mink by piercing the sac with a hypodermic needle and withdrawing the thick off-white emulsion.[17] The fluid, about 1.3 ml per animal, had a very strong, unpleasant odor characteristic of many volatile sulfides and thiols. After being centrifuged in a clinical centrifuge, the secretion separated into four layers. The upper layer - a light yellow oil - constituted twenty percent of the total secretion from the females and thirty-four percent from the males. A fatty layer, a yellowish aqueous solution, and a bottom fatty layer composed the remaining three layers. Bulb to bulb distillation of the oil at 0.02 mm with a bath temperature at 40° or below gave the volatile components amounting to about five per cent of the total oil.

Glc separation of the volatile compounds was achieved using a 12-ft x 1/4-in Teflon-lined aluminum column packed with 20% Apiezon-L on Chromosorb-W 60/80 with manual temperature programming from 150 to 230°. Two compounds were isolated. The most volatile compound gave an nmr spectrum (δ 1.65, s, 6H; δ 2.25, t, 2H; δ 3.4, t, 2H; in $CDCl_3$) identical to the spectrum of an authentic sample of 2,2-dimethylthietane (14). A mass spectrum of 14 gave fragmentation ions at m/e values of 102, 87, 74, 69, 68, 59, 56 (base peak), and 41 which are consistent with its structure. The less volatile component gave an nmr spectrum (δ 0.94, d, 6H;

δ 1.61, m, 3H; δ 2.75, t, 2H; in CDCl$_3$) identical to the spectrum of an authentic sample of diisopentyl disulfide (15).

Schildknecht et al. had previously found both 14 and 15 in mink scent, but they had also reported finding 3,3-dimethyl-1,2-dithiolane (16).[13] We were unable to find 16. Crump found 2-propylthietane (17) in male ermine scent.[14]

14   15

16   17

The non-volatile organic portions of both skunk and mink scent were also examined. In the skunk, this portion contains mostly nitrogen-containing aromatic compounds as judged by nmr and elemental analysis. Uv spectroscopy on substances partially purified by column and thin-layer chromatography suggested that they are mono- or di-substituted quinolines. In the mink, however, the non-volatile portion is composed of triglycerides.

CONCLUSIONS

So it seems that the docile skunk possesses large amounts of vile-smelling low molecular weight sulfur compounds which can be used to repell agressors whereas the mink, an agressive animal, has smaller amounts of similar but not identical sulfur compounds which it uses primarily for marking. The non-volatile portion of skunk scent may contain precursors of the volatile compounds or perhaps they are secondary metabolites produced in unrelated processes. In any event, they may add to the scent's repellent effectiveness. In the mink, however, the triglycerides apparently act as a fixative and decrease the evaporation rate of volatile compounds used in marking.

We would expect that examination of the anal sacs of other mustelids would reveal similar but not necessarily identical low molecular weight sulfur compounds. The quantity and nature of the scent should correlate with its use by the animal. In addition, a knowledge of how these compounds are produced might provide some insight into the taxonomic relationship among mustelids.[18,19]

ACKNOWLEDGMENTS

We thank the National Science Foundation for grant no. CHE 77-08983 allowing the purchase of an nmr spectrometer, Dr. Catherine E. Costello for obtaining the glc-ms data at MIT, and Mrs. Kathryn A. Reynolds for typing this manuscript.

REFERENCES

1. J. B. Harborne, Chemosystematics and coevolution, Pure and Appl. Chem. 49:1403 (1977).

2. D. Müller-Schwarze and M. M. Mozell, "Chemical Signals in Vertebrates," Plenum Press, New York (1977).

3. A. Kjaer, Low molecular weight sulphur-containing compounds in nature: a survey, Pure and Appl. Chem. 49:137 (1977).

4. G. Ohloff and I. Flament, The role of heteroatomic substances in the aroma compounds of food stuffs, in "Progress in the Chemistry of Organic Natural Products," vol. 36, W. Herz, H. Grisebach, and G. W. Kirby, eds., Springer-Verlag, Vienna (1979).

5. A. G. Singer, W. C. Agosta, R. J. O'Connell, C. Pfaffmann, D. V. Bowen, and F. H. Field, Dimethyl disulfide: an attractant pheromone in hamster vaginal secretion, Science 191:948 (1976).

6. J. W. Wheeler, D. W. von Endt, and C. Wemmer, 5-Thiomethylpentan-2,3-dione: a unique natural product from the striped hyena, J. Am. Chem. Soc. 97:441 (1975).

7. S. R. Wilson, M. Carmack, M. Novotny, J. W. Jorgenson, and W. K. Whitten, Δ^3-Isopentenyl methyl sulfide. A new terpenoid in the scent mark of the red fox (Vulpes vulpes), J. Org. Chem. 43:4676 (1978).

8. K. K. Andersen and D. T. Bernstein, Some chemical constituents of the scent of the striped skunk (Mephitis mephitis), J. Chem. Ecol. 1:493 (1975).

9. B. J. Verts, "The Biology of the Striped Skunk," The University of Illinois Press, Chicago (1967).

10. T. F. Aldrich, A chemical study of the secretion of the anal glands of Mephitis mephitica (common skunk) with remarks on the physiological properties of this secretion, J. Exp. Med. 1:323 (1896).

11. G. A. Fester and F. A. Bertuzzi, La secrecion del zorrino, Rev.
     Fac. Quim. Ind. Agr., Univ. Mac, litoral 5:85 (1937).

12. E. Beckmann, Ueber das Drüsensecret des Stinkdaches, Pharm.
     Zentralhalle Dsch. 37:557 (1896).

13. H. Schildknecht, I. Wilz, F. Enzmann, N. Grund, and M. Ziegler,
     Mustelan, the malodorous substance from the anal gland of the
     mink (Mustela vison) and the polecat (Mustela putorius),
     Ang. Chem. Int. Ed. Engl. 15:242 (1976).

14. D. R. Crump, The major malodorous substance from the anal gland
     of the stoat (Mustela erminea), Tetrahedron Lett. 5233 (1978)

15. K. K. Andersen and D. T. Bernstein, 1-Butanethiol and the
     striped skunk, J. Chem. Ed. 55:159 (1978).

16. V. E. Sokolov and I. M. Khorlina, Pheromones of mammals: study
     of the composition of volatile acids in vaginal discharge of
     mink (Mustela vison Bris.), Dokl, Akad, Nauk SSR 228:225
     (1976) in Chem. Abs. 85:60288q (1976).

17. D. T. Bernstein, Chemical constituents of the anal glands of the
     striped skunk (Mephitis mephitis) and the North American
     mink (Mustela vison), Ph.D. thesis, University of New Hamp-
     shire, Durham, New Hampshire, 1979.

18. G. E. Svendsen and J. D. Jollick, Bacterial contents of the
     anal and castor glands of beaver, J. Chem. Ecol. 4:563
     (1978).

19. E. S. Albone and G. C. Perry, Anal sac secretion of the red fox,
     Vulpes vulpes: volatile fatty acids and diamines: Implica-
     tions for a fermentation hypothesis of chemical recognition,
     J. Chem. Ecol. 2:101 (1975).

KEY WORDS

     Mustelids, striped skunk, mink, chemical communication

ON THE MECHANISM OF C-S BONDS FORMATION IN THE BIOSYNTHESIS OF BIOTIN

Andrée MARQUET [†], Adel Guirguis SALIB [††],

François FRAPPIER [†] and Georges GUILLERM [†]

[†] Laboratoire de Chimie Organique Biologique
Tour 44-45, Université P. et M. Curie,
4, Place Jussieu - 75230 PARIS CEDEX 05

[††] Biochemistry Department - Faculty of
Agriculture, Zagazig University
ZAGAZIG - EGYPT

The biosynthesis of biotin, 1, has already been intensively studied (1). All the intermediates in the biosynthetic pathway from pimelic acid to dethiobiotin, 2, are presently known and formed by classical biochemical reactions. But the mechanism of the conversion of dethiobiotin into biotin, a very unusual transformation, is still completely unknown.

In our first investigation (2) on biotin biogenesis, we have shown that the formerly proposed hypothesis of an unsatured intermediate of type (i) was highly improbable (scheme 1). These results were in agreement with those obtained separately by Parry and Kunitani with A. niger (3).

The hypothesis of an intermediate hydroxylation, (ii), for the functionalisation of the saturated carbons is now considered.

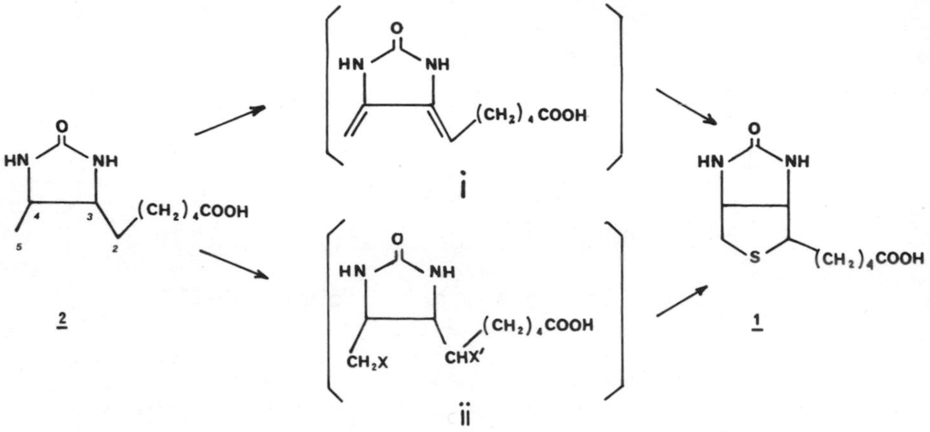

Scheme 1

The hydroxylation hypothesis

The three hydroxy compounds, 5, 6, 9, were prepared,
to be tested as biotin precursors. Their synthesis, starting from 4-methyl-5-(ω-carboethoxyvaleryl)-imidazolone-2,
3, (4) is summarized in scheme 2.

The aldehyde, 7, was obtained according to Zav'yalov
(5).

^{3}H labelled alcohols were easily prepared by reduction of their carbonyl precursors 4, 8 by NaB^{3}H$_4$.

The biosynthetic experiments were carried out with
E. coli C 124, a mutant which is unable to synthesize
dethiobiotin but which carries out the transformation
of dethiobiotin into biotin. If one of the above hydro-
xycompounds is an intermediate between dethiobiotin and
biotin, it should promote the growth of this mutant.

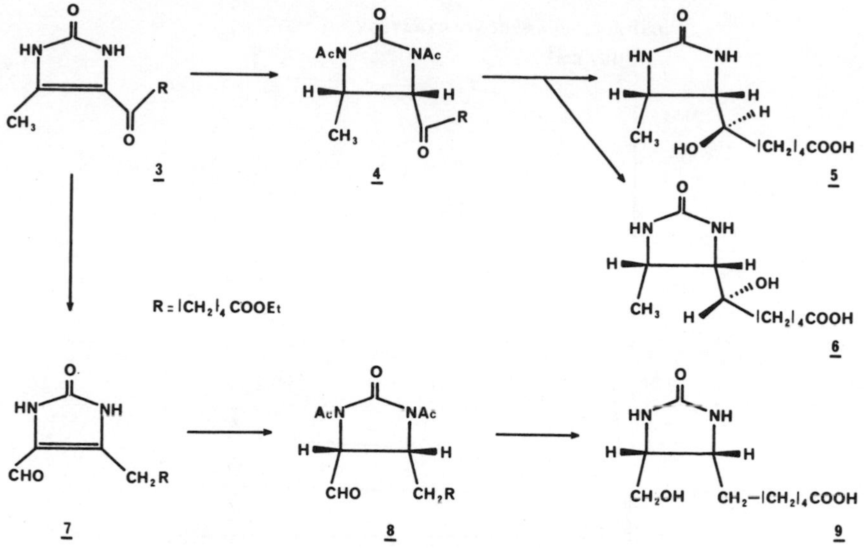

Scheme 2

The results are negative for the three compounds. No growth is observed with concentrations ranging from 5 to 200 ng/ml. However, negative results in growth experiments may be due to permeability problems and this point had to be checked before concluding that the alcohols are not intermediates.

It has been shown by Eisenberg (6) that biotin uptake in E. coli K12 proceeds through an active transport process characterized by the accumulation of biotin against a concentration gradient, and a marked dependence on temperature pH and energy source.

We checked that this process is also operating with E. coli C 124, the strain we use in these biosynthetic experiment (Fig. 1). Under the same conditions, no active transport process is observed for alcohols 5 and 6. With dethiobiotin and alcohol 9, the transport is temperature dependent, but the intracellular concentrations, higher than the initial ones, remain low compared to biotin (Fig. 1).

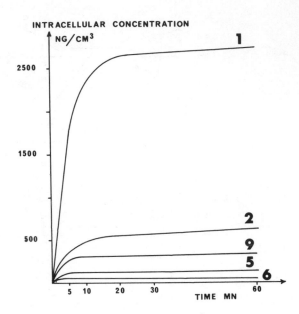

Fig.1 **Transport** in E. Coli C 124, at 37° C, from an uptake medium containing 75 NG/CM³ of Radioactive Compound.

It is however possible to let the hydroxy compounds enter the cells in sufficient amount, by a process similar to diffusion, by increasing the extracellular concentration. With dethiobiotin, under the normal growth conditions, a 1 ng/ml concentration in the incubation medium ensures a 15 to 25 ng/ml content in the intracellular space. A higher concentration of compounds 5, 6 and 9 respectively can be obtained if the exogeneous concentration is increased to 75 ng/ml.

The fact that no growth is observed under these conditions enables us to conclude that an intermediate hydroxylation is very unlikely. However, this hypothesis cannot be completely excluded since the conversion of dethiobiotin into biotin, which is certainly a multistep process, could involve a multienzymatic complex carrying out the whole transformation or part of it without releasing the intermediates (or incorporating a free intermediate from the medium). This point is of course very difficult to check.

Isolation of an intermediate between dethiobiotin and biotin

When resting cells of E. coli C 124 grown on biotin are incubated with ^3H and ^{14}C dethiobiotin, no radioactive biotin is produced but a new compound, X, can be isolated from the supernatant. It is purified by chromatography of its methyl ester on silicagel.

The corresponding acid promotes the growth of E. coli C 124 with an external concentration of 0.5 ng/ml similar to the concentrations used for dethiobiotin or biotin. The generation times are of the same order of magnitude. The conversion of X into biotin by growing cells of E. coli C 124 has been established. If a doubly labelled sample of dethiobiotin is incubated ($^3H/^{14}C$ = 6.70) the isolated X compound has the same $^3H/^{14}C$ ratio (6.68). This shows that, as expected (2), there is no degradation of the side chain and no loss of the hydrogens at positions 3 and 4 during its formation.

X contains sulfur as shown by another experiment carried out in a medium supplemented with ^{35}S sulfate.

Its structure elucidation is now underway.

Its knowledge will be very useful for further studies.

REFERENCES

1. M.A. Eisenberg in Advances in Enzymology, 38, 317-372 (1973)
2. G. Guillerm, F. Frappier, M. Gaudry and A. Marquet, Biochimie, 59, 119-121 (1977)
3. R.J. Parry and M.G. Kunitani, J. Amer. Chem. Soc., 98, 4024-4026 (1976)
4. R. Duschinsky and L.A. Dolan, J. Amer. Chem. Soc., 67, 2079-2084 (1945)
5. S.I. Zav'yalov, N.A. Rodionova and E.P. Gracheva Izv. Akad. Nauk. SSSR, Ser. Khim., 9, 2025-2028 (1972)
   S.I. Zav'yalov, O.M. Radoul, B.J. Gounar and N.A. Rodionova, Izv. Akad. Nauk. SSSR, Ser. Khim., 10, 2335-2337 (1972)
6. O. Prakash and M.A. Eisenberg, J. Bacteriol., 120, 785-791 (1974)

THE BIOSYNTHESIS OF LIPOIC ACID IN THE RAT

Silvestro Duprè, Giuseppe Spoto, Rosa Marina Matarese, Mariella Orlando* and Doriano Cavallini

Institute of Biological Chemistry, University of Rome, Centre of Molecular Biology, C.N.R., Rome and *Laboratorio di Patologia non infettiva, Istituto Superiore di Sanità, Rome, Italy

INTRODUCTION

Lipoic acid is widely distributed among microorganisms, plants and animals (1-4). The concentration of this co-factor in animal tissues is quite low (5). It is mainly found in a form bound covalently to the α-ketoacid dehydrogenase multienzyme complexes. Nutritional experiments with higher animals have failed to demonstrate a growth response to lipoic acid (6). Whether or not lipoic acid must be considered as a vitamin for higher organisms is still an open question. Metabolic disorders derived from lipoic acid deficiency are not reported in animals.

When ^{14}C-labeled lipoic acid is injected in the rat intraperitoneally or administered per os, the label is found to a large extent in the urine, associated to small amounts of lipoic acid and to partly identified catabolites (7). Part of the radioactivity has been found associated with tissues, mainly liver and muscle. Lipoic acid is found to be metabolized inside the mitochondria, and must therefore be taken up by these organelles (8). Administration of ^{35}S-lipoic acid led also to uptake of the label by liver mitochondria (9). In both cases however

lipoate (or a benzene-soluble metabolite thereof) seems
to be present in the tissue mainly as a not protein bound
form. Therefore it is not possible to argue, from the experimental data found in the literature, whether and to
what extent external supplemented lipoic acid is incorporated into the multienzyme complexes or if it is simply
metabolized, presumably through a β-oxidation-like pathway, at the level of mitochondria. Lipoic acid may in any
case cross the mitochondrial membrane (10). These data
are compatible with two different possibilities: i) lipoic
acid is synthesized in vivo by a non-mitochondrial system,
enters the membrane and is incorporated, via the well studied ATP-dependent lipoate activating system (whose existence in mammalian cells has been demonstrated (11), into
the α-ketoacid dehydrogenases complexes; ii) higher organisms are not able of synthesizing lipoic acid, and are
therefore dependent from an external supply, as for instance from the intestinal flora. They are able of incorporating preformed lipoic acid into the apoprotein. Data
reported in the experimental part of this paper are in
agreement with the first hypothesis.

Studies of incorporation of ^{14}C-labeled essential linoleic acid and non-essential oleic acid into cellular
components of young rats led Carreau et al. (12) to suggest a hypothetical biosynthetic pathway of lipoate, concerning the carbon backbone. No data are available about
the origin of sulfur atoms of lipoic acid in higher organisms.

More precise studies have been performed with microorganisms. Some strains, like Tetrahymena pyriformis,
Streptococcus cremoris, Pseudomonas putida and others,
have an absolute requirement for lipoic acid, and they are
used for useful and sensitive quantitative determination
methods. Other bacterial strains obviously are able to
grow on media containing, as sole sulfur source, sulfate,
cystine or methionine. In E. Coli the source of the sulfur moiety of biosynthesized lipoic acid appears to be
inorganic sulfate and methionine (13). The ability in
synthesizing lipoic acid seems to be widespread in microorganisms, and some accurate studies with ^{14}C-precursors
have been performed. 1-^{14}C-octanoic acid is incorporated

into lipoic acid as a unit, whereas 1-^{14}C-hexanoic acid
and 1,6-^{14}C-adipic acid are not incorporated when included
in the growth medium (14). It is interesting to note that
also 8-hydroxyoctanoate-1-^{14}C is not incorporated. Recently
the incorporation of specifically tritiated octanoic acid
and the stereochemistry of sulfur incorporation have been
studied by Parry and Trainor (15, 16) using E. Coli. They
confirmed that octanoic acid is the precursor of the carbon
backbone. The activation of the methylene groups (which
resembles the reaction leading from dethiobiotin to biotin)
and the formation of the C-S bond follow a yet unknown
mechanism. In any case neither biosynthetic intermediates
nor the actual precursor of sulfur atoms have been identified. It is interesting to note that E. Coli, grown on
^{35}S-lipoic acid containing medium, is able to incorporate
this compound as such in a protein bound, water insoluble
form (17). Incubation of ^{14}C-lipoic acid with rat liver
homogenate or mitochondrial preparations instead resulted
in the production of labeled CO_2, but it is not reported
to be incorporated. ^{35}S-lipoic acid is incorporated in
rat liver mitochondria, but most of it, in one form or
another, is lipide bound (9). Incorporation of ^{35}S-lipoic
acid to green algae into a lipide complex has also been
reported (18). All these data point to a biosynthetic
ability of higher organisms in respect to lipoic acid.

The studies we present here on the biosynthesis of
lipoic acid and particularly on the origin of the sulfur
atoms are performed in vivo with young Wistar rats and in
vitro with rat liver homogenates. Different ^{35}S-containing
probable precursors are assaied, and the labeled, protein-
bound lipoic acid is quantitatively evaluated after hydrolysis of the homogenate of the organ with 6 N HCl and
extraction with benzene, following with slight modifications the procedure of Mitra et al. (19). Cold lipoic
acid as carrier was added from the beginning, and the recovery of the whole procedure was evaluated, by a gaschromatographic method, to be about 30 percent.

In Vivo Experiments

Experiments in vivo with various ^{35}S-compounds have
shown that higher label concentration, after 24 hours

intraperitoneal injection of the precursor, was found in
liver and kidney. In the liver the incorporation was higher
for cysteine, followed by methionine and cystamine (fig.
1 A), thiosulfate labeled in the outer sulfur is not incorporated at all. These preliminary data (G. Spoto et
al., manuscript in preparation) indicate that lipoic
acid can not be considered as a vitamin, being synthesized in the living mammalian cell from sulfur containing
precursors. However these data do not imply necessarely
that cysteine is also the immediate donor of sulfur. In
fact these in vivo experiments do not take into account
the dilution of the label in intra- and extracellular
pools. Furthermore cysteine, cystamine and methionine are
involved in very different metabolic pathways, and have
possibly different turnovers and half-lives. The influence of these parameters and of the not known turnover
rate of bound lipoic acid may justify the very low incorporation figures. Incorporation data in vivo have been
obtained also with ^{14}C-labeled compounds, as cysteine,
octanoic acid, butyric acid and acetic acid. Results of
a single experiment for each compound are shown in fig.
1 B. Octanoic acid and its metabolic precursor acetate
seem to be incorporated. Also other labeled compounds are
found in chromatography, and, in the case of octanoic
acid, the separation from lipoic acid turned out to be not
complete. Incorporation figures for butyric acid and
cysteine are purely indicative, because, owing to the very
low amount of radioactivity recovered, it was not possible
to obtain chromatographic evidence for lipoic acid formation. Cysteine is in any case a good donor of sulfur
atoms, not of carbon atoms. As already discussed, octanoic acid was found to be a good precursor also for E.
Coli (14).

     In all experiments lipoic acid has been identified by
thin layer chromatography on silica gel plates, by using
some of five different solvent systems (20, 21). With
cysteine as precursor only lipoic acid has been found as
final extraction product. With both methionine and cystamine as precursors, more than one component is recovered.
One component, accounting for 30-70% of the total radioactivity, migrates like a pure lipoic acid sample. The
other radioactive spots have not been identified. These

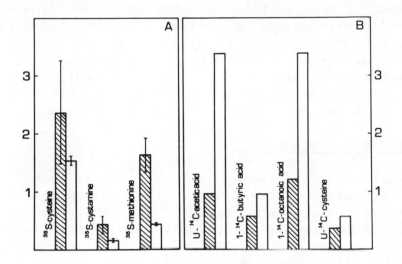

Fig. 1.   Incorporation of ^{35}S or ^{14}C of various precursors
into lipoic acid. Young Wistar rats (70-85 g) are injected intraperitoneally with labeled precursors. After 24
hours excised livers are homogenized in 6 N HCl (1:10),
refluxed for 4 hours and extracted with benzene (2x3 volumes). After washing with 0.01 N HCl and concentration
in vacuo, benzene is extracted with 20% $KHCO_3$ (2x3 vol);
the aqueous layer is acidified and extracted again with
2x3 vol benzene. The organic layer is concentrated in
vacuo: identification of lipoic acid is made with TLC on
silica gel plates with various solvent systems. Recovery
is given as follows:

$$\text{black bars} \quad = \quad \frac{mC_i \text{ extracted as lipoic acid}}{mC_i \text{ injected}} \times 10^5$$

$$\text{white bars A} \quad = \quad \frac{mC_i \text{ extracted as lipoic acid}}{g \text{ liver} \quad x \quad \frac{mC_i}{mmol} \text{ injected}} \times 10^7$$

$$\text{white bars B} \quad = \quad \frac{mC_i \text{ extracted as lipoic acid}}{g \text{ liver} \quad x \quad \frac{mC_i}{mmol} \text{ injected}} \times 10^8$$

A : ^{35}S-precursors, average of two experiments for
      each compound (⊢——⊣ = dispersion of data)

B : ^{14}C-precursors, single experiments.

results are compatible with two different interpretations.
i) Methionine and cystamine are incorporated not as well
as cysteine, and the heterogeneity of the benzene extract
is due to the presence of earlier stages of the biosynthetic pathway of lipoic acid; ii) cystamine and/or methionine are better precursors than cysteine, they are
incorporated faster into the lipoic acid molecule and the
other radioactive spots in the chromatography represent
degradation products of biosynthesized lipoate.

In Vitro Experiments

On the basis of the above in vivo experiments, in
vitro incubations at pH 7.6 with liver homogenates of
young rats were performed, with ^{35}S-cysteine as precursor.
The incorporation has been studied at various times and
results are given in fig. 2. The broken line represents
the radioactivity recovered in the final benzene extract
of a sample containing proteins inactivated by immersion
in boiling water for five minutes and incubated for 20
hours. Thin layer chromatography on silica gel plates
shows the presence, at short time, of only one component,
after longer incubation time and particularly after 20
hours incubation a second unidentified component appears.
The inactivated blank incubation mixture does not contain
lipoic acid. Incorporation values are much higher than
the in vivo experiments, about 10^5 times larger. This very
high increment may be explained by considering that cysteine is not any more diluted into pool(s). Also the
turnover rate of lipoic acid could be altered, being the
catabolism of lipoate apparently localized in the mitochondria.

The next step was a rough fractionation of the liver
homogenate into mitochondrial, microsomal and soluble
fraction. The crude mitochondrial and microsomal fractions
were prepared following the classical procedure (22);
the crude fractions were dissolved in 0.05 M Tris buffer,
pH 7.6. The supernatant soluble fraction was prepared,
after homogenization with 2 volumes of 0.02 M Tris buffer,
pH 7.6 containing 150 mM KCl and 5 mM $MgCl_2$, by centrifugation first at 7000xg for 20 min and then at 105,000xg
for 60 min. Table 1 reports, in terms of radioactivity

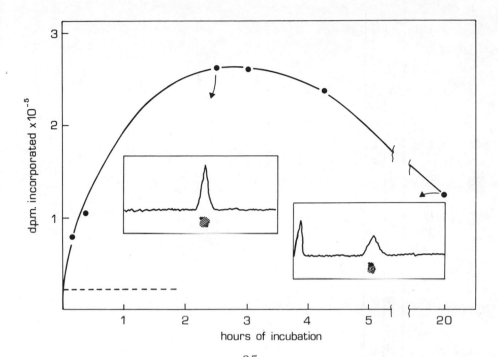

Fig. 2.   Incorporation of ^{35}S-cysteine into lipoic acid
after incubation with rat liver homogenate at
various times. Rat liver is homogenated with 3
volumes 0.02 M Tris buffer, pH 7.6; incubation
is performed with about 3 μC_i ^{35}S-cysteine
(110 mC_i/mmol), 145 mg protein at 37 °C. Incubations are stopped by adding HCl (6 N final)
and the mixtures are worked up as described in
fig. 1. Broken line gives the incorporation value of an incubation mixture at 20 hours with
proteins inactivated by immersion in boiling
water for 5 min: no lipoic acid is found with
TLC. Insets show TLC on silica gel plates (pe-
troleum ether 8: ether 2: acetic acid 1 as solvent) of incubation mixtures at 2.30 and 20
hours, monitored for radioactivity. Lined spots
are positive to the Folin reagent for disulfides
(23) and are due to lipoic acid added to the
homogenate as carrier.

Table 1. Incorporation of 35S from cysteine into lipoic acid with various sub-cellular fractions.

Fraction	Protein incubated mg	Label incorporation	
		$\dfrac{\mu Ci \text{ lipoic acid}}{\mu Ci \text{ incubated}}$	$\dfrac{\mu Ci \text{ lipoic acid}}{mg \text{ protein} \times \mu Ci \text{ incubated}} \times 10^3$
Homogenate	126.6	0.033	0.26
Cytosol	27.9	0.059	2.1
Mitochondrial	15.8	0.13	8.0
Microsomal	16.2	0.23	14.3
Cytosol + mitochondrial + microsomal	9.5 5.4 5.5	0.12	5.6
Cytosol + mitochondrial	13.9 7.9	0.039	1.8
Microsomal + mitochondrial	8.1 7.9	0.13	8.3
Cytosol + microsomal	13.9 8.1	0.12	5.5

Incubation for 3 hours at 37 °C, in 2 ml 50 mM Tris buffer, pH 7.6, with about 2.2 μCi 35S-cysteine (85 mCi/mmol). Reaction is stopped by adding HCl (6 N final); solutions are worked up as described in fig. 1.

recovered as lipoic acid in the final benzene extract,
results of the incubation of these three fractions with
^{35}S-cysteine. Incubations were performed with the homogenate as such and with the single fractions. Data obtained
by incubating a certain amount of the different fractions
two by two, and of all three fractions together are also
shown. These preliminary data indicate a higher activity
to be present in the microsomal fraction. The presence
of incorporating activity also in the other fractions
could be attributed simply to cross contaminations. Particularly the mitochondrial fraction may be contaminated
by microsomal proteins.

The conclusions arising from the above results may
be summarized as follows. Lipoic acid is not a vitamin for
the rat; it is synthesized by enzymatic systems present
in the hepatic cells, probably in the microsomal fraction.
Sulfur atoms of lipoic acid arise from cysteine, either
directly or through intermediates not yet identified.
Also cystamine and methionine can act, to a lesser extent,
as sulfur precursors, apparently via a metabolite shared
with cysteine. As far as the carbon backbone is concerned,
it does not originate from cysteine; the origin from acetate and, apparently, from octanoate seems to be in agreement with Reed's data for E. Coli (14).

References

1) M. Koike and K. Koike, "Lipoic acid" in:"Metabolism
    of sulfur compounds" ("Metabolic Pathways", vol. VII,
    III ed., D.M. Greenberg, ed.) pp. 87-99, Academic
    Press, New York (1975).
2) L.J. Reed, in "Comprehensive Biochemistry" (M. Flor-
    kin and E.H. Stotz, eds.), vol. 14, pp. 99-126, El-
    sevier, Amsterdam (1966).
3) U. Schmidt, P. Grafen, K. Altland and H.W. Goedde,
    Advan. Enzymol. 32:423-469 (1969).
4) L.J. Reed and D.J. Cox, Ann. Rev. Biochem. 35:57-84
    (1966).
5) P.C. Jocelyn, "Biochemistry of the SH group", p. 10,
    Academic Press, New York (1972).
6) E.L.R. Stokstad, H.P. Broquist and E.L. Patterson,
    Fed. Proc. 12:430 (1953).

7)   J.Y. Spence and D.B. McCormick, Arch. Biochem. Bio-
     phys. 174:13-19 (1976).

8)   E.H. Harrison and D.B. McCormick, Arch. Biochem.
     Biophys. 160:514-522 (1974).

9)   E.M. Gal and D.E. Razevska, Arch. Biochem. Biophys.
     89:253-261 (1960).

10) V.N. Totskii, Biochemistry (russ.) 41:894-903 (1976).

11) J.N. Tsunoda and K.T. Yasunobu, Arch. Biochem. Bio-
     phys. 118:395-401 (1967).

12) J.P. Carreau, D. Lapous and J. Raulin, Biochimie 59:
     487-496 (1977).

13) Y. Nose, personal communication, cited in "Metabolism
     of sulfur compounds" ("Metabolic Pathways", vol. VII,
     III ed., D.M. Greenberg, ed.), p. 91, Academic Press,
     New York (1975).

14) L.J. Reed, T. Okaichi and J. Nakanishi, Abst. Int.
     Symp. Chem. Natur. Prod., p. 218, Kyoto (1964).

15) R.J. Parry, J. Am. Chem. Soc. 99:6464-6466 (1977).

16) R.J. Parry and D.A. Trainor, J. Am. Chem. Soc. 100:
     5243-5244 (1978).

17) H. Nawa, W.T. Brady, M. Koike and L.J. Reed, J. Am.
     Chem. Soc. 82:896-903 (1960).

18) H. Grisebach, R.C. Fuller and M. Calvin, Biochim.
     Biophys. Acta 23:34-42 (1957).

19) S.K. Mitra, R.K. Mandal and D.P. Burma, Biochim. Bio-
     phys. Acta 107:131-133 (1965).

20) D.P. Kelly, J. Chromatogr. 51:343-345 (1970).

21) P.R. Brown and J.O. Edwards, J. Chromatogr. 43:515-
     518 (1969).

22) H.R. Mahler and E.M. Cordes "Biological Chemistry",
     II ed., p. 448, Harper and Row Int. Edition (1971).

23) K. Shinohara, J. Biol. Chem. 109:665-670 (1935).

SOME ASPECTS OF THE METABOLISM OF SULFUR-CONTAINING HETEROCYCLIC COFACTORS: LIPOIC ACID, BIOTIN, AND 8α-(S-L-CYSTEINYL)RIBOFLAVIN

Donald B. McCormick*

Divisions of Nutritional and Biological Sciences
Cornell University
Ithaca, New York 14853 (USA)

INTRODUCTION

The occurrence of sulfur in certain of the essential cofactors, viz. some vitamins and their derived coenzymes, has led to interesting avenues of investigation. Participation of the disulfide/sulfhydryl functions of lipoyl residues in transacylases is generally delineated, but details of the metabolism of the dithiolanyl ring are lacking. In the case of biotin, it is not clear what specific mechanistic advantage relating to carboxylations may be gained from the presence of the thiolanyl sulfur, though aspects of the formation and breakdown of this anchor point are being clarified. The converse is currently the situation with S-linked flavins where the effect on their redox properties is better understood than their metabolism.

It will be the purpose of this chapter to provide an overview of our knowledge on the catabolism of three S-containing heterocyclic cofactors, viz. lipoate, biotin, and the 8α-(S-L-cysteinyl)riboflavin, with emphasis on the fate of the molecular sulfur; others will deal with the more-biosynthetic aspects of lipoic acid and biotin.

LIPOIC ACID

The structure of natural α-(+)-lipoic acid (thioctic, 6,8-dithiooctanoic, or 1,2-dithiolane-3-valeric acid) is as follows:

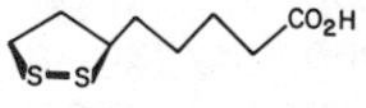

As has been summarized elsewhere,[1] the specific information now available on the metabolism of this cofactor largely derives from

*Present address: Dept. Biochem., Emory Univ., Atlanta, GA 30322.

studies made with the racemic [^{14}C]compound supplied to the bacterium *Pseudomonas putida* LP[2-6] and the rat.[7,8] The bacteria were usually cultured under aerobic conditions at 30°C in a neutral trace elements-salts solution with 0.4% *dl*-lipoate as sole source of carbon and sulfur and 0.2% NH$_4$NO$_3$ for nitrogen. Animals used were young, male Sprague-Dawley rats maintained on a pelleted commercial diet and administered lipoate at the level of 0.5 mg/100 g body weight.

Uptake, Utilization, and Metabolism in the Pseudomonad

In lipoate-grown cells that were preincubated for 15 minutes to deplete internal substrate (Fig. 1A), the rate and extent of uptake followed: acetate > octanoate > lipoate > lipoate thiolsulfinate.[3,5] However, preincubation of the cells with unlabeled acetate (Fig. 1B) markedly enhanced the uptake of either lipoate or its thiolsulfinate, reflecting the energy-dependent nature of the process.[5] Under these latter conditions, the poor uptake of dihydrolipoate was only modestly enhanced. The uptakes of these substrates were similar and very low when cells were grown in 0.4% glucose medium,[3] since the glyoxylate cycle, used by the pseudomonad efficiently to metabolize the acetyl-CoA derived from β-oxidation of lipoate and other alkanoates, was then subject to catabolite repression. Although the constitutive acyl-CoA synthetase remained fairly active, the two glyoxylate-cycle enzymes, isocitrate lyase and malate synthase, were found to be repressed by 94 and 85%, respectively, when compared to activities measured in extracts from acetate-grown cells, but was moderately active when octanoate or lipoate provided the C source.[3]

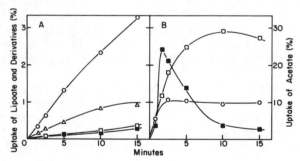

Fig. 1.   Uptake of ^{14}C-labeled acetate (○), octanoate (Δ), lipoate (□), and lipoate thiolsulfinate (■) by bacterial cells that were starved (A) or preincubated with 0.1 mM acetate (B).

The rates of substrate oxidation measured in washed-cell suspensions reflected good agreement between oxygen uptake and carbon dioxide production.[3] Octanoate was a more favorable substrate than lipoate. Among shorter chain fatty acids, oxidation rate decreased with decrease in chain length; however, acetate was still oxidized more rapidly than glucose, α-glyceryl phosphate, and succinate.[2] Among metabolites of lipoate, the thiolsulfinate was oxidized less readily than lipoate but, at higher concentrations, almost as well as acetate.[5] The bisnor (4,6-dithiohexanoate or 1,2-dithiolane-3-propano-

ate) and tetranor (2,4-dithiobutanoate or 1,2-dithiolane-3-carboxyl-
ate) catabolites had comparatively low rates of oxidation.[3] In intact or sonicated bacterial cells, production of $^{14}CO_2$ from [1-^{14}C]-
octanoate was faster than from [1,6-^{14}C]lipoate. These results indicate that a high rate of β-oxidation occurs in the lipoate-grown
pseudomonad, but the efficiency of this process is significantly decreased by the presence of the S-containing ring system, especially
after removal of the first acetyl-CoA. When the cells were grown on
lipoate, acyl-CoA synthetase and enzymes of the β-oxidative pathway
and glyoxylate cycle are used to activate, derive, and utilize acetate as the main growth substance. This is also reflected in the loss
of ^{14}C from [1,6-^{14}C]lipoate in the medium, which plateaued at approximately 50%, corresponding to the formation predominantly of the two-
carbon shorter bisnorlipoate with remaining label at position 4.[2,3]

The ability of the pseudomonad to derive S from lipoate, its metabolites, and other organic and inorganic salts has also been determined.[3-5] Best growth was obtained when $(NH_4)_2SO_4$ supplied both N
and S to the medium containing a readily oxidized C source, such as
octanoate (Fig. 2A).[3] Cystine also served as an excellent source of
S. Less satisfactory growth was obtained with lipoate as the C
source, as expected on the bases of poorer uptake and subsequent β-
oxidation. Growth became somewhat better when both S and C had to be
derived from lipoate with NH_4NO_3 to supply N. Moreover, d-, l-, or
dl-lipoate serve equally well as C and S sources. There was insufficient degradation of bisnor- and tetranorlipoate and of the thiolsulfinates of lipoate and bisnorlipoate even to permit growth. Among
synthetic analogs tested with NH_4NO_3 (Fig. 2B), both C and S could be
derived from 6,8-dithiononanoate (8-methyllipoate), which supported
growth as well as lipoate; however, 6,9-dithiononanoate, which is a
six-membered cyclic disulfide, was inactive. These results indicate
that the dithiolane, but not dithiane, ring can be degraded independent of β-oxidative cleavage of the side chain to yield C and that
the latter becomes more difficult after removal of the first acetyl-
CoA from a five-carbon side chain. The finding that much smaller
amounts of β-hydroxybisnorlipoate and tetranorlipoate than bisnor-

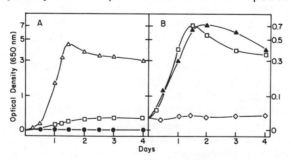

Fig. 2.   Growth of bacteria on 0.2% octanoate (△), lipoate (□), and
          catabolites (●) as C sources with 0.2% $(NH_4)_2SO_4$ (A); or on
          lipoate (□), 6,8-dithiononanoate (▲), and 6,9-dithiononano-
          ate (◇) as C and S sources with 0.2% NH_4NO_3 (B).

lipoate were found in culture filtrates of the lipoate-grown pseudo-monad supports these conclusions.[4] Growth experiments with the iso-lated and synthetic thiolsulfinates of lipoate and bisnorlipoate revealed that these compounds are "dead-end" metabolites, since they cannot adequately supply C and only poorly serve as S sources.[5]

Overall, the LP strain of *P. putida* appears unique in its ability not only to β-oxidize fatty acid chains, including the one of lipo-ate, but also degrade in ways not yet detailed sufficient quantities of the dithiolane ring to derive the S.

As mentioned, several metabolites of lipoate have been isolated from cultures of *P. putida* LP.[2-6] Procedures have involved the use of [^{14}C]lipoate labeled in positions 1 and 6 or 7 and 8 as source of C and S during growth, followed by solvent extractions and hydrophobic gel filtration or anion-exchange chromatographies of the culture supernatants. Purified compounds were characterized by chromato-graphic mobilities and spray reactions and by UV, IR, PMR, and mass spectrometries. When appropriate, comparisons were made with syn-thetic lipoic acid, the bisnor and tetranor analogs, and their thiol-sulfinates.[9,10] A flow scheme of lipoate catabolism based on com-pounds isolated (structures), as well as inferred (names in brack-ets), is shown below.

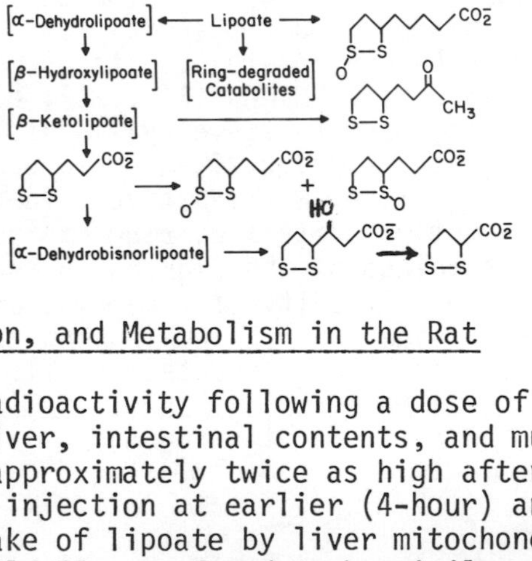

Uptake, Utilization, and Metabolism in the Rat

Localization of radioactivity following a dose of [1,6-^{14}C]lipoate was greatest in liver, intestinal contents, and muscle.[7] Retention of the label was approximately twice as high after per os adminis-tration than i.p. injection at earlier (4-hour) and later (24-hour) periods. The uptake of lipoate by liver mitochondria and subsequent β-oxidation to yield CO_2 was found to be similar to shorter-chain fatty acids, e.g. octanoate, rather than longer-chain ones, e.g. palmitate. Lipoate and octanoate were relatively insensitive to additions of *dl*-carnitine to supply the *l*-substrate for the long-chain acyltransferase upon which mitochondrial translocation of pal-mitate is dependent. Moreover, (+)-decanoylcarnitine inhibited the latter but had no effect on the short-chain compounds.

Rates of $^{14}CO_2$ formation from [1,6-^{14}C]lipoate with isotonic liver homogenates were expectedly depressed by the inclusion of equimolar unlabeled octanoate > palmitate. Also, the efficiency of oxidation of lipoate, greater in the intact animal, was considerably higher with the 1,6- than 7,8-labeled compound.[8] As was found with the pseudomonad, β-oxidative cleavage of the two-carbon terminus of the side chain is much more extensive than oxidative removal of the relatively resistant carbons at the dithiolane ring.

Following i.p. injection of dl-[1,6-^{14}C]lipoate, 56% of the radioactivity was recovered in urine within 24 hours.[8] When acidified and extracted with benzene, 92% remained in the aqueous phase, which contained at least ten [^{14}C]compounds when fractionated by anion-exchange chromatography (Fig. 3). Two of these, fractions 6 and 10, were identified as lipoate and β-hydroxybisnorlipoate, respectively. The 8% of urinary radioactivity extracted into benzene was fractionated by hydrophobic gel filtration (Fig. 4) followed by paper chromatography. Compounds identified were (1) lipoate, (2) bisnorlipoate, and (3) tetranorlipoate plus a keto compound.

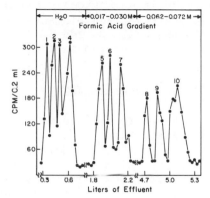

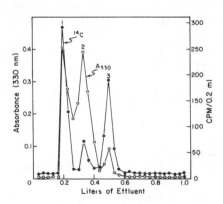

Fig. 3. Elution pattern of urinary [^{14}C]metabolites in the aqueous phase on Dowex 1-X2 (formate).

Fig. 4. Elution pattern of urinary [^{14}C]metabolites in the benzene phase on Sephadex LH-20 eluted with chloroform.

When the 81% of total radioactivity found in urine from rats that had received dl-[7,8-^{14}C]lipoate was similarly fractionated, 72% remained in the aqueous phase and 28% extracted into benzene. Elution patterns of metabolites were generally similar to those found with the 1,6-labeled compound. Overall, it appears that the rat, similar to *P. putida* LP, metabolized lipoate mainly via β-oxidation of the valeric acid side chain. In contrast to the catabolites found in culture media from the pseudomonad, however, a larger amount of more-polar metabolites are excreted by the rat. Much of this difference in amount rather than kind of metabolite is undoubtedly related to the different amounts of lipoate supplied to the organisms. The bacteria must derive the considerable C and energy needs, as well as lesser amounts of S, from lipoate, as can most easily be accomplished by the first β-oxidative cleavage to yield the benzene-soluble bisnorlipoate with only a little degradation of the dithiolane ring.

The animals derive such needs from ingested food but, at least, have
the nonspecific β-oxidative machinery to effect degradation of the
fatty acid portion of this S-containing cofactor.

BIOTIN

The structure of natural d-biotin (cis-tetrahydro-2-oxothieno[3,4-
d]imidazoline-4-valeric acid) is as follows:

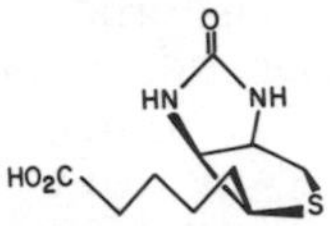

Similar to lipoate, most information on the metabolism of biotin, as
has been summarized elsewhere,[11,12] derives from our investigations
using the radioactive compound or homologs supplied as sole sources
of C, N, and S to a Pseudomonas sp.[13-20] or after administration of
small quantities to rats.[21-23] The bacteria were usually cultured
under aerobic conditions at 28°C in a neutral trace elements-salts
solution with 0.3% d-biotin as sole source of C, N, and S. Animals
used were young, male Sprague-Dawley rats fed biotin-sufficient or
-deficient diets and administered biotin at levels of 2 µg to 2 mg.

Uptake, Utilization, and Metabolism in the Pseudomonad

In biotin-grown cells that were preincubated to deplete internal substrate (Fig. 5), the initial rate of transport followed: biotin sulfone > biotin l-sulfoxide > biotin d-sulfoxide > biotin > dethiobiotin.
The greatest extents of uptake were achieved with the l-sulfoxide and
sulfone. Essentially the same behavior was found with cells grown on
d-sulfoxide, which is utilized after reduction to biotin.[18] These
results can be contrasted with those on rates of growth and loss of
radioactivity. The pseudomonad could not grow on the sulfone or l-
sulfoxide alone.[17,19] The sulfone lost no [14C] label from the carboxyl
or ureido carbonyl carbons,[14] although β-oxidation was initiated as
evidenced by isolation of radioactive β-hydroxybiotin sulfone from
culture filtrates of the organism grown on the [14C]sulfone plus unlabeled biotin.[19] The l-sulfoxide could be degraded in side chain and
ureido portions,[14] but again metabolic stricture was seen in accumulation of the β-hydroxy metabolite.[17] Growth and loss of ureido carbonyl label is slower on the d-sulfoxide than biotin (Fig. 6).[18] In
addition to the reduction of d-sulfoxide to form biotin and those
catabolites from side chain and ureido ring shown previously to arise
from the vitamin,[16,19] some further oxidation to the sulfones of
biotin and bisnorbiotin was also found.[18]

Isolation and physico-chemical characterization of the α-dehydro and
β-hydroxy compounds, methyl ketones derived from β-keto acids,[19] and
two- and four-carbon shorter catabolites[16] verified the early sequential β-oxidative cleavage of the valerate side chain of biotin. Also

we have established the similar breakdown of homo to nor to trisnor
analogs by particulate preparations of the pseudomonad,[15] and even
degradation of dethiobiotin to two- and four-carbon shorter side
chain catabolites by such other organisms as *Aspergillus niger.*[24]

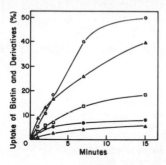 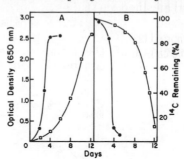

Fig. 5.   Bacterial uptake of [14]C-   Fig. 6.   Growth of bacteria (A)
          labeled *l*-sulfoxide   and loss of [14]C (B) on
          (o), sulfone (Δ), *d*-sul-   media containing 0.3%
          foxide (□), biotin (●),   ^{14}C-labeled biotin (●)
          and dethiobiotin (▲).   or the *d*-sulfoxide (□).

Degradation of the 2-ketoimidazole portion of biotin appears to occur
after initial β-oxidative removal of the first acetyl-CoA to form
bisnorbiotin but before metabolic alteration of the thiolane portion.
The tetranor catabolite was found metabolically inert and tended to
accumulate.[16]   Compounds with the bicyclic ring system of *d*-biotin but
with even-numbered acid side chains, i.e. *d*-homo- and *d*-norbiotins,
were not subject to loss of the ureido carbonyl;[14] they were β-oxidized
to the carboxymethyl of *d*-trisnorbiotin rather than the carboxyethyl
function of bisnorbiotin.[15]   Also, absence of the thiolanyl ring, i.e.
in dethiobiotin, did not allow significant degradation of the ureide
portion.[15]   Moreover, *d*-allobisnorbiotin, which must have been formed
by cleavage of the *cis*-ureido ring at the bridgehead carbon opposite
side chain, followed by de- and recarbamylations to form the *trans*
isomer, was found to have lost most of the original ureido ^{14}C but
retained ^{3}H in the side chain and thiolane portions.[20]   Some N as
carbamyl phosphate probably generated at this stage was used for biosynthetic steps, as free uracil containing relatively high radioactivity had been found to accompany metabolite formation.[16]   A second
oxidation at the bridgehead carbon adjacent the side chain allowed
release of urea, which retained the original ^{14}C label.[16]   The urea
was degraded to CO_2 and NH_3, which increased in the medium during
total degradation of biotin.[13]

Although some oxidation of biotin to form biotin sulfoxides had been
noted,[16,18] and the sulfones of biotin and bisnorbiotin were isolated
from cultures of the bacteria grown on the *d*-sulfoxide,[18] growth experiments indicated that the thiolanyl S must be at the level of sulfide to be extirpated.   As with lipoate, details of the metabolic
events that lead to release of S fragments are lacking; however, we
have recently isolated methyl thioacetate as a volatile metabolite.

This compound must result from the intact S plus adjacent methylene of the thiolane moiety or possibly via methylation of H_2S formed. Much of the sulfide could be accounted for as $(NH_4)_2S$, which accumulated in sufficient quantity to impart a yellowish color to vigorously aerated cultures. Both elemental sulfur and sulfate[13] also appeared in the culture medium.

A composite of the primary events and their general sequence with the total degradation of biotin is shown below.

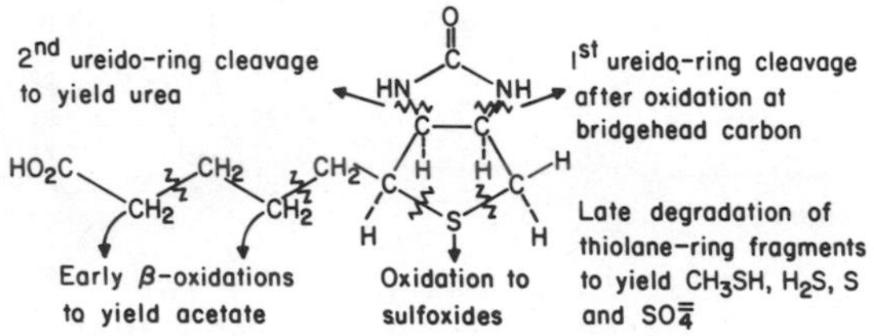

Uptake, Utilization, and Metabolism in the Rat

When [^{14}C-ureido]biotin was injected i.p. into rats, most was excreted in the urine within 24 hours.[21] The initial excretion rate decreased with decreasing amounts of biotin administered. Urine collected during days 0 to 5 and 11 to 14 from biotin-depleted, antibiotic-treated animals that received a single injection of 0.5 μg of [^{14}C]biotin/100 g body weight was fractionated for [^{14}C]catabolites. Similar fractionation was done for liver homogenates incubated with 1 μg/g tissue under conditions of limiting or excess O_2. Results, summarized in Table 1, demonstrate that most of the radioactivity can be accounted for by remaining biotin plus its d- and l-sulfoxides, bisnorbiotin, and the methyl bisnorbiotinyl ketone as a major component of neutral metabolites.

Table 1.   Comparison In Vivo and In Vitro of the Metabolism of Biotin in Rats

| [^{14}C]Compounds | Radioactivity Recovered (%)[a] | | | |
| | In Vivo (Urine) | | In Vitro (Homogenate) | |
	Early	Late	Excess O_2	Limited O_2
Neutral	3.4	trace	14.0[b]	4.0[b]
Biotin l-sulfoxide	2.5	trace	15.0	6.5
Biotin d-sulfoxide	6.6	trace	27.0	16.5
Bisnorbiotin	37.5	84.0	trace	4.0
Biotin	25.5	4.0	4.0	42.0
Total	75.5	88.0	60.0	73.0

[a]Percentage of total radioactivity prior to isolation.
[b]Includes urea as well as a neutral ketone.

The percent of the ^{14}C represented by bisnorbiotin increased relative to biotin as the total radioactivity in urine decreased exponentially with time. Aeration of homogenates expectedly increased the amount of sulfoxides. Also, trace amounts of $^{14}CO_2$ and [^{14}C]urea were detected in vitro but were not found with intact animals. The rat, therefore, has the ability to metabolize biotin via at least partial β-oxidation of the side chain and to effect some oxidation of the thioether sulfur to sulfoxides. Under physiological conditions, however, the metabolism of this vitamin in rats, unlike the pseudomonad, does not include extensive cleavage of the ring structure.

Additional studies[22] have shown that [^{14}C]biotin injected i.p. into rats as the avidin complex was excreted much more slowly than the free vitamin. Nevertheless, the biotin-avidin complex was dissociated in vivo and the released biotin excreted and metabolized to sulfoxides and bisnorbiotin. That liver has the capacity to cause such dissociation and metabolism was demonstrated. The rate and extent of excretion of bisnorbiotin is essentially the same as for biotin.[23] The more water-soluble tetranorbiotin, both d- and l-sulfoxides of biotin, and biotin sulfone were excreted even more rapidly.

8α-(S-L-CYSTEINYL)RIBOFLAVIN

The structure of an N-acylated 8α-(S-L-cysteinyl)riboflavin, which is known to occur at the FAD level within the active sites of mammalian monoamine oxidase and a *Chromatium* cyt c_{552} reductase, is as follows:

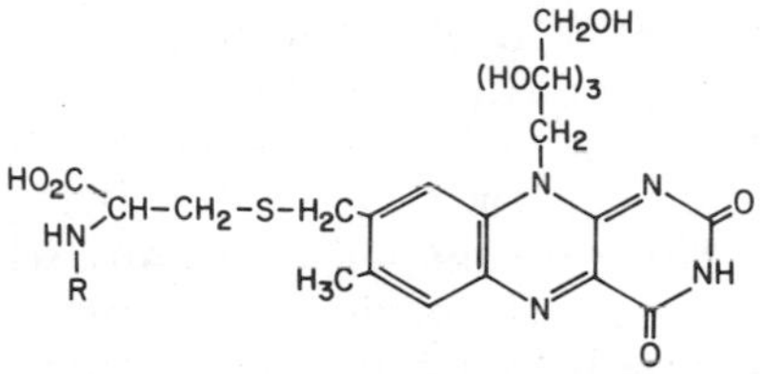

As there was no information on the absorption, metabolism, and excretion of 8α-(amino acid)riboflavins, which arise through digestion and by turnover of flavoproteins that contain them, studies were made on the S-cysteinyl-flavin in rats.[25] The flavin moiety was traced by radiolabeling the relatively stable 2-position in the isoalloxazine. The cysteinyl portion was N-acetylated to prevent the too facile sulfoxidation that occurs nonenzymatically in aerobic, aqueous solutions of the flavinyl compound with a free α-amino function.

The distribution of radioactivity in tissues and intestinal contents one hour after administration of 8α-[S-(N-acetyl)-L-cysteinyl]-D-[1-^{14}C]riboflavin is summarized in Table 2. That at least the flavinyl moiety must be absorbed from the gastrointestinal tract was indicated by the rapid and sequential appearance of a fraction of ^{14}C in those tissues, such as liver, that similarly reflect the uptake of [2-^{14}C]riboflavin. The rather rapid appearance of ^{14}C in kidney, especially after i.p. injection, was also typical for flavin localization and preceded excretion of a considerable fraction of the label. The considerable radioactivity in the small intestine and its contents after i.p., as well as per os, administration reflects enterohepatic circulation. Only traces of ^{14}CO$_2$ were exhaled within even

Table 2.   Distribution of ^{14}C One Hour After Administration of
       8α-[S-(N-Acetyl)-L-cysteinyl]-D-[2-^{14}C]riboflavin[a]

Material	Per os	I.P.
Blood	0.8	0.5
Liver	2.2	0.6
Kidney	0.9	4.3
Small intestine	5.7	4.9
Contents of small intestine	21.0	15.4
Cecum and large intestine	1.5	0.9
Contents of cecum and large intestine	0.9	0.7

[a]Values expressed as average % ^{14}C administered.

longer times after administration by either route.

At physiological levels, the cysteinyl flavin did not adequately replace or significantly antagonize the vitamin in deficient or normal
rats. Excretion of the N-acetylcysteinyl riboflavin and its metabolites was competitive with riboflavin; urinary loss was influenced by
the amount administered, either separately or together with vitamin.

Metabolic alteration of a portion of the amino acid-flavin was reflected by [^{14}C]compounds excreted in urine following per os administration. As characteristic of flavins, a considerable fraction
(55%) of the urinary ^{14}C could be extracted by phenol and redissolved
in water prior to chromatographic separation of components. Results
of anion-exchange column chromatography (Fig. 7) revealed that a considerable portion of the ^{14}C washed through, although N-acetylcysteinyl riboflavin and its sulfoxides were retained. This suggests
that the original carboxylate group has been lost.

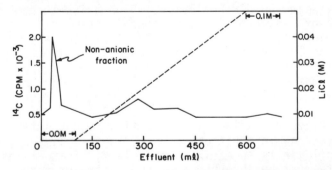

Fig. 7.   Elution pattern of phenol-extracted urinary [^{14}C]metabolites
         of 8α-[S-(N-acetyl)-L-cysteinyl]-D-[2-^{14}C]riboflavin on DEAE-
         cellulose (chloride).

The nonanionic fluorescent [^{14}C]metabolites derived from N-acetylcys-
teinyl-[^{14}C]riboflavin migrated on paper chromatograms in a manner
expected for decarboxylated rather than deacylated catabolites. This
was verified by absence of reactivities with both acid indicator and
ninhydrin sprays.

Considerable oxidation of the thioether linkage within the N-acetyl-cysteinyl-flavins led to formation of the more-fluorescent sulfoxide forms, which were formed as a result of incubations in vitro as well as occurred in urine. However, no significant deacylation of the N-acetyl function could be detected with homogenates found active on N-acetyl-L-cysteine. Purified acylase I cannot catalyze hydrolysis of the acetyl group from the flavin-attached amino acid.

ACKNOWLEDGMENTS

The work described was supported by Research Grants AM-04585 and AM-08721 from the National Institutes of Health, USPHS. Appreciation is also felt for coworkers who obtained most of the results on the metabolism of lipoate (J.C.H. Shih, M.L. Rozo, H.-H. Chang, and H.C. Furr), biotin (R.N. Brady, H. Ruis, S. Iwahara, H.C. Li, J.A. Roth, W.B. Im, M.N. Kazarinoff, and J.M. Westendorf), and the 8α-[S-(N-acetyl)-L-cysteinyl]riboflavin (C.P. Chia).

REFERENCES

1. E.H. Harrison and D.B. McCormick, Chemistry and physiology of nutrients and growth regulators: Riboflavin, *in*: "CRC Handbook of Nutrition and Food", M. Rechcigl, Jr., ed., CRC Press, Cleveland (1979), in press.
2. J.C.H. Shih, L.D. Wright, and D.B. McCormick, Isolation, identification, and characterization of a lipoate-degrading pseudomonad and of a lipoate catabolite, *J. Bacteriol.* 112:1043 (1972).
3. J.C.H. Shih, M.L. Rozo, L.D. Wright, and D.B. McCormick, Characterization of the growth of *Pseudomonas putida* LP on lipoate and its analogues: Transport, oxidation, sulphur source, and enzyme induction, *J. Gen. Microbiol.* 86:217 (1975).
4. H.-H. Chang, M.L. Rozo, and D.B. McCormick, Lipoate metabolism in *Pseudomonas putida* LP, *Arch. Biochem. Biophys.* 169:244 (1975).
5. H.C. Furr, H.H. Chang, and D.B. McCormick, Lipoate metabolism in *Pseudomonas putida* LP: Thiolsulfinates of lipoate and bisnorlipoate, *Arch. Biochem. Biophys.* 185:576 (1978).
6. H.C. Furr and D.B. McCormick, Bacterial catabolism of lipoic acid. Isolation and identification of a methyl ketone, *Internat. J. Vit. Nutr. Res.* 48:165 (1978).
7. E.H. Harrison and D.B. McCormick, The metabolism of dl-[1,6-^{14}C]-lipoic acid in the rat, *Arch. Biochem. Biophys.* 160:514 (1974).
8. J.T. Spence and D.B. McCormick, Lipoic acid metabolism in the rat, *Arch. Biochem. Biophys.* 174:13 (1976).
9. J.C.H. Shih, P.B. Williams, L.D. Wright, and D.B. McCormick, Properties of lipoic acid analogs, *J. Heterocycl. Chem.* 11:119 (1974).
10. H.C. Furr, J.C.H. Shih, E.H. Harrison, H.-H. Chang, J.T. Spence, L.D. Wright, and D.B. McCormick, Chromatographic and spectral properties of lipoic acid and its metabolites, *in*: "Methods in Enzymology, Vitamins and Coenzymes", D.B. McCormick and L.D. Wright, eds., Vol. 62D, pp. 129-135, Academic Press, New York (1979).

11. D.B. McCormick and L.D. Wright, The metabolism of biotin and analogues, *in*: "Comprehensive Biochemistry, Metabolism of Vitamins and Trace Elements", M. Florkin and E.H. Stotz, eds., Vol. 21, pp. 81-110, Elsevier, Amsterdam (1971).
12. D.B. McCormick, Biotin, *Nutr. Rev.* 33:97 (1975).
13. R.N. Brady, L.-F. Li, D.B. McCormick, and L.D. Wright, Bacterial and enzymatic degradation of biotin, *Biochem. Biophys. Res. Commun.* 19:777 (1965).
14. R.N. Brady, H. Ruis, D.B. McCormick, and L.D. Wright, Bacterial degradation of biotin. Catabolism of ^{14}C-biotin and its sulfoxides, *J. Biol. Chem.* 241:4717 (1966).
15. H. Ruis, R.N. Brady, D.B. McCormick, and L.D. Wright, Bacterial degradation of biotin. II. Catabolism of ^{14}C-homobiotin and ^{14}C-norbiotin, *J. Biol. Chem.* 243:547 (1968).
16. S. Iwahara, D.B. McCormick, L.D. Wright, and H.-C. Li, Bacterial degradation of biotin. III. Metabolism of ^{14}C-carbonyl-labeled biotin, *J. Biol. Chem.* 244:1393 (1969).
17. J.A. Roth, D.B. McCormick, and L.D. Wright, Bacterial degradation of biotin. IV. Metabolism of ^{14}C-carbonyl-labeled biotin *l*-sulfoxide, *J. Biol. Chem.* 245:6264 (1970).
18. W.B. Im, J.A. Roth, D.B. McCormick, and L.D. Wright, Bacterial degradation of biotin. V. Metabolism of ^{14}C-carbonyl-labeled biotin *d*-sulfoxide, *J. Biol. Chem.* 245:6269 (1970).
19. M.N. Kazarinoff, W.B. Im, J.A. Roth, D.B. McCormick, and L.D. Wright, Bacterial degradation of biotin. VI. Isolation and identification of β-hydroxy and β-keto compounds, *J. Biol. Chem.* 247:75 (1972).
20. W.B. Im, D.B. McCormick, and L.D. Wright, Bacterial degradation of biotin. Isolation and identification of *d*-allobisnorbiotin, *J. Biol. Chem.* 248:7798 (1973).
21. H.-M. Lee, L.D. Wright, and D.B. McCormick, Metabolism of carbonyl-labeled ^{14}C-biotin in the rat, *J. Nutr.* 102:1453 (1972).
22. H.-M. Lee, L.D. Wright, and D.B. McCormick, Metabolism, in the rat, of biotin injected intraperitoneally as the avidin-biotin complex, *Proc. Soc. Exp. Biol. Med.* 142:439 (1973).
23. H.-M. Lee, N.E. McCall, L.D. Wright, and D.B. McCormick, Urinary excretion of biotin and metabolites in the rat, *Proc. Soc. Exp. Biol. Med.* 142:642 (1973).
24. H.-C. Li, D.B. McCormick, and L.D. Wright, Metabolism of dethiobiotin in *Aspergillus niger*, *J. Biol. Chem.* 243:4391 (1968).
25. C.P. Chia, R. Addison, and D.B. McCormick, Absorption, metabolism, and excretion of 8α-(amino acid)riboflavins in the rat, *J. Nutr.* 108:373 (1978).

IMPLICATION OF ESSENTIAL FATTY ACIDS IN PYRUVATE DEHYDROGENASE ACTIVATION

J.Raulin, M.Launay, C.Loriette, D.Lapous & M.Bouchène
Nutrition Cellulaire & Lipophysiologie, C.N.R.S.
Université Paris 7, 2 place Jussieu, PARIS 05,
France

INTRODUCTION

Recent observations (Carreau et al.,1977) allowed a parallel to be drawn with the hypothesis concerning the possible participation of the linoleic acid in the pyruvate dehydrogenase complex (PDHc) .

The fatty acid belonging to the n 6 family, derived from linoleic acid, all have an "essential fatty acid" (EFA) biological activity. It is at present thought that this activity results from two major functions :

1/ EFAs are fundamental components for membrane construction; 2/ they are precursors of the prostaglandins and thromboxanes .

Studies carried out in our laboratory (Raulin et al.,1974) nevertheless suggest the need to look for other explanations which might explain the indispensable character of linoleic acid. The as yet unexplained increase in vivo in nuclear and mitochondrial DNA polymerase activity which was seen in adipose tissue enriched for linoleic acid (Launay et al.,1968,1969) has encouraged us to study the role of metabolic products of the EFAs in certain fundamental reactions of intermediary metabolism, other than that involving lipid components .

In order to examine the metabolic pathways of the various molecular products, we have first chosen to use,as tracer, uniformly radio-labelled linoleic acid. We have followed the fate of the radioactivity in water soluble and nucleo-

protein fractions after injection of $[U-^{14}C]$ linoleic
acid, $[U-^{14}C]$ oleic acid being used as control .

Analysis of the chloroform insoluble polar fractions, has
demonstrated the presence of several constituents showing
greater radioactivity after linoleic than after oleic acid
injection, the constituents being neither amino-acids, nor
prostaglandins, nor prostaglandin breakdown products.
Silica gel chromatography of the products obtained after
treatment of lipoic acid with 6 N HCl has demonstrated
that certain of the breakdown products show the same Rf
values as similarly chloroform insoluble, radioactive products obtained after treatment of the radio-labelled samples. The latter have been shown by the use of palladium
chloride to contain sulfur, thus suggesting that they are
derived by the action of 6 N HCl hydrolysis from lipoic
acid .

It does not, therefore, seem impossible that a relationship exists between the EFAs and lipoic acid, a hypothesis
supported by the appreciable improvement in the health of
rats on a lipoamide supplemented diet compared to the
strictly lipid deprived controls (Carreau et al.,1975).
According to this hypothesis, the EFAs would function as
lipoic acid precursors by a metabolic pathway as yet unknown, and which could be continued in animals by the
synthetic pathway described for E.Coli by Reed (1964)
and Parry (1977) . Confirmation of the existence of a
metabolic pathway used by linoleic acid derived fatty
acids, or in their absence by other fatty acids which it
has occasionally been suggested are essential, necessita-
tes additional experimentation .

The aim of the present experiment was to study the PDHc
activity in the livers and brains of developing rats born
to females kept on a fat-free diet from 10 days after mating. The question arises in the mechanism of action of
the supplemental linoleic acid given to the progeny after
weaning on PDHc activity. Any effect could obviously be
either direct, due to variation in the PDHc environment, or
could be secondary and due to the stimulation of lipoic
acid synthesis needed for further activation of the lipoyl
transacetylase and lipoamide dehydrogenase .

We have therefore determined lipoamide dehydrogenase activity in mitochondria prepared from the livers and brains
of some experimented animals .

EXPERIMENTAL

Chemicals, enzymes and diet: thiamine pyrophosphate chloride, CoA-SH, NAD$^+$, NADH, Lipoamide and sodium pyruvate were obtained from Sigma; lactate dehydrogenase and phosphotransacetylase from Boehringer; sodium [1-^{14}C] pyruvate (13.1 mCi/mmol) from Radiochemical Centre, Amersham (U.K.) The fat-free diet from U.A.R.,Villemoisson sur Orge,91360 Epiney (France) contained 30% casein, 63% sucrose, 2% cellulose, 1% salt mixture and 1% vitamin mixture.

Preparation of animals :Wistar strain rats weighing 250 g were used for this study. Females were kept on our stock colony diet until the 10th day of gestation. They were then transferred to a fat-free (FF) diet, remaining at this until their progeny was weaned, i.e. 42 days after mating. Progeny (male and female) were then randomly distributed into two groups :

- (FF) kept on a fat-free diet ,

- (LP) kept on a fat-free diet supplemented with 0.7% (by wt) sunflower oil .

Animals - mothers and progeny - were all pair fed, and kept on experimental diets until sacrificed .

Preparation of tissue homogenates and mitochondria :

As soon as possible after decapitation, the liver and brain were excised and weighed . The organs were then either homogenized or immediately frozen in liquid nitrogen. Homogenates were prepared according to Wieland et al. (1971) using a high-speed mechanical tissue desintegrator. A ninefold amount (v/w) of ice-cold 20 mM potassium phosphate buffer pH 7.0 containing 40% (v/v) glycerol was added, and the blender driven at top speed for 90 sec. Homogenates were either used immediately for enzyme assays or kept in liquid nitrogen. Mitochondria were prepared from the 900 rpm supernatant by sedimentation at 9,000 rpm in a Beckman J 21 centrifuge (rotor JA 20) .

Pyruvate dehydrogenase complex assays : PDHc activity was measured on crude liver and brain homogenates using an assay system adapted from Wieland et al.(1971) and Cremer & Teal (1974),and already described elsewhere (Raulin et al. 1979). Coenzyme A was omitted from the reaction blanks .

Lipoamide dehydrogenase assays : LADH activity was measured spectrophotometrically at 340 nm using a Unicam SP 1700 instrument equipped with a SP 1805 Programm Controller and an AR 25 Linear Recorder, on liver and brain mitochondrial fractions. The assay system used was that described by Reed & Willms (1966) and Ngo & Barbeau (1978) .

Lipid extraction and analysis :Lipids were extracted from
brains and livers according to Bligh & Dyer (1959) and
fatty acids were analysed by gas-liquid chromatography
using glass capillary columns of Carbowax 20 M, after
their conversion into methyl esters .

RESULTS

Fatty acid composition of maternal milk and organs

excised from the progeny

The fatty acid composition of lipids extracted from the
contents of stomach was determined for suckling rats of
the (FF) group, The concentration of linoleic acid was
rather high (23%) despite the fact that maternal diet was
shown to be strictly fat-free. In so far as the progeny's
fatty acid composition is concerned differences in the 2
different groups were not visible after 7-10 days. Only
traces of eicosatrienoic acid were observed in both (FF)
and (LP) livers and brains - except when rats presented
symptoms of deficiency: in this case, more than 5% of
C20 : 3 were found in the liver. Linoleic acid was low
in both (FF) and (LP) livers (approx. 5% with NS differences) and arachidonic acid did not change significantly
with either of the 2 diets. The proportion of non-EFAs
was as high in the (LP) as in the (FF) group of livers.

Increase in weight of developing rats

Animals in the (FF) group increased their weight very
slowly during the first 7 days after weaning (0.74 g per
day) and even more slowly later in experiment . When excised, the livers and brains were found to weigh much less
than the equivalent organs from the (LP) group of rats.
Ten days after weaning, wet weight differences were greater
than 100% for the livers (1.30 g vs 2.92 g) and approx.
20% for the brains (1.06 g vs 1.27 g) .

Pyruvate dehydrogenase activity

1/ Adult females

PDHc activity in the livers of females kept on an (FF)
diet from the 10th day after mating, then transferred to
the (LP) diet until 31-36 days after the end of suckling
was found significantly higher than when females were
kept on the (FF) diet. No significant differences were
found between PDHc activity in the (FF) and (LP) groups
of brains .

2/ Developing rats

As previously observed when PDHc was determined on maternal livers, the highest activity in developing rats was
found to be associated with the (LP) group livers. This
was independent of the number of days (7-10 days) post-
weaning on the diet. The differences between the (LP)
and the (FF) groups after 10 days on diets (469 vs 263
nmoles pyruvate consumed $g^{-1} min^{-1}$) was highly significant. PDHc activity was also significantly higher in the
(LP) group of brains (808 vs 341 nmoles pyruvate consumed $g^{-1} min^{-1}$) .

In the control group of rats kept on a stock colony diet
containing 4% lipids, no differences were found in PDHc
activity compared to the (LP) group of livers, but significant differences in PDHc brain activity were noted.
The results obtained with (FF) rats which presented several symptoms of deficiency suggest that while PDHc activity was still high in the liver, this activity was greatly
reduced in the brain compared to results obtained earlier.

Lipoamide dehydrogenase activity

The highest LADH activity measured on mitochondrial fractions were found to be associated with the (LP) group of
livers and brains. The differences between the (LP) and
the (FF) groups were approx. 12% in the brains and 45%
in livers of females kept 31 days on diets after the end
of suckling. Such a difference (17-20%) for LADH activity was also observed in mitochondria prepared from the
livers and brains of developing rats kept 7-10 days on
the two diets .

DISCUSSION

L.J. Reed pointed out a few years ago (1974) that in
view of its central position in metabolism, PDHc was a
likely candidate for metabolic regulation .

It has been proposed that the inhibition of pyruvate
oxidation by the NADH and acetylCoA produced in the fatty
acid oxidation resulted from direct inhibition of the
multienzyme complex (Nicholls et al.,1967). Several
publications have described a regulatory role for long-
chain fatty acylCoA esters (Erfle & Sauer, 1969; Loriette
et al., 1971; Batenburg & Olson, 1976; Waljtys-Rode,
1976; Scholtz et al., 1978) .

It was recently observed in our laboratory (Raulin et al.,

1979) that the PDHc activity for livers was very low when
animals were prepared as for the experiment described here
(born to females kept on a fat-free diet), and then transferred after weaning to a diet containing 20% fat. After
this treatment, PDHc activity was found to be even lower
than when rats were kept on the strictly fat-free diet .

In the experiment described here, we can exclude the possibility that the (LP) group of rats actually received
dietary lipids. They were only given limiting amounts of
linoleic acid (0.45%) to supplement the (FF) diet. This
was not enough to eliminate traces of eicosatrienoic acid
from the livers and brains, as expected when EFAs are
supplied for a longer period of time (Mead et al.,1956;
Holman, 1960). The very low "trienoic : tetraenoic acid"
ratio was the same for both the (FF) and (LP) groups of
rats. Nevertheless, the treatment with 0.45% linoleic
acid given in the (LP) diet, increased brain and liver
PDHc activity dramatically, even when administered for
as short time as 7 days .

We do not wish to draw definite conclusion as to the
exact role that the linoleic acid limiting supplement
played in these experiments. However, such observations
tend to support our previous hypothesis concerning the
possible close relationship between linoleic acid and
the PDH complex (Carreau et al.,1975,1977). It is also
interesting to notice that several destructive neurological diseases are characterized biochemically by abnormalities in PDHc (Clayton et al., 1974; Barbeau,1975;
Blass et al., 1976; Barbeau et al., 1976, Field et al.,
1977) and by a decrease in the HDL cholesterol ester
EFAs. Wurtman (1977) and Barbeau (1978, 1979) who have
initiated trials with lecithins (phosphatidylcholines
rich in linoleic acid) for five such diseases, have noted
an improvement following treatment in both neurological
and biochemical symptoms .

ABSTRACT

Wistar strain female rats were kept on a fat-free (FF)
diet from the 10th day after mating until weaning. Pyruvate dehudrogenase complex (PDHc) activity in the livers
and brains of their progeny was found to be very low when
the weaned rats (21 g body weight) were maintained 7-10
days on the same (FF) diet. PDHc activity increased dramatically in the (LP) group of livers and brains due to
the addition of a limiting amount (0.7%) of sunflower
oil in the (FF) diet, i.e. when the (LP) group of rats

were given limiting amounts (0.45%) of linoleic acid.
The highest lipoamide dehydrogenase (LADH) activity was
also associated with the (LP) group of brain and liver
mitochondrial fractions. Traces of eicosatrienoic acid
were still visible 7-10 days after weaning in both (LP)
and (FF) livers and brains.This suggests that the (LP)pro-
geny had not completely recovered from the maternal lipid
deprivation, and tends to support the hypothesis (Carreau
et al.,1975,1977) implicating EFAs in lipoamide synthesis and PDHc activation .

Aknowledgements : The author's study was supported in
part by grant (78.1.251.7) from the Institut National
de la Santé et de la Recherche Médicale (France) .

REFERENCES

Barbeau, A. (1975) Trans.Am.Neurol.Ass.,100,164-165.

Barbeau, A. (1978) J.Can.Sci.Neurol., 5,157-160.

Barbeau, A. (1979) T.I.B.S., pp.87-90.

Barbeau, A., Butterworth, R.F., Ngo, T.T., Breton, G.,
    Melancon, S., Shapcott, D., Geoffroy, G. and
    Lemieux, B. (1976) Can.J.Neurol.Sci., 3,379-388.

Batenburg, J.J. & Olson, M.S. (1976) J.Biol.Chem.,
    251,1364-1370.

Blass, J.P., Kark, R.A.P. & Menon, N.K. (1976)
    New Engl.J.Med.,295,62-67.

Bligh, F.G. & Dyer, W.J. (1959) Can.J.Biochem.Physiol.,
    37,911-917.

Carreau, J.P., Lapous, D. & Raulin, J. (1975)
    Comptes Rendus, Série D, 281,941-944.

Carreau, J.P., Lapous, D. & Raulin, J. (1977)
    Biochimie, 59,486-497.

Clayton-Love, W., Cashell, A., Reynolds, M. & Callaghan,N.
    (1974) Br.Med.J., 3,18-21.

Cremer, J.E. & Teal, H.M. (1974) Febs Letters, 39,17-20.

Erfle, J.D. & Sauer, F. (1969) Biochim.Biophys.Acta,
    178,44&-452.

Field, E.J., Joyce, G. & Smith, B.M. (1977) Neurology,
    214,113-127.

Holman, R.T. (1960) J. Nutrition, 70, 405-415.

Launay, M., Vodovar, N. & Raulin, J. (1968)
    Bull.Soc.Chim.Biol.,50,439-450.

Launay, M., Dauvillier, P. & Raulin, J. (1969)
    Bull.Soc.Chim.Biol.,51,95-104.

Loriette, C., Jomain-Baum, M. Macaire, I. & Raulin, J.
    (1971) Eur.J.Clin.Biol.Res., 16,366-372.

Mead, J.F., Slaton, W.H.(Jr) & Decker, A.B. (1956)
    J.Biol.Chem., 218,401-407.

Ngo, T.T. & Barbeau, A. (1978) Int.J.Biochem.,9,681-684.

Nicholls, D.G., Sheperd, D. & Garland, P.B. (1967)
    Biochem.J.,103,677-691.

Parry, R.J. (1977) J.Am.Chem.Soc., 99-19,6464-6466.

Raulin, J., Loriette, C., Launay, M., Lapous, D., Goureau-
    Counis, M.F., Counis, R. & Carreau, J.P. (1974) in:
    "The Regulation of Adipose Tissue Mass", J.Vague &
    J.Boyer,Ed., Excerpta Med.,Series 315, Elsevier Publ.,
    Amsterdam, pp.32-34 .

Raulin, J., Loriette, C., Launay, M., Lapous, D.,
    Bouchène, M. & N'Dalla, J. (1979) submitted for
    publication .

Reed, L.J. (1964) Int.Symp.Chem.Nat.Prod.,Kyoto,
    I.U.P.A.C., D-16, 218-220.

Reed, L.J. (1974) Acc.Chem.Res., 7, 40-46.

Scholtz, R., Olson, M.S., Schwab, A.J., Schwabe, U.,
    Noell, C. & Braun, W. (1978) Eur.J.Biochem.,86,519-530.

Walajtys-Rode, E.I. (1976) Eur.J.Biochem.,71,229-237.

Wieland, O., Siess, E., Schultze-Wethmar, F.H., v.Funcke,
    H.G. & Winton, B. (1971)Arch.Biochem.Biophys., 143,
    593-601.

Wurtman, R.J., Hirsch, M.J. & Growdon, J.H. (1977)
    Lancet, 2,68-69.

STIMULATION OF FATTY ACID METABOLISM BY PANTETH(E)INE

Kin-ya Kameda and Yasushi Abiko

Laboratory of Biochemistry,
Research Institute
Daiichi Seiyaku Co., Ltd.
Edogawa-ku, Tokyo 132, Japan

Introduction

Diabetes mellitus is a metabolic disease where lack of insulin action diminishes glucose utilization in the peripheral tissues. In compensation for this abnormal situation, fat is preferentially used as a sole energy source in the body. The metabolic shift to lipid utilization leads to hypertriglyceridemia accompanied by elevation of free fatty acid in blood and, in very advanced stages by elevation of ketone bodies including acetoacetate and β-hydroxy-butyrate in blood. Increased level of CoA and acyl CoA in the diabetic rat liver was reported by Smith et al. [1]. This seems to be a metabolic response to increased utilization of fatty acid in diabetic state and suggests increased requirement for CoA in diabetic tissues. It is, therefore, interesting to study the effect of some precursors of CoA on diabetic hyperlipidemia. The present paper deals with a favorable effect of pantethine on lipid metabolism in streptozotocin diabetic rats. Pantethine treatment has been found to reduce increased levels of serum triglycerides, free fatty acid and β-hydroxybutyrate in diabetic rats, but pantothenic acid had no effect. For getting insight into the mode of the specific action of pantethine, have been carried out some experiments in vitro including fatty acid oxidation and ketogenesis in rat liver and muscle tissues.

Methods and Materials

Male Sprague-Dawley rats weighing 150 g to 200 g were used throughout the experiment. Rats were made diabetic by a single intravenous injection of streptozotocin after overnight fasting.

Streptozotocin was dissolved in 0.1 M citrate buffer, pH 4.5, and administered in a volume of 2 ml per kg. body weight. From the preliminary study indicated in Table 1, a dose of streptozotocin at 45 mg/kg was selected as the best experimental conditions for inducing stable diabetic state. The lower dose gave a great variation among individuals and the higher doses resulted in high mortality. Under the selected conditions, all the rats showed typical symptoms of diabetesmellitus including hyperglycemia, polyphagia, polydipsia and polyurea accompanied by sugar cataract which was developed within 3 months. Control rats received the citrate buffer alone. Diabetic rats were maintained on a normal laboratory chow for 2 weeks. Then the diabetic rats were divided into two groups. One of them received normal diet for more 4 weeks and the other one received a diet containing 0.1 % (w/w) of pantethine. This concentration of pantethine corresponded to the dose of 200 mg/kg/day. Blood samples were taken from the tail vein at 2 and 6 weeks after streptozotocin injection for assay of serum lipids. Hepatic lipids and CoA and muscular CoA levels were also determined at 6 weeks after streptozotocin (Fig. 1). Triglyceride and phospholipid were determined by respective commercial kits (Wako Pure Chemical Industries Ltd., Osaka, Japan). Cholesterol was determined by the method of Zak [2].Free fatty acid was determined by a slight modification of the method of Itaya and Ui [3].Hepatic lipids were extracted by the method of Folch et al. [4]. CoA and ketone bodies were determined enzymatically by the method of Abiko et al. [5] and Williamson et al. [6], respectively. Acyl CoA was determined after hydrolysis to CoA according to Tubbs et al. [7] with a slight modification. For the experiment in vitro, the liver and the muscle tissue (quadricepts femoris) were excised off and homogenized with 9 volumes of 0.3 M mannitol containing 10 mM hydroxyethylpiperazine-N'-2-ethanesulfonic acid and 1 mM EDTA, pH 7.4. After centrifugation at 600 x g for 10 minutes, the supernatant fraction was used as the enzyme solution. Palmitate oxidation and ketogenesis from palmitoyl-

Table 1   Streptozotocin Diabetes in Male Sprague-Dawley Rats

Dose mg/kg, i.v.	Incidence (%) of hyperglycemia at 48 hr. >300 mg/dl (Fed)	Death rate (%) (1 week)	Note
40	100	0	Variation among individuals
45	100	10	Stable hyperglycemia, slight increase in body weight
50	100	40	Stable hyperglycemia High mortality
60	100	50	
70	100	60	
80	100	100	

Streptozotocin (Upjohn or Boehringer) : dissolved in citrate buffer pH 4.5
Slc : Sprague-Dawley rats (♂) 6 weeks old, 24 hr fasted

carnitine or palmitate were assayed according to the method of
Cederbaum et al. [8]. GTP-dependent fatty acid oxidation was
measured in a similar way to palmitate oxidation with exception
that GTP was used instead of ATP. Fatty acyl CoA synthetase activity
was determined by the method of Bar-Tana et al. [9]. Statistical
analysis of the results was performed by the Student t-test.

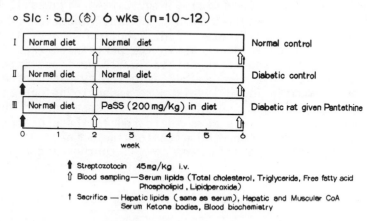

Fig. 1. Design of the experiments in vivo

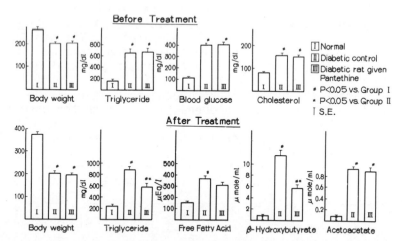

Fig. 2. Effect of pantethine treatment on serum lipids in
diabetic rats in vivo. The streptozotocin-induced diabetic
rats were given pantethine (group III, n=10) in diet for
4 weeks, being started at 2 weeks after streptozotocin
injection. Serum lipid levels were compared with those in
the normal (group I, n=12) and the diabetic control rats
(group II, n=10) before and after pantethine treatment.

Results

Effect of Pantethine on Diabetic Hyperlipidemia in vivo

At 2 weeks after streptozotocin injection, when pantethine treatment was just started, considerable elevation of serum triglyceride and cholesterol were observed in diabetic rats. Blood glucose concentration was about 400 mg/dl (Fig. 2, top figures). Pantethine treatment for 4 weeks resulted in significant reduction of serum triglyceride and β-hydroxybutyrate (Fig. 2,bottom figures). Free fatty acid level also tended to decrease in the pantethine-treated rats, but serum cholesterol and acetoacetate levels were not affected by the treatment. Blood glucose was not affected by pantethine. In the diabetic rats, hepatic lipid levels were found to be normal and not affected by pantethine. As shown in Table 2, hepatic and muscular CoA and long acyl CoA levels were increased in these diabetic rats. The increase was more pronounced with long acyl CoA and in the muscle tissue. Pantethine treatment caused further increase in CoA and acyl CoA.

Effect of Pantethine on Fatty Acid Oxidation in vitro

Effect of pantethine on palmitate oxidation was studied. Fig.3 shows the formation of radioactive CO_2 from 1-^{4}C-palmitate as a function of CoA concentration in the liver and muscle homogenate obtained from normal rats. Addition of pantethine at 40 μM significantly stimulated the oxidative reaction of palmitate, but pantethine itself was not active unless CoA was present in the reaction mixture, as can be clearly seen with the muscle preparation. The similar results were obtained with the tissue homogenates from the diabetic rats. The result of the experiment with the muscle preparation is illustrated in Fig. 4.

Fig.5 shows the effect of some pantethine derivatives on palmitate oxidation in the liver homogenates from both the normal and the

Table 2 CoA and Long Acyl CoA Levels in the Liver
and the Muscle Tissues

		CoA content (nmoles/g wet tissue)		
	Group	I	II	III
Liver	Free CoA	461 + 75.25	525 + 90.26	590 + 83.19
	Long acyl CoA	345 ∓ 45.28	471 ∓ 35.13#	513 ∓ 54.78
	Total CoA	825 + 90.68	968 + 47.80	1101 + 79.55
Muscle	Free CoA	98 + 26.12	106 + 13.45	164 + 48.0
	Long acyl CoA	225 ∓ 80.05	840 ∓ 258#	890 ∓ 145#
	Total CoA	332 + 54.21	960 + 123#	1158 + 97.8#

\# $p < 0.05$ versus Group I

diabetic rats. Phosphopantetheine was found to be most effective in stimulation of palmitate oxidation in both tissue preparations even in the absence of added CoA. Palmitoyl-S-pantetheine as well as pantethine was active only in the presence of added CoA. GTP-dependentpalmitate oxidation was also tested with the negative results of stimulation by these pantethine derivatives. When 1-^{14}C-octanoate was used as the substrate, however, formation of radioactive CO_2 was not affected at all by pantethine and palmitoyl-S-pantetheine, as shown in Fig.6.

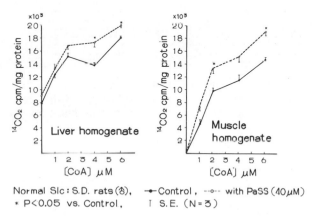

Normal Slc : S.D. rats (♂),   —•—Control, --o-- with PaSS (40μM)
* P<0.05 vs. Control,   ⊤ S.E. (N=3)

Fig.3. Effects of pantethine on palmitate oxidation in vitro. The reaction mixture contained 0.2 μmoles of sodium 1-^{14}C-palmitate, 2 μmoles of ATP, 20 nmoles of CoA, 2 μmoles of L-carnitine and 100 μl of rat liver or muscle homogenate from normal rats in a total volume of 2 ml. After incubation of the mixture at 37° for 30 minutes, CO_2 formed was determined.

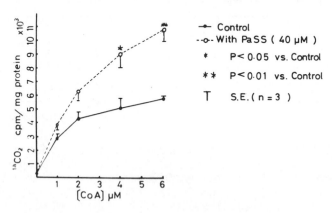

Fig.4. Effect of pantethine on palmitate oxidation in the muscle homogenate from diabetic rats. The experimental conditions were the same as in Fig.3, except for the muscle preparation from the diabetic rats.

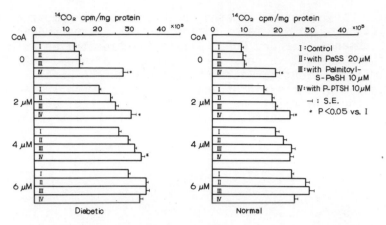

Fig.5. Effect of pantethine derivatives on palmitate oxidation in the normal and diabetic liver homogenates. The experimental conditions were the same as in Fig.3.

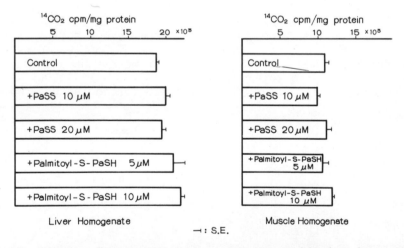

Fig.6. Effect of pantethine on octanoate oxidation in normal liver and muscle homogenates. For the liver preparation, the reaction mixture contained 0.2 μmoles of sodium 1-^{14}C-octanoate 2 μmoles of ATP, 0.8 μmoles of ADP and 100 μl of the liver homogenate in a total volume of 2 ml. For muscle preparation, the reaction mixture contained 0.2 μmoles of sodium 1-^{14}C-octanoate and 100 μl of the muscle homogenate in a total volume of 2 ml.

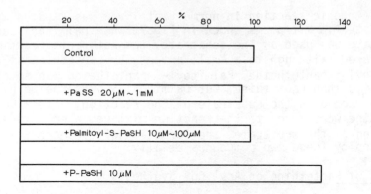

Assayed by the method of J. Bar-Tana et al.

Fig.7. Effect of pantethine derivatives on ketogenesis in the rat liver homogenate. The control reaction mixture contained 0.2 µmoles of a substrate, 2 µmoles of ATP, 0.4 µmoles of L-carnitine, 20 nmoles of CoA and 0.5 ml of the liver homogenate in a final volume of 2 ml (pH 7.4). When palmitoyl-L-carnitine or palmitoyl-S-pantetheine was the substrate, L-carnitine and CoA were omitted. After incubation at 37° for 30 min., aceto-acetate formed was determined.

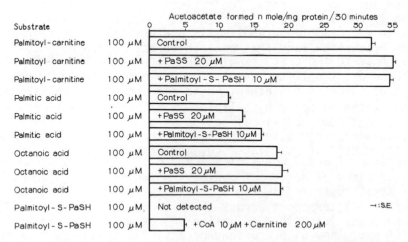

Fig.8. Effect of pantethine derivatives on palmitoyl CoA synthesis by rat liver microsomes. The microsomal acyl CoA synthetase activity was determined by the method of Bar-Tana et al. [9] with 0.2 M 1-^{14}C-palmitate as the substrate.

Effect of Pantethine on Ketogenesis in vitro

        Ketogenic reaction in normal rat liver was studied with various
substrates in vitro. As shown in Fig.7, when palmitoyl carnitine or
palmitate was used as the substrate, the ketogenic reaction was
stimulated, although to a small extent, by pantethine as well as
palmitoyl-S-pantetheine. Palmitoyl-S-pantetheine was somewhat more
effective than pantethine, but palmitoyl-S-pantetheine by itself
did not serve as the substrate of the reaction, unless CoA and
carnitine were added to the reaction mixture. When octanoic acid
was used as the substrate, the formation of acetoacetate was not
affected by these two compounds at all.

Effect of Pantethine on Acyl CoA Synthesis

        Synthesis of palmitoyl CoA was studied with the normal rat
liver microsomal fraction. As shown in Fig.8, palmitoyl CoA synthesis was stimulated by about 30 % by phosphopantetheine, but not by
pantethine or palmitoyl-S-pantetheine.

                              Discussion

        The present study revealed that pantethine treatment effective-
ly reduced serum triglyceride, β-hydroxybutyrate and free fatty acid
in diabetic hyperlipidemia of the rats. As reported by Smith et al.
[1], CoA and long acyl CoA levels were increased in the tissues
of the diabetic rats when compared with those of the normal rats.
The increase was especially pronounced in long acyl CoA levels and
in the muscle tissue. Pantethine treatment gave further increase in
CoA and acyl CoA contents. These findings may suggest that the increase in CoA and acyl CoA content is a metabolic response to diabetic state where lipid metabolism is preferentially accelerated for
energy production, and that pantethine has served as a CoA precursor
in response to increased requirement for CoA in the diabetic tissues.
        However, the latter assumption can not be the only mechanism
of action, because pantothenic acid did not exert so favorable
effect on the diabetic hyperlipidemia as pantethine (data not
shown). And some additional action of pantethine should be considered. This has been also supported by the present finding that
pantethine did not stimulate $^{14}CO_2$ formation from 1-^{14}C-palmitate in
the absence of added CoA (Figs.3 and 4). There may be several
possible sites of action of pantethine in the fatty acid metabolism
which are illustrated in Fig.9 : stimulation of 1) fatty acyl CoA
synthesis, 2) transport across the mitochondrial membrane, 3) intramitochondrial fatty acid oxidation, 4) oxidation of acetyl CoA in
the TCA cycle and 5) GTP dependent acyl CoA synthesis in mitochondria.
        Microsomal acyl CoA synthetase was recently reported to be
stimulated by phosphopantetheine by Bar-Tana et al. [9], which
we have reconfirmed here (Fig. 8). Pantethine was not effective
probably due to lack of a phosphorylating enzyme in this system.

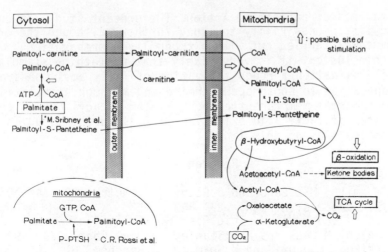

Fig.9 Metabolic map of fatty acid oxidation

Phosphopantetheine has been found to be the most active stimulant also in the overall palmitate oxidation to CO_2 (Fig.5). These findings suggest the major contribution of phosphopantetheine as an active principle to pantethine's action, because pantethine can be easily phosphorylated by pantothenate kinase in the cells [13].

The second site may include two possibilities: one is on the carnitine acyl transferase system and the other involves formation of palmitoyl-S-pantetheine and its non-enzymatic transport through the mitochondrial membranes followed by oxidation in β-oxidation system after conversion to palmitoyl CoA. The former possibility may be supported by the difference in effectiveness of pantethine between the palmitate and octanoate oxidation reactions (Fig.3, 4 and 6) and between the ketogenic reactions from these two substrates (Fig.7), because octanoic acid is freely permeable through the mitochondrial membranes. The latter possibility was based on the findings on the formation of acyl pantetheine in rat liver microsomes [10] and on the enzymatic interconversion between acyl CoA and acyl pantetheine [14] and on the much lower susceptibility of acyl pantetheine to β-oxidation than acyl CoA [14]. But this has been ruled out by the finding that palmitoyl-S-pantetheine did not serve as the substrate of the ketogenic reaction (fig.7).

The third and fourth possible sites may be excluded by the present results that acetoacetate formation from octanoic acid (Fig.7) and $^{14}CO_2$ formation from 1-^{14}C-octanoic acid (Fig.6) were not affected by pantethine under the conditions where palmitate oxidation was stimulated by pantethine. If either of these two oxidation cycles could be affected by pantethine, octanoate oxidation in either system had to be stimulated, because octanoic acid is non-enzymatically permeable through the mitochondrial membranes.

GTP dependent acyl CoA synthetase in mitochondria was recently reported to require phosphopantetheine as an essntial allosteric effector for activity [11]. This enzyme system thus might be one

of the targets of pantethine action. The present study, however, has
remained little possibility to study further, although the present
experiment was carried out with a crude liver preparation.

In conclusion, it is very likely that pantethine stimulated
fatty acid oxidation through activation of fatty acyl CoA synthetase
after conversion to phosphopantetheine and through stimulation of
the mitochondrial membrane transport system of acyl CoA which is
the key step of fatty acid oxidation [12]. Further studies are
now in progress along this line.

Reference

[1] C.M.Smith, M.L.Cano and J.Potyraj J.Nutr. 108 854 1978
[2] B.Zak Am.J.Clin.Path. 27 583 1957
[3] K.Itaya and M.Ui J.Lipid Res. 6 16 1965
[4] J.Folch, M.Lees and H.S.Stanley J.Biol.Chem. 226 497 1957
[5] Y.Abiko, T.Suzuki and M.Shimizu J.Biochem. 61 10 1967
[6] D.H.Williamson and J.Mellanby In " Methods of Enzymatic
Analysis " edited by H.U.Bergmeyer p. 454 1965 Academic Press
[7] P.K.Tubbs Biochem.J. 93 550 1964
[8] A.J.Cederbaum, C.S.Lieber and E.Rubin   A.B.B. 169 29 1975
[9] J.Bar-Tana, G.Rose and B.Shapiro Biochem.J. 129 1101 1972
[10] M.Sribney and J.L.Dove Lipids 13 422 1978
[11] C.R.Rossi, A.Alexander, L.Galzigna, L.Sartorelli and
D.M.Gibson   J.Biol.Chem. 245 3110 1970
[12] J.D.McGarry, P.H.Wright and D.W.Foster J.Clin.Invest. 55
1202 1975
[13] Y.Abiko   J.Biochem. 61 290 1967
[14] J.R.Stern   J.Am.Chem.Soc. 77 5194 1955

Key words

Pantethine, Phosphopantetheine, Palmitoylpantetheine, Diabetes
Fatty acid oxidation, Ketogenesis

EFFECTS OF A HYPOLIPIDEMIC DRUG (CLOFIBRATE) ON THE BIOSYNTHESIS AND DEGRADATION OF CoA IN RAT LIVER

Sverre Skrede

Institute of Clinical Biochemistry
Rikshospitalet, University of Oslo
Oslo, Norway

INTRODUCTION

It has recently been reported (1) that administration of the hypolipidemic drug clofibrate (ethyl-α-p-chlorophenoxyisobutyrate) to rats is followed by a great increase of the CoA-pool in their liver. The increase may amount to about 3-fold, and is mainly due to an increase of free CoA (1,2).

We have recently studied the content of CoA in different subcellular compartments in the liver of rats given clofibrate, and how different steps in the degradation and biosynthesis of CoA are affected by this drug (2).

MATERIALS AND METHODS

Male rats weighing about 135 g were given ordinary commercial rat food with (controls without) 0.3% clofibrate for 10-14 days. Liver subfractions were obtained by conventional tissue fractionation techniques, except that EDTA (5 mmol/l) and l-tartrate (25 mmol/l) were added to avoid degradation of CoA (3) during the procedure. Marker enzymes for the subfractions were assayed as described previously (2,3,4).

Total CoA was estimated by the CoA-dependent incorporation of radioactive carnitine into palmityl carnitine (5) or by a recycling phosphotransacetylase method (cf. 2).

Long-chain acyl CoA was assayed according to the method of Tubbs & Garland (6) with minor modifications (2), using palmityl-CoA as standard.

Degradation of CoA or dephospho-CoA was estimated at pH 7.4 as described previously (3).

Biosynthesis of CoA from pantothenate. Supernatants from livers homogenized in phosphate buffer (pH 7.2, 0.02 mmol/l) and dialyzed against the same buffer, were incubated according to Abiko et al (7), except that an ATP-regenerating system was present (2). In experiments with mitochondria, oligomycin was also added. In experiments with cold pantothenic acid, CoA was assayed as described above. When ^{14}C-pantothenic acid was used, CoA and its precursors was isolated by descending paper chromatography, usually with n-butanol-acetic acid-water (5:2:3, v/v), which gave good separation between CoA/dephosphoCoA, 4-phosphopantetheine and pantothenic acid (cf. 2).

Pantothenate kinase was assayed by the above system for "biosynthesis of CoA" (7), except that l-cysteine was omitted to avoid further metabolism of 4-phosphopantothenic acid.

Dephospho-CoA kinase was assayed at pH 8.5 as described previously (2).

Dephospho-CoA pyrophosphorylase. 4-phosphopanto-thenate was prepared from dephospho-CoA by incubation with commercial nucleotide pyrophosphatase, and isolated by affinity chromatography with Thiopropyl-Sepharose followed by preparative paper chromatography. Dephospho-CoA pyrophosphorylase was estimated at pH 8.5, in a system similar to that used for dephospho-CoA kinase.

RESULTS

Table 1 shows that treatment with clofibrate caused an increase of the total content of CoA in the liver by a factor of 2-3, and that long-chain acyl CoA increased to almost the same extent. The wet weight and protein content of the livers of clofibrate-treated rats was also higher than in the control animals.

Table 1. Clofibrate-induced increase of CoA in rat
         liver

The table shows arithmetical mean values from groups
of 17 normal rats and 17 rats treated with clofibrate.
The numbers refer to amounts in whole liver.

	Controls	Clofibrate-treated rats
Wet weight (g)	10.1	12.3
Total protein (q)	1.8	2.2
Total CoA (μmoles)	4.0	10.7
Long-chain acyl CoA (μmoles)	0.5	1.1

Liver subfractionation experiments (2) showed that
the protein content of all fractions was increased in
the animals treated with clofibrate - most pronounced
for the mitochondrial fraction. Also the amount of
peroxisomes was probably increased, since the amount of
catalase was rather high both in the mitochondrial and
in the lysosomal fraction. Total CoA (and long-chain
acyl CoA) was increased in all fractions - most in the
mitochondrial and lysosomal fractions, where about 5-
fold of the amounts in these fractions from the control
rats were found.

In table 2, the specific activities of the two
first steps in the degradation of CoA are shown for total
homogenate and tissue subfractions. There was no change
of their specific activities in total homogenate. A
decrease of CoA-degradation in the mitochondrial fraction in the clofibrate-treated group is thought to be
explained by·a change of the subfractionation pattern,
with less lysosomes in this fraction. A clofibrate-
induced increase of the degradation of CoA in the supernatant is thoughtto be due to greater fragility of the
lysosomes in such livers, since supernatant acid phosphatase was also increased.

In conclusion, the results of table 2 indicate that
changes in the degradation pathway cannot explain the
increase of CoA seen after treatment with clofibrate.

Table 2. Degradation of CoA and dephospho-CoA in liver
         total homogenate and subfractions from normal
         and clofibrate-treated rats

Fraction	Substrate	Normal rats	Clofibrate-treated rats
		(nmoles/mg of protein/min)	
Total homogenate	CoA	0.7	0.7
	Dephospho-CoA	3.9	3.9
Nuclear	CoA	0.8	0.8
	Dephospho-CoA	3.5	3.5
Mitochondrial	CoA	2.0	0.7
	Dephospho-CoA	1.4	0.8
Lysosomal	CoA	9.3	7.6
	Dephospho-CoA	7.8	2.3
Microsomal	CoA	1.1	1.0
	Dephospho-CoA	3.9	4.3
Supernatant	CoA	0.04	0.5
	Dephospho-CoA	1.0	0.8

The biosynthesis of CoA from pantothenic acid was
markedly increased in liver supernatant after administration of clofibrate. Both by estimating the final
product directly, or by the incorporation of ^{14}C-
pantothenic acid into CoA, rates about 3-fold of normal
were found (table 3).

Table 4 shows that this increase of biosynthesis cannot be explained by an increase of the main, supernatant pool of dephospho-CoA kinase. Table 4 also shows a mitochondrial pool of dephospho-CoA kinase, which we also demonstrated in earlier studies (4,5). This activity was also similar in the normal and the drug-exposed group.

Mitochondrial dephospho-CoA kinase (8) has a rather broad pH-optimum of about 8.5, and is dependent on Mg^{2+} (full activation is achieved at 0.5 mmol/l). Its K_m

Table 3. Increased biosynthesis of CoA from pantothenic acid in livers from clofibrate-treated rats

Liver fractions were incubated with cold (A) or ^{14}C-labelled (B) pantothenic acid as described in the Methods section and more detailed in ref. 2.

Group of rats	Method	
	A CoA formed (nmoles/mg of protein)	B ^{14}C-pantothenate activity incorporated into CoA (% of initial activity)
Normal	5.1	9.1
Clofibrate-treated	16.4	30.6

Table 4. Dephospho-CoA kinase in the supernatant and mitochondrial fractions from normal and clofibrate-treated rats

The formation of CoA from dephospho-CoA at pH 8.5 was estimated in the presence of l-tartrate (2). The mitochondria were washed twice with KCl (0.15 mol/l), and contained no activity of (supernatant) LD.

Fraction	Normal rats	Clofibrate-treated rats
	(nmoles/mg of protein/min)	
Supernatant	0.8	0.9
Mitochondria	0.5	0.6

for dephospho-CoA (0.01 mmol/l) and ATP (about 0.05
mmol/l) are both an order of magnitude less than reported by Abiko (8) for dephospho-CoA isolated from the
soluble fraction of the cell. The mitochondrial dephospho-CoA kinase is located in the inner mitochondrial
membrane (9).

Dephospho-CoA pyrophosphorylase was also found to
have a previously unrecognized mitochondrial pool (8).
Pantothenate kinase, on the other hand, could not be
detected in mitochondria (9).

Table 5 shows that pantothenate kinase in the liver
supernatant from clofibrate-treated animals had about
twice the activity of normal. We found feed-back inhibition of CoA at this step, as previously reported by
Abiko (8). The extent of this feed-back inhibition was
not significantly changed by treatment with clofibrate
(2).

Table 5. Increased activity of pantothenate kinase in
         the supernatant of liver from clofibrate-
         treated rats

^{14}C-pantothenic acid was incubated with dialyzed liver
supernatants as described in the Methods section and
ref. 2.

Group	4-phosphopantothenic acid formed (nmoles/mg of protein/min)
Normal rats	0.10
Clofibrate-treated rats	0.22

DISCUSSION

Effects of clofibrate on CoA in the liver. We could
confirm that treatment of rats with clofibrate will cause
an increase of CoA in the liver (1). Long-chain acyl-CoA
will also increase during such treatment, but not in
relation to total CoA (as per cent).

CoA and acyl-CoA were found to be increased in all
subcellular fractions, most in the mitochondria (2).
This is particularly interesting since the inner mito-

chondrial membrane is thought to be impermeable to CoA
(cf. 5), and made us perform studies on the possibility
of a separate mitochondrial system for biosynthesis of
CoA (see below).

Degradation of CoA in rats treated with clofibrate.
The first two steps in the cellular catabolism of CoA
are thought to be the removal of the ribose 3-phosphate
group by lysosomal acid phosphatase (4) and the cleavage
of the pyrophosphate bridge by plasma membrane nucleotide pyrophosphatase (3). None of these steps are significantly affected by treatment with clofibrate (2),
however, and cannot explain the changes in the CoA pool.
Clofibrate will influence the liver subfractionation
pattern, however, with less lysosomal acid phosphatase
in the mitochondrial fraction and more in the supernatant. Probably, the membrane structures are changed
by the drug, leading to increased "fragility" of the
particles.

Biosynthesis of CoA from pantothenic acid in the
supernatant fraction of the liver. All enzymes required
for the biosynthesis of CoA from pantothenic acid have
been isolated from the soluble fraction of the cell (8).
By using the in vitro system of Abiko (cf. 8) for the
synthesis of CoA from pantothenic acid, we found the
capacity for CoA-biosynthesis to be markedly increased
after treatment with clofibrate (2). Dephospho-CoA
kinase was not significantly (or only very slightly)
increased, but the activity of pantothenate kinase was
more than doubled. CoA will exert feed-back inhibition
of pantothenate kinase (8), and the extent of this
inhibition was not changed by clofibrate (2).

Mitochondrial biosynthesis of CoA. We have previously reported that dephospho-CoA kinase is present
in mitochondria from liver (2,5,9) and kidney (4). Even
though it has been generally thought that the inner
mitochondrial membrane is a permeability barrier for
CoA (cf. 5), the existence of a separate, mitochondrial
pathway for the biosynthesis of CoA has to our knowledge
not been suggested.

In this study, we have found that both enzymes
representing the last two steps in the biosynthesis of
CoA - dephospho-CoA pyrophosphorylase and dephospho-CoA
kinase - are present in the inner mitochondrial membrane.
Pantothenate kinase - representing the first step in
the biosynthesis of CoA - is not detectable in reason-

ably pure mitochondrial fractions, however.

We are now performing more detailed studies on the
mitochondrial pathway of biosynthesis of CoA, but it is
not clear at present which precursor this sequence of
reactions starts with. It has been settled, however,
that it is not the same as the initial precursor of
the biosynthetic pathway of the soluble fraction of the
cell, i.e. pantothenic acid.

SUMMARY

1. Clofibrate induces an increase of CoA (and long-
    chain acyl-CoA) in rat liver by a factor of 2-3.
2. The first two enzymatic steps in the catabolic
    pathway of CoA are not significantly affected by
    such treatment.
3. The in vitro biosynthesis of CoA from pantothenic
    acid in liver homogenate is increased 2-3 times
    after treatment with clofibrate, mainly due to an
    increase of pantothenate kinase.
4. A separate, mitochondrial pathway for the biosynthe-
    sis of CoA was found. It has the last two steps in
    common with the previously recognized pool in the
    soluble fraction of the cell, since dephospho-CoA
    pyrophosphorylase and dephospho-CoA kinase are both
    present in the inner mitochondrial membrane. Panto-
    thenate kinase - the first step in the cell sap
    pathway - is lacking in the mitochondria, however.
    Studies are in progress to detect the first pre-
    cursor for CoA in the mitochondrial biosynthetic
    pathway.

REFERENCES

1. M. J. Savolainen, V. P. Jauhonen and I. E. Hassinen,
    Effects of clofibrate on ethanol-induced modifica-
    tions in liver and adipose tissue metabolism: Role
    of hepatic redox state and hormonal mechanisms.
    Biochem. Pharmacol. 26: 425 (1977).
2. S. Skrede and O. Halvorsen, Increased biosynthesis
    of CoA in the liver of rats treated with clofibrate.
    Eur. J. Biochem. (in the press, 1979).
3. S. Skrede, The degradation of CoA: Subcellular
    localization and kinetic properties of CoA- and
    dephospho-CoA pyrophosphatase. Eur. J. Biochem. 38:
    401 (1973).
4. J. Bremer, A. Wojtczak and S. Skrede, The leakage
    and destruction of CoA in isolated mitochondria.

Eur. J. Biochem. 25: 190 (1972).

5. S. Skrede and J. Bremer, The compartmentation of
    CoA and fatty acid activating enzymes in rat liver
    mitochondria. Eur. J. Biochem. 14: 465 (1970).

6. P. K. Tubbs and P. B. Garland, Assay of coenzyme A
    and some acyl derivatives, Methods Enzymol. 13:
    535 (1969).

7. M. Shimizu and Y. Abiko, Biosynthesis of coenzyme A
    from pantothenate, pantethine and from S-benzoyl-
    pantethine in vitro and in vivo. Chem. Pharm. Bull.
    13: 189 (1965).

8. Y. Abiko, Metabolism of CoA, in Metabolic Pathways
    (D.M. Greenberg, ed.), 7: 1 (1975).

9. S. Skrede and O. Halvorsen (in preparation).

CLINICAL CHEMISTRY OF MERCAPTOPYRUVATE AND ITS METABOLITES

B. Sörbo, U. Hannestad, P. Lundquist, J. Mårtensson and
S. Öhman

Department of Clinical Chemistry
University of Linköping
S-581 85 Linköping, Sweden

INTRODUCTION

The catabolism of cysteine in the mammalian body occurs mainly through an oxidative pathway leading to sulfate and taurine with cysteine sulfinate as a common intermediate. There is, however, another catabolic pathway, which starts by transamination of cysteine to mercaptopyruvate[1]. The latter may then be reduced to mercaptolactate through the action of lactate dehydrogenase[2] or its sulfur may be transferred to sulfite or cyanide, giving thiosulfate and thiocyanate as products. The latter reactions are catalyzed by mercaptopyruvate sulfurtransferase[3]. Furthermore, thiosulfate may be converted to thiocyanate through the action of thiosulfate sulfurtransferase[3]. Mercaptolactate, thiosulfate and thiocyanate may thus be considered as metabolic products formed through a transaminative catabolic pathway of cysteine. This pathway has been established by in vitro experiments but very little is known about its function in human beings under normal and pathological conditions. Studies in this field have been hampered by lack of simple and specific methods for the determination of mercaptolactate, thiosulfate and thiocyanate but we have recently developed such methods applicable to the fairly low concentrations of these compounds found in normal human urine.

What follows is a brief description of these methods, which may be of interest also to the biochemist working in this field of sulfur metabolism.

DETERMINATION OF MERCAPTOLACTATE

Mercaptolactate in the form of its mixed disulfide with cysteine has been demonstrated as a constituent of normal human urine[4],
and increased amounts of this compound are excreted by subjects with
a rare metabolic defect, mercaptolactate cysteine disulfiduria[5].
This condition has recently been reported[6] to be caused by a deficiency of mercaptopyruvate sulfurtransferase. The fact that mercaptolactate is excreted as the mixed disulfide with cysteine is
probably explained by a rapid oxidation of intracellularly formed
thiols when they enter the extracellular space[7]. For statistical
reasons, thiols formed in minor amounts are thus converted to their
mixed disulfides with cysteine. Earlier methods for the determination of mercaptolactate cysteine disulfide in human urine made use
of the conventional amino acid analyzer. Although adequate for the
high concentrations of this compound present in urine from subjects
with mercaptolactate cysteine disulfiduria, they were not sensitive
enough for determinations of the low concentrations present in
normal human urine. We have now devised a sensitive gas chromatographic method[8] for the determination of mercaptolactate. The latter
is first liberated from the mixed disulfide by reduction, a "clean
up" by adsorption steps then follows, and mercaptolactate is then
converted to a derivative, suitable for gas chromatography. For the
reduction of the mixed disulfide we initially tried borohydride and
electrolytic reduction, but these methods were unsatisfactory. Quantitative reduction of the mixed disulfide was on the other hand
easily achieved by incubation of urine with an insoluble polymer
containing thiol groups (thiopropyl-Sepharose 6B). The "clean up"
procedure was based on the adsorption of the thiol compounds present
in the sample to an organomercurial adsorbent (p-acetoxymercuri-
aniline-Sepharose 4B). The adsorbed mercaptolactate could then be
eluted with an excess of cysteine. The latter must, however, be
eliminated before further analysis, by adsorption on a strongly
acid cation exchange resin in the hydrogen form. After these initial
steps, mercaptolactate was obtained as a dilute water solution,
which had to be concentrated before gas chromatography. Attempts to
concentrate these solutions by evaporation failed, but the problem
was solved by converting mercaptolactate to its dibenzyl derivative
by extractive alkylation. In this step mercaptolactate was extracted
as an ion pair with tetrabutylammonium to methylene chloride, containing benzyl bromide as the alkylating agent. The dibenzyl derivative thus formed had excellent gas chromatographic properties and
could be concentrated by evaporation from its methylene chloride
solution with no loss. The final gas chromatographic step was performed with flame ionization detection and OV-17 as the stationary
phase.

When the entire procedure was applied to normal human urine,
three major components were detected in the gas chromatograms. One
component had a retention time identical with that of the dibenzyl

derivative of mercaptolactate and its identity was verified by mass spectrometry. The two other components could conclusively be identified from their mass spectra as the benzyl derivatives of mercapto-acetate and N-acetylcysteine respectively. A bonus of our gas chromatographic method was thus that these two sulfur containing metabolites could be determined simultaneously with mercaptolactate. We may note here, that both mercaptoacetate and N-acetylcysteine have previously been reported to occur in normal urine as their mixed disulfides with cysteine[4],[9] but satisfactory methods for their determination have not been available up to now. Very little is known about the origin of these compounds, but N-acetylcysteine may arise as a side product from the biosynthesis of mercapturic acids[10]. Mercaptoacetate, on the other hand, could presumeably be formed by oxidative decarboxylation of mercaptolactate and should then be included with the metabolites formed by transaminative degradation of cysteine. However, no experimental evidence is available to support this hypothesis.

So far, nothing has been said about mercaptopyruvate as a possible constituent of human urine. Unfortunately, as mercapto-pyruvate is a fairly labile compound our gas chromatographic method was not applicable to the analysis of this compound. We made attempts to stabilize mercaptopyruvate by derivatization of its carbonyl group before analysis by the method just described. Some success was achieved by conversion to the ethoxime, and we observed in fact a small component on gas chromatograms from urine with a retention time identical with that of authentic mercaptopyruvate carried through the modified analytical procedure. However, the identity of this component has not been verified by mass spectrometry and the recovery of mercaptopyruvate is far from acceptable. The only conclusion to be drawn from these experiments is that if mercapto-pyruvate is present in normal human urine, its concentrations are much lower than those of mercaptolactate.

DETERMINATION OF THIOSULFATE

It has been known for a long time that thiosulfate is present in normal human urine, but reliable methods for the determination of fairly low concentrations have not been available. A method based on the precipitation of the nickel-ethylenediamine complex of thiosulfate followed by iodometric determination was reported[11] many years ago, but gives according to our experience unreliable results. We have earlier described[12] a simple colorimetric method for determination of thiosulfate, based on the cyanolysis of thiosulfate to thiocyanate by the action of cyanide and cupric ions followed by determination of thiocyanate as its ferric ion complex. Unfortunately, this method is not sensitive enough for direct application to urine and, furthermore, other urinary compounds interfere in the cyanolysis reaction. It may, however, be used for assay of the very high concentrations of thiosulfate found in urine from cases of the very rare metabolic defect sulfite oxidase deficiency[13].

We have recently[14] improved our original cyanolysis procedure for determination of thiosulfate by the introduction of certain chromatographic steps. The key to success was the observation that certain weakly basic anion exchange resins have a very strong affinity for thiocyanate. Apparently, thiocyanate is not bound to the resin by ion-exchange, as strong binding occurs even at a high pH, where the resin is uncharged. A short description of the method is as follows: Urinary compounds interfering with cyanolysis are first removed by chromatography on the weakly basic anion exchange resin AG 3, which does not separate thiosulfate from thiocyanate. Endogenous thiocyanate is then removed on a column of Lewatit MP7080, a resin with a high affinity for thiocyanate. The sample is then cyanolyzed in the presence of cupric ions, and the thiocyanate formed adsorbed on a second column of Lewatit MP7080. It is finally eluted with a small volume of an acid solution of ferric nitrate and determined colorimetrically as the ferric thiocyanate complex. The detection limit of this method is 1 μmol/l of thiosulfate. Although the method may appear complicated, its precision is adequate for clinical studies and no interference from compounds known to occur in urine has been encountered.

DETERMINATION OF THIOCYANATE

A number of methods for the determination of thiocyanate in body fluids have been reported but are according to our opinion not completely satisfactory. Methods based on the formation of the red-coloured ferric thiocyanate complex are thus fairly insensitive and unspecific. They are thus not applicable to urine, although they may under certain conditions be used for determination of serum thiocyanate. Other colorimetric methods are based on the König reaction. Thiocyanate is first converted to a cyanogen halide, which then reacts with pyridine to produce glutaconic aldehyde. The latter is finally converted to a dyestuff by coupling with a primary amine or a compound containing reactive methylen hydrogens. A well-known method[14] of this type used bromine as the halogenating agent, but the excess of bromine had to be removed by arsenite, as it otherwise interfered in the following coupling step. Furthermore, the highly carcinogenic compound benzidine was used to produce the dye-stuff. Although benzidine may be replaced by less carcinogenic compounds[15], the fairly toxic arsenite must nevertheless be employed as a reagent and the method gives inaccurate results when applied to urine. A more specific method, reported by Boxer and Richards[16], was based on the oxidation of thiocyanate to hydrogen cyanide, which was then separated from interfering compounds by aeration into sodium hydroxide and determined by a modification of the König reaction. Unfortunately this method is fairly laborious to perform and requires special glass equipment.

We have recently developed a simple and specific method[17] for determination of thiocyanate, based on our finding that the ion-exchange resin Lewatit MP7080 has a high affinity for thiocyanate. As earlier mentioned, the resin does not bind thiocyanate by an ion-exchange mechanism but by adsorption forces, probably related to the strong chaotropic effect of thiocyanate. This interpretation is supported by the fact that thiocyanate is displaced from the resin by other chaotropic ions such as perchlorate, trichloro-acetate and nitrate, but not by non-chaotropic ions such as chloride or sulfate (Table I).

Our method consists thus of a simple chromatographic separation of thiocyanate from interferring compounds by adsorption on a small column of Lewatit MP7080, followed by elution with sodium perchlorate. Thiocyanate is then determined by a new modification of the König reaction, where hypochlorite is used as the halogenating agent. Hypochlorite is the compound to be preferred for this purpose, as it instantaneously reacts with thiocyanate and an excess of hypo-chlorite does not interfere in the following coupling step. The method may also be used for determination of thiocyanate in serum or plasma without deproteinization of the sample.

Table 1. Elution of Thiocyanate
from Lewatit MP7080 by Anions

Anion	Thiocyanate eluted[a] %
ClO_4^-	100
Cl_3CCOO^-	92
NO_3^-	74
Cl^-	0
SO_4^{2-}	0

[a]Thiocyanate (0.1 μmol) was adsorbed on a 2.5x0.7 cm column of Lewatit MP7080 and eluted with 8 ml solution of the sodium salt of the anion studied (1 mol/l).

HUMAN STUDIES

We have so far used the methods described mainly to establish reference ("normal") values for later clinical studies on patients. As shown in Table 2, the compounds formed by transaminative catabolism of cysteine represent only a minor fraction (about 0.5%) of all sulfur compounds present in normal human urine. It should be noted that the values given for thiocyanate were obtained on non-smoking subjects, as tobacco-smoking results in an elevated thiocyanate excretion.

Clinical studies are in progress and we have focussed our interest on burned patients, receiving intravenous nutrition with amino acid mixtures. Increased values for the excretion of mercaptolactate and thiocyanate have been observed in these patients, but our data are of a too preliminary nature to warrant a more detailed presentation.

SUMMARY

We have developed methods for the analysis of mercaptolactate, thiosulfate and thiocyanate in urine. These compounds are supposed to be formed by a transaminative catabolic pathway of cysteine, with mercaptopyruvate as a common precursor. Mercaptolactate is determined by a gas chromatographic method, which also allows the simultaneous determination of mercaptoacetate and N-acetylcysteine. Thiosulfate and thiocyanate are determined by colorimeteric procedures after separation from interfering compounds by chromatographic step. Although the compounds studied represent only a small fraction of the sulfur compounds present in normal human urine, the methods developed may provide means to a better understanding of the metabolism of sulfur compounds in human beings under normal and pathological conditions.

Table 2. Reference Values for Urinary Excretion of Sulfur Compounds.

Compound	Excretion[a] μmol/24 h
Mercaptolactate	37.5 ± 15.5
Mercaptoacetate	9.3 ± 2.3
N-Acetylcysteine	30.6 ± 14.3
Thiosulfate	31.7 ± 12.8
Thiocyanate	44.0 ± 22.1[b]
Sulfate	$22,200 \pm 8,800$
Total sulfur	$27,200 \pm 9,800$

[a]Means $\pm$ SD from determinations on at least 20 adult subjects of both sexes
[b]Non-smokers

REFERENCES

1. T. Ubuka, S. Umemura, Y. Ishimoto and M. Shimomura, Transamination of L-cysteine in rat liver mitochondria, Physiol. Chem. Phys. 9:91 (1977).
2. E. Kund, The reaction of β-mercaptopyruvate with lactic dehydrogenase of heart muscle, Biochim. Biophys. Acta 25:135 (1957).
3. B. Sörbo, Thiosulfate sulfurtransferase and mercaptopyruvate sulfurtransferase, in: "Metabolism of Sulfur Compounds, Vol. VII", D. M. Greenberg, ed. Academic Press, New York (1975).
4. K. Kobayashi, Studies on the new metabolites of cystine: I. Orientation studies on the in vitro formation of S-(carboxymethylthio)cysteine and S-(2-hydroxy-2-carboxyethylthio)-cysteine from cystine and its related compounds, Physiol. Chem. Phys. 2:455 (1970).
5. A. Niederwisser, P. Giliberti and K. Baerlocher, β-Mercaptolactate cysteine disulfiduria in two normal sisters. Isolation and characterization of β-mercaptolactate cysteine disulfiduria, Ped. Res. 43:405 (1973).
6. V. E. Shih, M. M. Carney, L. Fitzgerald and V. Monedjikova, β-Mercaptopyruvate sulfur transferase deficiency. The enzyme defect in β-mercaptolactate cysteine disulfiduria, Ped. Res. 11:464 (1977).
7. J. C. Crawhall and S. Segal, The intracellular ratio of cysteine and cystine in various tissues, Biochem. J. 105:891 (1967).
8. U. Hannestad and B. Sörbo, Determination of 3-mercaptolactate, mercaptoacetate and N-acetylcysteine in urine by gas chromatography, Clin. Chim. Acta In press.
9. H. Kodama, T. Ikegami and T. Araki, N-monoacetylcystine in human urine, Physiol. Chem. Phys. 6:87 (1974).
10. R. M. Green and J. S. Elce, Acetylation of S-substituted cysteines by rat liver and kidney microsomal N-acetyltransferase, Biochem. J. 147:283 (1975).
11. J. H. Gast, K. Arai and F. Aldrich, Quantitative studies on urinary thiosulfate excretion by healthy human subjects, J. Biol. Chem. 195:875 (1952).
12. B. Sörbo, A colorimetric method for the determination of thiosulfate, Biochim. Biophys. Acta 23:412 (1957).
13. F. Irreverre, S. H. Mudd, W. D. Heizer and L. Laster, Sulfite oxidase deficiency: Studies of a patient with mental retardation, dislocated ocular lenses, and abnormal urinary excretion of S-Sulfo-L-cysteine, sulfite, and thiosulfate in urine, Biochem. Med. 1:187 (1967).
14. B. Sörbo and S. Öhman, Determination of thiosulfate in urine, Scand. J. Clin. Lab. Invest. 38:521 (1978).
15. W. N. Aldridge, The estimation of microquantities of cyanide and thiocyanate, Analyst (London) 70:474 (1945).

16. A. R. Pettigrew and G. S. Fell, Simplified colorimetric deter-
    mination of thiocyanate in biological fluids, and its applica-
    tion to investigation of the toxic amblyopias, <u>Clin</u>. <u>Chem</u>.
    18:996 (1972).
17. G. E. Boxer and J.C. Richards, Determination of thiocyanate in
    body fluids, <u>Arch</u>. <u>Biochem</u>. 39:292 (1952).
18. P. Lundquist, J. Mårtensson, B. Sörbo and S. Öhman, Method for
    determining thiocyanate in serum and urine, <u>Clin</u>. <u>Chem</u>.
    25:678 (1979).

SPECTRAL PROPERTIES OF PERSULFIDE SULFUR IN RHODANESE

C.Cannella, L.Pecci, F.Ascoli, M.Costa,
B.Pensa and D.Cavallini

Institute of Biological Chemistry, University
of Rome and Centre of Molecular Biology of
C.N.R., Rome - Italy

INTRODUCTION

Rhodanese (thiosulfate:cyanide sulfurtransferase
EC 2.8.1.1.) is a mitochondrial enzyme which transfer
sulfane sulfur of thiosulfate by a double displacement
mechanism involving a sulfur-substituted enzyme [1,2].
Sulfite is the first product released during the catalytic cycle while in the second step the sulfur bound
to the enzyme is transferred to cyanide with formation
of thiocyanate and restoration of the sulfur-free enzyme. As rhodanese is crystallized in the presence of
thiosulfate the crystalline enzyme is in the form of a
sulfur intermediate i. e. the sulfur-rhodanese. Under
this condition the transferable sulfur is covalently
bound to a cysteinyl residue and a persulfide group
E-S-SH is present [3]. This group exhibits a low intensity absorption band at 335 nm which is responsible for
the quenching of the tryptophan fluorescence due to a
non radiative energy transfer. The overlap between the
persulfide absorption and tryptophan emission is responsible of this phenomenon. The quenching was originally
attributed to a direct interaction of sulfur with a tryptophan residue presumed to be present in the active
site [4]. Volini and Wang [5] reported that upon sulfur release the enzyme undergoes a conformational change from
a β-type structure to a random coil involving the move-

ment of a particular group of aminoacid residues.

The knowledge of the x-ray structure of rhodanese, recently described by Ploegman et al.[6], can be of great help in the interpretation of the enzyme behaviour in solution. According to the crystallographic analysis, the rhodanese molecule is a single polypeptide chain of 293 aminoacids with the essential cysteine 247 located at the bottom of a pocket. The walls of this pocket are formed by hydrophobic residues on one side and by hydrophylic residues on the other side.

The present communication deals with some spectrophotometric properties of rhodanese which have been interpreted in the ligth of these new evidences.

MATERIALS AND METHODS

Crystalline bovine liver rhodanese was prepared by the method of Horowitz and De Toma[7] with the following modification in the third step. The acidification at pH 6 was omitted and replaced by gel filtration on a Sephadex G 75 column (2x200 cm) equilibrated with 10 mM phosphate buffer pH 7.6 containing 1 mM thiosulfate. The enzyme was finally crystallized at least twice. The crystalline rhodanese was stored in 1.8 M ammonium sulfate pH 7.9 containing 1 mM thiosulfate and dialyzed against 10 mM phosphate buffer pH 7.6 before use. Protein concentration was determined using the molar absorptivity $\varepsilon_{280} = 57.75 \times 10^3$. A molecular weight value of 33,000 was used[8], as recently obtained from the aminoacid sequence and the x-ray crystallographic data[6]. The enzyme activity was tested according to Sörbo[9].

Oxidized glutathione and bovine serum albumin were used to prepare the corresponding persulfide derivatives by incubation in 0.1 N NaOH for 90 min as reported by Cavallini et al.[10,11]. For the determination of differential extinction coefficients of persulfide group the decrease in optical densities at 335 nm and around 273 nm upon cyanide addition were correlated to the release of thiocyanate. This latter compound was determined by the method of Sörbo[9].

All spectra were taken in thermostated cells at 10°. Ultraviolet and visible spectra were measured using a Beckman 5260 spectrophotometer. Circular dichroism measurements were performed in a Cary 60 spectropolarimeter

with a 6002 attachment. Cylindrical cells, having optical
paths of 0.1 and 1 cm, were employed. The molar ellipti-
cities (deg. cm^2/dmole) are expressed either on a molar
basis $[\vartheta]_M$, or, below 250 nm, on the average aminoacid
residue concentration, $[\vartheta]_{aa}$.

RESULTS AND DISCUSSION

Fig.1 shows the spectral modification occurring in
the region from 400 to 240 nm when rhodanese is unloaded
of transferable sulfur by addition of cyanide. Besides
the bleaching of the absorption band at 335 nm, already
assigned to the persulfide group [3], two other modifications in the absorbance are evident at 290 and 276 nm,
which are better monitored by difference spectra. As reported in Fig.2 both acceptor substrates, cyanide and
sulfite, produce on the rhodanese molecule very similar
spectral changes which are almost completely reversible
by addition of thiosulfate. These data indicate that
also the absorbance at 276 nm can be ascribed to the
persulfide group even if it is difficult to evaluate the
extent of the enzyme-sulfur complex restoration because
of the negative absorbance due to the consumption of
thiosulfate in the sample cell only. For comparison the
absorption spectra of persulfurated model compounds have
been studied in the same region. Cleavage of the persulfide group of bovine serum albumin and glutathione by
cyanide produces difference spectra very similar to
those observed for rhodanese with two peaks at 335 and
around 273 nm (Fig.3).

These results allow to correlate these absorbance
changes with the presence of persulfide group in the
enzyme molecule. The differential molar extinction coefficients for sulfur-free and sulfur-containing rhodane-
se were found to be: $\Delta\varepsilon_{276} = 550$, $\Delta\varepsilon_{335} = 120$; these values are very close to those obtained for persulfurated
serum albumin and glutathione (see Table I).

A third spectral perturbation is induced at 290 nm
by cyanide treatment of rhodanese which is still present
upon addition of thiosulfate (see Fig.2). This finding
indicates that the cleavage of rhodanese persulfide
group by cyanide causes a perturbation of (a) tryptophanyl residue(s) which is not reversed by a sulfur
donor substrate. Since x-ray studies show that trypto-

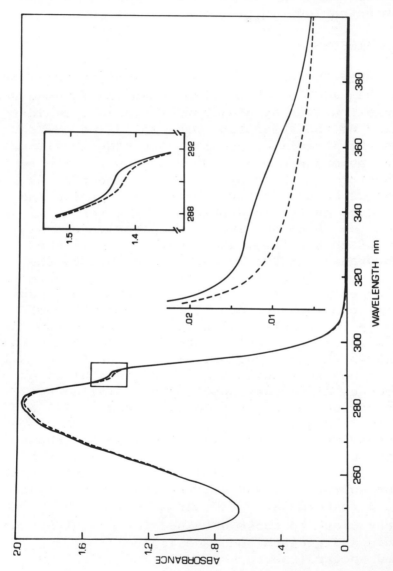

Fig.1 – Absorbance spectra of rhodanese. 34 μM sulfur–rhodanese in 10 mM phosphate buffer, pH 7.6 before (full line) and after addition of 0.34 mM cyanide (dashed line). Absorbance changes in the 310–400 nm region are evident when spectra were scanned with the ordinate scale expanded 50 times. Inset: absorbance modification between 288 and 292 nm (tryptophan region).

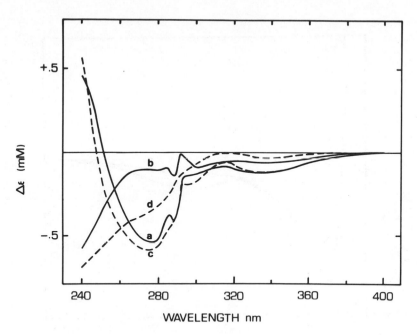

Fig.2 - Difference spectra of rhodanese in 10 mM phosphate buffer pH 7.6. Full lines: sulfur-rhodanese unloaded
with cyanide (a) and subsequently reconstituted with
thiosulfate (b) scanned against sulfur-rhodanese. Dashed
lines: c and d as a and b respectively but with the enzyme unloaded in the presence of sulfite. The molar exces-
ses of substrates added were (a) $CN^-/E=2.7$; (b) $S_2O_3^=/E=$
9.2; (c) $SO_3^=/E=9.2$; (d) $S_2O_3^=/E=50$. In the difference
spectra b and d thiosulfate was added in both sample and
reference cuvettes. All spectra were monitored at 10°C
with 35 µM enzyme in 1 cm light path cuvettes.

phan 35 is closely located to the cysteine bearing transferable sulfur the modification observed at 290 nm can
be ascribed to an irreversible perturbation of this residue when sulfur is removed from cysteine.

The structural modifications occurring in the rhodanese molecule during the catalytic cycle were also
studied by measuring the circular dichroism spectra. In

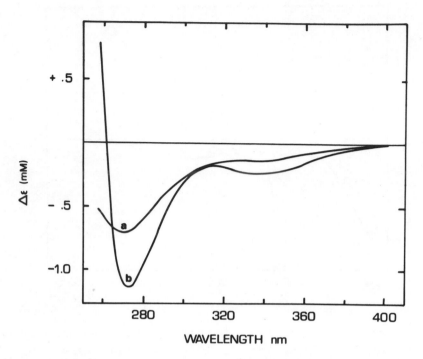

Fig.3 – Difference spectra of persulfurated model compounds in 0.1 N NaOH at 25°C. Curve a: 25 μM persulfurated bovine serum albumin treated with 320 molar excess cyanide against untreated sample; curve b: persulfurated glutathione, obtained by incubation of 0.67 μM oxidized glutathione in 0.1 N NaOH, treated with 10 molar excess cyanide against untreated sample.

figures 4 and 5, the CD in the range 380–190 nm are reported. The CD spectrum of the enzyme in the 190–250 nm region is typical of a β–structure with a negative peak at 208 nm. This spectrum and those obtained upon addition of cyanide or sulfite (Fig. 4 and 5 respectively) are in good agreement with those previously reported by Volini[5] for the native enzyme and the sulfite treated enzyme. They indicate that the removal of the persulfide group from the enzyme causes only a small increase in the ma-

Table I - Differential molar extinction coefficients of persulfide group in artificially persulfurated compounds and in sulfur-rhodanese.

persulfide derivative of:	$\Delta\varepsilon$	
	~ 273 nm	335 nm
bovine serum albumin	880	170
glutathione	1125	240
rhodanese	550	120

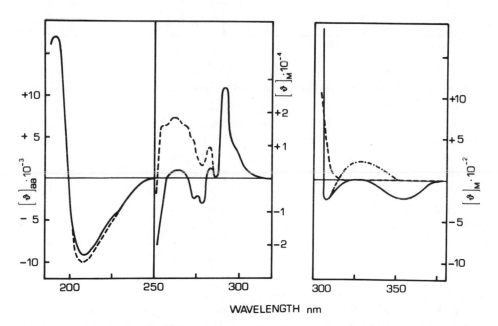

WAVELENGTH nm

Fig.4 - Circular dichroism spectra of rhodanese in 10 mM phosphate buffer pH 7.6 at 10°C.
Solid line: sulfur-rhodanese, dashed line: after treatment with 2 molar excess cyanide, dot-dashed line: reconstituted enzyme after addition of 10 molar excess thiosulfate. Unless indicated, CD spectra of reconstituted and native enzyme are indistinguishable. From left to right rhodanese concentration was: 14 μM (0.1 cm cell), 31 μM, 0.84 mM.

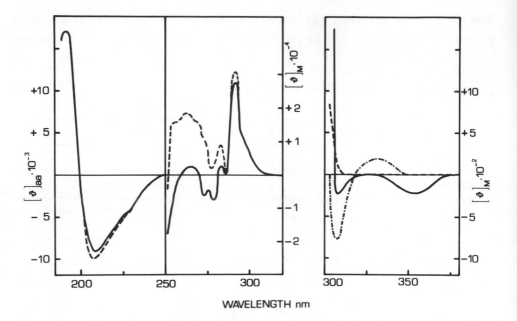

Fig.5 – Circular dichroism spectra of rhodanese in 10 mM
phosphate buffer pH 7.6 at 10°C.
Solid line: sulfur-rhodanese, dashed line: after treatment with 8 molar excess of sulfite, dot-dashed line:
reconstituted enzyme after addition of 50 molar excess
thiosulfate. Unless indicated, CD spectra of reconstituted and native enzyme are indistinguishable. From left
to right rhodanese concentration was: 14 μM (0.1 cm cell),
31 μM, 0.84 mM.

gnitude of the peak at 208 nm which is almost identical
with the two substrates. This result can be taken as a
proof of a change in the secondary structure; however
any conclusion on the extent of this change is not justi-
fied by the present experimental evidences. Recovery of
the original secondary structure is complete upon addition of thiosulfate as it is shown in the same figures.
Parallel and reversible changes in the CD spectra between
250-310 nm are also present. The bands present in this
region may arise from aromatic side chains [12]. The small
positive band at 255-270 nm, which can also be due to
the persulfide group (see absorption spectra), is largely increased upon unloading of the enzyme. The prominent

band at 292 nm and that at 282 nm, both positive, can be attributed to ^1L_b transitions of tryptophanyl residue. Their position probably corresponds to the ^1L_b band of a hydrogen bonded tryptophanyl group in a non polar environment [13]. This group could be identified with Trp 35 present in the active site of rhodanese. The increase of the CD positive band at 282 nm upon treatment with cyanide or sulfite indicates that a conformational change involving this tryptophanyl group occurs upon unloading. The band at 292 nm is slightly modified only upon addition of sulfite. Two small negative bands are present at 308 and 353 nm in the CD spectrum of the native enzyme. These bands, which disappear upon addition of cyanide or sulfite, can be tentatively related to the presence of the persulfide chromophore. A number of evidences substantiate this hypothesis: a) their disappearance in the presence of both acceptor substrates, which are known to react with the persulfide group; b) the absence of any other chromophore absorbing in this spectral region; c) the spectral changes occurring in absorption in the same region (see fig. 1 and 2) which can be attributed to the presence of persulfide group; d) the CD absorption spectra of oxidized and persulfurated glutathione, shown in Fig.6.

The oxidized glutathione shows the disulfide absorption band at 260 nm which is negative in CD. The persulfide glutathione in addition to the characteristic absorbance at 335 nm shows a strong increase of the absorbance at shorter wavelength. These bands appear to be better resolved in the CD spectrum where two positive bands at 330 and 280 nm and a negative band at 300 nm are present. All these features disappear when the persulfide group is cyanolyzed.

The restoration of the persulfide group in the rhodanese molecule by addition of thiosulfate lead to similar changes in absorption and CD spectra (Fig. 2,4 and 5). The optically active band at 308 nm present in the native molecule is still present together with a new positive band at 330 nm. These bands occur at similar wavelength where the persulfurated glutathione is optically active. The above mentioned irreversible perturbation of tryptophanyl residue(s), as a consequence of a modification of the active site, can explain also the different absorban-

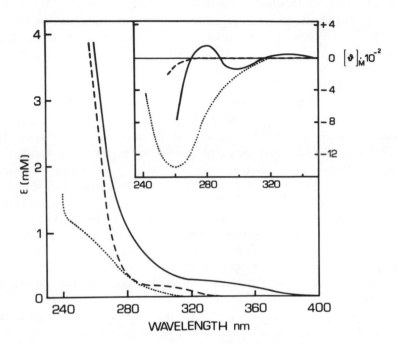

Fig.6 – Absorbance and circular dichroism of glutathione.
Dotted line: 2 mM oxidized glutathione in 5 mM NaOH;
solid line: persulfurated derivative after 90 min incubation in 0.1 M NaOH; dashed line: persulfide derivative
cyanolyzed with 10 molar excess cyanide. Circular dichroism spectra reported in the inset for comparison are
obtained in the same conditions.

ce and CD bands of persulfide group in reconstituted enzyme compared with sulfur-enzyme. This modification has
no effect on tryptophan fluorescence and does not influence the sulfur transferring ability of the enzyme. When
reconstituted rhodanese is crystallized recovery of the
native structure is complete.
      The results reported here are relevant from several
points of view. The persulfide group can be identified
by its absorptions (~273, 335 nm) and CD bands (280, 300

and 330 nm) either in model compounds or in the sulfur-rhodanese. Furthermore, unloading and loading of rhodanese are completely reversible processes, as far as the secondary structure and the catalytic activity are concerned. The perturbation of (a) tryptophanyl residue(s) of the enzyme upon unloading shown in the spectra reported in this paper, can be tentatively related with a modification at the tryptophan 35 present in the active site pocket.

Acknowledgment: the Authors wish to thank Dr R. Santucci for his help in performing the circular dichroism measurements.

REFERENCES

1. B.H. Sörbo, in:"Metabolic Pathways" D.M. Greenberg,ed. vol. 7:433-455 - Academic Press, New York (1975).
2. J. Westley, in:"Advan. Enzymol." vol. 39:327-368 (1973).
3. A. Finazzi Agrò, G. Federici, C. Giovagnoli, C. Cannella and D. Cavallini, Eur. J. Biochem. 28:89-93 (1972).
4. B. Davidson and J. Westley, J. Biol. Chem. 240:4463-9 (1965).
5. M. Volini and S-F. Wang, J. Biol. Chem. 248:7386-91 (1973).
6. J.H. Ploegman, G. Drent, K.H. Kalk and W.G.J. Hol, J. Mol. Biol. 123:557-94 (1978).
7. P. Horowitz and F. DeToma, J. Biol. Chem. 245:984-5 (1970).
8. J. Russell, L. Weng, P.S. Keim and R.L. Heinrikson, J. Biol. Chem. 253:8102-8 (1978).
9. B.H. Sörbo, Acta Chem. Scand. 7:1129-36 (1953).
10. D. Cavallini, G. Federici and E. Barboni, Eur.J.Biochem. 14:169-74 (1970).
11. G. Federici, S. Duprè, R.M. Matarese, S.P. Solinas and D. Cavallini, Int. J. Pept. Prot. Res. 10:185-9 (1977).
12. E.H. Strickland, C.R.C. Crit. Rev. Biochem. 113-76 (1974).
13. E.H. Strickland, C. Billups and F. Kaye, Biochemistry 11:3657-62 (1972).

BIOCHEMISTRY OF THIOCYSTINE

Rasul Abdolrasulnia* and John L. Wood
Department of Biochemistry
University of Tennessee
Center for the Health Sciences
Memphis, Tennessee 38163

INTRODUCTION

Thiocystine, bis-[2-amino-2-carboxyethyl]trisulfide, is a relatively stable trisulfide analog of cystine. The compound was first isolated by Fletcher and Robson[1] from hydrochloric acid hydrolysates of proteins which are rich in cystine. The trisulfide and a small amount of accompanying tetrasulfide were artefacts produced by acid-catalyzed exchange reactions between cystine and its decomposition products. Sandy et al.[2] isolated thiocystine, the trisulfide of glutathione, and the corresponding mixed trisulfide between glutathione and cysteine from Rhodopseudomonas spheroides. Also, Massey et al.[3] detected the presence of the trisulfide of glutathione in commercial samples of the compound.

Thiocystine was synthesized initially by the reaction of sulfur dichloride with N-acetylcysteine[1]. Thiocystine may also be prepared by the addition of sulfur to cysteine[4]. The compound can be prepared in good yield from the reaction of hydrogen sulfide and cystine monosulfoxide according to the methods of Savige et al.[5]

$$CySSCy \xrightarrow{peracetic} CyS^+SCy \xrightarrow{H_2S} CySSSCy + H_2O$$

where the middle species carries an O^- on the central sulfur.

---

*Present address: Department of Biochemistry
                   School of Medicine
                   Pahlavi University
                   Shiraz, Iran

483

Thiocystine has been estimated to be 10 times more soluble than cystine[1]. This makes separation of mixtures difficult. However, it can be separated by ion-exchange or electrophoresis. The compound is relatively stable in pure solution from pH 2-9. On standing in dilute solution, it slowly breaks down to cystine and sulfur. The process is accelerated in neutral or alkaline solutions by the presence of sulf-hydryl compounds.

$$CySSSCy \longrightarrow [CySS^- + CyS^+] \longrightarrow CySSCy + S^\circ$$

Szczepkowski and Wood[6] found that the action of cystathionase (L. homoserine hydrolase (deaminating), EC 4.2.1.15) on cystine produced thiocystine presumably from the intermediate product, thiocysteine, which was first detected by Cavallini and co-workers[4].

CH$_2$-S-S-CH$_2$ $\xrightarrow{\text{cystathionase}}$ CH$_2$SSH   CH$_3$
| | | |
CHNH$_2$ CHNH$_2$ CHNH$_2$ + C=O + NH$_3$
| | | |
COOH   COOH   COOH   COOH
     cystine   thiocysteine
                              (cysteine persulfide)

            NH$_2$ NH$_2$ NH$_2$
HS-CH$_2$-CH$_2$-COOH   +   HOOC-CH-CH$_2$-S-S-S-CH$_2$-CH-COOH
     cysteine   thiocystine
                        (cystine persulfide)

Thiocystine was also formed when cystine was incubated with cystathionase from R. spheroides (Wider de Xifra et al.[7]).

Szczepkowski and Wood[6] showed thiocystine can function as a substrate for rhodanese (thiosulfate:cyanide sulfur transferase, EC 2.-8.1.1). (Table I)

Table I. Thiocyanate formation from thiocystine and cyanide catalyzed by rhodanese[6].

Substrate	μmoles	μmole
Thiocystine	0.2	0.04
"	0.4	0.18
"	0.6	0.29
" (enzyme omitted)	0.6	0.03
Thiosulfate	0.6	0.03
"	2.0	0.14
"	4.0	0.25

Neuberger et al.[8] have shown that thiocystine (and other related trisulfides) can convert aminolevulinate synthetase of R. spheroides from a low activity to a high activity form. Wood and co-workers* found thiocystine, when injected intravenously into rats just prior to i.p. cyanide, will protect the animals against 3 times the LD_{50}. Also, the compound provides some protection against radiation. Thiocystine can replace cystine in the diet of growing rats.

All of these biological functions of thiocystine can be attributed to the persulfide-nature of the trisulfide. The term persulfide is used to describe compounds that can transfer sulfane sulfur to appropriate acceptors. On the basis of its similarity to thiosulfate in functioning as a substrate for rhodanese, other biological transfer reactions can be predicted for thiocystine.

RESULTS AND DISCUSSION

We have investigated the behavior of thiocystine in its transfer of persulfide sulfur (sulfane) in order to better understand its biochemical properties. When thiocystine was dissolved in a glycine phosphate buffer, pH 7.4, the solution was stable for at least 6 hours. However, when a trace of an electrophilic substance, such as cyanide, was added, the solution quickly became turbid and deposited free sulfur. The other product was cystine.

Figure 1 shows the time course of the reaction at pH 7.4 and at pH 9. Turbidity reached a maximum at 4-5 min at pH 9 and then declined as settling of elemental sulfur and slow formation of thiocyanate proceeded (Curve A). There was no thiocyanate formation at pH 7.4 indicating the absence of cyanolysis (Curve B). It is apparent that the catalysis of the decomposition by cyanide was a result of an initial scission of the trisulfide bond by cyanide to form a sulfhydryl compound.

$$CySSSCy + CN^- \longrightarrow CySS^- + CySCN$$
$$CySS^- \longrightarrow CyS^- + S^\circ$$

As reported by Szczepkowski and Wood[6], cysteine is active in catalyzing the decomposition of thiocystine. Mercaptosuccinate, utilized as a model mercaptan, accelerated the decomposition of thiocystine over approximately the same course as cyanide, but at a much lower concentration (Fig. 2,A).

$$CySSSCy + RS^- \rightleftharpoons CySS^- + RSSCy$$
$$CySS^- \longrightarrow CyS^- + S^\circ$$
$$RSSCy + CyS^- \longrightarrow RS^- + CySSCy$$

---

*Unpublished work

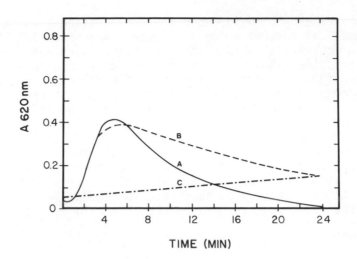

Fig. 1. Decomposition of thiocystine catalyzed
by cyanide or sulfite ion.
A: 4.4 mM thiocystine and 4.4 mM NaCN, pH 9.0.
B: 4.4 mM thiocystine and 17.8 mM NaCN, pH 7.9.
C: 4.4 mM thiocystine and 4.4 mM Na$_2$SO$_3$, pH 7.4.
"Reproduced with the permission from Bioorganic Chemistry"

In the absence of the thiol, virtually no turbidity occurred
during the period of 24 min. Thiocysteine was detected in the decomposition reaction by the addition of iodoacetic acid. Carboxy-
methylthiocysteine was isolated by electrophoresis and compared to an
authentic sample.

$$CySSSCy + 2 ICH_2COOH \xrightarrow{NaCN} CySSCH_2COOH$$
$$+ \quad + \quad 2NaI$$
$$(CySCH_2COOH)$$

The scission of trisulfide by sulfhydryl compounds is reversible.
Therefore, as shown by Szczepkowski and Wood, if one adds a disulfide, this reduces thiocysteine formation and hence, the amount of
turbidity observed. This is shown in Fig. 2 where thiocystine was
treated with mercaptosuccinic acid in the presence of oxidized glutathione or lipoic acid. Both compounds were effective in diminishing
the rate of thiocystine decomposition but lipoic acid, which is a good
acceptor of persulfide sulfur, was more efficient.

Additional evidence for the intermediate formation of the persulfide compound, thiocysteine, was obtained by spectrophotometry.
Fig. 3 shows the development of the characteristic absorption of the

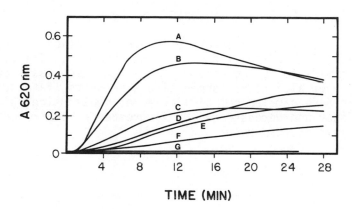

TIME (MIN)

Fig. 2.   Effect of disulfide on decomposition of 4.4 mM
thiocystine catalyzed by 0.1 mM mercaptosuccinate.
A: thiocystine and mercaptosuccinate at pH 7.4.
B: A + 4.4 mM GSSG.
C: A + 4.4 mM lipoic acid.
D: thiocystine and mercaptosuccinate at pH 8.6.
E: D + 4.4 mM GSSG.
F: D + 4.4 mM lipoic acid.
G: 4.4 mM thiocystine at pH 7.4.
"Reproduced with the permission of Bioorganic Chemistry"

persulfide group, which absorbs weakly between 330-350 nm, when thiocystine was incubated with cysteine.

    As noted above, the system approaches equilibrium due to the re-
verse reaction of the product cystine with thiocysteine (Curve A).
The decrease in the absorption peak with time (Curve B) predicts the
instability of the persulfide species. Addition of cyanide to the
medium caused immediate loss of the characteristic absorption of the
persulfide group. The resulting absorption spectrum was identical
with that of thiocystine (Curve C).

Rhodanese is a sulfhydryl enzyme and it is tempting to formulate the reaction of rhodanese with thiocystine in terms of sulfhydryl function.   The reaction should be formulated as follows:

$$CySSSCy \; + \; EnzS^- \; \longrightarrow \; EnzSSCy \; + \; CySS^-$$

$$EnzSSCy \; + \; CySS^- \; \longrightarrow \; EnzSS^- \; + \; CySSCy$$

However, there is no evidence for the formation of an intermediate disulfide between the enzyme and cysteine.   The first identifiable intermediate is the enzyme persulfide which was shown by Finazzi-Agro et al.[9] to be formed also when thiosulfate acted on the enzyme.   We have carried out similar experiments using thiocystine as a sulfur donor and demonstrated by fluorescence spectroscopy that an enzyme persulfide form results (Fig. 4).

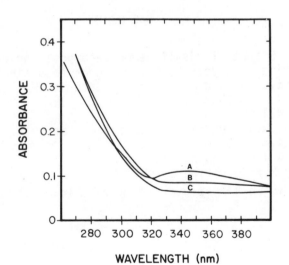

Fig. 3.   Development of thiocysteine.
A: 0.3 mM thiocystine and 0.5 mM cysteine,
     pH 8.6 after 15 min.
B: A after 2.5 hours.
C: 0.3 mM thiocystine.
Curve C was also obtained when A was made
0.3 mM in cyanide.

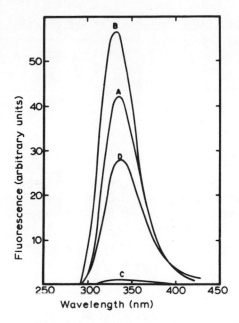

Fig. 4.   Fluorescence spectra of reaction of
rhodanese with thiocystine.
A: 5 μM rhodanese (sulfur-containing form).
B: 50 μM KCN added to A.
C: 3.7 μmoles solid thiocystine added to B.
D: 8 μmoles KCN added to C.
Excitation at 278 nm.

Abdolrasulnia and Wood[10] showed that labeled thiocystine transferred only sulfane sulfur to rhodanese when the persulfide was
formed.

The molecular specifications for the molecule which reacts with
the enzyme are fairly strict at pH 7.4. Thus, as shown in Table II,
homothiocystine does not function as well as thiocystine as a substrate, although diacetylthiocystamine is better at pH 8.6. Other
trisulfides show little or no activity.

A comparison of the rates of decomposition of the trisulfides
in the presence of mercaptosuccinate (Tables II and III) shows a
similarity between the instability of the compounds at pH 7.4 and
their substrate activity.

Table II. Trisulfide compounds as substrates for rhodanese.

Substrate	(μmol SCN$^-$ per min)			
	pH 7.4		pH 8.6	
	(a)	(b)	(c)	(d)
Thiocystine	0.77	1.38	2.03	0.14
Homothiocystine	0.16	0.896	——	0.08
Diacetylthiocystamine	0.096	2.680	——	0.02
Trithiodiacetic acid	0.024	0.096	——	0.01
Trithiodipropionic acid	0.0	——	0.096	0.0
Ethyl trisulfide	0.0	——	0.64	0.01

(c) medium was 25% ethanol.   (d) enzyme omitted.

Table III. Decomposition of Trisulfides.

	Agent μmoles	Minutes before Haze	
		pH 7.4	pH 9.0
Trithiodipropionic acid			
Mercaptosuccinate	0.25	none	none[3]
NaCN	80	150[2]	none[1,3]
Homothiocystine			
Mercaptosuccinate	0.05	8	15
NaCN	80	8.5	none[1]
N,N'-Diacetylthiocystamine			
Mercaptosuccinate	0.05	14	6
NaCN	50	1	——
Thiocystine			
Mercaptosuccinate	0.05	1	2
NaCN	20	4	2[1]

[1] strong SCN$^-$ test
[2] weak SCN$^-$ test
[3] sulfur precipitate in 15 hrs.
Incubations were done at room temperature in a volume of
2.4 ml containing 20 μmoles of trisulfide. Solutions were
observed for 20 min.

One can predict from these findings that thiocystine can function as a sulfur donor in all systems which thiosulfate can. Furthermore, since the means for synthesizing thiocystine is present in biological systems, thiocystine can be postulated as a storage form for persulfide sulfur.

REFERENCES

1. J. C. Fletcher and A. Robson, The occurrence of bis-(2-amino-2-carboxyethyl)trisulfide in hydrolysates of wool and other proteins, Biochem. J. 87:553 (1963).

2. J. D. Sandy, R. E. Davies and A. Neuberger, Control of 5-aminolaevulinate synthetase activity in Rhodopseudomonas spheroides. A role for trisulfides, Biochem. J. 150:245 (1975).

3. V. Massey, C. H. Williams, Jr., and G. Palmer, The presence of S-containing impurities in commercial samples of oxidized glutathione and their catalytic effect on the reduction of cytochrome c, Biochem. Biophys. Res. Commun. 42:370 (1971).

4. D. Cavallini, C. DeMarco, B. Mondovi, and B. S. Mori, The cleavage of cystine by cystathionase and the transfulfuration of hypotaurine, Enzymologia 22:167 (1960).

5. W. E. Savige, J. Eager, J. A. Maclaren, and C. M. Roxburgh, The S-monoxides of cystine, cystamine and homocystine, Tetrahedron Lett. 44:3289 (1964).

6. T. W. Szczepkowski and J. L. Wood, The cystathionase-rhodanese system, Biochim. Biophys. Acta 139:469 (1967).

7. E. A. Wider de Xifra, J. D. Sandy, R. C. Davies, and A. Neuberger, Control of 5-aminolaevulinate synthetase activity in Rhodopseudomonas spheroides, Philos. Trans. Roy. Soc. London, Ser. B, 237:79 (1976).

8. A. Neuberger, J. D. Sandy, and G. H. Tait, Control of aminolaevulinate synthetase in Rhodopseudomonas spheroides, Enzyme 16:79 (1973).

9. A. Finazzi Agro, G. Federici, C. Giovagnoli, C. Cannella and D. Cavallini, Effect of sulfur binding on rhodanese fluorescence, Eur. J. Biochem. 28:89 (1972).

10. R. Abdolrasulnia and J. L. Wood, Transer of persulfide sulfur from thiocystine to rhodanese, Biochim. Biophys. Acta 567:135 (1979).

SUBCELLULAR COMPARTMENTATION AND BIOLOGICAL FUNCTIONS

OF MERCAPTOPYRUVATE SULPHURTRANSFERASE AND RHODANESE

Aleksander Koj

Institute of Molecular Biology,
Jagiellonian University,
Grodzka 53, 31-001 Krakow, Poland

According to present knowledge transfer of bivalent or sulphane sulphur from thiosulphate, thiocystine, 3-mercaptopyruvate and some other donors is catalyzed by at least three different enzymes: rhodanese /EC 2.8. 1.1/, thiosulphate reductase /no EC number/ and mercapto- pyruvate sulphurtransferase /EC 2.8.1.2/ /for references see Westley, 1973; Koj et al.,1975/. The mechanism of their action is not uniform: rhodanese operates by a double-displacement mechanism /cf. Westley, 1973/, mercaptopyruvate sulphurtransferase by a sequential reaction /Jarabak and Westley,1978/, while the mode of action of thiosulphate reductase is not elucidated.

Transformations of these sulphur compounds in the animal cell are regulated by variable concentrations of metabolites and enzymes in appropriate compartments. Thiosulphate enters cells only slowly, and studies of Crompton et al./1974/ showed that its transport to mitochondria is catalyzed by a dicarboxylate carrier. Little is known about other substrates of sulphurtrans- ferases such as thiocystine, 3-mercaptopyruvate, gluta- thione or lipoic acid, but presumably their transloca- tions between subcellular compartments require active transport. Fig.1 shows a proposed model of distribution of principal sulphurtransferases and related enzymes, as well as some of their natural substrates. The scheme is certainly oversimplified and not complete but it represents the background of our investigations.

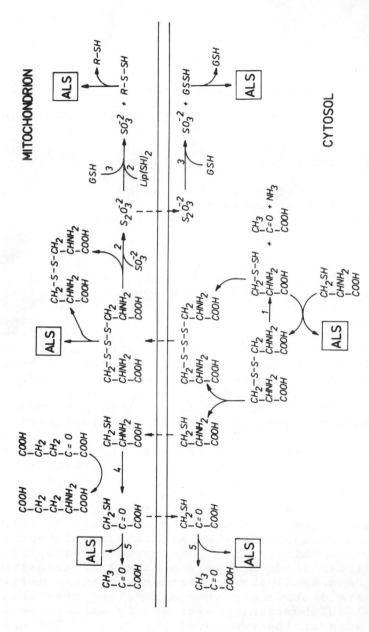

Fig.1. Principal enzymes involved in the formation of acid labile sulphide /ALS/ in the mitochondria and cytosol of rat liver; 1 – cystathionase, 2 – rhodanese, 3 – thiosulphate reductase, 4 – cysteine aminotransferase, 5 – mercaptopyruvate sulphurtransferase.

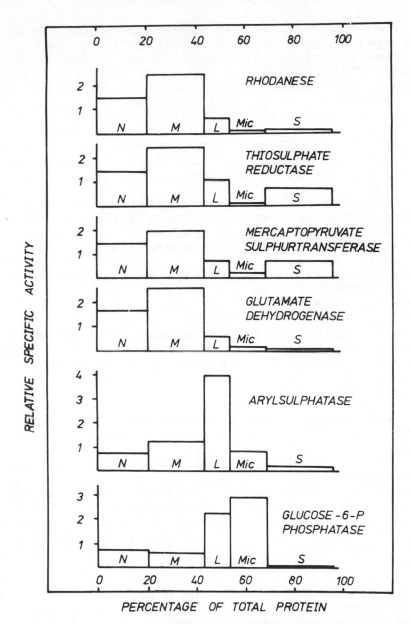

PERCENTAGE OF TOTAL PROTEIN

Fig.2. Distribution patterns of three sulphurtransferases and some marker enzymes in rat liver /after
Koj et al.,1975/. N - nuclear fraction; M -
mitochondrial fraction; L - lysosomal fraction;
Mic - microsomal fraction; S - final supernatant.

The common metabolite appearing during the action
of several enzymes is acid labile sulphide /ALS/ which
may serve for reconstitution of some iron-sulphur
proteins /Finazzi-Agro et al.,1971; Bonomi et al.,1977;
Taniguchi and Kimura, 1974; Volini et al.,1977/. Other
well established biological functions of sulphurtransferases are associated with detoxication of cyanide
/Sörbo, 1957; Auriga and Koj, 1975/ and sulphide
/Szczepkowski and Wood, 1967/, or with participation in
the redox reactions and biosynthesis of sulphur amino
acids /for references see Westley, 1973/.

Since the pioneering studies of DeDuve et al./1955/
it was assumed that rhodanese is associated with the
mitochondrial fraction. We analyzed the accurate distribution of all three sulphurtransferases in rat liver
homogenate /Koj et al.,1975/. As shown in Fig.2 rhodanese
and glutamate dehydrogenase exhibit one maximum - in the
mitochondrial fraction - while thiosulphate reductase
and mercaptopyruvate sulphurtransferase show a bimodal
distribution with a significant proportion of activity
recovered in the final supernatant. This can be explained
either by the presence of independent mitochondrial and
cytosolic isoenzymes, or by their easier release during
homogenization, especially if they were present in the
intermembrane space.

To test the latter possibility mitochondria were
further fractionated with digitonin and the enzymes
determined in intramitochondrial compartments. As shown
in Fig.3 over 80% of the activity of all three sulphurtransferases was recovered in the matrix fraction,
similarly to malate and glutamate dehydrogenases. There
is no indication that the sulphurtransferases are
located either in the intermembrane space or in the
membranes, since their activities in these fractions
are the same as those of the two marker enzymes for the
matrix.

Until now it has not been elucidated what is the
origin of mitochondrial sulphurtransferases, although
majority of mitochondrial proteins is synthesized on
cytoplasmic ribosomes and later transferred to the
organelle /for references see Schatz and Mason, 1974/.
Recently Marra and co-workers /1978/ studied selective
permeation of labelled aspartate aminotransferase isoenzymes into mitochondria in vitro. Following their
ideas we labelled crystalline bovine liver rhodanese
with ^{131}I and incubated it with fresh preparation of

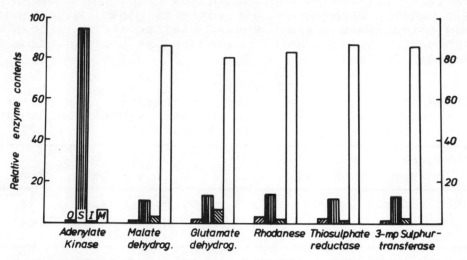

Fig.3. Distribution of some enzymes in submitochondrial
       fractions of rat liver /modified from Koj et al.,
       1975/. O - outer membrane; S - intermembrane
       space; I - inner membrane; M - matrix.

homologous mitochondria. In a control experiment identical amount of bovine serum albumin was labelled and
added to the mitochondrial suspension. When the mitochondria were briefly centrifuged we found that significantly higher proportion of radioactivity was always
associated with the sediment in the case of rhodanese
than in case of albumin. However, binding of labelled
enzyme to mitochondria may be non-specific since it is
not inhibited by "cold" rhodanese, and it is not directly
related to the amount of mitochondrial protein in the
incubation mixture. It is possible that rhodanese is
synthesized as a precursor with higher affinity to mitochondrial membranes. Direct proof of the transfer of
cytoplasmatically synthesized rhodanese into mitochondria
will require immunochemical studies with the monovalent
antiserum to the enzyme.

     In the hope that some clue of the bimodal distribu-
tion of mercaptopyruvate sulphurtransferase lies in the
structure of the enzyme we isolated cytoplasmic and
mitochondrial forms of this transferase from rat liver
/Kasperczyk et al.,1977/. However, as shown in Table 1
both forms were found to have similar molecular weights,
and after electrofocusing in sucrose gradient enzymic

activity was always located in the fractions between pH
5.1 and 6.7 with three main activity peaks. Mitochondrial
and cytoplasmic enzymes showed similar affinity toward
typical substrates as indicated by the corresponding K_m
values /Table 1/.

Table 1.
Comparison of some molecular and catalytic parameters of
mitochondrial and cytosolic mercaptopyruvate sulphurtrans
ferase from rat liver /after Kasperczyk et al.,1977/

Source of enzyme	Molecular weight	Isoelectric point	K_m /mM/	
			3-mp	2-me
Mitochondria	34,000	5.3, 5.8, 6.4	7.4	152
Cytosol	33,000	5.3, 5.8, 6.4	7.7	155

3-mp = 3-mercaptopyruvate; 2-me = 2-mercaptoethanol

At the same time we found considerable differences
among rat tissues in the ratio of mitochondrial to cytosol total activities of mercaptopyruvate sulphurtransferase /Koj et al.,1977/. In the intestine less than 30%
of cellular enzyme was found within mitochondria; heart
and transplantable Morris hepatoma showed almost equal
distribution between cytosol and mitochondria while in
the case of kidney over 80% of the enzyme was localized
in the mitochondrial fraction /Fig.4/. During these
experiments we observed that estimations of cellular
compartmentation of sulphurtransferases are usually
encumbered with an error due to leakage of mitochondrial
proteins during tissue homogenization and centrifugation.
However, it may be assumed that any activity of glutamate
dehydrogenase or rhodanese found in the cytosol derives
from mitochondria. On this ground a suitable correction
is obtained by calculating the relative amount of rhodanese or glutamate dehydrogenase in the total volume of
cytosol in respect of total mitochondrial fraction. The
experimental correction found in our experiments with
rat tissues ranged from approximately 5% in soft tissues
/brain, testis/ up to 40% in compact tissues /heart,
kidney/. In general a fairly good agreement was observed
between the values based on either rhodanese or glutamate
dehydrogenase.

Recently we have compared subcellular compartmentation of rhodanese and mercaptopyruvate sulphurtransferase
in the liver of seven vertebrate species. After calcula-

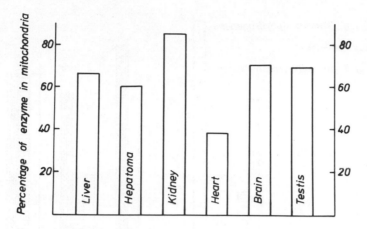

Fig.4. Cellular compartmentation of mercaptopyruvate
sulphurtransferase in some rat tissues /modified
from Kasperczyk et al.,1977/.

ting specific activities of mercaptopyruvate sulphurtransferase /units of enzyme per mg protein/ we found
very low values for fish, amphibia and reptilia in
comparison with birds and mammals /Fig.5/. On the other
hand, the highest specific activities of rhodanese were
observed in rat, hamster and frog. A high rhodanese
activity in frog tissues was already observed by us
/Koj and Frendo, 1967/ and by Schievelbein et al./1969/.
Further studies are required to elucidate whether a high
rhodanese activity in the frog is related to sex of the
animal, food intake and season of the year.

The distribution of both sulphurtransferases in
the subcellular compartments of the liver of different
animal species is better compared after calculating
total activities of the enzymes found in those fractions.
To our surprise we found that in lower vertebrates rhodanese activity does not parallel glutamate dehydrogenase distribution in this respect that much more rhodanese occurs in the cytosol fraction. After applying a
correction for mitochondrial leakage as based on the
presence of glutamate dehydrogenase in cytosol a rather
clear picture emerged /Fig.6/. In trout approximately
80% of rhodanese is localized extramitochondrially,
while in frog and axolotl the trend becomes reversed.

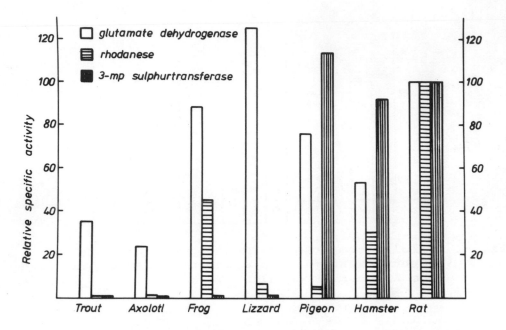

Fig.5. Specific activities of glutamate dehydrogenase,
       rhodanese and mercaptopyruvate sulphurtransfe-
       rase in mitochondrial fractions from the liver
       of seven vertebrate species. The values found
       in rat liver were assumed as 100%.

In lizzard and pigeon almost 90% of rhodanese is confi-
ned to mitochondria, and finally in rat and hamster over
99% of the enzyme is within the organelle. In case of
mercaptopyruvate sulphurtransferase no such clear relationship exists, frog being the principal exception.
However, it can be stated that poikilothermic animals,
such as trout, axolotl and lizzard have majority of
this enzyme in cytosol, while in pigeon, hamster and
rat two-third to three-quarter of mercaptopyruvate
sulphurtransferase is localized in mitochondria.

    Looking at this picture one is tempted to speculate
that in the process of vertebrate evolution sulphurtransferases of the liver greatly increased in activity
and underwent translocation from the soluble cytoplasm
into the mitochondrial compartment. Perhaps this is
related to organization of the electron transport chain
and importance of oxidative phosphorylation in homoio-
thermic animals. However, these suppositions must be
verified by more extensive and systematic studies.

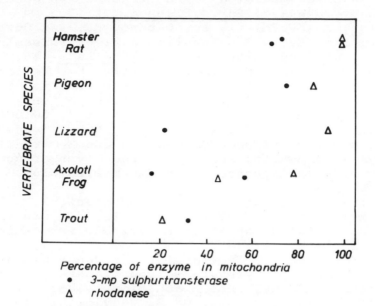

Fig.6. Cellular compartmentation of rhodanese and
mercaptopyruvate sulphurtransferase in the liver
of seven vertebrate species /after Dudek et al.,
1979/. The values plotted are corrected for
enzyme leakage from mitochondria.

        Acknowledgements: Our studies on properties and
biological functions of sulphurtransferases were supported by the National Science Foundation /USA/ through
the M. Sklodowska-Curie Fund established by contributions of the United States and Polish governments
/Grant No OIP 75-2980, principal investigator A.Koj,
US consulting scientist J.L.Wood/.

REFERENCES

Auriga,M., and Koj, A., 1975, Protective effect of
    rhodanese on the respiration of isolated mito-
    chondria intoxicated with cyanide, Bull.Acad.Pol.
    Sci., Ser.Sci.Biol., 23:305.

Bonomi, F., Pagani, S., Cerletti, P., Cannella C., 1977, Rhodanese-mediated sulfur transfer to succinate dehydrogenase, Eur.J.Biochem.,72:17.

Crompton, M., Palmieri, F., Capano, M., and Quagliariello, E., 1974, The transport of thiosulphate in rat liver mitochondria, FEBS Lett.,46:247.

DeDuve, C., Pressman, B.C., Gianetto, R., Wattiaux, R., and Appelmans, F., 1955, Tissue fractionation studies. Intracellular distribution patterns of enzymes in rat liver tissues, Biochem.J.,60:604.

Dudek, M., Frendo, J., and Koj, A., 1979, Subcellular compartmentation of rhodanese and 3-mercaptopyruvate sulphurtransferase in the liver of some vertebrate species, Comp.Biochem.Physiol. - in the press.

Finazzi-Agro, A., Cannella, C., Graziani, M.T., and Cavallini, D., 1971, A possible role for rhodanese: the formation of "labile" sulphur from thiosulphate, FEBS Lett.,16:172.

Jarabak, R., and J. Westley, 1978, Steady state kinetics of 3-mercaptopyruvate sulphurtransferase from bovine kidney, Arch.Bioch.Bioph.,185:458.

Kasperczyk, H., Koj, A., and Z. Wasylewski, 1977, Similarity of some molecular and catalytic parameters of mitochondrial and cytosolic mercaptopyruvate sulphurtransferase from rat liver, Bull. Acad.Pol.Sci.,Ser.Sci.Biol.,25:7.

Koj, A., and Frendo, J., 1967, Oxidation of thiosulphate to sulphate in animal tissues, Folia Biol./Krakow/, 15:49.

Koj, A., Frendo, J., and Wojtczak, L., 1975, Subcellular distribution and intramitochondrial localization of three sulphurtransferases in rat liver, FEBS Lett.,57:42.

Koj, A., Michalik, M., and Kasperczyk, H., 1977, Mitochondrial and cytosol activities of three sulphurtransferases in some rat tissues and Morris hepatomas, Bull.Acad.Pol.Sci.,Ser.Sci.Biol.,25:1.

Marra, E., Doonan, S., Saccone, C., and Quagliariello, E., 1978, Studies on the selective permeation of radioactivelly labelled aspartate aminotransferase isozymes into mitochondria in vitro, Eur.J.Biochem., 83:427.

Schatz, G, and Mason, T.L., 1974, The biosynthesis of mitochondrial proteins, Ann.Rev.Biochem.,43:51

Schievelbein, H., Baumeister, R., and Vogel, R., 1969, Comparative investigations on the activity of thiosulphate sulphurtransferase, Naturwissenschaften,56:416.

Sörbo, B., 1957, Sulphite and complex-bound cyanide
    as sulphur acceptors for rhodanese, Acta Chem.
    Scand.,11:628.
Szczepkowski, T.W., and Wood, J.L., 1967, The cystathio-
    nase-rhodanese system, Biochim.Biophys.Acta,139:
    469.
Taniguchi, T., and Kimura, T., 1974, Role of 3-mercapto-
    pyruvate sulphurtransferase in the formation of
    the iron-sulphur chromophore of adrenal ferredoxin,
    Biochim.Biophys.Acta,364:284.
Volini, M., Craven, D., and Ogata, K., 1977, Rhodanese
    iron protein association in bovine liver extracts,
    Biochem.Biophys.Res.Comm.,79:890.
Westley, J., 1973, Rhodanese, Adv.Enzymol.,39:327.

SEMISYNTHETIC CEPHALOSPORINS. III.[1] SYNTHESIS AND STRUCTURE ACTIVITY RELATIONSHIPS OF NOVEL ORALLY ACTIVE 7-[4-HYDROXY-3-(SUBSTITUTED METHYL)PHENYL]-ACETAMIDO-3-CEPHEM-4-CARBOXYLIC ACIDS

Abraham Nudelman[*], Abraham Patchornick, Eva Karoly-
Hafely and Frida Braun
Department of Organic Chemistry, The Weizmann Institute
of Science, Rehovot, Israel
Richard C. Erickson
Merrel Research Center, Merrell-National Laboratories,
Division of Richardson, Merrell Inc., Cincinnati, Ohio
43215, U.S.A.

The majority of the semisynthetic cephalosporins are not effective when administered orally. The main exceptions are cephalexin and cephradine, and recently reported compounds such as cefaclor[2], cefadroxyl[3] and cefatrizine[4]. In view of the fact that oral activity is observed primarily in derivatives of α-amino aryl acetic acids, we designed our synthetic course on the basis of modification of the aryl part of a readily available α-amino acid, in such a way that initial functionalization of the aromatic ring would provide a "handle", which could in turn be easily converted into a variety of functional groups. The second important criterion was the necessary optical activity of the amino acids. In order to avoid the resolution of each individual compound thus obtained, it was deemed important to start with an optically active amino acid whose modification would proceed without racemization. These two requirements were fulfilled upon chloromethylation of D-α-amino-4-hydroxyphenyl acetic acid 1. This reaction proceeded in high yield, and the isolated product 2

TABLE I.   Novel Amino Acids.

	Nu	Reagent
A	$-N_3$	NaN_3
B	-SCN	KSCN
C	-OH	$NaHCO_3$
D	$-SCH_3$	$NaSCH_3$
E	-S-(1-methyltetrazol-5-yl)	NaS-(1-methyltetrazol-5-yl)
F	$-OCH_3$	CH_3OH
G	$-S-C(=NH)-NH_2$	$H_2N-C(=S)-NH_2$
H	-S-C(imidazoline)	HN-C(=S)-NH (imidazolidinethione)
I	$-SO_3H$	Na_2SO_3
J	$-S-C(=NH)-NH-C(=NH)-NH_2$	$H_2N-C(=S)-NH-C(=NH)-NH_2$
K	$-SC(=S)OEt$	$Na-SC(=S)OEt$
L	-H	$H_2/Pd/C$

was devoid of side products derived from poly-chloromethylation, N-formylation, polymerization, etc. Amino acid 2 was in turn readily converted to a series of novel optically active acids 3, primarily by simple nucleophilic displacement of the benzylic chloride (Table I). The most convenient method for the preparation of the methoxy derivative 3F involved initial conversion to the methoxy ester 4 upon reflux in methanol. Facile hydrolysis of 4 gave the desired 3F.

$$\underline{2} \xrightarrow{CH_3OH} HO-\left\langle\text{aryl, } CH_2OCH_3\right\rangle-CH\text{-}\overset{O}{\overset{\|}{C}}OCH_3 \cdot HCl \xrightarrow{NaOH} \underline{3F}$$

with NH_2 substituent

4

Some of $\underline{3}$ were obtained as mixtures with equivalent amounts of sodium chloride upon lyophilization of the aqueous solution in which the reactions were carried out. They were used subsequently without further purification. Compound $\underline{3L}$ was obtained by hydrogenolytic reduction of the chloride in a Parr hydrogenator. Selective oxidations of $\underline{3D}$ gave sulfoxide $\underline{7}$ and sulfone $\underline{8}$.

$$HO-\left\langle CH_2\text{-}\overset{O}{\overset{\|}{S}}\text{-}CH_3\right\rangle-CH\text{-}CO_2H \xleftarrow{NaIO_4} \underline{3D} \xrightarrow{H_2O_2} HO-\left\langle CH_2SO_2CH_3\right\rangle-CH\text{-}CO_2H$$

with NH_2

7   **8**

The conversion of $\underline{3}$ to the corresponding N-tert-butoxy-carbonyl (Boc) derivatives $\underline{9}$ proceeded smoothly with compounds $\underline{3A}$, $\underline{3C}$, $\underline{3D}$, $\underline{3E}$, $\underline{3F}$, $\underline{3K}$, $\underline{3L}$, $\underline{7}$ and $\underline{8}$ by standard procedures[5] where the preferred reagent used was di-tert-butyldicarbonate $\underline{10}$[5]. The attempted conversion of amino acids $\underline{3B}$, $\underline{3G}$, $\underline{3H}$ and $\underline{3J}$ to the Boc-derivatives failed in view of the easy concomitant hydrolysis of the RS-substituent under the basic reaction conditions to give $\underline{9C}$

$$HO-\left\langle CH_2SR\right\rangle-CH\text{-}CO_2H \quad \left[(CH_3)_3 CO\overset{O}{\overset{\|}{C}}\right]_2 O \quad HO-\left\langle CH_2OH\right\rangle-CH\text{-}CO_2H$$

with NH_2; NH-$\overset{O}{\overset{\|}{C}}OC(CH_3)_3$

$\underline{3B}$, $\underline{3G}$, $\underline{3H}$, $\underline{3J}$ $\underline{9C}$

$$R = -CN, \quad -\overset{NH}{\overset{\|}{C}}\text{-}NH_2, \quad -\overset{NH}{\overset{\|}{C}}\text{-}NH\text{-}\overset{NH}{\overset{\|}{C}}\text{-}NH_2, \quad \text{(imidazoline ring)}$$

The Boc-derivative of $\underline{3I}$ could not be isolated and unambiguously identified due to the strong acidic conditions necessary to free the sulfonic acid group from its sodium salt, which caused extensive decomposition of the Boc-protecting group.

The coupling of the acids $\underline{9}$ to the respective 7-amino-cephems[6] $\underline{10}$ was carried out by the general mixed anhydride procedure

of Spencer et al.[7] using i-butyl chloroformate or pivaloyl
chloride.

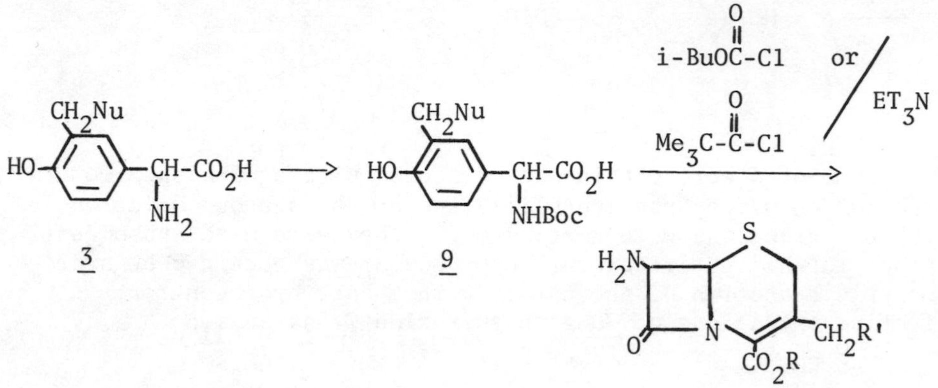

$$\underline{3} \longrightarrow \underline{9} \xrightarrow{\text{i-BuOC-Cl or Me}_3\text{C-C-Cl / ET}_3\text{N}}$$

10a (R=H)

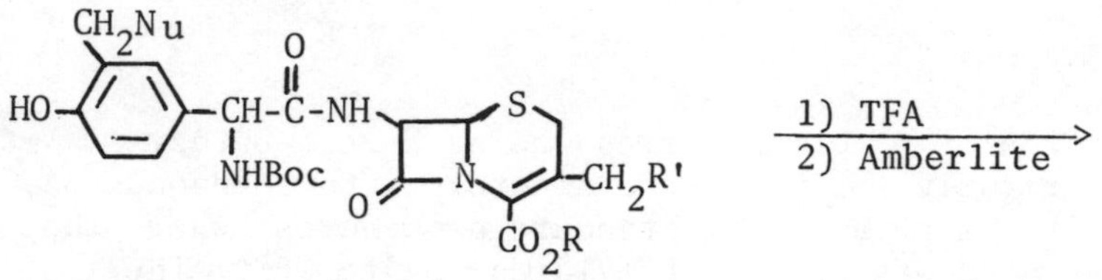

11a (R=H)

$$\xrightarrow{\text{1) TFA}}{\text{2) Amberlite}}$$

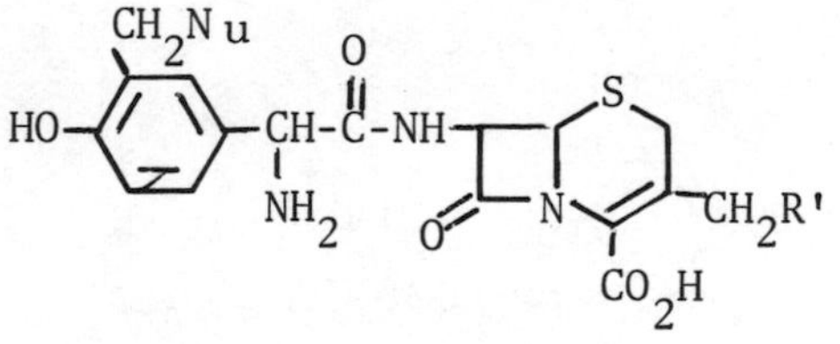

12

R'= -H,

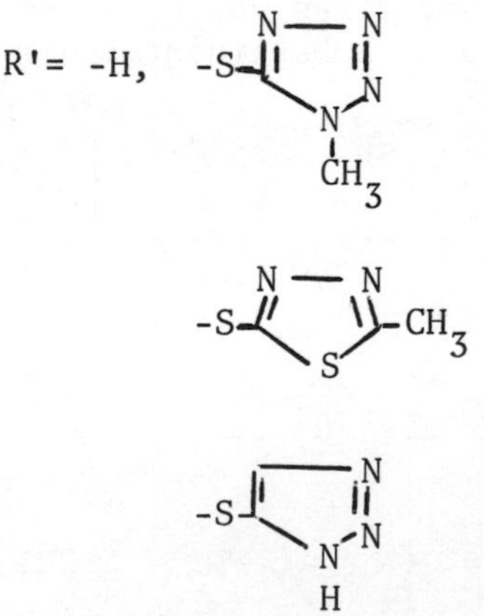

Alternatively coupling could be carried out with the esters
10B[8] in the presence of N-ethoxycarbonyl-2-ethoxy-1,2-dihydroquin-
oline (EEDQ).

TABLE II. Compartive In Vivo Activities ED_{50} (mg/Kg)

Cephalosporins 12		S PNEUM D137		S PYOG ST139		E COLI ES 59		S AUREUS M 238		S SCHOTT SA 27		S AUREUS M 290		S PNEUM K 39	
Nu	R'	S.C	P.O	S.C	P.O	S.C	P.O	S.C	P.O	S.C	P.O	S.C.	P.O	S.C.	P.O
N_3	CT			0.27	1.99	4.6	10				>80	<2.5	22.4		
N_3	CTD	5.0	14.2	0.5	2.6	12.9	13.6								
CH_3O	CT	1.25	8.4	0.11	0.55	4.6	13.5	29.6		<5	9.3			9.4	24.2
CH_3O	CTD	1.25	5	0.17	0.39	11.1	40	2.9	8.9	8.25	23.8			28.1	45.7
CH_3S	CT	1.25	5.8	0.13	0.54	4.6	6.6	20	>40	1.25	14.8	3.46		9.1	31
CH_3S	CTD	4.3	6.7	0.42	1.4					19	>40				
Cepha-lexin	Average values	36	31.8	2	1.2	3.4	7.3	2	5	6.5	13.5	1.35	1.83	41.3	13.5

$R' = \ -S-$ (CTD: 5-methyl-1,3,4-thiadiazol-2-yl, N–N / S / CH_3) ; $-S-$ (CT: 1-methyl-tetrazol-5-yl, N–N / N=N / CH_3)

$$\underline{9} \ + \ \underline{10B} \xrightarrow[\text{EEDQ}]{} \quad \underline{11B}$$

$$R=C(CH_3)_3 \qquad\qquad\qquad R=C(CH_3)_3$$

The final products $\underline{12}$ were obtained upon treatment of $\underline{11}$ with trifluoroacetic acid followed by neutralization with basic Amberlite IR-45 resin.[6]

The effective doses (ED_{50}) of a selected group of cephalosporins $\underline{12}$ against a variety of gram-positive and gram-negative bacteria are presented in Table II. The values obtained by subcutaneous (S.C.) as well as oral (P.O.) administration are compared to those of cephalexin. In general it is apparent that similar activities to those of cephalexin are observed, with somewhat higher potency against St. Pneumoniae D137 and St. Pyogenes ST139. Presently these compounds are undergoing a more extensive microbiological as well as toxicollogical evaluation.

In summary a variety of novel optically active aryl amino acids have been prepared coupling of these to various 7-aminocephems has given a family of cephalosporins whose activity, based on primary evaluation is comparable to that of cephalexin, further investigation into these compounds is being pursued.

REFERENCES

1. A. Nudelman, F. Braun, E. Karoly and R.C. Erickson, J. Org. Chem., 43:3788 (1978).
2. R.R. Chauvette and P.A. Pennington, J. Med. Chem., 18:403 (1975).
3. R.E. Buck and K.E. Price, Antimicrob. Agents Chemother., 11: 324 (1977).
4. P. Actor, J.V. Uri, L. Phillips, C.S. Sachs, J.R. Guarini, I. Zajac, D.A. Berges, G.L. Dunn, J.R.E. Hoover and J.A. Weisbach, J. Antibiot., 28:594 (1975).
5. V.F. Pozdnev, Khim. Prir. Soed., 764 (1974).
6. G.L. Dunn, J.R.E. Hoover, D.A. Berges, J.J. Taggart, L.D. Davis, E.M. Dietz, D.R. Jakas, N. Yim, P. Actor, J.V. Uri and J.A. Weisbach, J. Antibiot., 29:65 (1976).
7. J.L. Spencer, E.H. Flynn, R.W. Roeske, F.Y. Siu and R.R. Chauvette, J. Med. Chem., 9:746 (1966).
8. J. Stedman, J. Med. Chem., 9:444 (1966).

AMBIGUITIES IN THE ENZYMOLOGY OF SULFUR-CONTAINING COMPOUNDS

D. Cavallini, G. Federici, S. Duprè, C. Cannella and

R. Scandurra

Institute of Biological Chemistry of the University
of Rome, Rome, Italy

In the course of this meeting we have occasionally met reactions involving a variety of sulfur-containing compounds carried out by systems or enzymes also used for different purposes. In this final lecture we should like to point out how this phenomenon is much more general than believed, thus indicating the occurrence of a real ambiguity in the enzymology of sulfur-containing compounds. This fact became to the attention of one of us early in 1956, when cystamine (I) and lanthionamine (II) were assayed as substrates for diamine oxidase (1). It was known at that time that diamine oxidase was an enzyme of broad specificity, being able to use as substrates diamines having different carbon chain lengths and also diamines with one of the amino groups substituted with an imidazole or a guanido group (2). However it was not known whether a substantial change of the chemical composition would have impaired the enzymatic attack on the substrate. It was therefore a surprise to find an enzyme unable to recognize the substitution of a methylene carbon of the substrate with one or two atoms of sulfur. Diamine oxidase uses not only I and II as substrates but also lanthionamine sulfone, although at a lower rate than I and II (1). The interest of this finding was increased by the observation that the rate of the oxidation of I and II was in the range of that of the traditional substrates and that the oxidation was followed by the cleavage of the intermediate cystaldimine (IV) leading to the formation of compounds of biological interest like thiocysteamine (VI), hypotaurine (VII), thiotaurine (VIII) and taurine (IX). (3-7). It is evident that diamine oxidase uses as reacting points the nitrogen-containing groups while the composition of the carbon chain between them could be radically modified to the point of being partially substituted with sulfur atoms without impairing the enzymatic attack.

Cysteinesulfinic acid (XI) and cysteic acid (XII) are also known to take profit from number of enzymes used for other purposes. In this case the group producing the ambiguity is the sulfinic or the sulfonic group which simulate the carboxyl group of the carbon analogs glutamic and aspartic acids. XI and XII are able to substitute glutamic and aspartic acids in the common transaminase reac-

$$CH_2-NH_2$$
$$|$$
$$CH_2$$
$$|$$
$$S$$
$$|$$
$$S$$
$$|$$
$$CH_2$$
$$|$$
$$CH_2-NH_2$$

Cystamine I

$$CH_2-NH_2$$
$$|$$
$$CH_2$$
$$|$$
$$S$$
$$|$$
$$CH_2$$
$$|$$
$$CH_2-NH_2$$

Lanthionamine II

$$CH_2-NH_2$$
$$|$$
$$CH_2$$
$$|$$
$$SO_2$$
$$|$$
$$CH_2$$
$$|$$
$$CH_2-NH_2$$

Lanthionamine
sulfone III

Cystaldimine
IV

$$CH_2-SH$$
$$|$$
$$CH_2-NH_2$$

Cysteamine V

$$CH_2-SSH$$
$$|$$
$$CH_2-NH_2$$

Thiocysteamine
VI

$$CH_2-SO_2N$$
$$|$$
$$CH_2-NH_2$$

Hypotaurine
VII

$$CH_2-SO_2SH$$
$$|$$
$$CH_2-NH_2$$

Thiotaurine
VIII

$$CH_2-SO_3H$$
$$|$$
$$CH_2-NH_2$$

Taurine IX

tions (8–11). The possible occurrence of a specific transaminase for XI (12) is yet to be substantiated by the isolation of the pure enzyme. We had the opportunity to use XI as substrate for mitochondrial aspartate amino transferase purified up to homogeneity thus confirming that this enzyme actually transaminates XI irrespective of the possible occurrence of a specific enzyme. Other sulfur containing substrates for this transaminase are: (XIII to XV), cysteine-S-sulfonate (13), alanine thiosulfonate (14), and serine-O-sulfate (15). It is remarkable that while the enzyme uses XV as substrate, it is inactive towards serine-O-phosphate. The fate of the transamination products is quite different. In fact, while pyruvate-β-sulfonate produced by the transamination of cysteic acid is a stable compound (9), the transamination products of the other substrates release: sulfite XI and thiosulfate XIII and XIV. The behaviour of serine sulfate (XV) is quite interesting

and merits consideration. This compound is substrate of purified
aspartate amino transferase, however, when incubated in the absence
of the keto acid acceptor it is α,β-eliminated yielding ammonia,
sulfate and pyruvate. Aminoacrylate, which is produced in the course of this reaction as an intermediate inactivates the enzyme by
reacting with sulfhydryl and other groups of the active site. This
type of inhibition, frequently called suicide reaction (15), occurs

$$CH_2-SH$$
$$CH-NH_2$$
$$COOH$$

Cysteine X

$$CH_2-SO_2H$$
$$CH-NH_2$$
$$COOH$$

Cysteinesulfinic
acid XI

$$CH_2-SO_3H$$
$$CH-NH_2$$
$$COOH$$

Cysteic acid
XII

$$CH_2-SSO_3H$$
$$CH-NH_2$$
$$COOH$$

Cysteine-S-sulfonate
XIII

$$CH_2-SO_2SH$$
$$CH-NH_2$$
$$COOH$$

Alanine thiosul-
fonate XIV

$$CH_2-O-SO_3H$$
$$CH-NH_2$$
$$COOH$$

Serine-O-sulfate
XV

also with other substrates and other enzymes (16). Another point of
interest is that thiosulfate (17) and other S-containing compounds
add spontaneously to the intermediate aminoacrylate thus removing
the inhibition. Incidentally this procedure not only represents a
way to protect these enzymes from their suicidal inclinations but
is also an alternative way to produce a number of sulfur containing
compounds. Thus cysteine-S-sulfonate (and consequently cysteine upon
the reductive removal of sulfite) is produced in the presence of
thiosulfate (17), thialysine in the presence of cysteamine (16).

Purified bacterial aspartate β-decarboxylase has been found
to use also XI in a manner very similar to that observed in the case
of aspartate. The final products in the case of XI are alanine and
sulfite (18). To our knowledge XIII, XIV and XV have not been assayed
as possible substrates for this enzyme and we will not be surprised
if they should function too.

Another point of interest related to the enzymatic similarity
of the sulfur analogs of aspartic and glutamic acids is an observation made some years ago but not published. Living rats injected
with XI or XII excrete in the urine an amount of glutamic and aspartic acids much higher than under normal conditions. This finding indicates a competition at the level of the kidney reabsorbing system
between XI or XII and glutamic or aspartic acid. If so it is evident

that the ambiguity occurs also at the level of the transport systems.

Still debated is the question of the occurrence in animal tissues of decarboxylases specific for the decarboxylation of XI and XII
distinct from glutamate decarboxylase. A number of reports lend support to the occurrence of a single enzyme operating the decarboxylation of XI and XII, although at different rates (19-21). Some dis-
agreement, however, has appeared at this regard (22).

$$CH_2\text{--}SH \qquad\qquad CH_2SO_2H \qquad\qquad CH_2\text{--}SO_3H$$
$$CH_2 \qquad\qquad\qquad\quad CH_2 \qquad\qquad\qquad\quad CH_2$$
$$CH\text{--}NH_2 \qquad\qquad\quad CH\text{--}NH_2 \qquad\qquad\quad CH\text{--}NH_2$$
$$COOH \qquad\qquad\qquad COOH \qquad\qquad\qquad COOH$$

Homocysteine	Homocysteine	Homocysteic
XVI	sulfinic acid	acid XVIII
	XVII	

To our knowledge the possibility that glutamate decarboxylase could
operate also the decarboxylation of XI and XII has not been eliminated. Actually we have been surprised to find deleted out of the
official Enzyme Nomenclature the name of cysteinesulfinic acid decarboxylase, formerly listed under the number 4.1.1.29, and now
listed as a side property of glutamate decarboxylase registered
under the number 4.1.1.15 (23). Decarbosylation of glutamate and
that of XI and XII in rat brain have actually been reported to behave similarly under a set of various conditions (24), on the other
hand the variation of the ratio of the decarboxylation rate of glutamate compared with that of XI of rat brain with age has been claimed as an indication of the occurrence of different enzymes (25).
The occurrence of different forms of glutamate decarboxylase in various tissues (26) complicates the solution of this problem. Bacterial glutamate decarboxylase has been found inactive on XI, XII and
XVIII while it is very active on homocysteinesulfinic acid (XVII).
The inactive compounds however exhibit a strong competitive inhibition on the decarboxylation of XVII (24). This finding cannot be
used to understand the mechanism of the animal decarboxylase, because taurine is not an important metabolite in the bacterial kingdom and bacteria could have developed enzymes with different fina-
lities. A preferential activity toward XVII compared with XI is
also exhibited by the beef liver glutamate dehydrogenase (27). XVII
seems therefore a better simulator of glutamic acid than the lower
homolog (XI). Another sulfur containing simulator of glutamate for
the dehydrogenase is carboxymethylcysteine (XIX). While the product
of the oxidation of XVII is the respective sulfinyl keto acid, in
the case of XIX the occurrence of a sulfur atom in β promotes further reactions and thioglycolic acid has been detected among the
final products (28). Snake venom L-aminoacid oxidase is inactive

with the L forms of aspartic and glutamic acid while it uses aminoa-
dipic and aminopimelic acids as substrates. It is interesting to
point out here that the sulfur containing dicarboxylic acids XIX,
XX and XXI are good substrates for this enzyme thus mimicking more
the higher homologues of glutamic acid than glutamic acid itself
(29).

$$
\begin{array}{ccc}
& COOH & COOH \\
COOH & | & | \\
| & CH_2 & CH_2 \\
CH_2 & | & | \\
| & S & CH_2 \\
S & | & | \\
| & CH_2 & S \\
CH_2 & | & | \\
| & CH_2 & CH_2 \\
CH-NH_2 & | & | \\
| & CH-NH_2 & CH-NH_2 \\
COOH & | & | \\
& COOH & COOH
\end{array}
$$

Carboxymethyl   Carboxymethyl   Carboxyethyl
cysteine XIX   homocysteine   cysteine XXI
                              XX

Cystathionase is an enzyme using as substrate not only cystathionine but also a number of compounds structurally related to
cystathionine like cystine (XXIII), lanthionine (XXIV) and djenko-
lic acid (XXV) (30-32). It also cleaves homoserine (31), thus acting on a sulfur free compound. This appears therefore as a reverse

| Cystathionine XXII | Cystine XXIII | Lanthionine XXIV | Djenkolic acid XXV |

$$
\begin{array}{cccc}
& & & COOH \\
& & & | \\
COOH & COOH & & CH-NH_2 \\
| & | & & | \\
CH-NH_2 & CH-NH_2 & COOH & CH_2 \\
| & | & | & | \\
CH_2 & CH_2 & CH-NH_2 & S \\
| & | & | & | \\
S & S & CH_2 & CH_2 \\
| & | & | & | \\
CH_2 & S & S & S \\
| & | & | & | \\
CH_2 & CH_2 & CH_2 & CH_2 \\
| & | & | & | \\
CH-NH_2 & CH-NH_2 & CH-NH_2 & CH-NH_2 \\
| & | & | & | \\
COOH & COOH & COOH & COOH
\end{array}
$$

ambiguity, i.e., an enzyme involved in the cleavage of a number of
sulfur compounds acting on a sulfur free substrate. However the Km
for homoserine is so low compared with that for the other substrates that this appears the physiological substrate. A point to stress
is that this enzyme, although of broad specificity, is very poorly

active on homocysteine, which is very similar to homoserine while
it cleaves cysteine into pyruvate, NH_3 and H_2S (33). This intriguing finding was explained when it was observed that the substrate
for this reaction is not cysteine but cystine present in the samples of cysteine. Cystine is cleaved by cystathionase into ammonia,
pyruvate and thiocysteine, the latter being than reconverted into
cystine by the excess of cysteine (34, 35). It is evident that sulfhydryl and hydroxyl groups are not enzymically similar. It is very
interesting to remark at this point how sulfur in the form of sulfhydryl groups is subject to variable degree of specificity. In
fact sulfhydryls are non-specifically oxidized to the disulfide level by a number of metal ions, by metal containing proteins and by
other catalysts, provided they have the appropriate redox potential.
In the contrary the oxidation of cysteine and cysteamine to the
respective sulfinates requires two distinct oxygenases, although
the reaction is formally very similar (36, 37). On the other hand
the oxidation of dihydrolipoic acid requires also a specific dehydrogenase.

$$
\begin{array}{ccc}
\text{CH}_2\text{-NH}_2 & \text{CHO} & \text{COOH} \\
| & | & | \\
\text{CH}_2 & \text{CH}_2 & \text{CH}_2 \\
| & | & | \\
\text{CH}_2 & \text{CH}_2 & \text{CH}_2 \\
| & | & | \\
\text{SO}_2\text{H} & \text{SO}_2\text{H} & \text{SO}_2\text{H}
\end{array}
$$

Homohypotaurine	Semialdehyde	Succinic mono-
XXVI	XXVII	sulfinic analog
		XXVIII

Cystathionine β-synthetase has a broad specificity with respect to the sulfur compound to be added to the serine moiety and
also with respect to serine. Serine can be replaced by O-acetylserine, cyanoalanine, chloroalanine and even by cysteine, while cysteamine, mercaptoethanol, thioglycolic acid, sulfide, cysteine and
other thiol compounds could replace homocysteine (38, 39). This
β-replacement is so fast that it has been used to prepare a number
of cysteine-S-derivatives by simply passing the appropriate reagents through a column of a bacterial synthetase bound to a rigid
support (40). The table reported in the next page lists some of the
substrates used by this enzyme.

The products of the decarboxylation of homocysteine-sulfinic
acid and homocysteic acid are respectively homohypotaurine (XXVI)
and homotaurine. The biological significance of these two compounds
is not yet known nor whether they are produced from homocysteine in
the animal body. However, their similarity to GABA has stimulated
a number of investigations on the relationship of these compounds
at the enzymatic level. The system GABA transaminase succinic semi-
aldehyde dehydrogenase, which converts GABA into succinic acid has

CYSTATHIONINE-β-SYNTHASE
(Serine sulfhydrase)
EC 4.2.1.22

Main reaction: Serine + Homocysteine ⟶ Cystathionine

Serine can be   Homocysteine can   S-containing products detected by
replaced by: be replaced by: using appropriate substrates:

O-acetylserine   H₂S   Cysteine
β-Cl-alanine   Mercaptoethanol   OH-ethyl cysteine
β-CN-alanine   Cysteamine   Aminoethyl cysteine
Cysteine   Cysteine   Lanthionine
Various alkyl-S-   Methyl-mercaptan   Methylcysteine
Cysteines   Various mercaptans   S-alkyl cysteine

General reaction: β-X-Ala + HS-R ⟶ Cys-R + XH

where X = OH, SH, CN, CH₃COO, S-R

Proposed new name: β-X-Ala sulfurtransferase.

been found operative also in converting XXVI into the monosulfinic analog of succinic acid XXVIII with a mechanism identical to that known for GABA (41). The physiological role of this reaction is unknown at present.

Du Vigneaud was among the first to investigate the biological and biochemical behaviour of an aminoacid having part of the carbon chain replaced by sulfur. Thienylalanine was used as a sulfur analog of phenylalanine (XXIX) and it was found that XXIX was an inhibitor of rat growth by competing with phenylalanine (42). Aminoethylcysteine, the sulfur analog of lysine (thialysine XXX) was prepared

Thienylalanine XXIX   Thialysine XXX   Selenalysine XXXI

a number of years ago (43) and more recently was the selenium analog XXXI (44). Thialysine is so strictly related to lysine that aminoethylation of cysteine residues in proteins is frequently used in order to increase the points of cleavage of proteins by trypsin (45). In our laboratory thialysine was prepared with the aim to test its possible role as an intermediate between cysteine and ethanolamine in the biosynthesis of cysteamine. We did not obtain clear cut results on this point, however we were able to obtain evidence on the ability of the living rat to produce cystamine and taurine from thialysine (46). More recently it has been demonstrated that certain tissues, in particular kidney extracts, are able to decarboxylate lanthionine, producing thialysine which in its turn is metabolized to taurine (47). Whether this could be taken as a possible alternative for the production of taurine through the cysteamine pathway by those tissues without the cysteinesulfinic pathway, is now under investigation.

The similarity of thialysine with lysine prompted a series of investigations aimed at establishing how far the delicate mechanism of protein biosynthesis could distinguish the two compounds and whether thialysine could function as an antagonist of lysine in animals and bacteria. In recent studies it was found that thialysine and also selenalysine are used by the protein synthesizing system of E. coli, that of rat liver and that prepared from rabbit reticulocites (48, 49). Both the lysine analogs are in fact activated by

aminoacyl–tRNA synthetase, are transferred to $tRNA^{lys}$ and are incorporated into polypeptides. Furthermore both the analogs act as competitive inhibitors of lysine in all these reactions.

           Thiaproline XXXII   Selenaproline XXXIII

Thiazolidine carboxylic acid (thiaproline XXXII) and selenazolidi-
ne carboxylic acid (selenaproline XXXIII) which are respectively
the S containing and the Se containing analogs of proline, behave
similarly (50, 51). These results indicate that even the very precise system of protein biosynthesis is unable to distinguish between
an S or Se atom and a methylene carbon in the substrate. The ability
of S and Se substituted amino acids to mimick the carbon containing
analogs suggested an attempt to use this kind of competition in
those cases where protein biosynthesis is very active. Thus it was
found that thialysine inhibits the replication of the Mengovirus in
mammalian cell cultures, although the growth of the Vescicular Sto-
matitis Virus was unaffected (52). An attempt to use selenalysine
as an antagonist of lysine in order to depress the growth of can-
cerous cells, however, has not given promising results up to now.

    The competition between sulfur and selenium in the biological
field is a well known phenomenon and sulfur amino acids with sulfur
replaced by selenium have been occasionally detected (53). In our
laboratory we have found that rhodanese is able to use selenosulfa-
te in the place of thiosulfate producing selenocyanide as the final
product (54). As an intermediate the seleno–containing enzyme is
formed which is the analog of the sulfur charged enzyme when thiosulfate is the substrate. In the case of the selenoenzyme the typical absorption at 330 nm detected in the active site as the result
of the persulfide formation (55) is shifted to 375 nm and is even
more pronounced, indicating the formation of a perselenosulfide
group (54). Whether to the selenorhodanese form could be ascribed
the role of the intermediate for the transfer of selenium from inorganic to organic compounds of biological interest is a stimulating
approach worth further investigation.

    As a conclusion of this short survey on the enzymatic behaviour
of sulfur compounds it appears that sulfur and selenium are able,
to a certain extent, to replace each other and both to replace carbon, being poorly distinguished by a number of enzymes which, otherwise, exhibit definite specificity towards other properties of the
substrate. Another relevant point to remark, which could be the con-

sequence of the first conclusion, is the observation that the large variety of organic and inorganic sulfur compounds occurring in living organisms are produced by a relatively low number of specific enzymes. Many sulfur compounds arise in fact by the action of enzymes used also for reactions not involving sulfur. Since it is frequent that the enzymatic product of a compound containing sulfur is more labile than the product of the original substrate, the product may undergo further non enzymatic changes to other compounds. This fact helps explain why a limited number of specific enzymes are able to produce such large number of sulfur compounds as those found in living organisms.

REFERENCES

1. D. Cavallini, C. De Marco, and B. Mondovì, The oxidation of cystamine and other sulfur-diamines by diamine-oxidase preparations. Experientia. 1956, 12; 377.
2. B.A. Zeller, Diamin-oxidase, Adv. Enzymol. 1942, 2: 93.
3. D. Cavallini, C. De Marco, and B. Mondovì, Cystaldimine: the product of oxidation of cystamine by diamine-oxidase. Biochim. Biophys. Acta, 1957, 24: 353.
4. C. De Marco, G. Bombardieri, F. Riva, S. Duprè and D. Cavallini, Degradation of cystaldimine, the product of oxidative deamination of cystamine. Biochim. Biophys. Acta, 1956, 100:89.
5. D. Cavallini, C. De Marco, and B. Mondovì, The enzymic conversion of cystamine and thiocysteamine into thiotaurine and hypotaurine. Enzymologia, 1961, 23:101.
6. G. Rotilio, B. Mondovì, and D. Cavallini, Oxidation of cystamine by diamine oxidase. Italian J. Biochem., 1966, 15:250.
7. S. Duprè, F. Granata, G. Federici, and D. Cavallini. Hydroxyamine compounds produced by the oxidation of diamines with diamine oxidase in the presence of alcohol dehydrogenase. Physiol. Chem. & Phys., 1975, 7, 517.
8. P.P. Cohen, Transamination with purified enzyme preparations. J. Biol. Chem., 1940, 136, 565.
9. S. Darling. Cysteic acid transaminase. Nature, 1952, 170: 749.
10. F. Chatagner, B. Bergeret, T. Sejourné, and C. Fromageot. Transamination et desulfination de l'ac. cysteinesulfinique. Biochim. Biophys. Acta, 1952, 9: 340.
11. F.J. Leinweber and K.J. Monty. Microdetermination of cysteinesulfinic acid. Anal. Biochem., 1962, 4: 252.
12. M. Recasenses, M.M. Gabellec, G. Mack and P. Mandel. Comparative study of miscellaneous properties of cysteinesulfinate transaminase and glutamate transaminase in chick retina homogenate. Neurochemical Res., 1978, 3: 27.
13. M. Coletta, S.A. Benerecetti, and C. De Marco. Enzymic transamination of S-sulfocysteine. Italian J. Biochem., 1961, 10:244.
14. C. De Marco and M. Coletta. Thiosulfate production during transamination of alaninethiosulfonic acid. Biochim. Biophys.Acta

1961, 47: 257.

15. R.A. John and P. Fasella. The reaction of L-serine-O-sulfate
     with aspartate aminotransferase. Biochemistry, 1969, 8: 4477.

16. T.S. Soper and J.M. Manning. β-elimination of β-alosubstrates
     by D aminoacid transaminase associated with inactivation of the
     enzyme. Trapping of a key intermediate in the reaction. Bioche-
     mistry, 1978, 17: 3377.

17. D. Cavallini, G. Federici, F. Bossa and F. Granata. The protec-
     tive effect of thiosulfate upon the inactivation of aspartate
     aminotransferase by aminoacrylic acid producing substrates.
     Eur. J. Biochem., 1973, 39: 301.

18. K. Soda, A. Novogrodsky and A. Meister. Enzymatic desulfina-
     tion of cysteinesulfinic acid. Biochemistry, 1964, 3: 1450.

19. J.G. Jacobsen, L.L. Thomas and L.H. Smith. Properties and di-
     stribution of mammalian L-cysteinesulfinate carboxy-lyase.
     Biochim. Biophys. Acta, 1964, 85: 103.

20. M.C. Guion-Rain and F. Chatagner. Rat liver cysteinesulfinate
     decarboxylase: some observation about substrate specificity.
     Biochim. Biophys. Acta, 1972, 276: 272.

21. G. Federici, U. Tomati, L. Santoro and C. Cannella. Attività
     della L-cisteinsolfinico e della L-cisteico decarbossilasi nel
     fegato e rene di mammiferi. Boll. Soc. Ital. Biol. Sper., 1973,
     49: 675.

22. J. Lin, R.H. Demeio and R.M. Metrione. Purification of rat li-
     ver cysteinesulfinate decarboxylase. Biochim. Biophys. Acta,
     1971, 250: 558.

23. Enzyme Nomenclature. Recommendations of the IUPAC and the IUB.
     Elsevier Publ. Co. Amsterdam, 1973, 272.

24. B. Jolles-Bergeret and M.H. Vaucher. Decarboxylation of DL-ho-
     mocysteinesulfinic acid in rat brain. J. Neurochem., 1973, 20:
     1797.

25. S.S. Oja, M.L. Karvonen and P. Lähdesmäki. Biosynthesis of tau-
     rine and enhancement of decarboxylation of cysteinesulfinate
     and glutamate by the electrical stimulation of rat brain sli-
     ces. Brain Res., 1973, 55: 173.

26. D.T. Whelan, C.R. Scriver and F. Mohyuddin. Glutamic acid de-
     carboxylase and γ-aminobutyric acid in mammalian kidney. Natu-
     re, 1969, 224: 916.

27. B. Jolles-Bergeret. Desamination oxidative de l'acide L-homo-
     cysteinesulfinique par la L-glutamodeshydrogenase de foie de
     boeuf. Inhibition de l'enzyme apr l'acid L-homocysteique. Bio-
     chim. Biophys. Acta, 1967, 146: 45.

28. A. Fiori, M. Costa and E. Barboni. Oxidation of S-carboxymethyl-
     cysteine by glutamate dehydrogenase. Physiol. Chem. & Phys.,
     1972, 4: 457.

29. A. Rinaldi, M.B. Fadda and C. De Marco, Oxidative deamination
     of carboxyethyl-cysteine and carboxymethyl-homocysteine. Phy-
     siol. Chem. & Phys., 1978, 10: 47.

30. D. Cavallini, C. De Marco, B. Mondovì and B.G. Mori. Cleavage
     of cystine by cystathionase and the transulfuration of hypotau-

rine. Enzymologia, 1960, 22: 161.

31. Y. Matsuo and D.M. Greemberg. A crystalline enzyme that clea-
    ves homoserine and cystathionine. J. Biol. Chem., 1959, 234:
    516.

32. D.M. Greenberg, P. Mastalerz and A. Nagabhushanam. Mechanism
    of djenkolic acid decomposition by cystathionase. Biochim.
    Biophys. Acta, 1964, 81: 158.

33. D. Cavallini, B. Mondovì, C. De Marco and A. Scioscia Santoro.
    The mechanism of desulfhydration of cysteine. Enzymologia,1962,
    24: 253.

34. B. Mondovì, A. Scioscia Santoro and D. Cavallini. Further evi-
    dence on the identity of cystathionase and cysteine desulfhy-
    drase. Arch. Biochem. Biophys., 1963, 101: 363.

35. D. Cavallini, B. Mondovì, C. De Marco and A. Scioscia Santoro.
    Inhibitory effect of mercaptoethanol and hypotaurine on the de-
    sulfhydration of cysteine by cystathionase. Arch. Biochem.
    Biophys., 1962, 96: 456.

36. D. Cavallini, G. Federici, G. Ricci, S. Dupre, A. Antonucci and
    C. De Marco. The specificity of cysteamine oxygenase. FEBS-Let-
    ters, 1975, 56: 348.

37. J.L. Lombardini, T.P. Singer and P.D. Boyer. Cysteine oxygenase.
    J. Biol. Chem., 1969, 244: 1172.

38. A.E. Braunstein, E.V. Goryachenkova, E.A. Tolosa, I.H. Wileha-
    rdt and   Yefremova. Specificity and some other properties of
    liver serine sulfhydrase: evidence for its identity with cy-
    stathionine β-synthase. Biochim. Biophys. Acta, 1971, 242:247.

39. J. Kraus, S. Packman, B. Fowler and L.E. Rosemberg. Purifica-
    tion and properties of cystathionine β-synthase from human li-
    ver. J. Biol. Chem., 1978, 253: 6523.

40. I. Willhardt and P. Hermann. Enzymatic synthesis of labelled
    sulphur containing amino acids. Abstracts of the Warsaw Meeting
    on Low Molecular Weight Sulphur Containing Natural Products.
    Warsaw 1976, p. 41.

41. D.G. De Gracia and B. Jollès-Bergeret. Sulfinic and sulfonic
    analogs of γ-aminobutyric acid and succinate semialdehyde,
    new substrates for the aminobutyrate aminotransferase and the
    succinate semialdehyde dehydrogenase of Pseudomonas Fluorescens.
    Biochim. Biophys. Acta, 1973, 315: 49.

42. M.F. Ferger and V. Du Vigneaud. The antiphenylalanine effect of
    thielylalanine for the rat. J. Biol. Chem., 1949, 179: 61.

43. D. Cavallini, C. De Marco, B. Mondovì and G.F. Azzone. A new
    synthetic sulfur-containing amino acid, S-aminoethyl cysteine.
    Experientia, 1955, 11:61.

44. C. De Marco, A. Rinaldi, S. Dernini and D. Cavallini. The syn-
    thesis of 2-aminoethyl-2-amino-3-carboxyethyl selenide (selena-
    lysine). A new analogue of lysine. Gazzetta Chimica Italiana,
    1975, 105: 1113.

45. R.D. Cole. S-aminoethylation, in "Methods in Enzymology" S.P.
    Colowick and N.O. Kaplan, eds., Academic Press, New York,1967,
    XI: 315.

46. D. Cavallini, B. Mondovì and C. De Marco. A preliminary report on the metabolism of S-aminoethylcysteine by the rat in vivo. Biochim. Biophys. Acta, 1955, 18: 122.

47. In press.

48. C. De Marco, V. Busiello, M. Di Girolamo and D. Cavallini. Selenalysine and protein synthesis. Biochim. Biophys. Acta., 1976, 454: 298.

49. M. Fàbry, P. Hermann and I. Rychlìk. Poly(A)-directed formation of thialysine oligopeptides in E. coli cell-free system. Collection Czechoslov. Commun., 1977, 42: 1077.

50. C. De Marco, V. Busiello, M. Di Girolamo and D. Cavallini. Selenaproline and protein synthesis. Biochim. Biophys. Acta 1977, 478: 156.

51. A. Antonucci, C. Foppoli, C. De Marco and D. Cavallini. Inhibition of protein synthesis in rabbit reticulocytes by selenaproline. Bull. Mol. Biol. and Med., 1977, 2: 80.

52. A. Scioscia Santoro, D. Cavallini, A.M. Degener, R. Perez-Bercoff and G. Rita. Effect of L-aminoethyl-cysteine, a sulfur analogue of L-lysine, on virus multiplication in mammalian cell cultures. Experientia, 1977, 33: 451.

53. I. Rosenfeld and O.A. Beath. Selenium: geobotany, biochemistry, toxicity and nutrition. Academic Press, New York, 1964, p. 133.

54. C. Cannella, L. Pecci, A. Finazzi Agrò, G. Federici, B. Pensa and D. Cavallini. Selenium binding to beef-kidney rhodanese. Eur. J. Biochem., 1975, 55: 285.

55. A. Finazzi Agrò, G. Federici, C. Giovagnoli, C. Cannella and D. Cavallini. Effect of sulfur binding on rhodanese fluorescence. Eur. J. Biochem., 1972, 28: 89.

CONTRIBUTORS

Abdolrasulnia, R.

Department of Biochemistry
School of Medicine
Pahlavi University
Shiraz, Iran

Abiko, Y.

Laboratory of Biochemistry,
Research Institute
Daiichi Seiyaku Co., Ltd.
Edogawa-ku, Tokyo 132,
Japan

Andersen, K.K.

Department of Chemistry,
University of New Hampshire
Durham, N.H. 03824, U.S.A.

Andreoli, V.

Department of Biochemistry,
The Medical School,
University of Perugia,
Via del Giochetto,
06100 Perugia, Italy

Antonucci, A.

Institute of Biological
Chemistry, University of
Rome and Centre of Molecu-
lar Biology, C.N.R. Rome,
Italy

Ascoli, F.

Institute of Biological
Chemistry, University of
Rome and Centre of Molecu-
lar Biology, C.N.R. Rome,
Italy

Benn, M.   Department of Chemistry,
                                  The University, Calgary,
                                  Alberta T2N 1N4, Canada

Bernstein, D.T.   Department of Chemistry,
                                  University of New Hampshire
                                  Durham, N.H. 03824, U.S.A.

Bielawska, H.   Department of Chemistry,
                                  University of Warsaw, Poland

Bouchène, M.   Nutrition Cellulaire and
                                  Lipophysiologie, C.N.R.F.
                                  Université Paris 7, 2 place
                                  Jussieu, Paris 05, France

Braun, F.   Department of Organic Che-
                                  mistry, The Weizmann Insti-
                                  tute of Science, Rehovot,
                                  Israel

Braunstein, A.E.   Institute of Molecular Biolo-
                                  gy, USSR Academy of Sciences,
                                  Moscow B-334, U.S.S.R.

Bukin, Y.V.   Cancer Research Center of
                                  USSR Academy of Medical Scien-
                                  ces, Moscow, U.S.S.R.

Cannella, C.   Institute of Biological Che-
                                  mistry of the University of
                                  Rome, Rome, Italy

Cantoni, G.L.   Laboratory of General and Com-
                                  parative Biochemistry, Natio-
                                  nal Institute of Mental Health
                                  Bethesda, Maryland 20205, U.S.A.

Cavallini, D.   Institute of Biological Che-
                                  mistry, University of Rome,
                                  Rome, Italy

Cacciapuoti, G.   Department of Biochemistry,
                                  II Chair, University of Naples
                                  First Medical School, Naples,
                                  Italy

Cartenì-Farina, M.   Department of Biochemistry,
II Chair, University of Naples
First Medical School, Naples,
Italy

Chatagner, F.   Laboratoire de Biochimie,
96 Boulevard Raspail,
75006 Paris, France

Chiang, P.K.   Laboratory of General and Comparative Biochemistry,
National Institute of Mental
Health, Bethesda, Maryland
20205,U.S.A.

Ciliberto, G.   Istituto di Chimica Biologica,
II Facoltà di Medicina e Chirurgia, Università di Napoli,
Via S. Pansini 5, 80131 Napoli,
Italy

Cimino, F.   Istituto di Chimica Biologica,
II Facoltà di Medicina e Chirurgia, Università di Napoli,
Via S. Pansini 5, 80131 Napoli,
Italy

Colonna, A.   Istituto di Chimica Biologica,
II Facoltà di Medicina e Chirurgia, Università di Napoli,
Via S. Pansini 5, 80131 Napoli,
Italy

Consalvi, V.   Institute of Biological Chemistry, University of Rome,
Rome, Italy

Costa, M.   Institute of Biological Chemistry, University of Rome and
Centre of Molecular Biology
of C.N.R., Rome, Italy

Coward, J.K.   Department of Pharmacology,
Yale University School of Medicine, New Haven,
Connecticut 06510,U.S.A.

Datko, A.H.	National Institute of Mental Health, Laboratory of General and Comparative Biochemistry Bethesda, Maryland 20205,U.S.A.
De Luca, G.	Department of Biochemistry, University of Messina, Italy
De Marco, C.	Centro di Biologia Molecolare del C.N.R., Istituto di Chimica Biologica, Università di Roma, Roma, Italy
Di Giorgio, R.M.	Department of Biochemistry, University of Messina, Messina, Italy
Di Girolamo, M.	Centro per lo studio degli Acidi Nucleici del C.N.R., Istituto di Fisiologia Generale, Università di Roma, Roma, Italy
Dupré, S.	Institute of Biological Chemistry of the University of Rome, Rome, Italy
Erickson, R.C.	Merrel Research Center, Merrel-National Laboratories, Division of Richardson, Merrel Inc., Cincinnati, Ohio 43215, U.S.A.
Federici, G.	Institute of Biological Chemistry, University of Rome, Rome, Italy
Frappier, F.	Laboratoire de Chimie Organique Biologique, Tour 44-45, Université P. et M. Curie, 4 place Jussieu, 75230 Paris Cedex 05, France
Galletti, P.	Department of Biochemistry, II Chair, University of Naples First Medical School, Naples, Italy

Gambacorta, A.

Laboratory for the Chemistry
of Molecules of Biological
Interest, C.N.R., Naples,
Italy

Gamboa, A.

Centro de Investigaciones en
Fisiología Celular, Universi-
dad Nacional Autónoma de México, México 20,D.F., México

Gaull, G.E.

Department of Human Develop-
ment and Nutrition, New York
State Institute for Basic
Research in Mental Retardation,
Staten Island, New York 10314,
U.S.A.

Giovanelli, J.

National Institute of Mental
Health, Laboratory of General
and Comparative Biochemistry
Bethesda, Maryland 20205,U.S.A.

Goryachencova, E.V.

Insitute of Molecular Biology,
USSR Academy of Sciences,Moscow
B-334, U.S.S.R.

Grue-Sørensen, G.

Department of Organic Chemistry,
The Technical University of
Denmark, 2800 Lyngby,Denmark

Guillerm, G.

Laboratoire de Chimie Organique
Biologique,Tour 44-45, Univer-
sité P. et M. Curie, 4 place
Jussieu-75230, Paris Cedex 05,
France

Hannestad, U.

Department of Clinical Chemistry, University of Linköping,
S-581 85 Linköping, Sweden

Hashimoto, Y.

Kobe Women's College of Pharmacy
Motoyama-Kitamachi, Higashinada-
ku, Kobe, Hyogo 658, Japan

Heird, W.C.

Department of Pediatrics, Columbia University College of Physicians and Surgeons, New York,
New York 10032, U.S.A.

Hoskin, F.C.G.	Department of Biology, Illinois Institute of Technology, Chicago, Illinois 60616,U.S.A
Huxtable, R.J.	Department of Pharmacology, University of Arizona,Health Sciences Center, Tucson,Arizona 85724, U.S.A.
Iwanow, A.	Department of Chemistry, University of Warsaw, Poland
Jacobsen, J.G.	Medical Department M, Odense University Hospital, DK-5000 Odense C, Denmark
Jellum, E.	Institute of Clinical Biochemistry, Rikshospitalet,Oslo and Olso Sanitetsforening, University Rheumatism Hospital, Oslo, Norway
Kameda, K.	Laboratory of Biochemistry, Research Institute, Daiichi Seiyaku Co., Ltd. Edogawa-ku, Tokyo 132, Japan
Karoly-Hafely, E.	Department of Organic Chemistry The Weizmann Institute of Science, Rehovot, Israel.
Kato, A.	Kobe Women's College of Pharmac Motoyama-Kitamachi, Higashinada ku, Kobe, Hyogo 658, Japan
Kelstrup, E.	Department of Organic Chemistry, The Technical University of Denmark, 2800, Lyngby, Denmark
Kim, S.	The Fels Research Institute and Department of Biochemistry, Temple University School of Medicine, Philadelphia,Pa.19140 U.S.A.
Kjær , A.	Department of Organic Chemistry, The Technical University of Denmark,2800, Lyngby, Denmark

Kochakian, C.D.	Laboratory Experimental Endocrinology University of Alabama, Medical Center, Birmingham, Alabama 35294, U.S.A.
Koj, A.	Institute of Molecular Biology, Jagiellonian University, Grodzka 53, 31-001 Krakow, Poland
Kontro, P	Department of Biomedical Sciences, University of Tampere, Box 607, SF-33101, Tampere 10, Finland
Lapous, D.	Nutrition Cellulaire & Lipophysiologie, C.N.R.S., Université Paris 7, 2 place Jussieu, Paris 05, France
Launay, M.	Nutrition Cellulaire & Lipophysiologie, C.N.R.S., Université Paris 7, 2 place Jussieu, Paris 05, France
Lombardini, J.B.	Department of Pharmacology and Therapeutics, Texas Tech University School of Medicine, Lubbock, Texas 79430, U.S.A.
Loriette, C.	Laboratoire de Biochimie, 96 boulevard Raspail, Paris, France
Lundquist, P.	Department of Clinical Chemistry, University of Linköping S-581 85 Linköping, Sweden
Luthy, J.	Institut fur Toxikologie der Eidgenössischen Technischen Hochschule und der Universität Zürich, CH-8603 Schwerzenbach, Switzerland
Macaione, S.	Department of Biochemistry, University of Messina, Italy

Malloy, M.H.   Department of Human Develop-
                                 ment and Nutrition, New York
                                 State Institute for Basic Re-
                                 search in Mental Retardation,
                                 Staten Island, New York, 10314
                                 and Department of Pediatrics,
                                 Columbia University College
                                 of Physicians and Surgeons,
                                 New York, New York 10032, U.S.A

Marquet, A.   Laboratoire de Chimie Organique
                                 Biologique, Tour 44-45, Univer-
                                 sité P. et M. Curie, 4 place
                                 Jussieu-75230 Paris, France

Mårtensson, J.   Department of Clinical Chemi-
                                 stry, University of Linköping
                                 S-581 85 Linköping, Sweden

Matarese, R.M.   Institute of Biological Che-
                                 mistry, University of Rome,
                                 Rome, Italy

McCormick, D.B.   Department of Biochemistry,
                                 Emory University, Atlanta,
                                 GA 30322, U.S.A.

Meister, A.   Department of Biochemistry,
                                 Cornell University, Medical
                                 College, 1300 York Avenue,
                                 New York, New York 10021,U.S.A.

Mozzi, R.   Department of Biochemistry,
                                 The Medical School, University
                                 of Perugia, Via del Giochetto,
                                 06100   Perugia,Italy

Mudd, S.H.   National Institute of Mental
                                 Health, Laboratory of General
                                 and Comparative Biochemistry,
                                 Bethesda, Maryland 20205,U.S.A.

Munthe, E.   Institute of Clinical Bioche-
                                 mistry, Rikshospitalet, Oslo
                                 and Oslo Sanitetsforening,
                                 University Rheumatism Hospital
                                 Oslo, Norway

Nisticò, G.   Department of Pharmacology,
                                    University of Messina, Italy

Noonan, P.K.   Department of Biology,Illinois
                                    Institute of Technology,
                                    Chicago, Illinois 60616,U.S.A.

Nudelman, A.   Department of Organic Chemi-
                                    stry, The Weizmann Institute
                                    of Science, Rehovot, Israel

Öhman, S.   Department of Clinical Chemi-
                                    stry, University of Linköping
                                    S-581 85 Linköping, Sweden

Øgaard Madsen, J.   Department of Organic Chemi-
                                    stry, The Technical University
                                    of Denmark, 2800 Lyngby,Denmark

Oja, S.S.   Department of Biomedical Scien-
                                    ces, University of Tampere,
                                    Box 607, SF-33101, Tampere 10,
                                    Finland

Oliva, A.   Department of Biochemistry,
                                    II Chair,University of Naples
                                    First Medical School, Naples,
                                    Italy

Orlando, M.   Laboratorio di Patologia non
                                    Infettiva, Istituto Superiore
                                    di Sanità, Rome, Italy

Orlov, E.N.   Cancer Research Center of
                                    U.S.S.R. Academy of Medical
                                    Sciences, Moscow, U.S.S.R.

Paik, W.K.   The Fels Research Institute and
                                    Department of Biochemistry,
                                    Temple University School of
                                    Medicine, Philadelphia, Pa.
                                    19140,U.S.A.

Paolella, G.   Istituto di Chimica Biologica,
                                    II Facoltà di Medicina e Chi-
                                    rurgia, Università di Napoli,
                                    Via S. Pansini 5, 80131 Napoli,
                                    Italy

Pasantes-Morales, H.   Centro de Investigaciones en
                              Fisiología Celular, Universi-
                              dad Nacional Autónoma de Mé-
                              xico, México 20, D.F.,México

Patchornick, A.   Department of Organic Chemi-
                              stry, The Weizmann Institute
                              of Science, Rehovot, Israel

Pecci, L.   Institute of Biological Che-
                              mistry, University of Rome
                              and Centre of Molecular Bio-
                              logy of C.N.R., Rome, Italy

Pensa, B.   Institute of Biological Che-
                              mistry, University of Rome
                              and Centre of Molecular Bio-
                              logy of C.N.R., Rome,Italy

Pierre, Y.   Laboratoire de Biochimie,
                              96 boulevard Raspail, 75006
                              Paris, France

Politi, L.   Institute of Biological Che-
                              mistry, University of Rome,
                              Rome, Italy

Porcellati, G.   Department of Biochemistry,
                              The Medical School, University
                              of Perugia, Via del Giochetto,
                              06100 Perugia,Italy

Prota, G.   Istituto di Chimica Organica
                              e Biologica, Università di
                              Napoli, Via Mezzocannone 16,
                              Napoli, Italy

Rassin, D.K.   Department of Human Develop-
                              ment and Nutrition,New York
                              State Institute for Basic
                              Research in Mental Retardation,
                              Staten Island, New York 10314,
                              U.S.A.

Raulin, J.   Nutrition Cellulaire & Lipo-
                              physiologie, C.N.R.S.,Univer-
                              sité Paris 7, 2 place Jussieu,
                              Paris 05, France

Ricci, G.   Institute of Biological Che-
                                   mistry, University of Rome
                                   and Centre of Molecular Biolo-
                                   gy of C.N.R., Rome, Italy

Ruszkowska, J.   Department of Chemistry, Uni-
                                   versity of Warsaw, Poland

Salib, A.G.   Department of Biochemistry,
                                   Faculty of Agriculture, Zaga-
                                   zig University, Zagazig, Egypt

Salvatore, F.   Istituto di Chimica Biologica,
                                   II Facoltà di Medicina e Chi-
                                   rurgia, Università di Napoli,
                                   Via S. Pansini 5, 80131 Napoli,
                                   Italy

Santoro, L.   Institute of Biological Chemi-
                                   stry, University of Rome and
                                   Centre of Molecular Biology of
                                   C.N.R., Rome, Italy

Scandurra, R.   Institute of Biological Chemi-
                                   stry, University of Rome, Rome,
                                   Italy

Schmidt, S.Y.   Berman-Gund Laboratory for
                                   the Study of Retinal Degenera-
                                   tions, Harvard Medical School,
                                   243 Charles Street, Boston, Mas-
                                   sachusetts 02114, U.S.A.

Skrede, S.   Institute of Clinical Biochemi-
                                   stry, Rikshospitalet, Univer-
                                   sity of Oslo, Oslo, Norway

Sörbo, B.   Department of Clinical Chemi-
                                   stry, University of Linköping
                                   S-581 85 Linköping, Sweden

Spoto, G.   Institute of Biological Chemi-
                                   stry, University of Rome and
                                   Centre of Molecular Biology of
                                   C.N.R., Rome, Italy

Sturman, J.A.   Developmental Neurochemistry
                                  Laboratory, Department of
                                  Pathological Neurobiology,
                                  Institute for Basic Research
                                  in Mental Retardation, Staten
                                  Island, New York 10314,U.S.A.

Traboni, C.   Istituto di Chimica Biologica,
                                  II Facoltà di Medicina e Chi-
                                  rurgia, Università di Napoli,
                                  Via S. Pansini 5, 80131 Napoli,
                                  Italy

van der Horst, C.J.G.   Laboratory of Biochemistry of
                                  Reproduction, Oudwijk 11, 3581
                                  TE Utrecht,The Netherlands

Wood, J.L.   Department of Biochemistry,
                                  University of Tennessee,Center
                                  for the Health Sciences, Memphis
                                  Tennessee 38163,U.S.A.

Wróbel, J.T.   Department of Chemistry, Uni-
                                  versity of Warsaw, Poland

Yamaguchi, K.   Department of Medical Chemistry
                                  Osaka Medical College 2-7,
                                  Daigaku-machi, Takatsuki, Osaka,
                                  Japan

Zappia, V.   Department of Biochemistry,
                                  II Chair, University of Naples
                                  First Medical School, Via Costan-
                                  tinopoli 16, 80138 Napoli,Italy